AF527771

ENVIRONMENTAL ENGINEERING

ENVIRONMENTAL ENGINEERING

Arvind Kumar
Shishir Suman Mishra

2024

Daya Publishing House®
A Division of
Astral International Pvt. Ltd.
New Delhi – 110 002

First Published, 2010

Reprinted, 2024

ISBN: 978-81-7035-644-8 (HB)

Published by : **Daya Publishing House®**
A Division of
Astral International Pvt. Ltd.
– ISO 9001:2015 Certified Company –
4736/23, Ansari Road, Darya Ganj,
New Delhi-110 002
Ph. 011-43549197, 23278134
E-mail: info@astralint.com
Website: www.astralint.com

Printed at : **Replika Press Pvt. Ltd.**

Preface

Environmental Engineering is the application of science and engineering principles to improve the environment (air, water, and soil resources), to provide healthy water, air, and land for human habitation and for other organisms, and to remediate polluted sites. It involves water and air pollution control, recycling, waste disposal, and public health issues as well as a knowledge of environmental engineering law. It also includes studies on the environmental impact of proposed construction projects.

Environmental engineers conduct hazardous-waste management studies to evaluate the significance of such hazards, advise on treatment and containment, and develop regulations to prevent mishaps. Environmental engineers also design municipal water supply and industrial wastewater treatment systems as well as being concerned with local and worldwide environmental issues such as the effects of acid rain, ozone layer depletion, water pollution and air pollution from automobile exhausts and industrial sources.

Ever since people first recognized that their health and well-being were related to the quality of their environment, they have applied thoughtful principles to attempt to improve the quality of their environment. The ancient Harappan civilization utilized early sewers in some cities. The Romans constructed aqueducts to prevent drought and to create a clean, healthful water supply for the metropolis of Rome. In the 15th century, Bavaria created laws restricting the development and degradation of alpine country that constituted the region's water supply.

Modern environmental engineering began in London in the mid-19th century when Joseph Bazalgette designed the first major sewerage system that reduced the incidence of waterborne diseases such as cholera. The introduction of drinking water treatment and sewage treatment in industrialized countries reduced waterborne diseases from leading causes of death to rarities.

In many cases, as societies grew, actions that were intended to achieve benefits for those societies had longer-term impacts which reduced other environmental qualities. One example is the widespread application of DDT to control agricultural pests in the years following World War II. While the agricultural benefits were outstanding and crop yields increased dramatically, thus reducing world hunger substantially, and malaria was controlled better than it ever had been, numerous species were brought to the verge of extinction due to the impact of the DDT on their reproductive cycles. The story of DDT as vividly told in Rachel Carson's "Silent Spring" is considered to be the birth of the modern environmental movement and the development of the modern field of "environmental engineering."

Conservation movements and laws restricting public actions that would harm the environment have been developed by various societies for millennia. Notable examples are the laws decreeing the construction of sewers in London and Paris in the 19th century and the creation of the U.S. national park system in the early 20th century.

Briefly speaking, the main task of environmental engineering is to protect public health by protecting (from further degradation), preserving (the present condition of), and enhancing the environment.

Professor (Dr.) Arvind Kumar

Dr. Shishir Suman Mishra

Contents

Chapter 1

The Geneva Experiment: An End or a New Beginning in Relation to the Environment

K.P. Ramakrishnan[1], Alok Kumar[2] and Anil Mittal[3]
[1]Department of B.C.A. and Engineering, Maharaja College, Rajasthan University, Jaipur – 302 004, Rajasthan
[2]Managing Director, Computer Point Foundation, Jaipur, Rajasthan
[3]Managing Director, Exponent Computer Center, Jaipur, Rajasthan

Keywords: *CERN, LHC, Micro black holes, Strangelets, Protons.*

Introduction

Large Hadron Collider (LHC) was switched on 10th September 2008, at the European Organization for Nuclear Research at Geneva located at the border of Swiss and France. Its purpose is one of the first big experiments for which it will be employed: to "hunt for a particle that gave the universe its form."

In early April of last year, a welder worked on the interconnections of magnets in the last sector of the Large Hadron Collider at the European Organization for Nuclear Research. Putting to an expensive test the theory of the U.K. based scientist Peter Higgs, the experiment at CERN, as the European research center is known by its French acronym, is designed, Reuters notes, to "smash together particles at close to the speed of light." In doing so by means of the Large Hadron Collider, which has been described as the largest and most complex machine ever made and as the world's most powerful particle accelerator to date, scientists will try to recreate the conditions that existed a millionth of a millionth of a second after the Big Bang, or the birth of the universe. By smashing particles together at unprecedented energies, the scientists at CERN hope to answer questions such as why some subatomic particles are heavier than others, and how particles were formed in the first place. The experimenters will also try to locate and identify a subatomic particle Higgs has theorized was the essential agent in the emergence of what we think of as matter today. With its vast labyrinth of tunnels and equipment stretching for miles, the complex has the feel of a cathedral to science. There, scientists have embarked on their biggest experiment ever, the hunt for a particle that gave the universe its form. Its scientific name is the Higgs Boson, but because it is so fundamental in shaping the universe, others have called it the "God particle". It is a particle that is supposed to endow other fundamental particles with mass. Without it, there would be no gravity, no universe as we know it. No one has seen it, but physicists have invoked it because it is the simplest explanation for how the universe evolved.

Materials and Methods

The European Organization for Nuclear Research (CERN) known as the world's largest particle physics laboratory, situated in the northwest suburbs of Geneva on the Franco-Swiss border, established on 29 September 1954 by 12 countries in Western Europe. The organization has twenty European member states, and is currently the workplace of approximately 2600 full-time employees, as well as some 7931 scientists and engineers (representing 500 universities and 80 nationalities). CERN's main function is to provide the particle accelerators and other infrastructure needed for high-energy physics research. Numerous experiments have been constructed at CERN by international collaborations to make use of them. The main site at Meyrin also has a large computer centre containing very powerful data processing facilities primarily for experimental data analysis, and because of the need to make them available to researchers elsewhere, has historically been (and continues to be) a major wide area networking hub member states' contributions to CERN for the year 2008 totaled CHF 1 billion (approximately €664 million, US$1 billion). The acronym was retained for the new laboratory after the provisional council was dissolved, even though the name changed to the current Organisation Europèene pour la Recherche Nuclèaire (European Organization for Nuclear Research) in 1954. According to Lew Kowarski, a fonner direc10r of CERN, when the name was changed, the acronym could have become the awkward CERN, and Heisenberg said but the acronym can still be CERN even if the name is not. Soon after its establishment, the work at the laboratory went beyond the study of the atomic nucleus, into higher-energy physics, an activity which is mainly concerned with the study of interactions between particles. Therefore the laboratory operated by CERN is commonly referred to as the European laboratory for particle physics which, better describes the current research being performed at CERN (source: www.wikipedia.com).

Most of the activities at CERN are currently directed towards building a new collider, the Large Hadron Collider (LHC) and the experiments for it. The LHC represents a large-scale, worldwide scientific cooperation project. The LHC tunnel is located 330 feet underground, in the region between the Geneva airport and the nearby Jura Mountains. It uses the 27 km circumference circular tunnel previously occupied by LEP which was closed down in November 2000 (Figure 1.1). CERN's existing PS/SPS accelerator complexes will be used to pre-accelerate protons which will then be injected into the LHC. This accelerator will generate vast quantities of computer data, which CERN will stream to laboratories around the world for distributed processing (making use of a specialized GRID infrastructure, the LHC Computing Grid).

In April 2005, a trial successfully streamed 600 MB per second to seven different sites across the world. If all the data generated by the LHC is to be analyzed, then scientists must achieve 1,800 MB per second before 2008. As of August 2008 the majority of the LHC ring is now cooled to the 1.9 K (–271.25°C) operating temperature. The initial particle beams were injected in August 2008, the first

Figure 1.1: Showing LHC Tunnel Located 330 Feet Underground

attempt to circulate a beam through the entire LHC was at 0830Z on 10 September, 2008, and the first high-energy collisions are planned to take place after the LHC is officially unveiled, on 21 October 2008. The Large Hadron Collider (LHC), which will provide the setting in which the particle-smashing activity will take place, has a circumference of 27 kilometers (about 17 miles). It lies 100 meters (nearly 330 feet) below ground, snaking into both French and Swiss territory. At full power, trillions of protons will race around the LHC accelerator ring 11,245 times a second, travelling at 99.99 per cent the speed of light.

More about what's under the hood at CERN: "When two beams of protons collide (inside the LHC), they will generate temperatures more than 100,000 times hotter than the heart of the sun, concentrated within a miniscule space. Meanwhile, the cooling system that circulates superfluid helium around the LHC's accelerator ring keeps the machine at minus 271.3 degrees Celsius minus 456.34 degrees Fahrenheit)." The Photon Multiplicity Detector (PMD) which is the part of the ALICE project in the LH C was developed at the Variable Energy Cyclotron Centre in Kolkata, which is a body of the Department of Atomic Energy. To give you an idea, everything that you see around you, including yourself can be reduced to atoms. Now, atoms can be further broken up into neutrons, protons and electrons. Neutrons and protons together form the nucleus of an atom. But what makes up neutrons and protons? That's where quarks come into the picture. These are subatomic particles held together by gluons and form the nucleus of an atom.

In nature, quarks are always found bound together in groups, and never in isolation, because of a phenomenon known as confinement. These groups of quarks are called hadrons. Now, when beams of protons smash together at almost the speed of light there will be such a high concentration of energy that a form of matter called quark-gluon plasma will be created. In this phase, for a brief period of time, a large number of free quarks and gluons can exist. That was how things were just after the Big Bang. The machines sent from Kolkata were fitted in the LHC by June. Experts from IIT-Mumbai, Punjab University, Jammu University, Institute of Physics, Bhubaneswar, TIFR, Mumbai, and Rajasthan University worked, together to develop the PMD.

Member Nations of CERN

Belgium, Denmark, Germany, France, Greece, Italy, Norway, Sweden, Switzerland, Netherlands, United Kingdom, India, Spain, Portugal, Finland, Poland, Hungary, Chezk Republic, Slovakia, Bulgaria, European Comission, Russia, Japan, Israel, Turkey, UNESCO and United States.

Results and Discussion

Concerns have been raised in the media, on the Internet and through the law courts about the safety of the particle physics experiments planned to take place at the Large Hadron Collider (LHC), the world's largest and most powerful particle accelerator to date. The claimed dangers of the LHC particle collisions, which are scheduled to begin during the week of 22 September 2008, include doomsday scenarios involving the production of stable micro black holes and the creation of hypothetical particles called strangelets. To address such concerns, CERN mandated a group of independent scientists to review these scenarios. In a report issued in 2003, they concluded that, like current particle experiments such as the Relativistic Heavy Ion Collider (RHIC), the LHC particle collisions pose no conceivable threat. A second review of the evidence commissioned by CERN was released in 2008. Table 1.1 gives all the events regarding this experiment. The report, prepared by a group of physicists not involved in the LHC experiments, reaffirmed the safety of the LHC collisions in light of further research conducted since the 2003 assessment. It was reviewed and endorsed by a CERN committee of 20 external scientists and by the Executive Committee of the Division of Particles and Fields of the American Physical Society, and was later published in the peer-reviewed Journal of Physics G. The report concludes that any doomsday scenarios at the LHC are ruled out because the physical conditions and events that will be created in the LHC experiments occur naturally in the universe without hazardous consequences.

In 2007, CERN mandated a group offive particle physicists not involved in the LHC experiments, the LHC Safety Assessment Group (LSAG), to monitor the latest concerns about the LHC collisions, namely John Ellis, Gian Giudice, Michelangelo Mangano ahd Urs Wiedemann, of CERN, and Igor Tkachev of the Institute for Nuclear Research in Moscow. On 20 June, 2008, in light of new experimental data and theoretical understanding, the LSAG issued a report updating the 2003 safety review, in which they reaffirmed and extended its conclusions that "LHC collisions present no danger and that there are no reasons for concern" (Ellis *et al.*). The LSAG report was then reviewed by CERN's Scientific Policy Committee (SPC), a group of external scientists that advises CERN's governing body, its Council. The report was reviewed and endorsed by a panel offive independent scientists, Peter Braun-Munzinger, Matteo Cavalli-Sforza, Gerardus't Hooft, Bryan Webber and Fabio Zwirner, and their conclusions were unanimously approved by the full 20 members of the SPC.

CERN operates a network of six accelerators and a decelerator. Each machine in the chain increases the energy of particle beams before delivering them to experiments or to the next more powerful accelerator. Currently active machines are:

Two linear accelerators generate low energy particles for injection into the Proton Synchrotron. The 50 MeV Linac2 is for protons, and the 4.2 MeV/u Linac3 is for heavy ions. The Proton Synchrotron

Booster increases the energy of particles generated by the proton linear accelerator before they are transferred to the other accelerators; The Low Energy Ion Ring (LEIR) accelerates the ions from the ion linear accelerator, before transferring them to the Proton Synchrotron (PS). Figure 1.2 showing the map of Large Hadron Collider. This accelerator was commissioned in 2005, after having been reconfigured from the previous Low Energy Anti-proton Ring (LEAR); The 28 GeV Proton Synchrotron (PS), built in 1959 and still operating as a feeder to the more powerful SPS; The Super Proton Synchrotron (SPS), a circular accelerator with a diameter of 2 kilometres built in a tunnel, which started operation in 1976. It was designed to deliver an energy of 300 GeV and was gradually upgraded to 450 GeV. As well as having its own beamlines for fixed-target experiments, it has been operated as a proton-antiproton collider, and for accelerating high energy electrons and positrons which were injected into the Large Electron-Positron Collider (LEP). From 2008 onwards, it will inject protons and heavy ions into the Large Hadron Collider (LHC); The On-Line Isotope Mass Separator (ISOLDE), which is used to study unstable nuclei. Particles are initially accelerated in the PS Booster before entering ISOLDE. It was first commissioned in 1967 and was rebuilt with major upgrades in 1974 and 1992 ;The Antiproton Decelerator (AD), which reduces the velocity of antiprotons to about 10 per cent the speed of light for research into antimatter.

Table 1.1: Showing All Events Related to CERN Since 1991

Years	*Events*
1991	CERN's Council endores proposed Large Hardron Collider
1993	The U.S. Congress cancels the Superconducting Super Collider which was to have been the world's largest collider after five year of construction in Texas costing US$2 billion
1997	After agreeing to provide significant financial contributions to the Lhc, the United States becomes an observer at CERN council.
1998	Civil Engineering for the Lhc gets underway, with parts manufactured around the world
2000	LEP is shut down after running for 11 years to make way for the LHC. First LHC parts from the United States arrive.
2001	A computer grid project to connect tens of thousands of computers worldwide is launched to handle data from LHC.
2003–2008	Installation of major LHC equipments in the tunnel
September 10, 2008	Launch of LHnC after a outlay of US$10 billion
September 11, 2008	Shutdown of LHC after a transformer fails
September 18, 2008	LHC is re-cooled to hear absolute zero after transformer is replaced.
September 19, 2008	Apparent electrical fault in a separate superconducting area causes a large leak of liquid helium into the tunnel. the LHC is shutdown for at least two months so that its temperature can be raised to permit repairs.

Source: www.msnbc.com.

LHC Experiments

Six experiments are currently being built, and will be running on the collider; each of them will study particle collisions under a different point of view, and with different technologies. Construction for these experiments required an extraordinary engineering effort *i.e.* ATLAS: A Toroidal LHC Apparatus; CMS: Compact Muon Solenoid; LHCb: LHC-beauty; ALICE: A Large Ion Collider

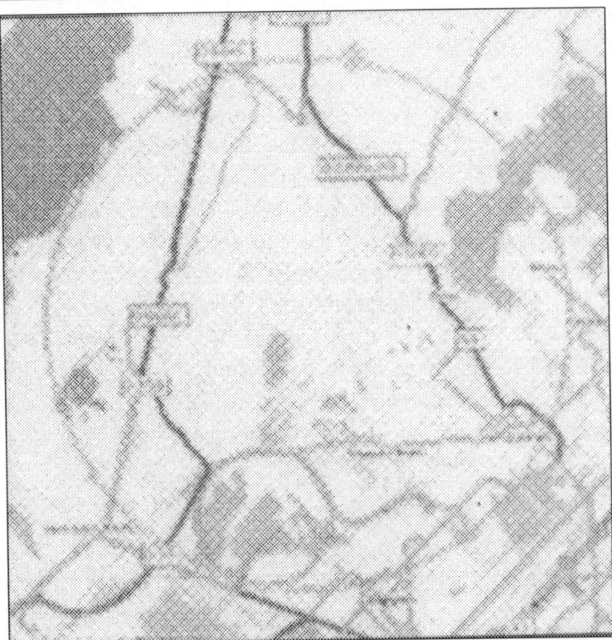

Figure 1.2: Map of the Large Hadron Collider at CERN

Experiment; TOTEM: Total Cross Section, Elastic Scattering and Diffraction Dissociation; LHCf: LHC-forward.

Safety Issues Regarding this Experiment

Drawing from research performed to assess the safety of the RHIC collisions, the LHC Safety Study Group, a group of independent scientists, performed a safety analysis of the LHC, and released their findings In the 2003 report Study of Potentially Dangerous Events, During Heavy-Ion Collisions at the LHC. The report concluded that there is "no basis for any conceivable threat". Several of its arguments were based on the predicted evaporation of hypothetical micro black holes by Hawking radiation and on the theoretical predictions of the Standard Model with regard to the outcome of events to be studied in the LHC. One argument raised against doomsday fears was that collisions at energies equivalent to and higher than those of the LHC have been happening in nature for billions of years apparently without any hazardous effects, as ultra-high-energy cosmic rays impact Earth's atmosphere and other bodies in the universe.

Micro Black Holes

Although the Standard Model of particle physics predicts that LHC energies are far too low to create black holes, some extensions of the Standard Model posit the existence of extra spatial dimensions, in which it would be possible to create micro black holes at the LHC at a rate on the order of one per second. (Aurèlien *et al.*). According to the standard calculations these are harmless because they would quickly decay by Hawking radiation.

Strangelets

Strangelets are a hypothetical form of strange matter that contains roughly equal numbers of two lightest types of quarks, the up and down quarks and strange quarks and are more stable than ordinary nuclei. If strangelets can actually exist, and if they were produced at LHC, they could conceivably initiate a runaway fusion process in which all the nuclei in the planet were converted to strange matter, similar to a strange star. The probability of the creation of strangelets decreases at higher energies. Furthermore, models indicate that strangelets are only stable or long-lived at low temperatures. Strangelets are bound at low energies (in the range of 1–10 MeV), (Ellis *et al.*) whilst the collisions in the LHC release energies in the range of 1–4 TeV. The second law of thermodynamics precludes the formation of a cold condensate that is an order of magnitude cooler than the surrounding medium. This can be illustrated by the example of trying to form an ice cube in boiling water.

New Species and Uncontrolled Explosion

As the experiment will come to an end, the evolution of new species may take place. If it will happen then there is danger of other diseases and some endemics also. The scientists hasn't given any clue of new species yet but there are assumptions of evolution. There is also a danger of uncontrolled explosion, in case if this might happen then no one can assume how much loss will occur.

Conclusion

$E = mc^2$, the well known equation given by 'Sir Albert Einstein' is going to prove a helping hand in solving the mystries of the formation of universe. This experiment can create a very tiny universe and then all the questions will be answered but a very small mistake will ruin the whole Earth. This search will answer some of the following questions: How the universe was created? What is 'Dark matter'? Mystry behind the 'God Particles'? and some other also related with the universe. So, LHC experiment will be a boon to our present queries or a disaster to our life, there is no answer to this question at present. Only "Time" can give all the answers.

Acknowledgements

The first author is grateful to Professor Arvind Kumar, Pro- Vice-Chancellor, S.K.M. University, Dumka and Dr. S.S. Mishra, Reader in Physics, K.K.M. College, Pakur for constant guiding, encouragements and blessings.

References

Aurèlien, Barran, and Julien, Grain, 2004. The case for mini black holes. *CERN Courier*, CERN.

Ellis, John, Giudice, Gian, Mangano, Michelangelo, Tkachev, Igor and Wiedemann, U, (LHC Safety Assessment Group) 2008. The LSAG's "Review of the safety of LHC Collisions. *Journal of Physics G: Nuclear and Particle Physics*. 35: 115004 (18 pp). doi: 10.1088/0954-3899/35/11/115004.arXiv:0806.3414. CERN record.

Phillip, Schewe F., Ben, Stein, and James, Riordon, 2001. Physics News Update. *The American Institute of Physics Bulletin of Physics News*, Number 558.

Website: www.msnbc.com

Website: www.wikipedia.com

Chapter 2

Remote Sensing Applications in Identifying Suitable Site for Solid Waste Disposal in Visakhapatnam, Andhra Pradesh, India

M. Vijaya Chandran[1], K. Soumya Mohan[2], T. Bragi Reddy[3] and E.U.B. Reddi[4]

**[1]Research Scholar, [2]Final Year Student, [3]Associate Professor, [4]Professor
Department of Environmental Sciences, Andhra University, Visakhapatnam, A.P.**

ABSTRACT

Unplanned solid waste management strategies have given rise to many problems world wide. These unwanted ill effects can be minimized by proper planning strategies. This study would deal with, how Remote Sensing and Geographical Information System can be used as a decision support tools for planning solid waste management. Keeping public health, economy groundwater, Environmental Sustainability and environmental sensitivity in consideration a site suitability selection criterion was followed in selecting suitable waste disposal site. An Integrated Remote Sensing and Geographical Information System (GIS) approach was used to model a design for the Visakhapatnam city future solid waste disposal. From the final site suitability analysis a site near Lankelapalem which is approximately 10 kilometers away from Gajuwaka towards Anakapalli and another site between Saripalli and Jangalapalem which is 3 kilometers north of Pendurti towards S.kota were identified as suitable sites for future solid waste disposal.

Introduction

With rapid urbanization and ever increasing population growth there has been a substantial increase in the generation of solid waste and contamination of air, water and land resources. The solid waste from different municipalities, not managed properly, have been creating problems for human health and environment. Some of the solid waste have been proved to be extremely toxic and infectious. The uncontrolled dumping of such wastes have not only brought about increasing number of incidents of health hazard but also causing the surface and groundwater contamination and thus posing serious environmental threat to the human beings.

Majority of the municipal authorities, not only in our State but also in the country, have not taken the MSW management as their priority agenda.

Study Area

The present study was carried out in Visakhapatnam, "the city of destiny". On account of rapid industrialisation, there has been significant migration into the city. It has experienced rapid industrialisation with the growth of major industries, including steel, petroleum refining and fertilizer. With the formation of "Greater Visakhapatnam" in 2005 the city's development is set for a quantum leap.

With a population of 14.5 lakhs in 2006 and a land area of 540 square kilometers, Visakhapatnam is country's largest city in terms of land and Andhra Pradesh's second largest urban agglomeration in population.

Objective of the Study

The present study would deal with, how Geographical Information System (GIS) and Remote sensing techniques can be used as a decision support tool for planning waste management. Geographic Information Systems (GIS) are ideal for preliminary site selection studies because it can manage large volumes of spatially distributed data from a variety of sources and efficiently store, retrieve, analyze and display information (Siddiqui, 1996). Using Remote sensing and GIS for site selection not only increases the objectivity and flexibility but also ensures that a large amount of spatial data can be processed in a short time.

Materials and Methods

Materials Used

Survey of India toposheet (65 O/1, 65 O/2 and 3, 65 O/5, 65 O/6, and 65 O/9), IRS LISS III and pan data. Eradas imagine.

The information from all these mediums has to be brought on a common platform where they can be linked together with each other. As the information is about the same location, they can be correlated. This will form a complete set of data with both spatial and non-spatial information. This would fulfill the basic principle of GIS to relate the information spatially.

The preparation of database includes scanning of the city ward map, digitizing and then geo-referencing it to prepare primary data. Maximum data was tried to be incorporated in the GIS data base. Also there was some information which could not be added in the database but it was important which had some local waste related information, which were used as the background information and were quite helpful for framing the guidelines for the work proposal.

Then the base map, drainage map, road map-raster image and the contours of Visakhapatnam city was digitized from the toposheet to enter the spatial information in vector form. Different information was spatially located on the maps in point, line and polygon features. These were later imported into Environmental Systems Research Institute (ESRI) ARCGIS.

The second step involves the processing of the satellite data using ERDAS IMAGINE image processing software. Masks of known water body, settlements and greenbelt areas were created by using spectral characteristics. Road networks were manually digitized using visual interpretations. And Land Use/Land Cover Map and Waste land map was created from them.

All the buffering operations and final selection of the appropriate sites was done in ESRI ARCVIEW GIS Spatial Analyst extension software.

Solid Waste disposal Site Selection Criteria

A number of essential factors are considered in locating landfill sites. Such factors include both physical and social environments. Me Kechnie *et al.* (1983) documented six factors that constitute these essential factors: Topography, Climate, Surface and Groundwater Hydrology, Geology, Cover material (Land Cover), and Land Uses. Due to data constraints, we used Topography, Surface hydrology, Existing settlements, Sensitive area and Land development (Roads etc.) of the area as guides to site selection. Criteria were specified to assume that dumpsite would be outside the buffer zone of the hydrology, forested areas, roads and existing housing.

Figure 2.1: Map Showing Major Road, Settlements, Forested Area, Endangered Species and Sensitive Zone

These criteria were:

1. Must not be located within 160 meters of a wetland
2. Must not be located within 160 meters of a sensitive area.
3. Must not be located within 200 meters of an existing waste site.
4. Must not be located within 1000 meters of an existing settlement.
5. Must be located within 300 meters of major road.
6. No landfill should be constructed within critical habitat area.

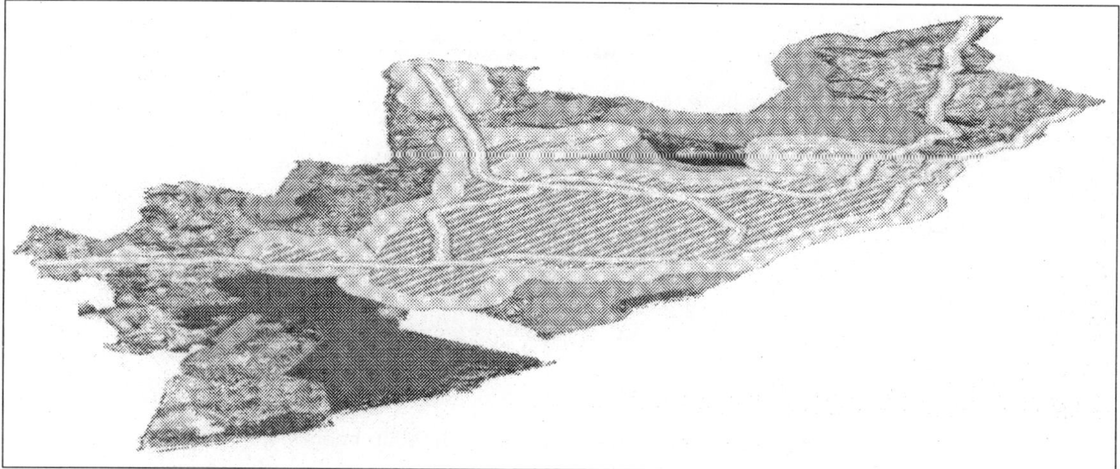

Figure 2.2: Map Showing Buffer of Roads, Settlements, Sensitive Area

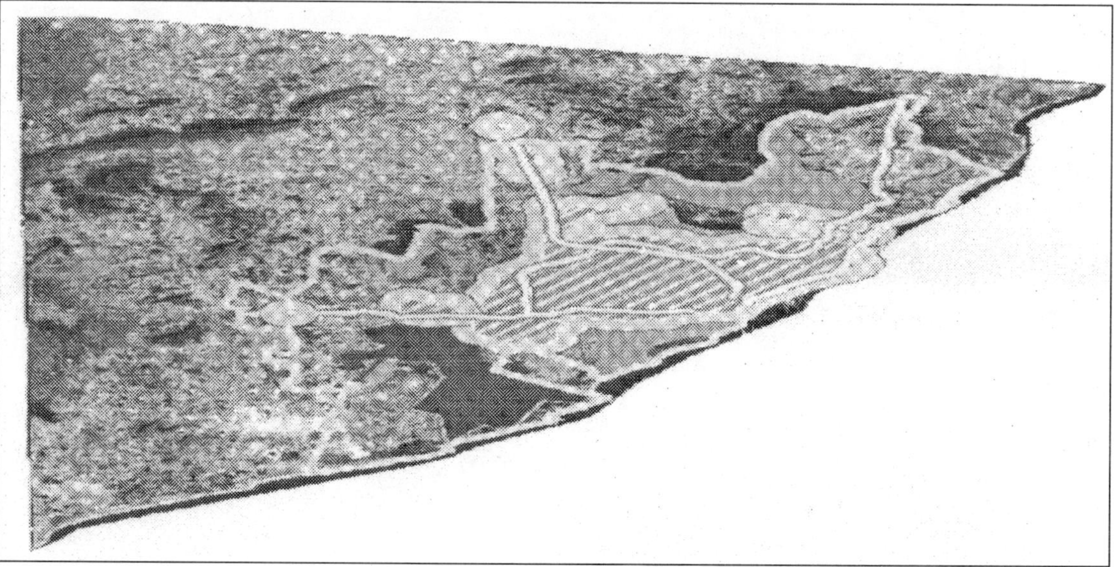

Figure 2.3: Map Showing the Proposed Site for Landfill

7. No landfill should be constructed within 3048 m (10,000 ft) of any airport.
8. Not be located on or within 500 m of sites of historic importance.

Results

Results shown in Figures 2.1–2.6.

Conclusion

This study has shown how Geographical Information System and Remote Sensing techniques may be used to protect the health and safety of citizens and shows how city planning managers may

Figure 2.4: Proposed Site–1a on Satellite Image

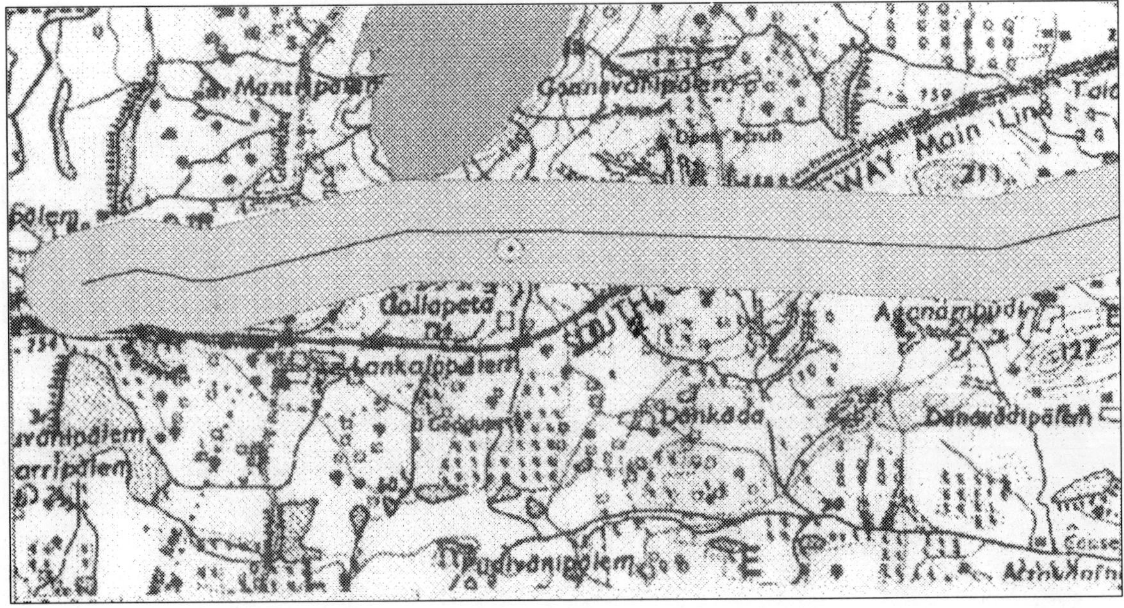

Figure 2.5: Proposed Site–1b on Toposheet

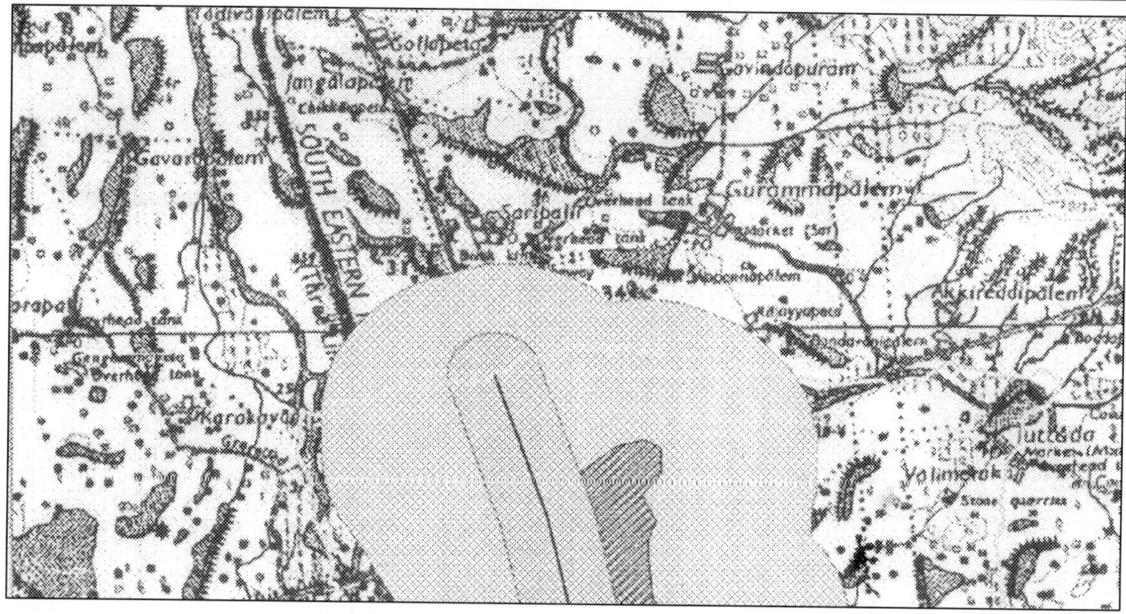

Figure 2.6: Proposed Site–2 on Toposheet

minimise the impacts of waste disposal on the environment and the economy. A point made in this study is that site planning should minimize impacts by siting the waste disposal facility under a list of rough criteria which specify that the site minimise impacts on local populations.

From the final suitability map, the following sites are identified for waste disposal:

1. Site near Lankalapalem which is approximately 10 kilometers away from Gajuwaka towards Ankapalli Highway.
2. Site between Saripalli and Jangalapalem which is 3 kilometers north of Pendurthi, towards S. Kota.

References

Colon, M. and Fawcett, B., 2005. Community-based household waste management: Lessons learnt from EXNORA's 'zero waste management' scheme in two South Indian cities. *Habitat International*, Elsevier.

CPHEEO, 2000. *Manual on Municipal Solid Waste management.* Central Public Health and Environmental Engineering, New Delhi, India, pp. 219–227.

Lerner, B.R., 2002. Managing yard wastes: Clippings and composts. In: *Solid Waste Management Series*, Indiana, USA.

Lunkapis, G.J., 2004. GIS as decision support tool for landfills siting. In: *Map Asia Conference.*

Ministry of Environment and Forest, 2000. *Notification on Municipal Solid Waste (Management and Handling) Rules, India*, pp. 3.

Narayan, L., 1999. *Remote Sensing in its Applications*. University Press, India.

Reddy, Anji. *Geoinformatics on Environmental Management*.

Singh, A.L. and Rehman, A., 2002. Malaria and related environmental issues in India: A case study of Aligarh city. *Geo Journal*.

Solano, E., Ranjitha, S.R., Barlaz, M.A. and Brill, E.D., 2002. Life-cycle based solid waste management. I: Model development. *Journal of Environmental Engineering*.

Suri, B., 1996. Polythene problems. In: *The Tribune*.

Tinmaz, E. and Demir, I., 2005. Research on solid waste management systems to improve existing situation in Corlu Town of Turkey. In: *Science Direct*.

Wilson, D.C., Velis, C. and Cheesemam, C., 2005. Role of informal sector recycling in waste management in developing countries. In: *Habitat International, Science Direct*.

Chapter 3

Time Series Model Application for Rainfall Forecasting of Saidapet Raingauge Station in Tamil Nadu

P. Srinivas[1], S. Thayumanavan[2], N. Rajesh[3] and K. Pavan Kumar[3]

[1]Assistant Professor, Department of Civil Engineering, R.V.R. and J.C. College of Engineering, Chowdavaram, Guntur – 522 019

[2]Former Professor and Head, Department of Civil Engineering, Anna University, Chennai – 600 025

[3]Junior Research Fellow, Centre for Water Resources, I.S.T., J.N.T. University, Hyderabad – 500 085

ABSTRACT

Water is an important resource for the development of any community and especially for a developing country like India it should be used judiciously. Hence, utilization of rainwater which otherwise flows into ocean and not useful, gains significance. Effective and efficient operation of water resource systems frequently demand the prediction of hydrological sequences like rainfall. Hence, forecasting models are necessary to predict the future rainfall based on past and current information. The objective of this study is to develop a time series model to forecast the rainfall. The Chembarambakkam reservoir, which feeds the Chennai metropolis, is controlled by Saidapet raingauge station. The analysis of Saidapet rainfall data for the 100 years (1901–2000) was done. ARMA model of order (2, 2) was fitted by using 84 years data (1901–1984) and checked its validity by using 16 years data (1985–2000), by comparing fitted values with actual rainfall data. Then forecast were made for the 25 years, (2001–2025). From the moving average values, a cyclical trend was not clearly visible but it may be inferred that successive droughts recurred once in a decade. Hence, necessary steps are to be taken for optimum water management and utilization.

Keywords: *Rainfall, Prediction, Fitted values, Forecast values, ARMA Model.*

Introduction

Nature has bestowed many bounties on its worldly creations and we are blessed with the same. Rainfall is one among them. Rainfall is the major means by which fresh water is made available. Water is an important resource for the development of any community and especially for a developing country like India it should be used judiciously. In view of the vital importance of water, the principal constituent for sustenance of all living things in the universe and considering its increasing scarcity for maintaining ecological balance, economic and developmental activities of all kinds, the planning and management of this resource and its optimal, economical and equitable use has become a matter of utmost urgency. Hence, utilization of rainwater which otherwise flows into ocean and not useful, gains significance. Makridakis *et al.* (1978) states that effective and efficient operations of water resource systems frequently demand the prediction of hydrological sequences like rainfall. Hence, forecasting models are necessary to predict the future rainfall based on past and current information. Before being able to forecast hydrological sequences, models have to be found which describe past data adequately. Rainfall data is of utmost importance to hydrologists as it forms the basis of many hydrological studies. Rainfall prediction will tell us whether the coming years will be a good, bad or average rainfall years. Accordingly contingency measures can be planned well in advance, tide over difficulties in augmenting water for the use of the teeming population of the Metropolis.

Some of the earlier works in forecasting of hydrological events are listed below:

A real-time river flow forecasting model, SMAR-AR is used for forecasting the Blue Nile river flows at Eldeim station near the Sudenese-Ethiopia border by Shamseldin *et al.* (1999). Box-Jenkins modeling of time series data can be improved and simplified by adhering to contemporary modeling procedures, proposes by Hipel *et al.* (1977). Mohan *et al.* (1991) in their paper highlights that forecasting models are necessary to predict inflows into a reservoir that are helpful in making decisions regarding releases from it. Bertoni *et al.* (1991) in their article, discussed conceptual rainfall runoff model and precipitation forecasting model. Online hydrologic forecasting is applied by Bras (1976) for short-term forecasting of discharges into a river basin and to forecast rainfall over small urban areas. Queen's University Forecasting Method (QUFM) developed by Watt *et al.* (1982) is a general forecasting method since it incorporates both the deterministic and stochastic components, the relative weights of each varying with application.

All the earlier models detailed here have used Box-Jenkins models, for forecasting. Each one have their own adjustments and assumptions to suit their study. The above details are taken as guidance for our study. Box-Jenkins (1976) describe a family of linear stochastic models that are commonly referred as either Box-Jenkins or ARMA models.

The objective of the present study is to develop a time series model to forecast the rainfall for the next twenty five years. Watershed area feeding the storage reservoirs serving the Chennai Metropolis is the study area. The feeding reservoir Chembarambakkam is controlled by Saidapet rainguage station.

So, Saidapet rainfall data for the past 100 years (1901–2000) is obtained and used for forecasting rainfall for the next twenty-five years (2001–2025).

Materials and Methods

Time Series Models

These models use past data as the basis for estimating future results. Past patterns are analyzed and attempts are made to compensate for these patterns in the forecasting process.

There are various types of time series models. Some of them are listed below:

1. Single Exponential Smoothing Model (SES)
2. Auto Regressive Moving Average Model (ARMA)
3. Auto Regressive Integrated Moving Average Model (A RIMA)
4. Decomposition Model
5. Winter's Method
6. Adaptive Filtering
7. Census II Method

Time Series Forecasting

Time series forecasting is the process of obtaining information from time series data for forecasting purposes. There are two basic approaches of time series forecasting. The self-projecting approach and the cause-and-effect approach. The self-projecting time series forecasting approach uses only the time series data of the activity to be forecasted to generate forecasts. The cause-and-effect time series forecasting approach derives forecasts on the basis of establishing relationship between the time series to be forecasted and one or more other series that influence or cause the first time series. The Box-Jenkins forecasting method used in the present study is a self-projecting time series forecasting method.

Box and Jenkins Methods

This is one of the most sophisticated time series techniques as it is capable of handling almost any pattern of data. It usually provides the most accurate short-range forecasts of any model and even enables statistical tests to be applied to each forecast along with confidence intervals. The objective of this method is to end up with a specific formula for the forecast values that replicates the pattern in the series as closely as possible and also produce accurate results. The phases involved in this process are

1. Model Identification Phase
2. Model Estimation and Validation Phase
3. Model Forecasting Phase
4. Model Updating Phase

Box and Jenkins Model Building Process

The methodology adopted to develop a Box-Jenkins time series model for forecasting purpose is as follows:

1. Plot the original series. Series is said to be stationary if it tends to wander more or less uniformly about mean. Otherwise non-stationary is implied.

2. Non-stationary series can be converted to stationary series by computational process known as regular differencing.
3. Find the Auto correlations and the Partial auto correlations (Partials) of the original series.
4. Examine the Auto correlations, partials for the patterns, which reveal the type of model (AR/MA/ARMA) and its order.
5. Identify the tentative model and parameters needed to construct the model.
6. The best or optimum values for the selected model parameters are determined so that the model generates fitted values, which are as close as possible to the original series values.
7. Certain diagnostics are provided to check the validity of the model and suggest alternative models.
8. The selected and estimated model is used to generate forecasts of the time series.

The methodology is shown as a flowchart in Figure 3.1.

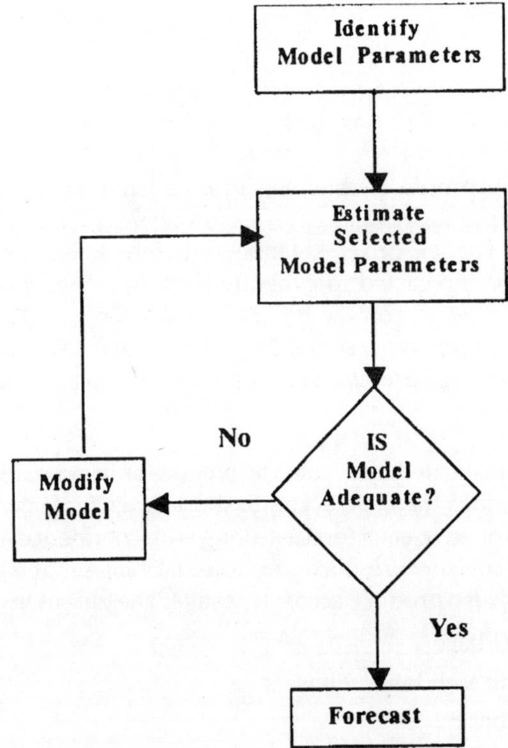

Figure 3.1: Time Series Model Building Process

Auto Regressive Moving Average (ARMA) Model

The ARMA model of order (p, q) is given by the equation as follows:

$$Xt = \Phi_1 X_{t-1} + \Phi_2 X_{t-2} + \text{- - - -} + \Phi_p X_{t-p} + e_t - \theta_1 e_{t-1} - \theta_2 e_{t-2} \text{- - - - - - -} \theta_q e_{t-q} \qquad (1)$$

where,

Φ_1, Φ_2 ------- Φ_p are Auto Regressive parameters

θ_1, θ_2 ------- θ_q are Moving Average parameters

e_t is random error

The choice of p and q are determined by Auto correlation coefficients calculated for the data. The parameters are estimated by following equations, which are obtained after solving Yule-Walker equations.

Analysis of Data

The analysis of Saidapet rainfall data was done for the past 100 years, (1901–2000) which are plotted in Figure 3.2. Auto correlations and Partial Autocorrelations of the original series were found out for corresponding lags. Autocorrelogram and Partial Autocorrelogram are shown in Figures 3.3 and 3.4 respectively. Autocorrelogram and Partial Autocorrelogram clearly indicate that the order of Model can be ARMA (2, 2).

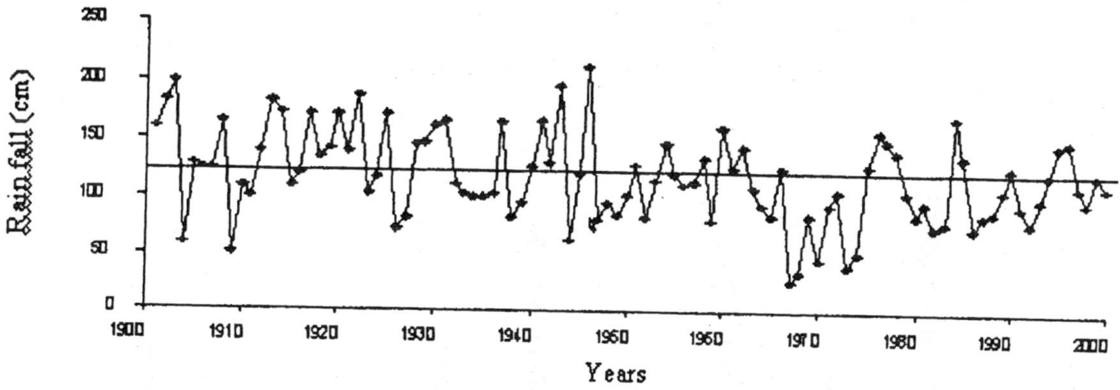

Figure 3.2: Saidapet Rainfall Data

Forecast with Box-Jenkins ARMA Model

A Box-Jenkins ARMA model of order (2, 2) was fitted to the time series data. Rainfall data for past 100 years (1901–2000) was obtained. 84 years data was taken to develop the ARMA model. Last 16 years were used to check the validity of model by comparing the fitted values with the actual rainfall values.

The developed ARMA model of order (2, 2) is given by the equation.

$$X_t = 0.38\, X_{t-1} + 0.15\, X_{t-2} + e_t + 0.27 e_{t-1} + 0.12\, e_{t-2} \tag{2}$$

Rainfall values and fitted values using ARMA (2,2) for the years 1985–2000 are shown in Figure 3.5.

Results and Discussion

Rainfall data of Saidapet for the past 100 years, (1901–2000) was taken for study. ARMA model of order (2, 2) was fitted by using 84 years data (1901–1984) and checked its validity by using 16 years

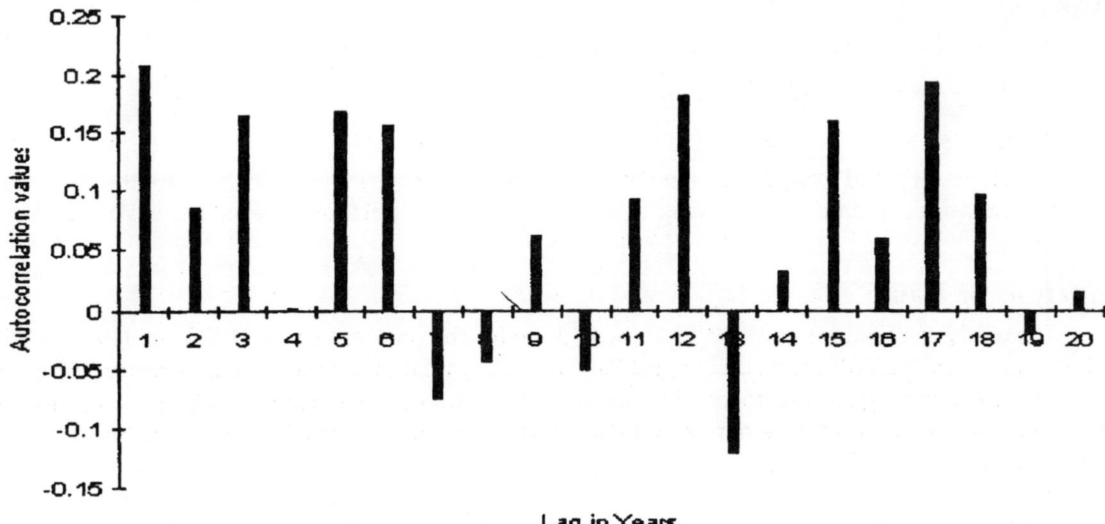

Figure 3.3: Autocorrelogram

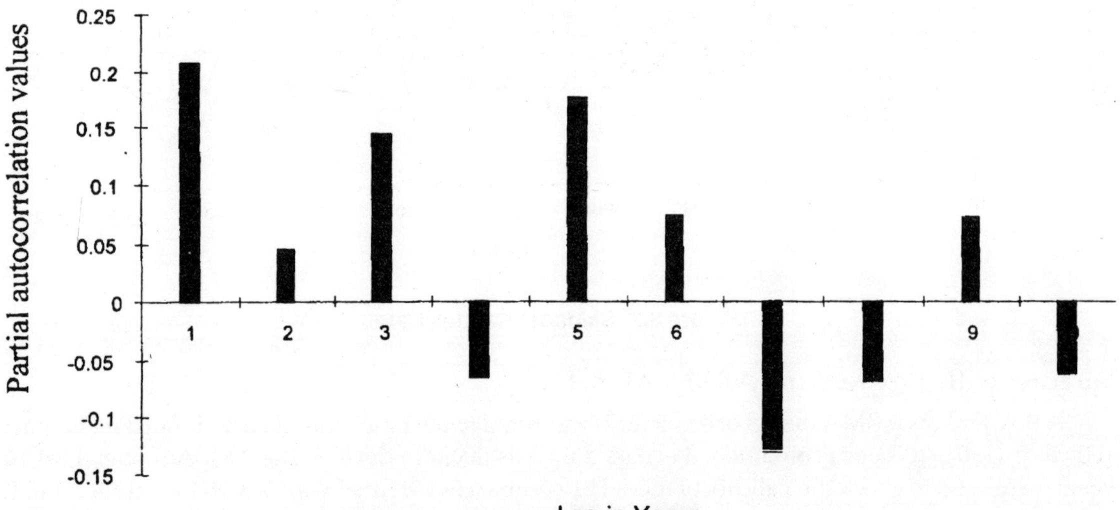

Figure 3.4: Partial Autocorrelogram

data (1985–2000), by comparing fitted values with actual rainfall data. Then forecast was made for the next 25 years, (2001 –2025). Forecast values for the rainfall using ARMA (2, 2) for the next 25 years, (2001–2025) is tabulated in Table 3.1. The variation of forecast values with the values of actual rainfall for the first five years (2001–2005) are tabulated in the Table 3.2. The plot of forecast values from year 2001–2025 is shown in Figure 3.6.

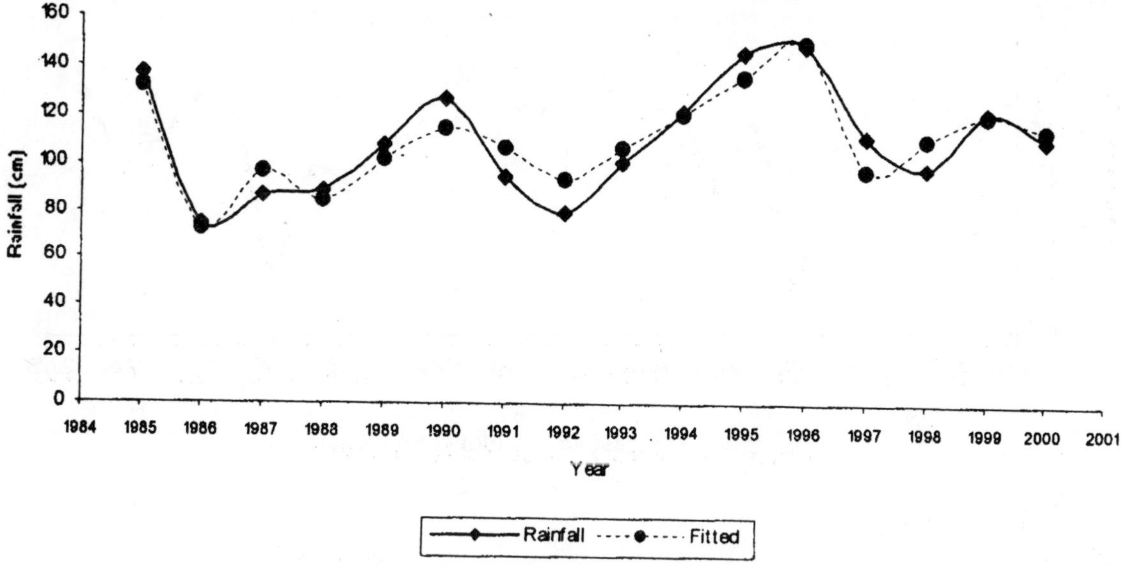

Figure 3.5: Comparison of Fitted Values with Rainfall Values

Table 3.1: Forecast of Rainfall Data for Saidapet Raingauge Station

Year	*Forecast Values Using ARMA (2,2) (cm)*	*Year*	*Forecast Values Using ARMA (2,2) (cm)*
2001	127.02	2014	143.83
2002	133.28	2015	146.92
2003	137.33	2016	131.08
2004	131.77	2017	123.41
2005	118.84	2018	106.72
2006	71.44	2019	111.46
2007	126.33	2020	77.06
2008	106.18	2021	132.59
2009	130.5	2022	138.15
2010	133.59	2023	122.98
2011	142.79	2024	114.14
2012	150.35	2025	138.92
2013	132.97		

The plot of fitted values and forecast values from 1985–2025 are shown in Figure 3.7. Forecast values plotted were matching fairly with the fitted values. Figure 3.8 shows the plots of rainfall including fitted and forecast values for 125 years. 3 year moving average and 5 year moving average for 125 years were computed. From the plots and moving average values, a cyclical trend was not clearly visible but it may be inferred that successive droughts recurred once in a decade. There is not

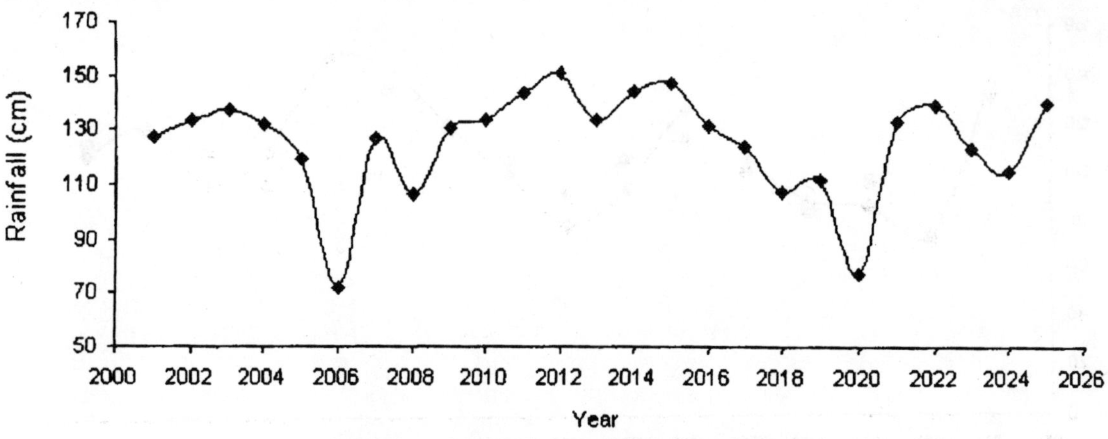

Figure 3.6: Forecast Values Plot (2001–2025)

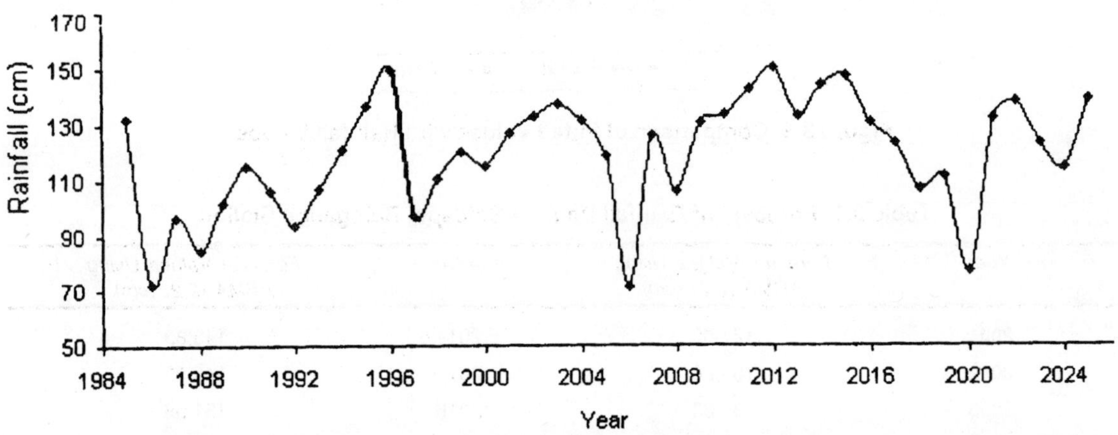

Figure 3.7: Rainfall Data of Saidapet (Both Fitted and Forecast) (1985–2025)

much deviation for the Standard Errors calculated for 100 years and 125 years. It reveals that forecast values were well fitted with actual values.

Table 3.2: Comparison of Forecast Values and Actual Rainfall for Saidapet Raingauge Station

Year	*Forecast Values Using ARMA (2,2) (cm)*	*Actual Values of Rainfall (cm)*	*Variation (cm)*
2001	127.02	125.72	1.30
2002	133.28	135.96	2.68
2003	137.33	133.67	3.66
2004	131.77	127.74	4.03
2005	118.84	122.76	3.92

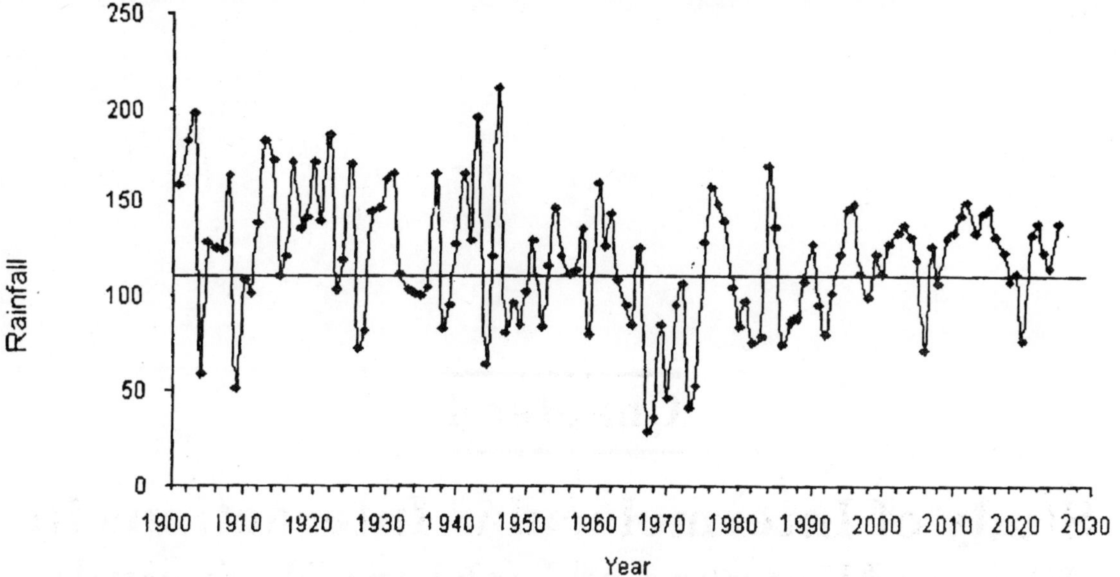

Figure 3.8: Rainfall Data of Saidapet (Actual data, Fitted Values, Forecast Values) (1901–2025)

Conclusions

The ARMA (2, 2) model was selected as the best fit from the analysis. Box and Jenkins method is best suited for long range forecasts. By combining this method with Linear regression, forecast was made for next 25 years. Since, droughts may be expected once in a decade there is a need to have better water management. Hence, necessary steps are to be taken for optimum water management and utilization.

References

Bertoni, Juan Carlos, Tucci, Carlos Eduardo and Clarke, Robbin Thomas, 1991. Rainfall based real-time flood forecasting. *Journal of Hydrology*, 131: 313–339.

Box, G.E.P. and Jenkins, G.M., 1976. Time Series analysis forecasting and control. *Holden Day.*

Bras, Rafel L., 1976. Short-term forecasting of rainfall and runoff. In: *Real-time Forecasting and Control of Water Resources Systems*. Pergamon Press, pp. 125–137.

Hipel, Keith W., McLeod, Angus Ian and Lennox, W.C., 1977. Advances in Box-Jenkins modeling: Model construction. *Water Resources Research*, 13: 567–575.

Makridakis, S., Wheelwright, S.C. and Mcgee, V.E., 1978. *Forecasting Methods and Applications*. John Wiley and Sons.

Mohan, S., Raman, H. and Premganesh, G., 1991. A comparative study on models for forecasting inflows. *Journal of Indian Water Resources Society*, 11(3): 19–22.

Shamseldin, Asaad Y., Abdoand, Gamal, M. and Elzein, Abderhman S., 1999. Realtime flood forecasting on the Blue Nile River. *Water International*, 24: 39–45.

Watt, W.E and Nozdryn, M.J., 1982. *Real-time Flood Forecasting for Flood Damage Reduction.* Water Resources Publication, pp. 551–571.

Chapter 4

Study of Intermolecular Interactions in Binary Mixtures of Lithium Hydroxide and Lithium Sulphate at 303K

G. Sudarmozhi[1]* and K. Ramanathan[2]
[1]Department of Physics, Muthayammal Engineering College, Rasipuram, T.N.
[2]Department of Physics, Sacs MAVMM Engineering College, Madurai, T.N.

ABSTRACT

In this study the ultrasonic velocity and density of the mixture of lithium hydroxide (LiOH) and lithium sulphate (Li_2SO_4) in wide range of concentration are determined in the frequency of 3MHz at 303K using ultrasonic interferometer technique. The ultrasonic parameters like adiabatic compressibility, intermolecular free length, acoustical impedance, molar sound velocity, molar compressibility and molar volume have been calculated by using experimental velocity and density data. The ion-solvent interactions at wide range of concentrations are studied and it shows the complex nature at 0.6 mole fraction of lithium sulphate (Li_2SO_4). The observed variations with different concentration of the mixture are discussed.

Introduction

Measurement of ultrasonic velocity has been widely employed in understanding the nature of molecular interaction in liquids (Varma and Singh, 1988; Prasad and Naidu, 2003; Bhandakkar *et al.*, 2003) and it could be used to investigate solute-solvent interaction in the mixture of various

* E-mail: sudarmozhi@gmail.com

proportionate. Ultrasound has been extensively used to determine the ion-solvent interaction in aqueous solutions containing electrolytes. In the solution of ionic solute the attraction between the solute and solvent is essentially of ion-dipole type. The ion- dipole interaction depends mainly on ion size and polarity of the solvent. During the dissolution of electrolyte in a solvent, a volume contraction is caused due to interaction between ions and solvent molecules and in solution, other acoustical properties may also be influenced (Rajendran *et al.*, 1999; Rajendran and Benny, 1993). The acoustical parameters such as adiabatic compressibility (β), intermolecular free length (L_f), acoustical impedance (Z), molar volume (V), molar compressibility (B) and molar sound velocity (R) have been calculated from the measured values of ultrasonic velocity and density in various concentrations at 303K. In this study, the ion-solvent interactions in aqueous solutions are studied and the formation of complex nature at a definite concentration is also investigated.

Experimental Details

1N solutions of lithium hydroxide and lithium sulphate are prepared separately by dissolving the salts of Otto kemi make in double distilled water and they are allowed for complete ionization. The appropriate volume of solutions have been taken and then mixed to form a desired concentration. The ultrasonic velocity of various concentrations are measured by using M-83 multi frequency interferometer of Mittal's make at 303K. The accuracy of velocity measurement is ±2 parts in 10^4. The temperature of the mixture is maintained to an accuracy of ±0.05K in an electronically controlled thermostatic water bath. Density of the mixtures are measured by using specific gravity bottle of 5ml with an accuracy of ±0.1 kgm^{-3}.The viscosity of the mixture is estimated using Ostwald's viscometer and the mass measurement is performed with a single pan digital balance. For density and velocity measurements, the experiment is performed twice and then the values are averaged. The calculated values are given in Table 4.1.

Table 4.1: Values of Ultrasonic Velocity, Density, Acoustic Impedance, Adiabatic Compressibility, Intermolecular Free Length and Molar Volume at 303K

Mole Fraction of		*U*	*ρ*	*Z(10⁶)*	*β (10⁻¹⁰)*	*L_f*	*V*
LiOH (X_1)	*Li_2SO_4 (X_2)*	*ms^{-1}*	*kgm^{-3}*	*$kgm^{-2}sec^{-1}$*	*$kg^{-1}mS^2$*	*Å*	*m^3mole^{-1}*
1.0000	0.0000	1590.0	1101.8	1.7522	3.5890	0.3788	0.0380
0.8980	0.1000	1599.0	1107.8	1.7717	3.5299	0.3757	0.0455
0.8000	0.2000	1611.6	1114.9	1.7963	3.4557	0.3717	0.0530
0.7064	0.3034	1621.8	1121.9	1.8188	3.3919	0.3683	0.0610
0.6000	0.4000	1628.4	1128.4	1.8364	.3.3450	0.3658	0.0676
0.5000	0.5000	1644.0	1134.9	1.8659	3.2599	0.3611	0.0748
0.4000	0.6000	1632.0	1141.4	1.8621	3.2906	0.3628	0.0819
0.3004	0.7012	1632.0	1148.5	1.8752	3.2677	0.3615	0.0890
0.2000	0.8000	1638.0	1155.5	1.8935	3.2242	0.3591	0.0958
0.1002	0.8997	1650.0	1163.0	1.9189	3.1583	0.3554	0.1025
0.0000	1.0000	1662.0	1170.0	1.9445	3.0942	0.3518	0.1093

Computation

Using ultrasonic velocity and density values, the acoustical parameters like acoustic impedance (Z), intermolecular free length (L_f), adiabatic compressibility (β), molar volume (V), molar compressibility (B) and molar sound velocity (R) are calculated by using the following relations (Srivastava *et al.*, 1983).

$$\text{Acoustic impedance (Z)} = U\rho$$

$$\text{Adiabatic compressibility } (\beta) = (U^2\rho)^{-1}$$

$$\text{Intermolecular free length } (L_f) = (\beta)^{1/2}K$$

$$\text{Molar volume (V)} = \frac{M_{eff}}{\rho}$$

$$\text{Molar sound velocity (R)} = V(U)^{1/3}$$

$$\text{Molar compressibility (B)} = \frac{M}{\rho}\beta^{-1/7}$$

Where K is Jacobson's temperature dependent constant (Jacobson, 1975) and its value for the present calculation has been taken as 200×10^{-8} at 303 K, M_{eff} is the effective molecular weight.

Results and Discussion

In Table 4.1, the experimental values of density, ultrasonic velocity of the mixtures of electrolytic solutions as a function of concentrations are given.

Variation in ultrasonic velocity in any solution generally indicates molecular association in it. The ultrasonic velocity shows increase with increase in mole fraction of lithium sulphate (X_2) up to 0.5 mole fraction. But it drops down at 0.6 and 0.7 mole fraction. Then it increases again with X_2. The change in velocity depends on the intermolecular free length or the decrease in intermolecular free length causing increase in velocity.

The increase in mole fraction of lithium sulphate weakens the molecular forces and hence the change in velocity is observed at 0.6 mole fraction. Polar molecules can interact with non-polar molecules through dipole induced dipole interaction and with polar molecules through dipole-dipole interaction (Kannappan *et al.*, 2003). When the salts are dissolved in water, the ions are splitted into Li^+, SO_4^{2-} and OH^- ions and they are coordinated with water molecules. Since the charge to mass ratio of SO_4^{2-} ions are small, it induces higher order in the water structure (Ramanathan and Ravichandran, 2004). The increase in structural order of water may cause more collision and hence the value of adiabatic compressibility is decreased with the increase in velocity. A drop in velocity at 0.6 mole fraction of lithium sulphate results in weakening of intermolecular forces and it is concluded that a significant interaction between solute and solvent takes place and it coincides with the previous work (Ramanathan and Ravichandran, 2004; Tabhane and Patki, 1985).

The density values at different concentrations against the mole fraction of lithium sulphate are shown in Figure 4.2. It is seen that the density values have increased linearly with the concentration of Li_2SO_4 where as it is inversely changed with LiOH. As well, the viscosities of various concentrations are increased correspondingly with density as shown in Table 4.2.

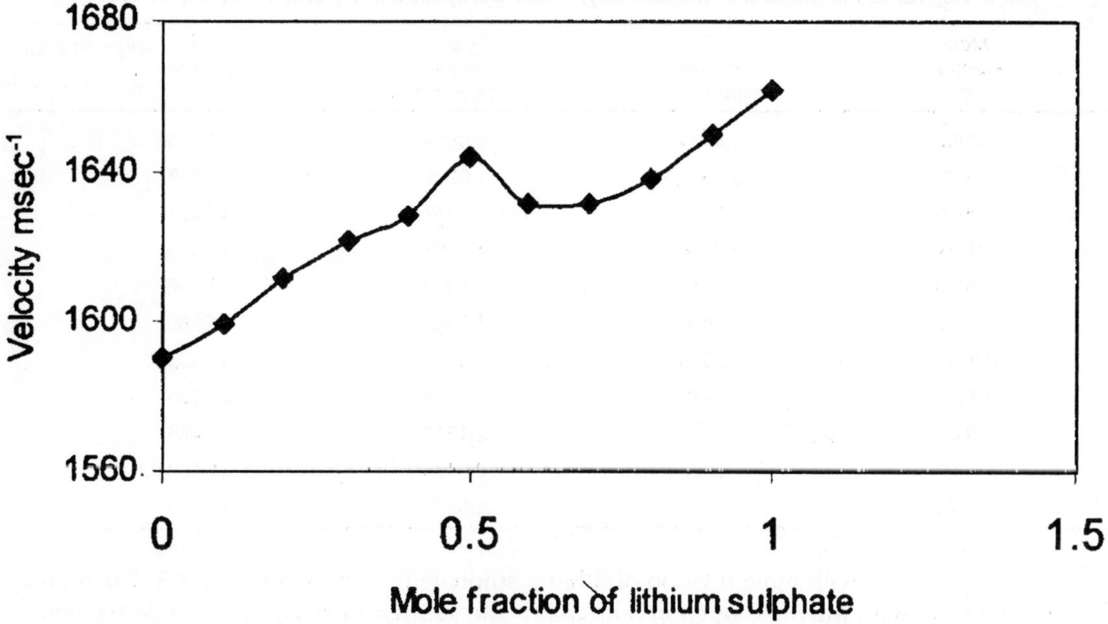

Figure 4.1: Variation of Ultrasonic Velocity with Mole Fraction of Lithium Sulphate

1200
1160
1120
1080
Density Kgm-3
0 0.2 0.4 0.6 0.8 1 1.2
Mole fraction of lithium sulphate

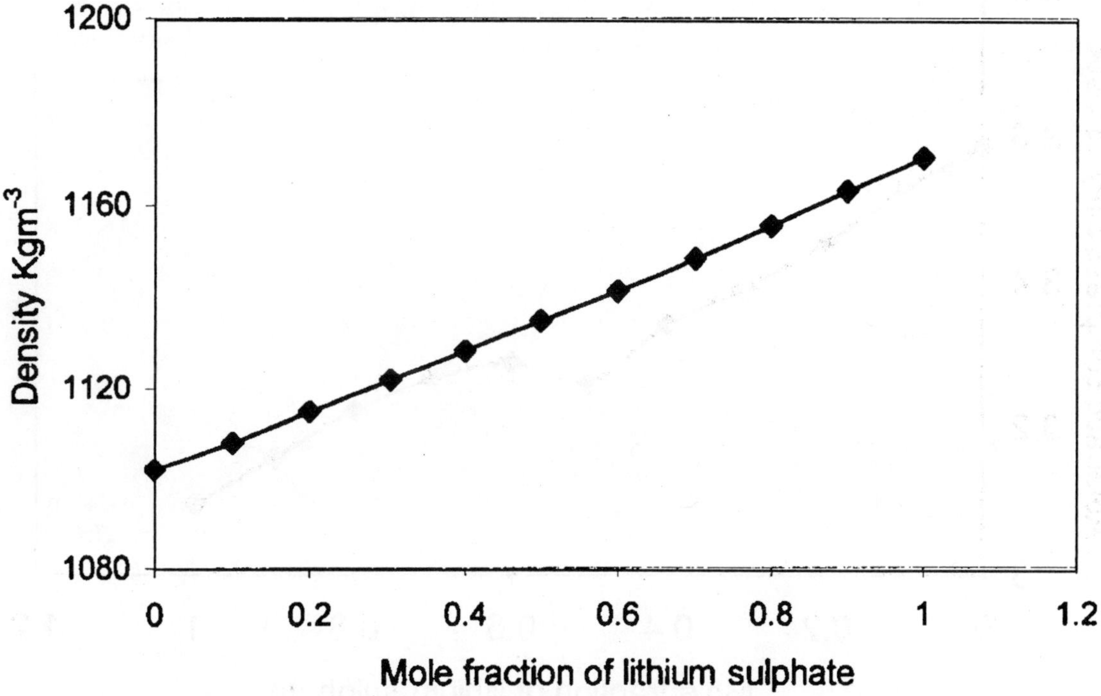

Figure 4.2: Variation of Density with Mole Fraction of Lithium Sulphate

Table 4.2: Values of Molar Sound Velocity, Molar Compressibility and Viscosity at 303K

Mole Fraction X_2	R m^3/mol $(m/s)^{1/3}$	B m^3/mol $(N/m^2)^{-1/7}$	η 10^{-4} N sm^{-2}
0.0000	0.4424	0.8519	17.680
0.1000	0.5307	1.0225	17.947
0.2000	0.6197	1.1947	18.231
0.3034	0.7148	1.3787	18.518
0.4000	0.7932	1.5309	18.798
0.5000	0.8806	1.7002	19.080
0.6000	0.9618	1.8591	19.538
0.7012	1.0452	2.0223	20.009
0.8000	1.1264	2.1810	20.485
0.8997	1.2082	2.3404	21.153
1.0000	1.2914	2.5030	21.999

The variation of β with mole fraction of lithium sulphate is shown in Figure 4.3. The β value changes inversely with the mole fraction and attains the sudden increase at 0.6 mole fraction of lithium sulphate. Then it starts decreasing. This indicates the unstable nature of ions in this solution.

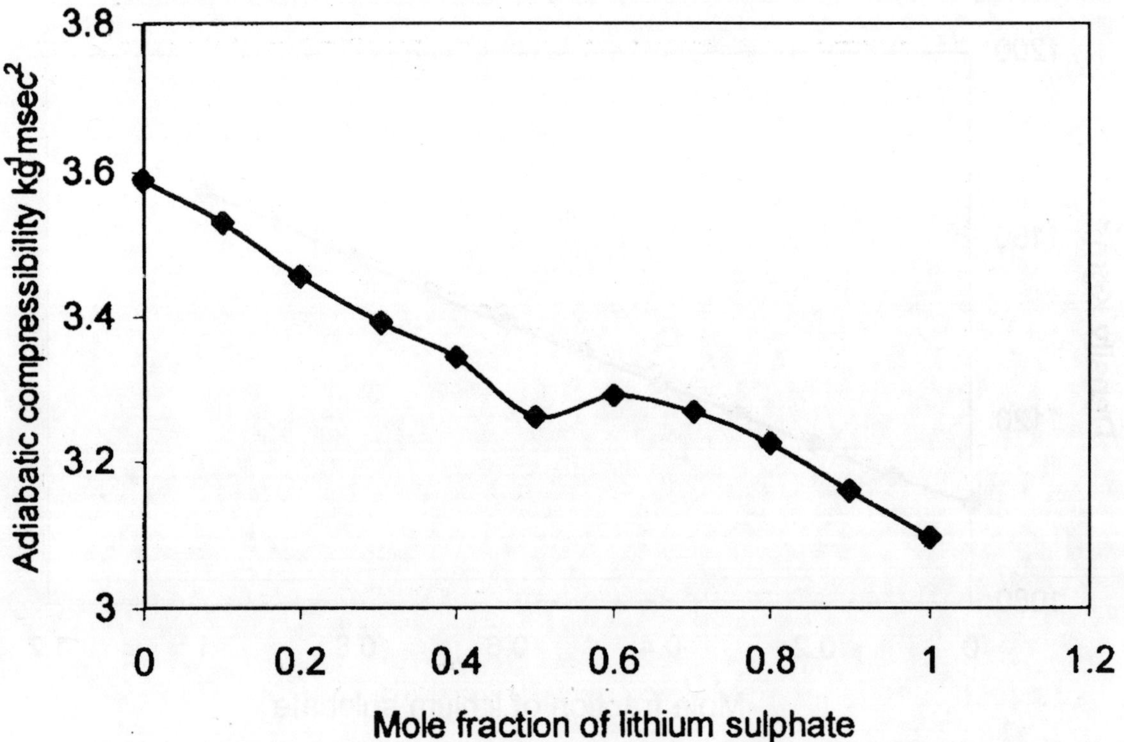

Figure 4.3: Variation of Adiabatic Compressibility with Mole Fraction of Lithium Sulphate

The decrease in value of β and Lf due to increase in mole fraction of lithium sulphate is explained on the basis of the fact that the ions in solution are surrounded by a layer of solvent molecules firmly bound and oriented towards ions. The orientation of the solvent molecules around the ion is due to the influence of electrostatic field of ions and it causes increase in internal pressure, lowering the compressibility of the solution except the mole fraction at 0.6. The complex formation in an ionic system is due to the ionic association.

Acoustic impedance increases gradually and it gets down at 0.6 mole fraction of lithium sulphate (Figure 4.4). Then it increases again and changes inversely with L_f. Impedance is almost reciprocal of β. The sudden drop in acoustic impedance is due to the complex formation of ions. Both the parameters R and B increase with the increase in concentration of lithium sulphate.

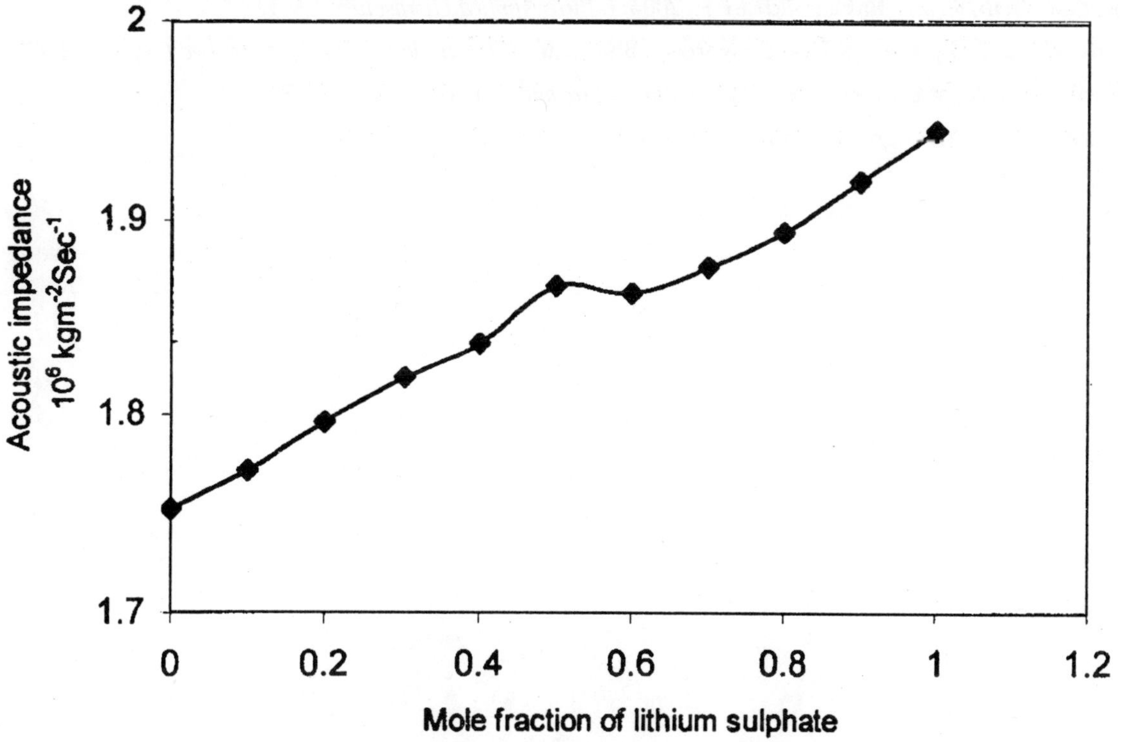

Figure 4.4: Variation of Acoustical Impedance with Mole Fraction of Lithium Sulphate

Conclusion

The study of 1N aqueous solutions of lithium hydroxide and lithium sulphate mixed salt solution states that the ion-association and formation at the concentrations of 0.6 and 0.7 mole fraction of lithium sulphate shows a specific complex nature and it reveals the instability of molecules in these solute and solvent molecules of binary liquids.

Acknowledgement

One of the authors (GS) would like to thank the principal and management for their encouragement and providing laboratory facilities.

References

Bhandakkar, V.D., Tabhane,V.A. and Ghosh, Sharda, 2003. *Indian J. Pure and Applied Physics*, A1: 849.

Jacobson, B., 1975. *Acta. chem. Scand.*, 34: 121.

Kannappan, V., Xavier Jesuit Raja, S. and Jeyasanthi, R., 2003. *Indian J. Pure and Applied Physics*, 41: 690.

Prasad, Ravindra K. and Naidu, P. Subramanyam, 2003. *Indian J. Pure and Applied Physics*, 41: 686.

Rajendran, V., Palanivelu, N. and Chaudhuri, B.K., 1999. *J. Acoust. Soc., India*, 27: 249.

Rajendran, V. and Benny, Christopher Newton, 1999. *Acoust. Lett.*, 17: 33.

Ramanathan, K. and Ravichandran, S., 2004. *J. Pure Applied Ultrasonics*, 26: 12.

Srivastava, T.N., Singh, R.P. and Swaroop, Bhagwan, 1983. *Indian J. Pure and Applied Physics*, 21: 67.

Tabhane, V.A., and Patki, B.A., 1985. *Indian J. Pure and Applied Physics*, 23: 58.

Varma, R.P. and Singh, Anuradha, 1988. *Indian J. Pure and Applied Physics* 26: 606.

Chapter 5

Communication between Microcontroller to Microcontroller for Control Applications

Ranjna Bohre, Ashish Verma*, Alka Dubey, R.S. Verma, Shalini Jain, and Fazal Noorbasha

Department of Physics and Electronics, Dr. Harisingh Gour University, Sagar – 470 003, M.P.

ABSTRACT

A system is design using UART for Microcontroller to Microcontroller communication. UART is used for receiving and transmitting the LASER beam. A LASER source is used to interface between them. The communication over the distance of about five kilometer is established using LASER beam. The receiving signal will be the input signal to the microcontroller. An algorithm is written along with flowchart for programming these microcontrollers.

Keywords: UART, LASER, Microcontroller, Programme, Communication.

Introduction

A system is design and developed to trace the signal from the distance of nearly five kilometer and the above. In this system two UARTs are used to make the communication link between the areas of

* E-mail: vermaashish31@rediffmail.com

nearly five kilometer. LASER beam is used as transmitting signal from one UART to the other. This LASER beam will hit on the receiving part of the UART. The output of the UART will further control the microcontroller. This microcontroller already programmed for control application.

System Description

The microcontroller to microcontroller communication circuit is as shown in Figure 5.1. In this communication link transmitting signal is LASER beam. As soon as LASER beam from the transmitter of UART is hit on the receiving part of the UART, connected microcontroller turn ON. This microcontroller control farther interfaced circuits.

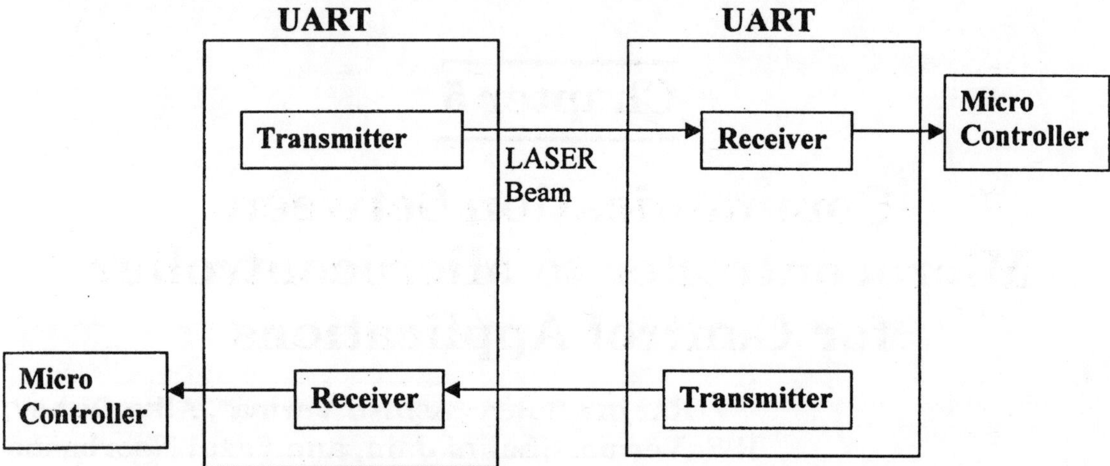

Figure 5.1: Block Diagram of Microcontroller to Microcontroller Communication

Microcontroller (μc) Interface

In this communication circuit UART is MAX232 as shown in Figure 5.2. This UART is 16pin IC. We can transmit the signal from pin 9 and taking the output from pin 7. Figure 5.3 shows the 89c51 Microcontroller, which is used in this communication circuit. This Microcontroller has the 16pins. The transmitted signal *i.e.* LASER beam is strike on the receiving part *i.e.* the sensor of the UART and the signal is to turn ON the connected microcontroller and interfaced to the other control applications as shown in Figure 5.4.

Application Software

The basic flow chart is used for Microcontroller to Microcontroller communication is as shown in Figure 5.5. In this software we first initialized the ports of Microcontroller. As well as signal *i.e.* LASER beam is transmitted from the UART and hit on the receiver of the other UART, connected microcontroller gets the signal to turn ON' the micro controller. The microcontroller is already programmed for control application and hence control the interfaced circuit.

Results and Discussion

In the present study we conclude that two microcontrollers, which are nearly five kilometer apart, are linked by LASER beam using UARTs and microcontrollers in serial communication.

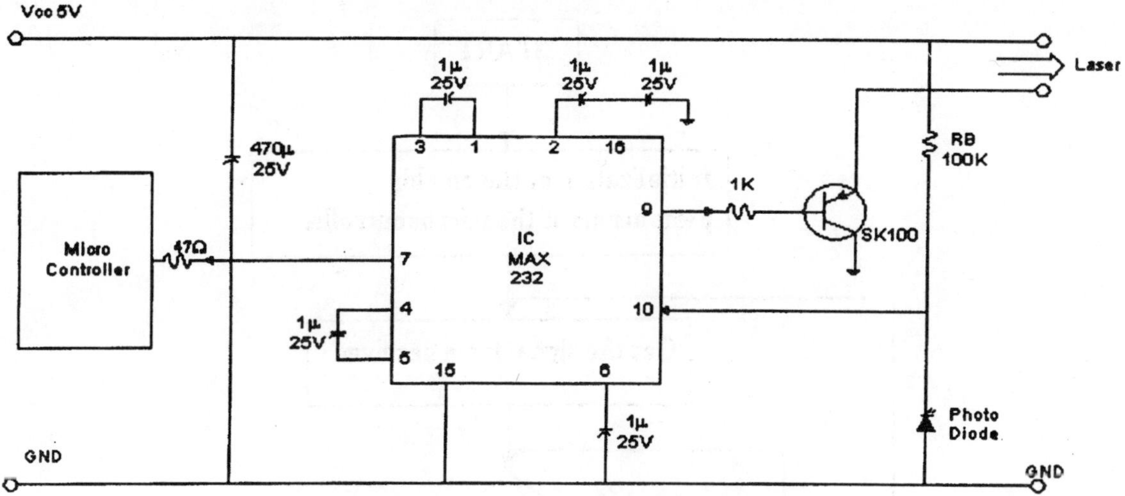

Figure 5.2: Circuit Diagram of UART (MAX 232)

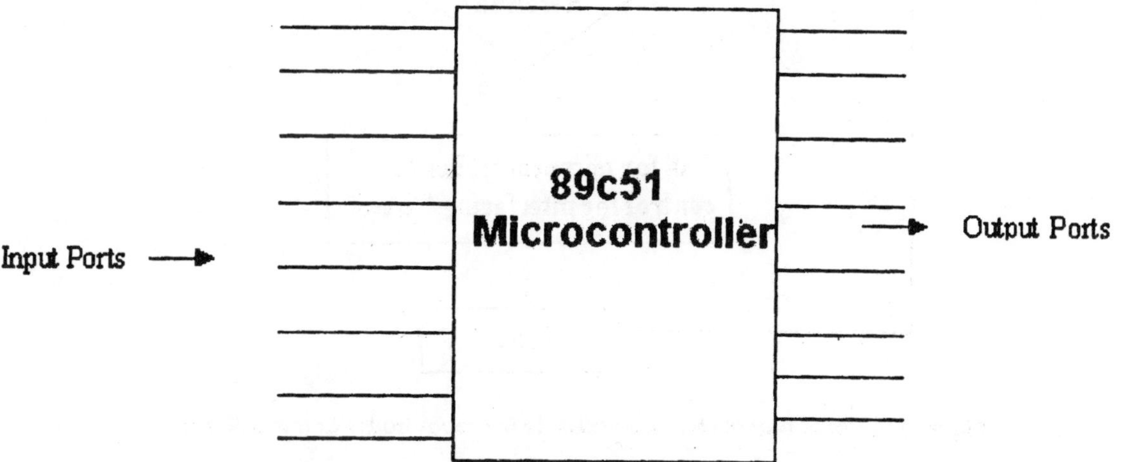

Figure 5.3: Block Diagram of 89c51 Microcontroller

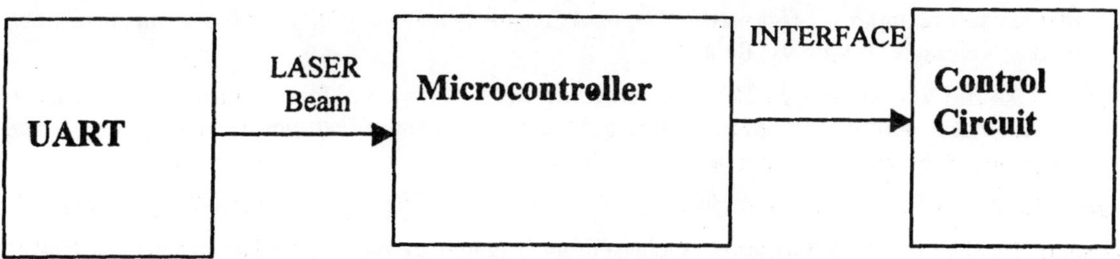

Figure 5.4: Block Diagram of Microcontroller Interface

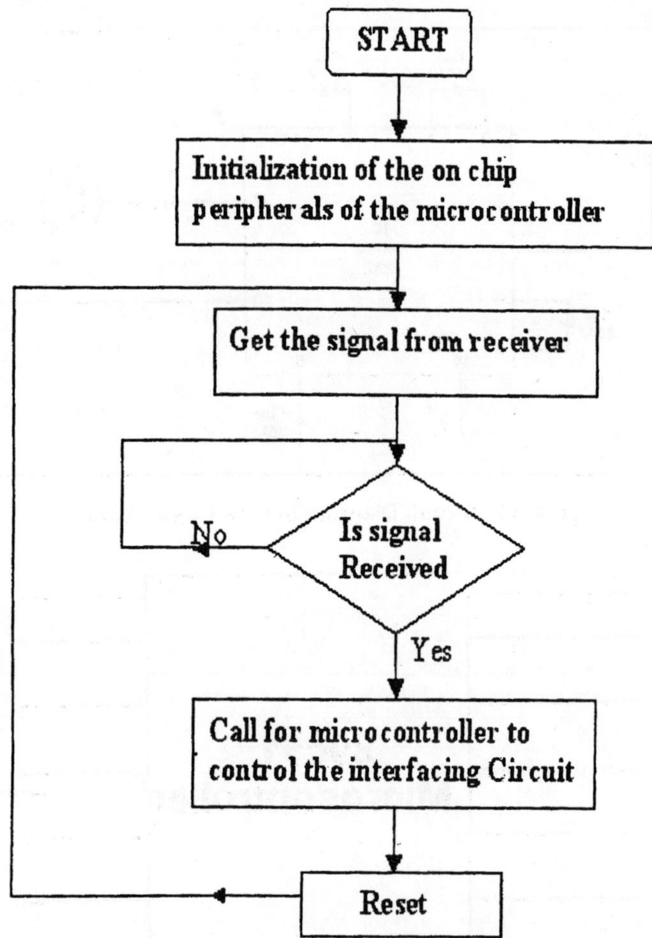

Figure 5.5: Flow Chart of Microcontroller to Microcontroller Communication

References

Bohre, Ranjna and Verma, Ashish, 2005. The receiving and transmitting system using USART for railway safety paper is presented in *National Conference on Emerging Technology and Application* held at Sourastra University, Rajkot.

Dubey, Alka and Verma, Ashish, 2006. A microcontroller based system for reducing sound pollution by automobiles in urban area presented in *National Seminar on Pollution and its Effect on Urban Environment*, Benazeer College held on 25–26 November.

Kamal, Raj 2005. *Microcontrollers: Architecture, Programming, Interfacing and System Design*. Pearson.

Kenneth, J. and Hintz, 1992. Merging C and assembly language in microcontroller application. *Journal of Microcomputer Application.*

Mazidi, Muhammad Ali, 2005. *The 8051 Microcontroller and Embedded System.* Pearson.

Peatman, John B., 2005. *Design with PIC-Microcontroller*. Pearson.

Sankar, K.S., 2000. *EFY*, pp. 46.

Singh, B.P., 1994. *Microprocessor and Microcontroller*. Galgotia Pub. Pvt. Ltd..

Chapter 6

Biosorption of Chromium (III) from Aqueous Solutions by Neem Sawdust

V. Vinodhini and Nilanjana Das

Environmental Biotechnology Division, School of Biotechnology, Chemical and Biomedical Engineering, VIT University, Vellore – 632 014, Tamil Nadu

ABSTRACT

Batch biosorption of chromium (III) from an aqueous solution was investigated using Neem sawdust. The biosorption capacities of Neem sawdust for Cr (III) was studied using biosorbent dosages (1–9 g/l), initial metal concentrations (10–150 mg/l) and solution pH values (2–8) in batch mode. At optimum biosorbent dosage (2 g/l) and initial solution pH (pH 5); Neem sawdust recorded maximum chromium (III) uptakes of 294.11 mg/g. The biosorption rate was fast and most of the processes were completed within 1 hr followed by slow attainment of equilibrium. The chromium removal increased with the increase in temperature up to 40°C and then started decreasing. The adsorption data were fitted to Langmuir and Freundlich isotherm models. Sorption equilibrium exhibited better fit to Langmuir isotherm.

Keywords: *Pollution, Biosorption, Chromium (III), Sorption equilibrium.*

Introduction

In recent years, increasing awareness of water pollution and its far reaching effects has prompted concerted efforts towards pollution abatement. Among the different heavy metals, chromium is a common and very toxic pollutant introduced into natural waters from a variety of industrial wastewaters (Donmez and Aksu, 2002). Chromium occurs most frequently as Cr (VI) or Cr (III) in aqueous solutions (Dakiky *et al.*, 2002). The two major sources of contamination are electroplating, metal finishing industries (hexavalent chromium) and tanneries (trivalent chromium). Removal technology of heavy metal ions by biosorption is appeared to be more promising technology than the conventional methods for removing metal ions in the viewpoint of environmental protection. Considerable attention has been paid on the recovery and removal of heavy metal ions from aqueous systems by using various biosubstances or natural products (Volesky, 1990; Nakajima, 2002; Stephen and Fergus, 1987).

The purpose of this study was to investigate the sorption of Cr (III) onto Neem sawdust particles. Experiments were done in a batch system and the metal ions sorption was investigated with respect to initial pH, Contact time, amount of biomass, temperature, initial metal ion concentration. The adsorption equilibrium study was conducted using the Langmuir and Freundlich isotherm models.

Materials and Methods

Biosorbent Preparation

The Neem sawdust was obtained from local saw mill. After washing with deionized water and drying at 60°C in an oven, the residues were ground and sieved into different fractions.

Batch Sorption Experiments

The experiments were performed in 250 ml Erlenmeyer flasks containing 100 ml of Cr (III) solution in rotary shaker at 120 rpm. To study the effect of pH on sorption, the pH of the chromium solution was adjusted to the values in the pH range of 2–8 by the addition of 0.1M HNO_3/NaOH prior to the experiment. The influence of sorbent dose was studied with biosorbent concentration varying from 1.0 to 9.0 g. The time dependent experiments were carried out by shaking the adsorption mixture at various predetermined intervals of 5, 10, 15, 20, 25, 30, 45, 60, 90, 120, 180, 240 and 300 minutes. The temperature of chromium (III) solution was adjusted in the range of 10 to 50°C. The metal biosorption potential was determined at different initial metal ion concentrations (10–150 mg/l), fixed biosorbent dose of 0.2 g in 100 ml of metal solution. The mixture was agitated for about 2 hrs at 120 rpm to allow the sorbent to adsorb the metal ions until the solution reached equilibrium. After agitation, the reaction mixtures were filtered through whatman no.1 filter paper and the filtrates were analyzed by Atomic Absorption Spectrophotometer (Varian Spectra 240) for determining the concentration of Cr (III) ions. The biosorption constants were calculated by using Langmuir and Freundlich isotherms. Based on Mass balance, the amount of metal ion adsorbed per gram of biomass was calculated using the following equation:

$$Q = \frac{(C_0 - C)}{M} \times V$$

where,

Q: Amount of metal ion biosorbed per gram of biomass, mg g^{-1};

C_o: Initial metal ion concentration, mg L^{-1}

C: Final metal ion concentration, mg L^{-1}

M: Weight of biomass in the reaction mixture, g

V: Volume of the reaction mixture.

Results and Discussion

Effect of pH

Since the solution pH can significantly influence the removal of heavy metals, it is therefore an important condition for adsorption of Cr (III) ions. Experiments over a range of pH values (2–8) with 100 mg/L of chromium solution shown in Figure 6.1 reveal that biosorption capacity of chromium (III) is maximum at pH 5. The pH dependency of biosorption efficiency could be explained by the functional groups involved in metal uptake and the metal chemistry (Matheickal, 1999). At low pH values, protons occupies most of the biosorption sites on the biomass surface and Cr(III) ions could be sorbed because of electric repulsion with the protons on the biomass. When the pH values increased, biosorbent surface were more negatively charged and the biosorption of the metal ions with positive charge (Cr^{3+}) process reached maximum around pH 5. Decrease in biosorption at higher pH is due to the formation of soluble hydroxylated complexes of the metal ions and their competition with active sites, and as a consequence, the retention would decrease again. Previous studies also reported that the maximum biosorption efficiency of Cr (III) was observed at pH 5 (Chojnacka, 2005; Van and Viraragavan, 2003)

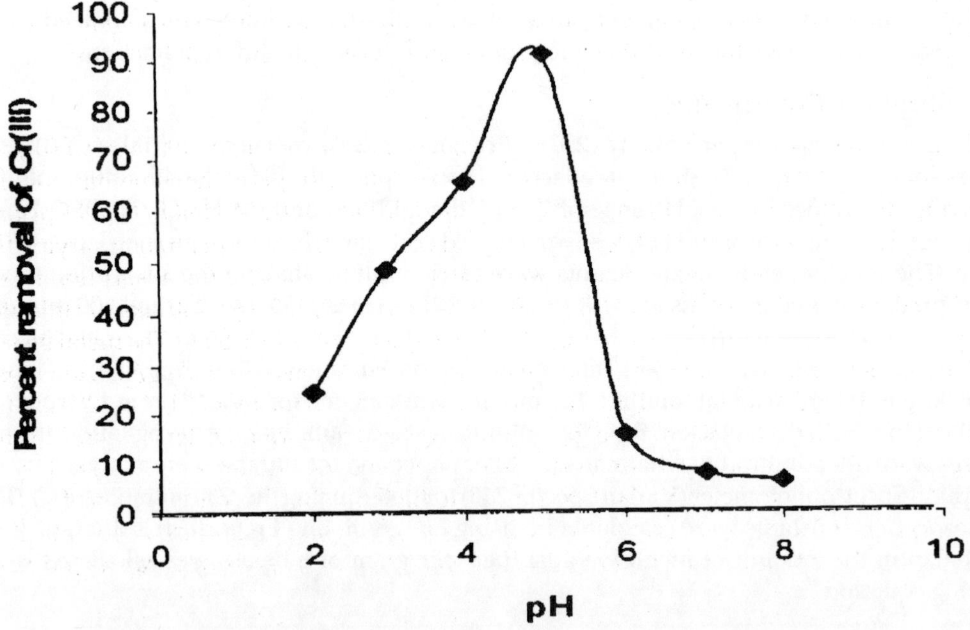

Figure 6.1: Effect of pH on Biosorption of Chromium (III) by Neem Sawdust

Effect of Contact Time

The equilibrium time required for the biosorption of Cr (III) on Neem sawdust with 2 g/l of the biosorbent at different time intervals was studied (Figure 6.2). The removal rate of Cr (III) increases with the increase of the adsorption time. However, it remains constant after an equilibrium time of 60 min which indicates that the adsorption tends toward saturation at 60 min. It was observed that the rate of removal of metal ions increases with increase in contact time to some extent. Further increase in contact time does not increase the uptake due to deposition of metal ion available on the adsorption sites on adsorbent surface. (Kaur *et al.*, 2008)

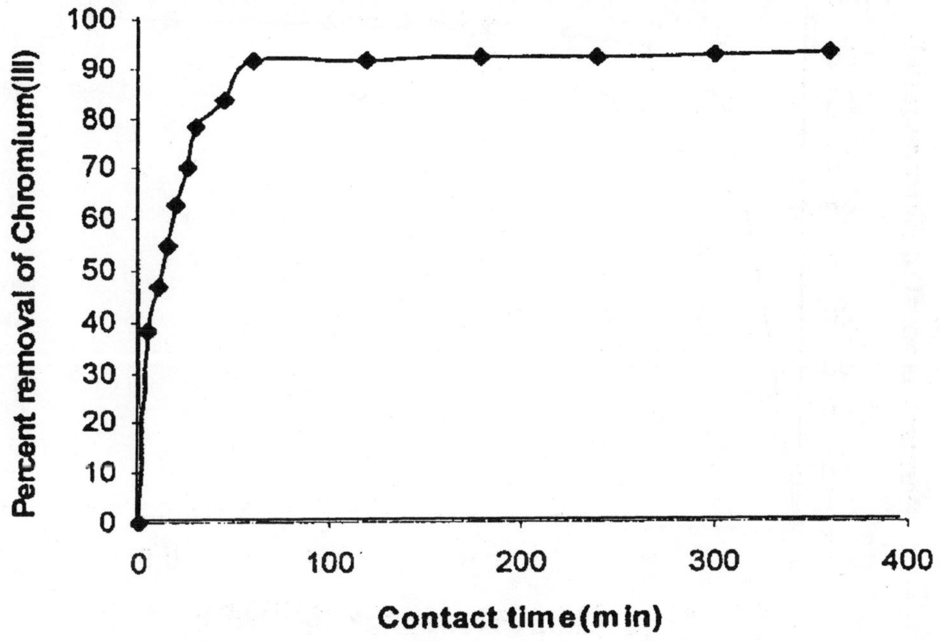

Figure 6.2: Effect of Contact Time on Biosorption of Chromium (III) by Neem Sawdust

Effect of Biomass Dosage

The effect of biosorbent dosage on the removal of chromium (III) is shown in Figure 6.3. Amount of sorbent was varied from 0.1 to 0.9 g in 100 ml and equilibrated for 60 min at an initial Cr (III) ion concentration of 100 mg/l. The results indicate that the per cent removal of Cr (III) increases with increase in the amount of adsorbent and efficiency for Neem sawdust was 95 per cent for removal of Cr (III). The removal of metal ions increased with the increase in biosorbent concentration and attained equilibrium after 4 g/l of adsorbent dosage for the chromium (III) ions. It can easily be inferred that the per cent removal of metal ions increases with increasing weight of the sawdust. This is due to the greater availability of the exchangeable sites or surface area at higher concentration of the adsorbent (Yu *et al.*, 2003).

Effect of Temperature

The sorption of chromium increased slightly with the increase in temperature up to 40°C and then started decreasing (Figure 6.4). The temperature higher than 40°C caused a change in the texture of the biomass and thus reduced its sorption capacity. Similar results were reported in case of *S. cerevisiae* (Goyal, 2003). It was found that the metal biosorption of Cr (III) by *S. cerevisiae* increased with the increasing temperature in the range of 25–45°C. They explained that higher temperature would lead to higher affinity of sites for metal binding sites on the yeast.

Effect of Initial Metal Ion Concentration

The effect of Cr (III) concentration on the sorption by the Neem sawdust was investigated by varying the concentration 10 to 150 ppm at a pH of 5 for 60 min equilibrium time (Figure 6.5). Cr (III) removal by Neem sawdust was 294.11 mg/g. In the case of low concentrations, the ratio of the initial

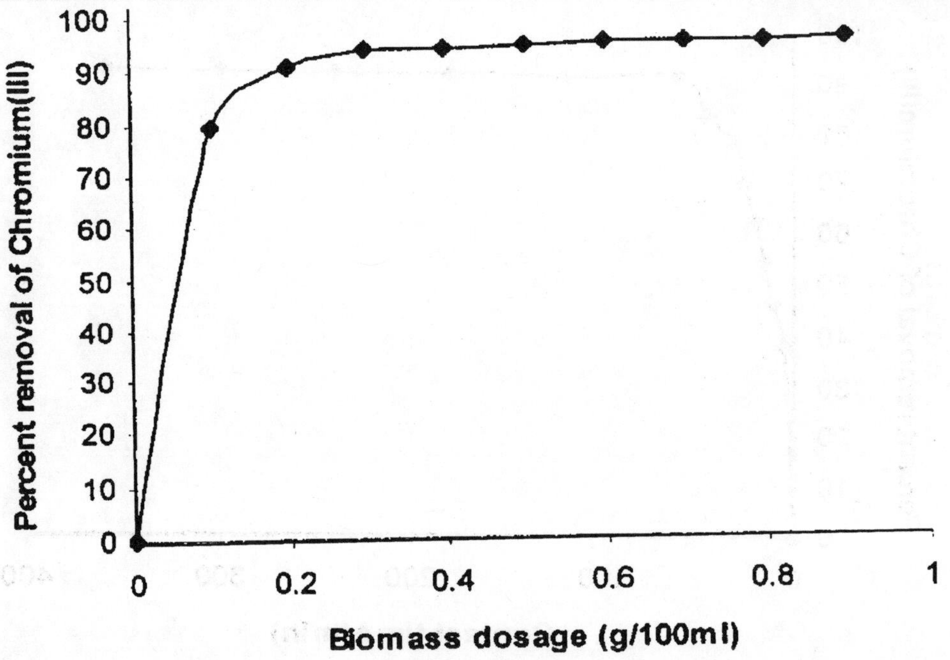

Figure 6.3: Effect of Biomass Dosage on Biosorption of Chromium (III) by Neem Sawdust

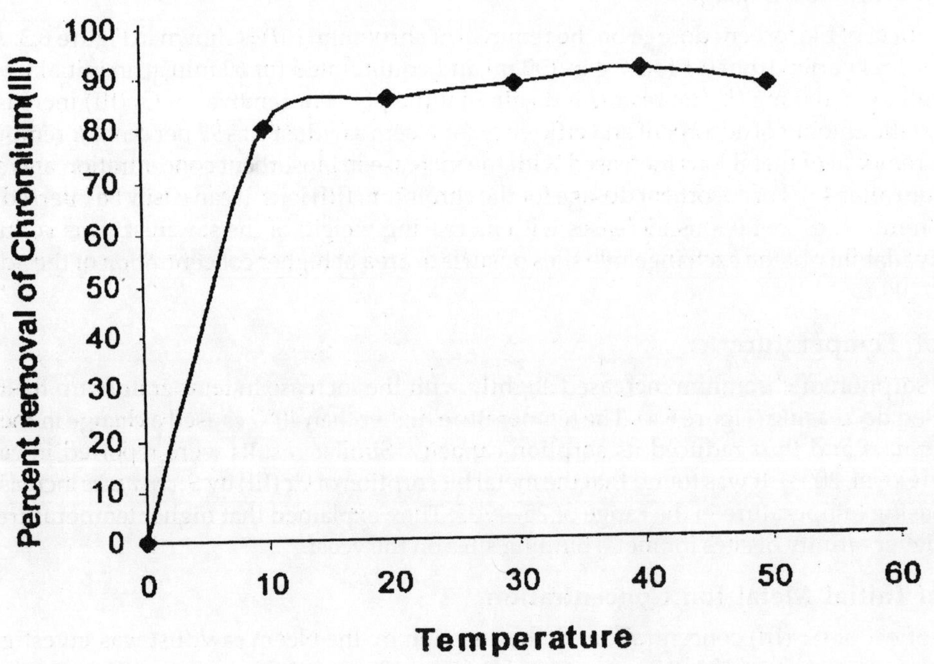

Figure 6.4: Effect of Temperature on Biosorption of Chromium (III) by Neem Sawdust

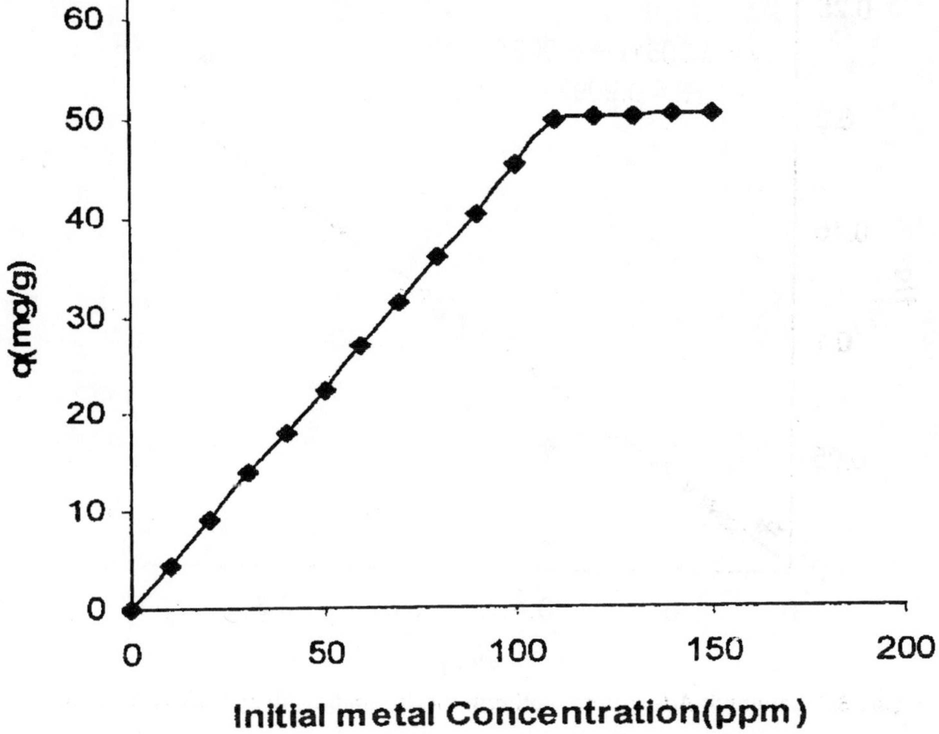

Figure 6.5: Effect of Initial Metal Concentration on Biosorption of Chromium (III) by Neem Sawdust

number of moles of metal ions to the available surface area is larger and subsequently the fractional adsorption becomes independent of initial concentrations. However, at higher concentrations the available sites of adsorption become fewer, and hence, the percentage removal of metal ions depends upon the initial concentration.

Adsorption Isotherms

Equilibrium data, commonly known as adsorption isotherms, are basic requirements for the design of adsorption systems. Classical adsorption models (Langmuir and Freundlich) are used to describe the equilibrium between adsorbed metal ions on the Sawdust (q_{eq}) and metal ions in solution (C_{eq}). The equilibrium established between adsorbed component on the biosorbent and unadsorbed component in solution. can be represented by adsorption isotherms. The most widely used isotherm equation for modeling equilibrium is the Langmuir equation which is valid for monolayer sorption onto a surface with a finite number of identical sites which are homogeneously distributed over the sorbent surface (Popuri *et al.*, 2007). The Freundlich expression is an empirical equation based on sorption on a heterogeneous surface suggesting that binding sites are not equivalent and/or independent. The Langmuir and Freundlich adsorption isotherms of Cr (III) ions for Neem sawdust was shown in Figures 6.6 and 6.7. The values of the Langmuir and Freundlich constants for the biosorption of Chromium (III) by Neem sawdust are presented in Table 6.1. The data fits well to Langmuir isotherm. The maximum adsorption capacity of Neem sawdust was 294.11 mg/g.

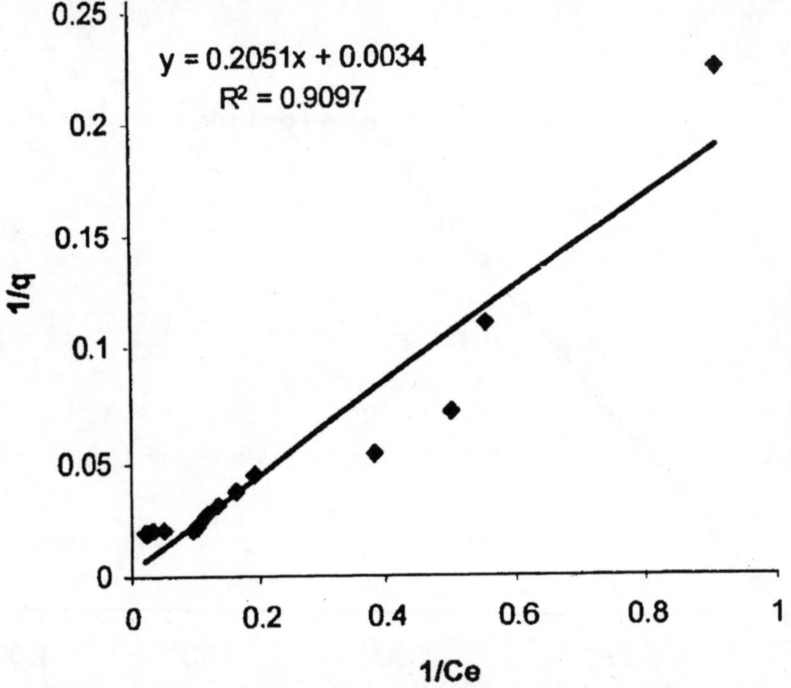

Figure 6.6: Langmuir Adsorption Isotherm for Chromium (III) by Neem Sawdust

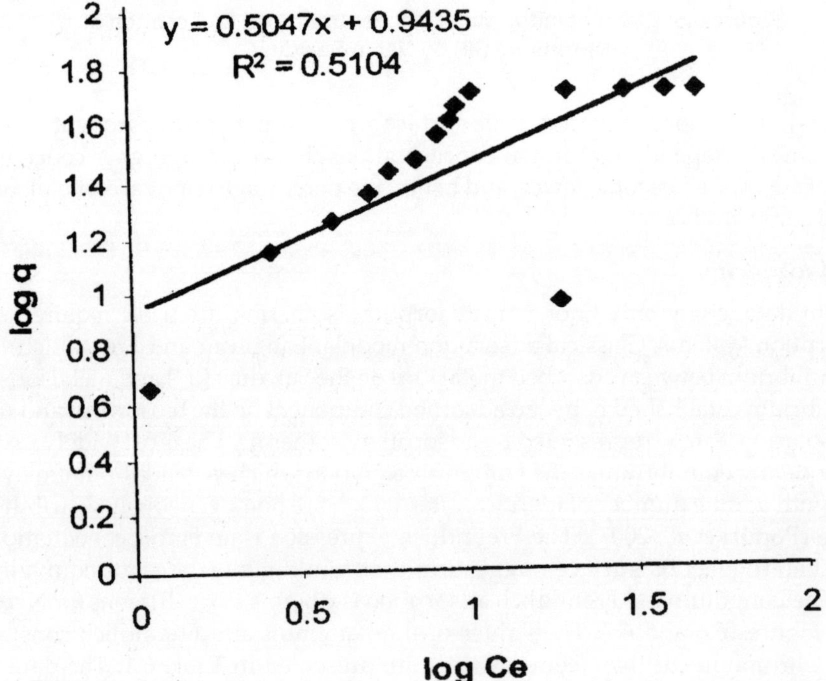

Figure 6.7: Freundlich Adsorption Isotherm for Chromium (III) by Neem Sawdust

Table 6.1: Linear Regression Data for Langmuir and Freundlich Isotherms for Chromium (III) Biosorption by Neem Sawdust

Biosorbent	*Langmuir Parameters*			*Freundlich Parameters*			*Best Fitted Model*
	Qmax (mg/g)	*b (L/mg)*	R^2	*N*	*Kf*	R^2	
Neem Sawdust	294.11	0.2051	0.9097	1.9813	8.780	0.5047	Langmuir

The results obtained have shown that Neem sawdust is an excellent biosorbent for the removal of chromium (III) ions. The maximum uptake of chromium (III) ions occurs at pH 5. Increase in the amount of biosorbent increases the per cent removal of the metal ions. The data fits well to Langmuir isotherm and the maximum adsorption capacity of Neem sawdust was 294.11 mg/g. Therefore, the present study clearly demonstrates the possibility of usage of Neem sawdust as suitable alternatives for the removal of chromium (III) ions from wastewater.

References

Chojnacka, K., 2005. Biosorption of Cr (III) ions by eggshells. *J. Hazard. Mater.*, 121: 167–173.

Dakiky, M., Khami, A., Manassra, A. and Mereb, M., 2002. Selective adsorption of chromium (VI) in industrial wastewater using low cost abundantly available adsorbents. *Advances in Environmental Research*, 6: 533–540.

Donmez, D. and Aksu, Z., 2002. Removal of chromium (VI) from saline wastewaters by Dunaliella species. *Process Biochemistry*, 38: 751–762.

Goyal, N., Jain, S.C. and Banerjee, D.C., 2003. Comparative studies on the microbial adsorption of heavy metals. *Adv. Environ. Res.*, 7: 311–319.

Kaur, S., Walia, T.P.S. and Mahajan, R.K., 2008. Comparative studies of zinc, cadmium, lead and copper on economically viable adsorbents. *J. Environ. Eng. Sci.*, 7: 83–90.

Matheickal, Q., Yu, J.T. and Woodburn, G.M., 1999. Biosorption of cadmium from aqueous solutions by pretreated biomass of marine alga *Durvillaea potatorum*. *Water Res.*, 33: 335–342.

Nakajima, A., 2002. Electron spin resonance study on the vanadium adsorption by Persimmon tannin gel. *Talanta*, 57: 537–544.

Popuri, S.R., Jammala, A., Reddy, K.V.N.S. and Abburi, K., 2007. Biosorption of hexavalent chromium using tamarind (*Tamarindus indica*) fruit shell: A comparative study. *Electronic Journal of Biotechnology*, 10: 2007.

Stephen, P. and Fergus, C., 1987. Mineral binding characteristics of lignin, guar gum, cellulose, pectin and neutral detergent fiber under simulated duodenal pH conditions. *J. Food Sci.*, 52: 1414–1419.

Volesky, B., 1990. *Biosorption of Heavy Metals*. CRC Press, Boca Raton.

Yan, G. and Viraraghavan, T., 2003. Heavy metal removal in aqueous solution by fungus Mucor rouxii. *Water Res.*, 78: 243–249.

Yu, L.J., Shyam, S., Shukla, Kenneth, L., Dorris, Shukla, A. and Margrave, J.L., 2003. Adsorption of chromium from aqueous solutions by maple sawdust. *Journal of Hazardous Matter*, 100: 53–63.

Chapter 7

Mass Transfer Studies in Hybrid Rotating and Reciprocating Perforated Plate Bubble Column

S. Dhanasekaran*, V. Vijayagopal and T. Karunanithi
Mass Transfer Laboratory, Department of Chemical Engineering, Faculty of Engineering and Technology, Annamalai University, Annamalai Nagar – 608 002, Tamil Nadu

ABSTRACT

Gas-Liquid mass transfer studies have been carried out in a 10 cm diameter Hybrid Rotating and Reciprocating Perforated Plate Bubble Column. Absorption of oxygen from air into deoxygenated water system is used in this study. The effect of agitation level, superficial air velocity, superficial liquid velocity and plate spacing on volumetric mass transfer co-efficient is studied and found to be significant. The volumetric mass transfer co-efficient (kLa) is correlated in terms of agitation level and gas holdup as

$$k_La = 0.0001(Af)^{0.8}(e)^{0.8}$$

and analyzed statistically. The values of k_La obtained by the correlation are found to concur with the experimental values.

Keywords: *Hybrid rotating and reciprocating perforated plate bubble column, Gas-liquid contact, Gas holdup, Bubble column, Interfacial area, Volumetric mass transfer co-efficient, Reciprocating plate column, Stirred tank reactor.*

* E-mail: sudha040506@yahoo.co.in

Introduction

Gas-liquid contact operation has a considerable industrial importance. The development of this technology has increased as a result of the growth of interest in biotechnology in general and in fermentation processes in particular (Gagnon *et al.*, 1998]. In order to accelerate mass transfer between gas and liquid phases, reactors in which the gas is dispersed as fine bubbles into the liquid are often used. Many type of reactors namely bubble columns, stirred tank reactors, Reciprocating plate columns, pulsed plate columns and liquid pulsating plate columns etc., have been studied extensively. The stirred tank reactors (Archis *et al.*, 2002) normally provide highly efficient contact by means of an external agitation but a single such reactor normally provides no more than one equilibrium stage. Bubble columns operated counter currently offer the possibility of multistage performance. An additional advantage of a tall bubble column is the increased solubility of gas at the base of the column (liquid exit). This is due to the higher hydrostatic pressure in accordance with Henry's law. Axial mixing in the liquid phase of the bubble columns generally can not be ignored; the axial dispersion coefficients in the absence of internals increase approximately as the 4/3 power of column diameter (Baird and Rice, 1975). Thus as the scale of the column is increased, the advantages of the counter current operation (relative to a stirred tank absorber) is reduced (Ramo Rao and Baird, 1988). Axial mixing in bubble columns is largely due to the radial non-uniformity in two phase flow pattern. It can be reduced by careful vertical alignment of the column (Tinge and Drinkenbird, 1986) by inserting baffles, packing or partitions and by agitation whether rotary or reciprocatary. The insertion of baffles and the provision of agitation also have additional beneficial effects of increasing the interfacial area. It is due to increased holdup and decreased bubble size (Mersmann *et al.*, 1988; Maruoka and Brauer, 1989; Midoux and Charpentier, 1984 and Brauer, 1988). If an absorption process is accompanied by a fast chemical reaction, the available interfacial area per unit volume will be the main design parameter since it is much more affected directly by the gas holdup and agitation level (Veljkovic and Skala, 1988, 1989). In order to increase the overall mass transfer rate between the phases, it is necessary to increase the liquid film mass transfer coefficient and/or interfacial area. This is connected with extra cost usually as external energy input. It is advisable that an increase in the mass transfer rate can be achieved in more suitable way by increasing interfacial area rather than liquid film mass transfer coefficient. An advancement of bubble column is Karr's Reciprocating Plate Column (KRPC). It is a type of multi phase reactor in which agitation is provided by a set of perforated plates moving up and down through a central shaft in the column. By introducing a small amount of external energy through the reciprocating plates, uniform gas dispersion is created and a frequent renewal of the interfacial area is resulted along with increased interfacial mass transfer. The main advantages of this reciprocating plate column are large gas holdup (Mustapha and Thibault, 1993), nearly uniform bubble size, large interfacial area (Yang *et al.*, 1986) and reduced axial mixing. Although the studies on reciprocating plate columns (Mustapha and Thibault, 1993; Brauer and Sucker, 1979; Ramo Rao *et al.*, 1983, 1986, and 1988 Veljkovic and Skala, 1983, 1989; Yang *et al.*, 1986) and its application are quite abundant as a gas-liquid contactor, there has been considerable general interest in the effects of rotary and reciprocatary flow on gas-liquid mass transfer.

The present study lies in combining the effects of bubble column, stirred tank reactor and reciprocating plate column. This resulted in Hybrid Rotating and Reciprocating Perforated Plate Bubble Column. Recently we have reported the results of hydrodynamic parameters (Dhanasekaran and Vijayagopal, 2007) namely Gas holdup, bubble diameter and interfacial area for this hybrid column. In a continuing study, the objective of the present investigation is to study the mass transfer parameter *i.e.*, volumetric mass transfer coefficient (k_La) of this hybrid column. Since the knowledge of the volumetric mass transfer coefficient or individual parameter such as the liquid film mass transfer coefficient 'k_L' and specific interfacial area 'a' is of essential importance for the designing gas -liquid contracting equipment.

Experimental Setup

The experimental setup and its ancillary connections are shown schematically in Figure 7.1. The column is 10.0 cm in internal diameter and 120 cm in height flanged acrylic column. The plate stack height is 100 cm, extending from a position of 10 cm above the base of the column. The plate is of 0.2 cm thickness plastic sheet with circular perforation. The plates are mounted on a 1.0 cm diameter acrylic drive shaft with equal spacing. Copper sleeves are used to the arrangement the plate spacing. The plate stack is driven in vertical and horizontal hybrid motion by the bevel gear arrangement, attached to a variable speed DC motor. A frequency controller FC is used to control the speed. The amplitude A (Half of the full reciprocation stroke) could be adjusted up to 1 cm. The drive frequency could be maintained up to 3.5 sec^{-1}. The deoxygenated water is supplied from a tank to the top of the plate stack through a globe valve and rotameter by the pump. Two solenoid valves in the deoxygenated water line one at the inlet and the other at the outlet permits shut-off during holdup measurements. A globe valve permits adjustment of the solution outlet to control the interface level IL at the top of the column. Air is fed to the column from the air compressor through the pressure regulator and rotameter. The surge tank is used to avoid the pressure fluctuations. The air line is also provided with a solenoid valve which enables to measure holdups. Air leaves the top of the column at atmospheric pressure.

Experimental Procedure

Mass transfer parameters are measured for the absorption of oxygen from air into deoxygenated water at room temperature (27±1°C). Distilled water in a 100 L feed tank was deoxygenated by the addition of sodium sulphite. The amount of sodium sulphite is 10 gm/L is added so as to bring the oxygen concentration down from 0.5 to 2 mg/L and trace amount of copper sulphate was also added as a catalyst. The water surface in the feed tank was covered with a circular wooden floating lid to prevent uptake of oxygen from the ambient air. The dissolved oxygen concentration of distilled water at 27±1°C is found to be 8.0±0.11 mg/L (Metcalf and Eddy, 1995). The dissolved oxygen concentration measurement is carried out in PE 138 water quality analyzer (Elico Limited). Henry's law constant was taken from Foust. Alan *et al.* (1980) and used to calculate the equilibrium dissolved oxygen concentration C*. For the study of Effect of agitation level on volumetric mass transfer co-efficient, the agitation level is varied with the fixed rate of both gas and liquid phases. After the interface level reaches the steady state, gas holdup is measured and solution samples are collected at the outlet of the column for the dissolved oxygen concentration measurement. Like that the effect of superficial gas and liquid velocity on volumetric mass transfer co-efficient is studied with maintaining the other parameters held constant.

The mass balance of sulphite in the liquid phase gives a basis for calculating the average value of kLa in the column for steady-state continuous counter current flow condition is (Dejan Skala and Vlada Veljkovic, 1988)

$$k_L a = \frac{U_L(C_{.in} - C_{.out})}{2LC*} \tag{1}$$

The values of $k_L a$ under steady-state conditions are calculated on the basis of the following assumptions (Dejan Škala and Vlada Veljkovic, 1988)

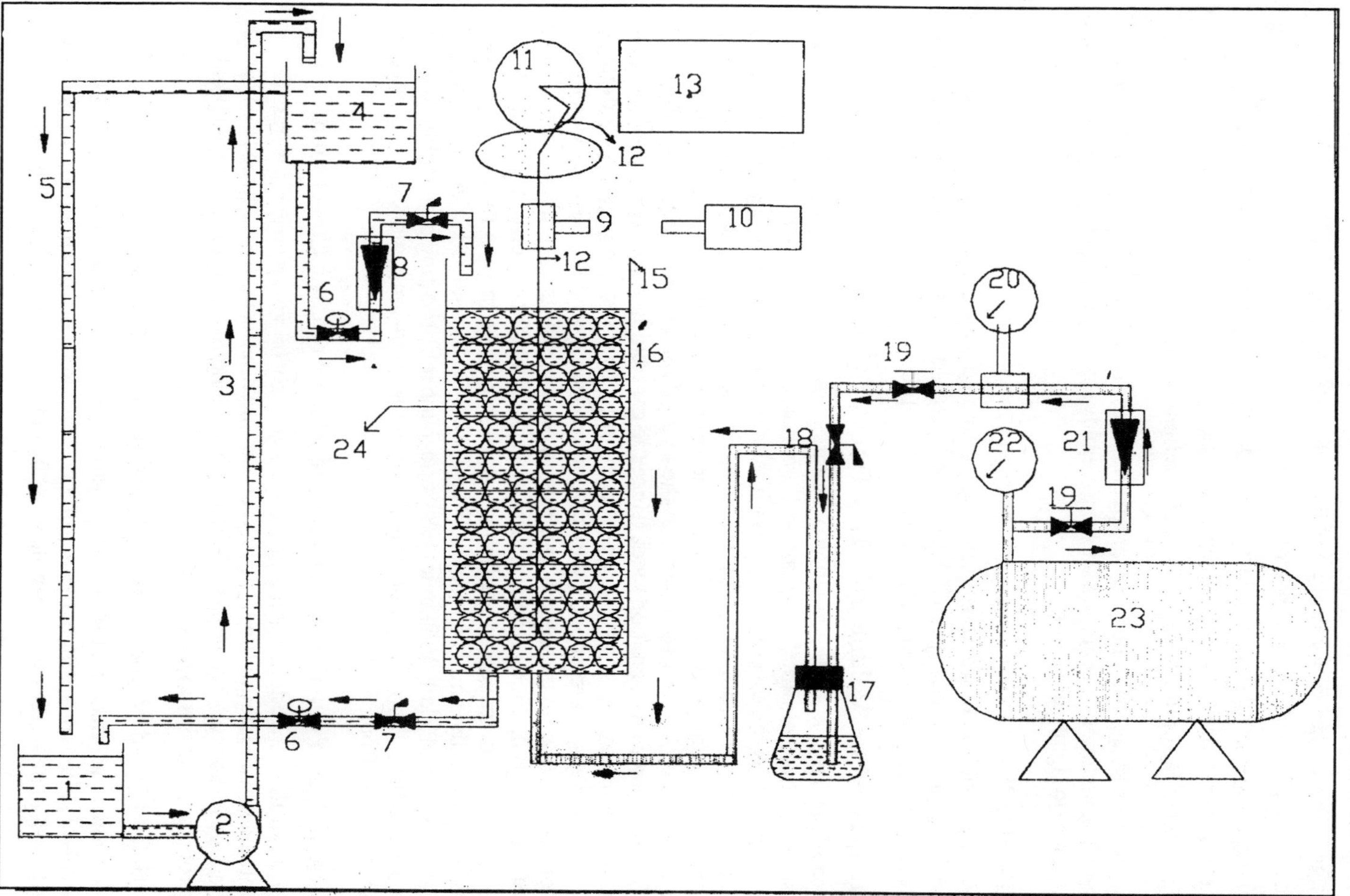

Figure 7.1: Experimental Setup: Hybrid Rotating and Reciprocating Perforated Plate Bubble Column

1: Sump; 2: Pump; 3: Water Line; 4: Overhead Tank; 5: Water Recirculation; 6: Globe Valve (Water); 7: Solinoid Valve (Water); 8: Rotameter (Water); 9: Sensor Point for Rotation; 10: RPM Indicator; 11: Bevel gear arrangement; 12: Connecting Rod; 13: DC-Motor; 14: Central Shaft; 15: Column (Acrylic); 16: Perforated Plates; 17: Surge Tank; 18: Solinoid Valve (Air); 19: Needle Valve (Air); 20: Pressure Regulator; 21: Rotameter (Air); 22: Pressure Gauge; 23: Air Compressor; 24: Gas Bubbles.

1. Plug flow of the phases,
2. The change in the oxygen concentration of the gas phase is negligible,
3. Gas holdup along the column is constant,
4. The equilibrium dissolved oxygen concentration is a constant with respect to height,
5. The gas phase resistance to mass transfer is negligible,
6. The enhancement factor under experimental conditions is 1.

The fourth assumption meant that the mean arithmetic value of oxygen solubility at the bottom and top of the column. The assumption enables one to considerably simplify the model for the calculation of k_La while introducing no significant error.

Results and Discussion

The value of k_La in hybrid rotating and reciprocating perforated plate bubble column is expected to be dependent on the agitation level, superficial gas velocity, superficial liquid velocity, perforation diameter, plate spacing and physico-chemical properties of the phases.

Effect of Agitation Level on Volumetric Mass Transfer Co-efficient

Figures 7.2a–c is shown that k_La is the function of agitation level. This is to be expected from the strong effect of increased gas holdup. This increased gas holdup results more dispersion of gas bubbles in the liquid. Based on the dispersion of gas bubbles with respect to the agitation level, the column is classified into three regions namely mixer-settler region (upto the agitation level of 1.5 cm/s), transition region (agitation level of 1.5 cm/s) and emulsion region (above the agitation level of 1.5 cm/s). This is discussed in detailed manner in the hydrodynamics part (Dhanasekaran and Vijayagopal, 2007). With increasing agitation level, (since the gas holdup decreased upto transition region) the increase of volumetric mass transfer co-efficient upto the transition region is moderately small. In the mixer-settler region, corresponding to lower intensities of agitation level, oxygen mass transfer depends on the aeration. In this region, this hybrid column seems to a bubble column with irregular shapes of bubbles. With increasing agitation level at a constant superficial gas and liquid velocity, the breakage process of dispersed gas bubbles starts to be more pronounced and intensive oxygen mass transfer occurs in the emulsion region. The same effect is observed in the reciprocating plate column by Dejan Skala and Vlada Veljkovic (1988) and in the reciprocating plate bubble column by Baird and Rama Rao (1988). From the graph, it is found in the emulsion region, that the effect of agitation level on k_La is very strong.

Effect of Superficial Gas Velocity on Volumetric Mass Transfer Co-efficient

Figures 7.3a–c show the dependence of superficial gas velocity on volumetric mass transfer co-efficient with agitation level and liquid flow rate held constant. The effect is found to be linear. With increasing superficial gas velocity, k_La increases due to higher gas holdup and the enhanced breakup of bubbles during their passing through the plate orifices.

Effect of Superficial Liquid Velocity on Volumetric Mass Transfer Co-efficient

Figures 7.4a–c shows the effect of superficial liquid velocity on volumetric mass transfer Co-efficient. At low levels of agitation, k_La increases slightly with liquid flow rate and at higher agitation levels the effect is found to be significant. A similar conclusion is reached earlier by Rama Rao and

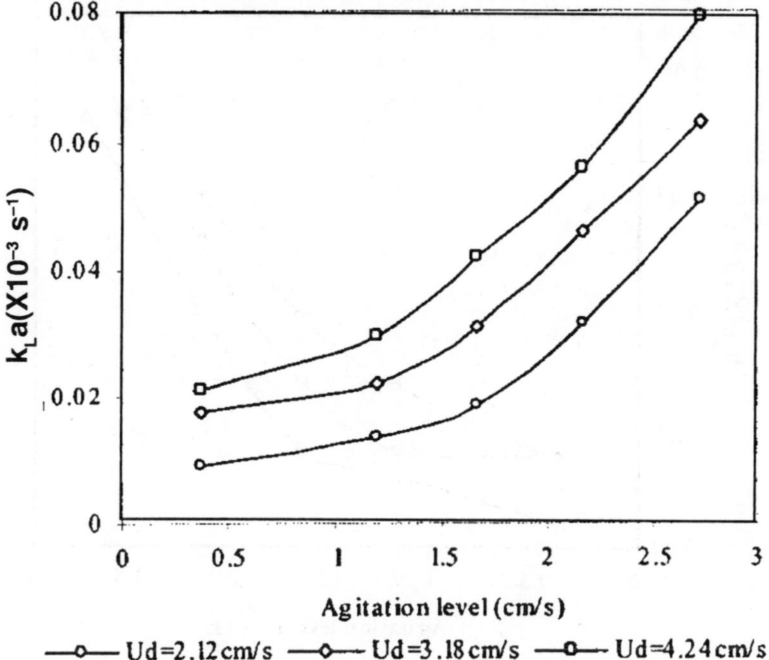

Figure 7.2a: Effect of Agitation Level on Volumetric Mass Transfer Co-efficient, U_c = 0.035 cm/s, d_p = 0.8 cm, S_p = 4.5 cm, n_p = 10

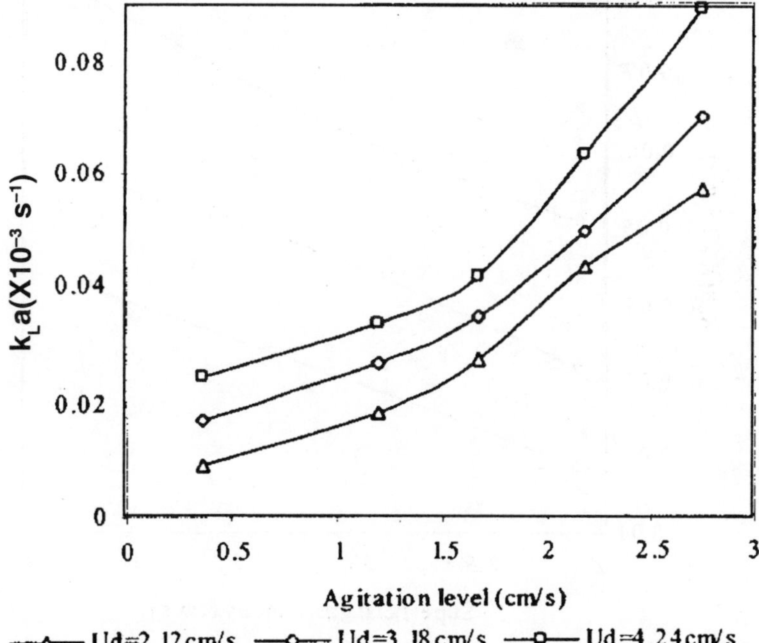

Figure 7.2b: Effect of Agitation Level on Volumetric Mass Transfer Co-efficient, U_c = 0.07 cm/s, d_p = 0.8 cm, S_p = 4.5 cm, n_p = 10

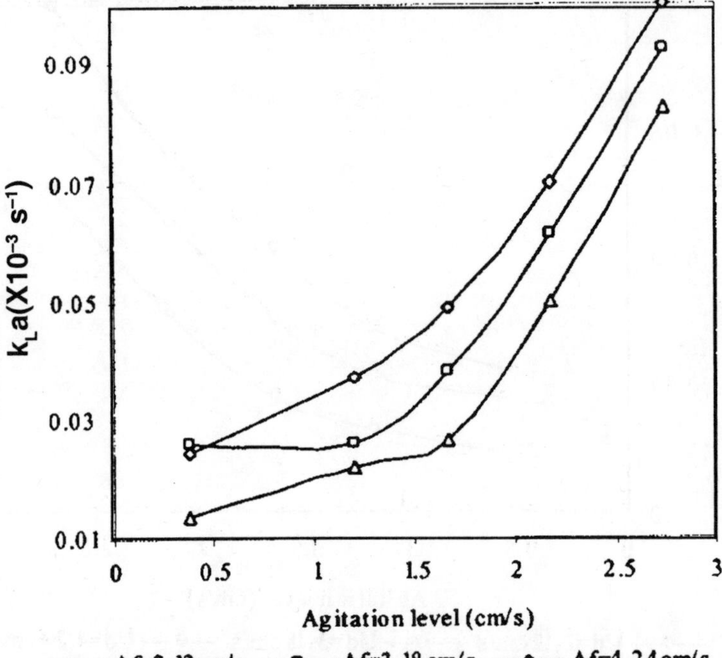

Figure 7.2c: Effect of Agitation Level on Volumetric Mass Transfer Co-efficient, U_c = 0.106 cm/s, d_p = 0.8 cm, S_p = 4.5 cm, n_p = 10

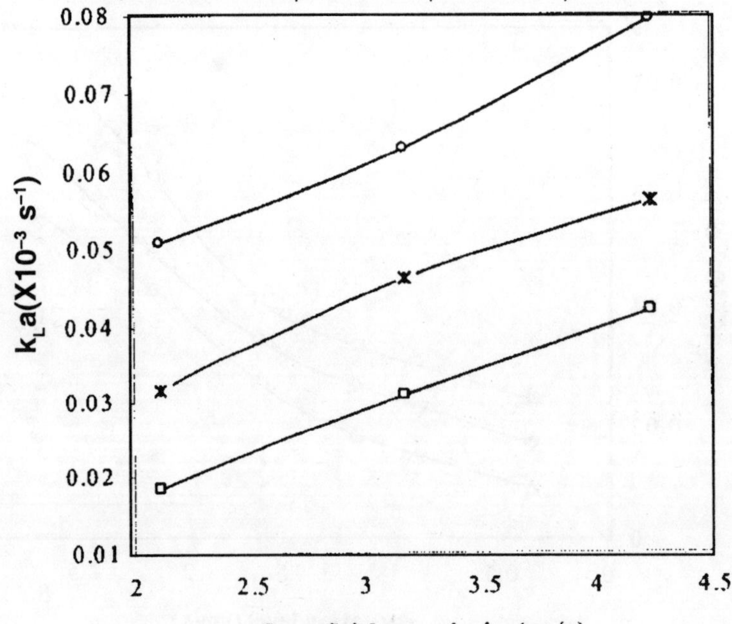

Figure 7.3a: Effect of Superficial Gas Velocity on Volumetric Mass Transfer Co-efficient, U_c = 0.035 cm/s, d_p = 0.8 cm, S_p = 4.5 cm, n_p = 10

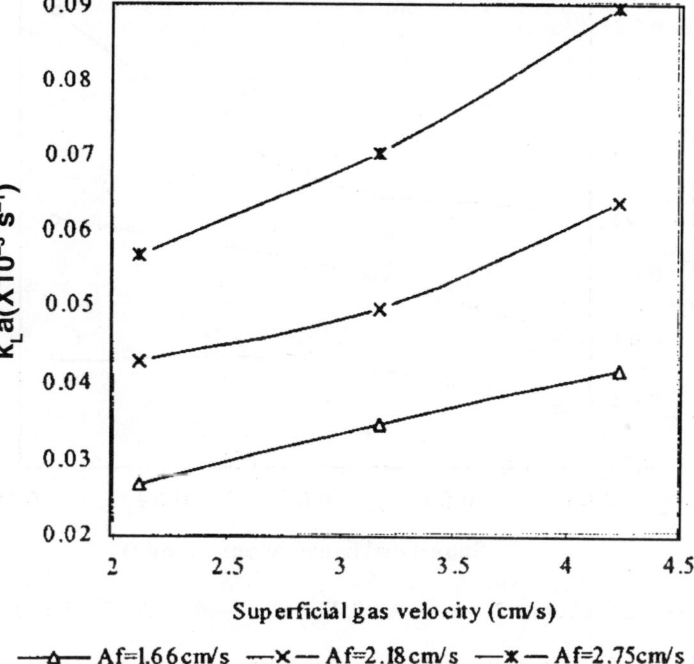

Figure 7.3b: Effect of Superficial Gas Velocity on Volumetric Mass Transfer Co-efficient, U_c = 0.07 cm/s, d_p = 0.8 cm, S_p = 4.5 cm, n_p = 10

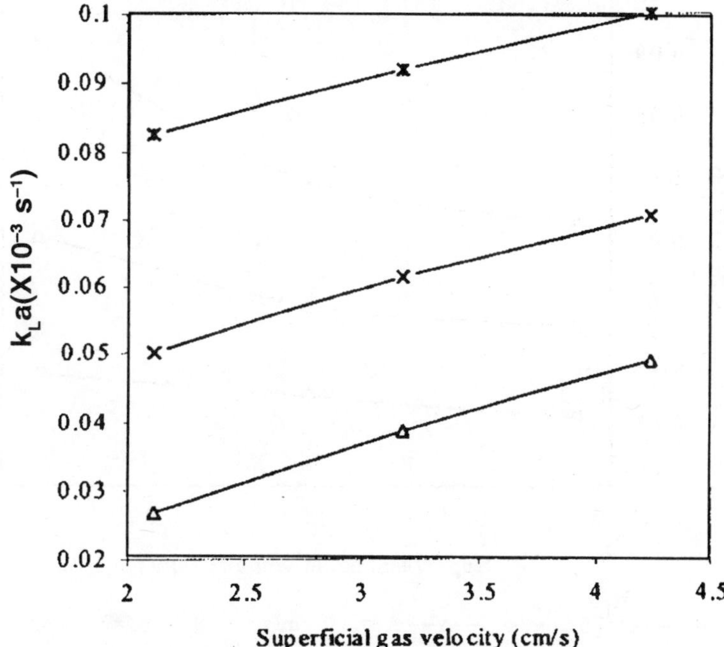

Figure 7.3c: Effect of Superficial Gas Velocity on Volumetric Mass Transfer Co-efficient, U_c = 0.106 cm/s, d_p = 0.8 cm, S_p = 4.5 cm, n_p = 10

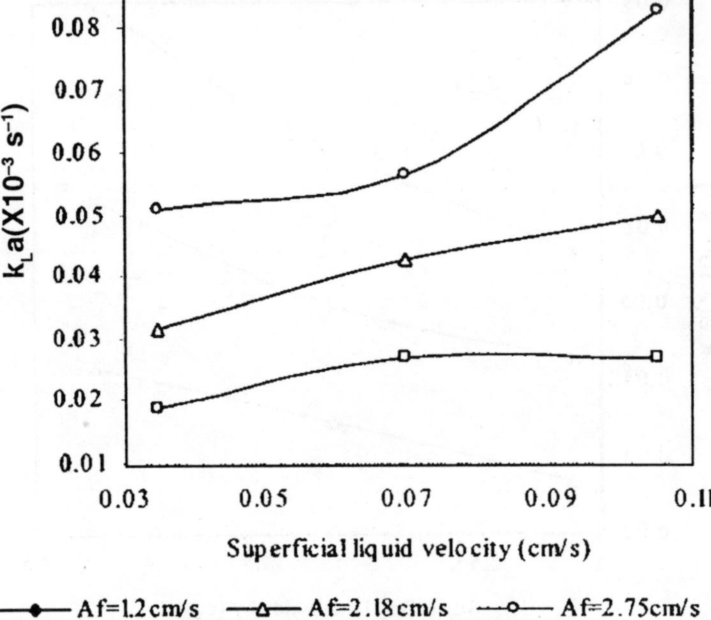

Figure 7.4a: Effect of Superficial Liquid Velocity on Volumetric Mass Transfer Co-efficient, U_d = 2.12 cm/s, d_p = 0.8 cm, S_p = 4.5 cm, n_p = 10

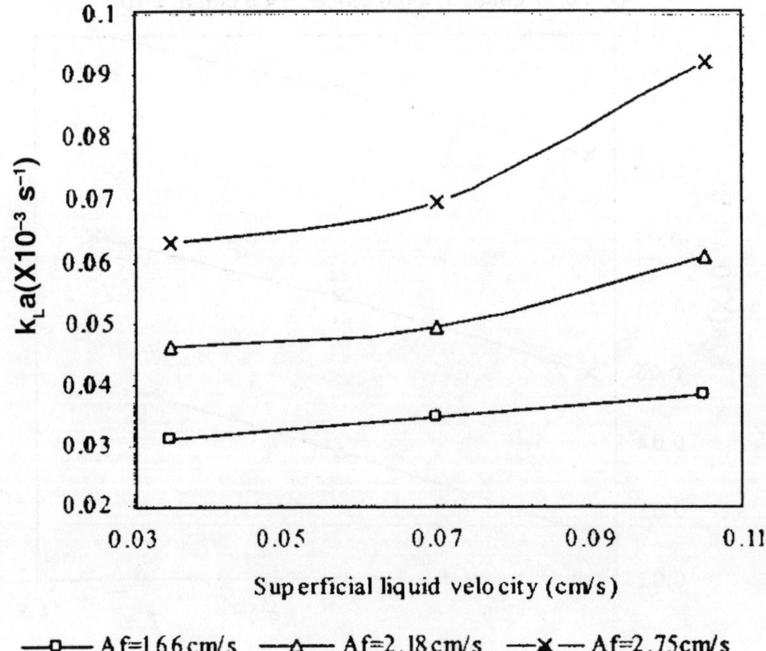

Figure 7.4b: Effect of Superficial Liquid Velocity on Volumetric Mass Transfer Co-efficient, U_d = 3.18 cm/s, d_p = 0.8 cm, S_p = 4.5 cm, n_p = 10

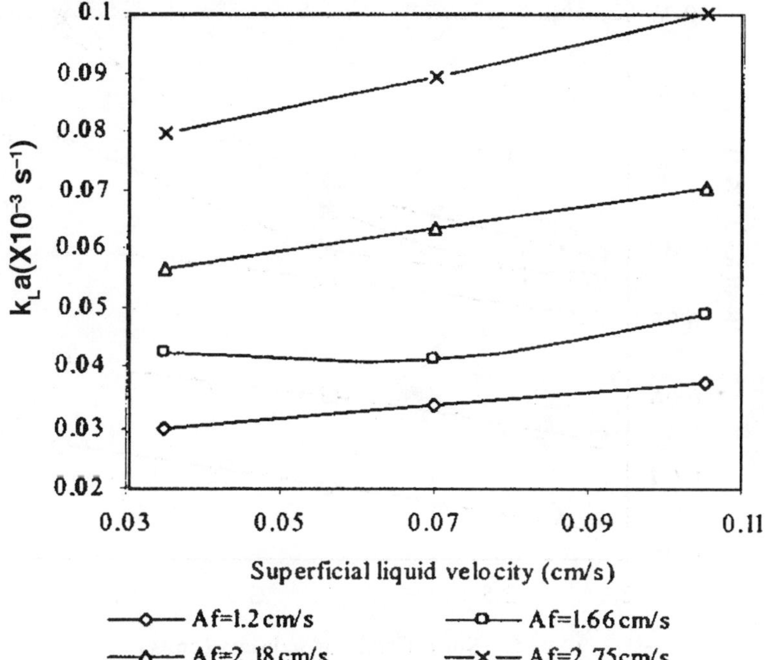

Figure 7.4c: Effect of Superficial Liquid Velocity on Volumetric Mass Transfer Co-efficient, U_d = 4.24 cm/s, d_p = 0.8 cm, S_p = 4.5 cm, n_p = 10

Baird (1988) and by Dhanasekaran and Vijayagopal (2007) regarding the effect of agitation level upon holdup.

Effect of Plate Spacing on Volumetric Mass Transfer Co-efficient

The column used in this study is very similar in geometrical characteristics, and only the effect of spacing between the plates, *i.e.*, of the number of plates, on $k_L a$ has been considered. As can be seen in Figure 7.5, for a given agitation level and superficial gas velocity, a higher value of ha is obtained with larger number of plates (spacing between the plates reduced). It has been already discussed in hydrodynamics part (Dhanasekaran and Vijayagopal, 2007) that the gas holdup is increased with increase in number of plates.

Correlation for Determination of Volumetric Mass Transfer Co-efficient

From the gas-liquid mass transfer investigation it is found that, the volumetric mass transfer co-efficient is depends on agitation level, Plate spacing, superficial liquid velocity, superficial gas velocity, liquid phase density and liquid phase viscosity.

$$k_L a = f(Af, S_p, U_d, U_L, \rho_L, \mu_L) \tag{2}$$

In addition to that, it is found from the previous hydrodynamic investigations (Dhanasekaran and Vijayagopal, 2007), the gas holdup have a linear dependence with all the above said parameters except the parameter Af. The gas holdup is first goes to decrease and then increases with increasing agitation level. It seems to be a non-linear dependence.

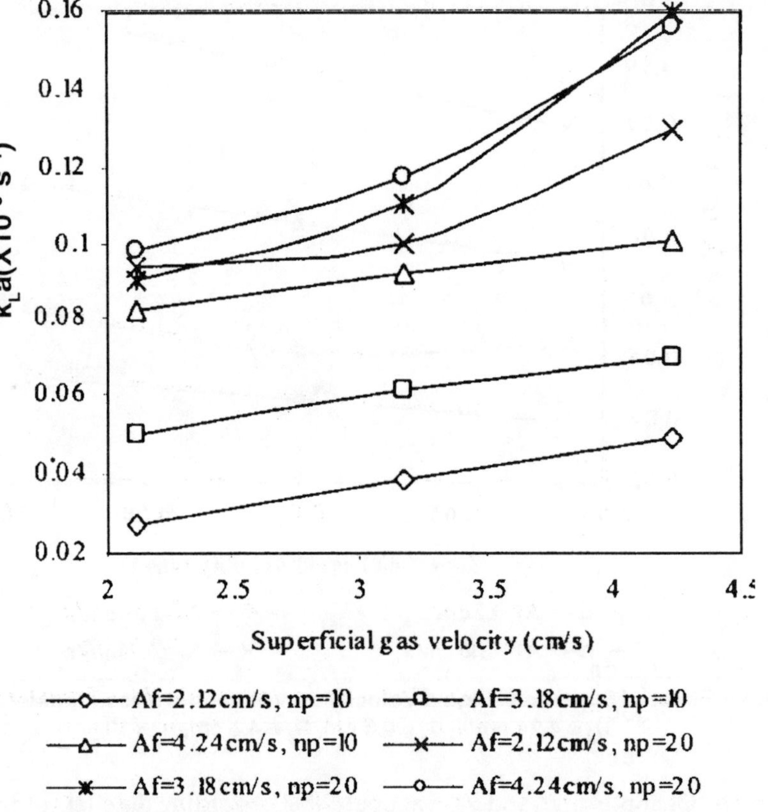

Figure 7.5: Effect of Plate Spacing on Volumetric Mass Transfer Co-efficient, U_c = 0.105 cm/s, d_p =0.8 cm

Therefore, the Oxygen mass transfer in this hybrid column could be correlated in terms of agitation level and gas holdup itself.

$$k_La = f(Af,\varepsilon) \tag{3}$$

Regression analysis is performed for the evaluation of constants and co-efficient. The regression co-efficient is found to be 0.9506. The standard deviation of percentage error is 6.5.

$$k_La = 0.0001(Af)^{0.8}(\varepsilon)^{0.8} \tag{4}$$

For checking the significance of the correlation, the Null hypothesis is used. The F value for the regression is 767. The table value of F for (3, 42) degree of freedom at 5 per cent probability level obtained from statistical table is 2.84. Since the calculated value of F is greater than the table value, Ho is rejected. It means that the developed correlation 5 is significant with 95 per cent confidence level. In Figure 7.6 the measured values of k_La is compared with calculated values and found to concur.

where,

$$\varepsilon_G = 0.035\left[\frac{Af}{U_L}\right]^{-0.068}\left[\frac{U_d}{U_L}\right]^{-0.378}\left[\frac{d_p.U_L.\rho_L}{\mu_L}\right]^{-0.364}\left[\frac{d_p}{S_P}\right]^{0.095} \tag{5}$$

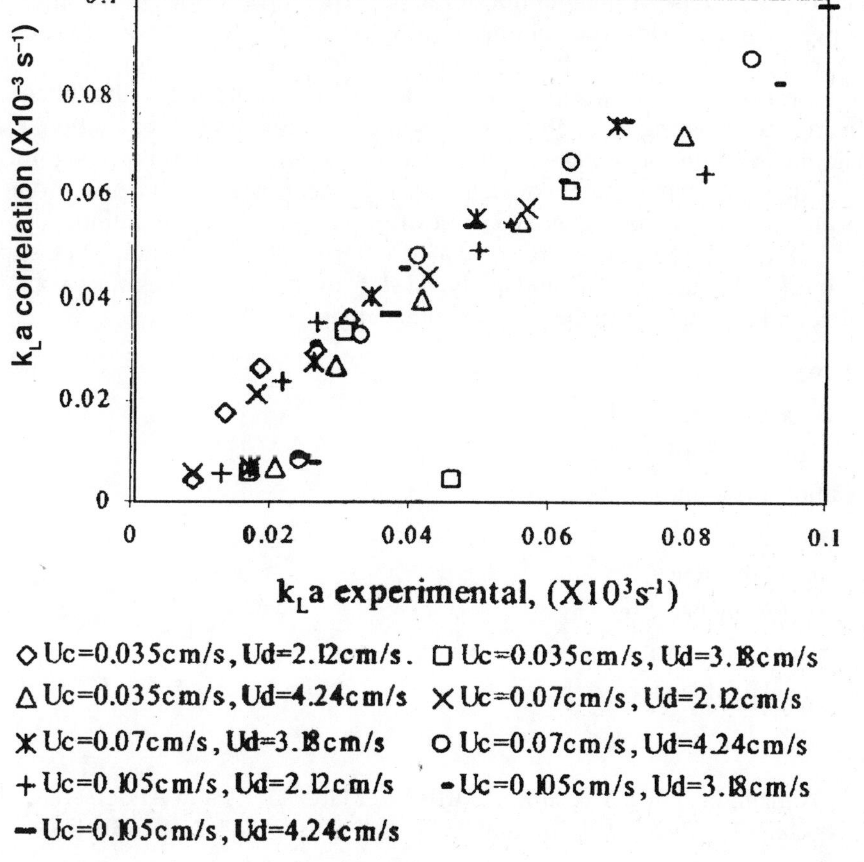

Figure 7.6: Comparison of Measured Values of $k_L a$ with Values Calculated from Correlation 4

For mixer-settler region to the transition region.

$$\varepsilon_G = 0.019\left[\frac{Af}{U_L}\right]^{-0.456}\left[\frac{U_d}{U_L}\right]^{-0.235}\left[\frac{d_p.U_L.\rho_L}{\mu_L}\right]^{-0.65}\left[\frac{d_P}{S_P}\right]^{0.56} \tag{6}$$

For transition region to the emulsion region.

And for interfacial area,

$$a = \frac{4.53}{d_P}\left[\frac{U_d}{U_L}\right]^{0.13}\left[\frac{Af}{U_L}\right]^{0.41}\left[\frac{S_P}{d_P}\right]^{-0.67}\left[\frac{d_p.U_L.\rho_L}{\mu_L}\right]^{-0.88}\left[\frac{d_p.U_L{}^2.\rho_L}{\sigma_L}\right]^{-0.2} \tag{7}$$

Conclusion

The slow chemical absorption of oxygen by sodium sulphite method is used for measuring $k_L a$ in hybrid rotating and reciprocating perforated plate bubble column. The volumetric mass transfer co-efficient k_La is found to depend on agitation level, superficial gas velocity and superficial liquid velocity. The effect of agitation level on volumetric mass transfer co-efficient is found to be strong. The effect of superficial gas velocity on volumetric mass transfer co-efficient is found to be linear. With increasing superficial gas velocity, k_La increases due to higher gas holdup and the enhanced breakup of bubbles during their passing through the plate orifices. At low levels of agitation, k_La increases slightly with liquid flow rate and at higher agitation levels the effect is found to be significant The co-efficient kLa also depends on the number of plates, *i.e.*, the spacing between the plates. When the other operating conditions are the same, higher kLa values are obtained with larger number of plates due to the higher gas holdup. The ha values are correlated in terms of the agitation level, superficial air velocity and superficial liquid velocity and analyzed statistically. The values of kLa obtained by the correlation are found to concur with the experimental values.

Nomenclature

- a: interfacial area, cm^2/cm^3
- A: Amplitude of reciprocation, cm
- Af: Agitation level, cm/s
- f: Frequency, s^{-1}
- n_p: Number of the plates
- S_p: Space between the plates, cm
- U_d: Superficial gas velocity, cm/s
- U_c: Superficial liquid velocity, cm/s

References

Alan, Foust S., Wenzel, Leonard A., Clump, Curtis W., Maus, Louis and Andersen, L. Bryce, 1980. *Principles of Unit Operations*, 2nd Edn. John Wiley and Sons, p. 734, Appendix D–3.

Baird, M.H.I. and Rice, R.G., 1975. Axial dispersion in large unbaffled columns. *Chem. Eng. J.*, 9: 171–174.

Baird, M.H.I. and Rama Rao, N.V., 1988. Characteristics of a countercurrent reciprocating plate bubble column II: Axial mixing and mass transfer. *Can. J. Chem. Eng.*, 66: 222–231.

Brauer, H. and Sucker, D., 1979. Biological wastewater treatment in a high efficiency reactor. *Ger. Chem. Eng.*, 2: 77–86.

Brauer, H., 1988. Mass transfer machines: The road to better design of equipment in the processing industries. *Int. Chem. Eng.*, 28(2): 207–219.

Dhanasekaran, S. and Vijayagopal, V., 2007. Hydrodynamic investigations in hybrid rotating and reciprocating perforated plate bubble column. *J. Curr. Sci.*, 10(2): 907–917.

Gagnon, Helene, Lounes, Mustapha and Thibault, Jules, 1998. Power consumption and mass transfer in agitated gas-liquid columns: A comparative study. *Can. J. Chem. Eng.*, 76: 379–389.

Lee, Dong-Hyun, Grace, John R. and Epstein, Norman, 2000. Gas holdup for gas velocities in a gas-liquid concurrent upward-flow system. *Can. J. Chem. Eng.*, 78: 1006–1010.

Lounes, M., Audet, J., Thibault, J.and Leduy, A., 1995. Mass transfer in a reciprocating plate bioreactor. *Bioprocess Eng.*, 13: 1–11.

Meikap, Bhim C., Kundu, Gautam and Biswas, Manindra N., 2002. Prediction of dispersed phase holdup in a modified multi-stage bubble column scrubber. *Can. J. Chem. Eng.*, 80: 306–312.

Maruoka, Y., and Brauer, H., 1989. Fluid dynamics and mass transfer in a multistage rotating disk reactor. *Int. Chem. Eng.*, 29(4): 577–615.

Mersmann, A., Voit, H., and Zeppenfeld, R., 1988. Do we need mass transfer machines. *Int. Chem. Eng.*, 28(1): 1–13.

Metcalf and Eddy, 1995. *Wastewater Engineering: Treatment, Disposal and Reuse*, 3rd Edn. Tata McGraw-Hill Publishing Company Limited, p. 1258, Table E–1.

Midoux, N and Charpentier, J.-C., 1984. Mechanically agitated gas-liquid reactors, Part 1: Hydrodynamics. *Int. Chem. Eng.*, 24(2): 49–287.

Midoux, N and Charpentier, J.-C., 1984. Mechanically agitated gas-liquid reactors, Part 2: Interfacial Area. *Int. Chem. Eng.*, 24(2): 452–481.

Miyanami, K., Tojo, K., Minami, I. and Yano, T., 1978. Gas-liquid mass transfer in a vibrating disk column. *Chem. Eng. Sci.*, 33: 601–608.

Lounes, Mustapha and Thibault, Jules, 1993. Hydrodynamics and power consumption of a reciprocating plate gas-liquid column. *Can. J. Chem. Eng.*, 71: 497–506.

Rama Rao, N.V., Srinivas, N.S. and Varma, Y.B.G., 1983. Dispersed phase holdup and drop size distribution in reciprocating plate column. *Can. J. Chem. Eng.*, 61: 168–177.

Rama Rao, N.V. and Baird, M.H.I., 1986. Gas-liquid pressure drop studies in a reciprocating plate column. *Can. J. Chem. Eng.*, 64: 42–47.

Rama Rao, N.V. and Baird, M.H.I., 1988. Characteristics of a countercurrent reciprocating plate bubble column, I: Holdup, pressure drop and bubble diameter. *Can. J. Chem. Eng.*, 66: 211–221.

Skala, Dejan and Veljkovic, Vlada, 1988. Mass transfer studies in a gas-liquid reciprocating plate column-1: Liquid phase volumetric mass transfer coefficient. *Can. J. Chem. Eng.*, 66: 192–199.

Tinge, J.T. and Drinkenburg, A.A.H., 1986. The influence of slight departures from vertical alignment on liquid dispersion and gas hold-up in a bubble column. *Chem. Eng. Sci.*, 41: 165–169.

Veljkovic, V. and Skala, D., 1986. Hydrodynamic investigations of gas-liquid contacting in a reciprocating plate column. *Can. J. Chem. Eng.*, 64: 906–914.

Veljkovic, V. and Skala, D., 1988. Mass transfer characteristics in a gas-liquid reciprocating plate column II: Interfacial area. *Can. J. Chem. Eng.*, 66: 200–210.

Veljkovic, V. and Skala, D., 1989. On the possibility of general prediction of gas holdup in reciprocating plate column. *Can. J. Chem. Eng.*, 67: 1015–1018.

Yang, N.S., Shen, Z.Q., Chen, B.H. and McMillan, A.F., 1986. Pressure drop, gas holdup and interfacial area for gas-liquid contact in Karr columns. *Ind. Eng. Chem. Proc. Des. Dev.*, 25: 660–664.

Yawalkar, Archis A., Pangarkar, Vishwas G. and Beenackers, Anthony A.C.M., 2002. Gas hold-up in stirred tank reactors. *Can. J. Chem. Eng.*, 80: 158–166.

Chapter 8

Biodiversity and Carbon Sequestration, Two Important Values of the Wetlands of West Bengal

Malabika Biswas, Bipal Kumar Jana, Mrinmoy Majumder, Soumyajit Biswas, Pankaj Kumar Roy, Arunabha Majumder and Asis Mazumdar

School of Water Resources Engineering, Jadavpur University, Kolkata

ABSTRACT

Wetlands cover 3 per cent of the Indian landmass, or nearly 100 000 sq. km (Olson *et al.*, 1983). Wetlands in India harbor a vast variety of life forms that are a part of the complex food of these transitional ecosystems. About 320 species of bird are associated with Wetlands of India. Apart from birds, the wetlands support a diverse population of plants and animals including 150 species of amphibian. Wetlands are the habitat of some of the world's endangered and threatened flora and fauna. Wetland ecosystems are cradles of biological diversity, providing the water and primary productivity upon which countless species of plant and animal depend for survival. They support high concentrations of birds (especially waterfowl), mammals, reptiles, amphibians, fish and invertebrate species. Levels of species diversity do, however, vary considerably between different wetland ecosystems: some lakes display high levels of diversity, whereas others support little life. Unlike terrestrial ecosystems, the richness of freshwater biodiversity is still poorly known. Identification and classification of wetland species is hampered by the fact that many species may spend part of their life cycle in both freshwater and marine ecosystems. Wetland ecosystems are also valuable regarding carbon sequestration. The aquatic plants are very much important as well as terrestrial plants for carbon sequestration. The magnitude of carbon storage depends upon wetland type and size, vegetation, depth of wetland soils, groundwater levels, nutrient levels, pH value and other factors. The net carbon sequestering versus carbon release roles of wetlands are complex and change over time. The conversion of CO_2 to carbonate minerals through sediment formation by a series of reactions may contribute significantly to CO_2 sequestration during a long period.

Keywords: *Wetlands, Biodiversity, Utilization, Carbon sequestration, Carbon stock, Conservation and participatory management.*

Introduction

The National Biodiversity Strategy and Action Plan (NBSAP) process is one of the largest environmental planning exercises being carried out in the world and certainly the largest ever in India. Initiated by the Government of India's Ministry of Environment and Forest funded by GEF through UNDP and technically executed by the NGO Kalpavriksh, the NBSAP in India is being developed through a participatory planning process involving all major stakeholders (Kothari, 2000).

The NBSAP process entails the assessment and stock of biodiversity related information at various levels including distribution of endemic and endangered species and site-specific threats and pressures with an ultimate aim to develop Strategy and Action Plans (SAP's) at the following five levels as below:

1. Local and Regional Strategy and Action Plan (LSAP's) for a few related in the country, *e.g.*, Karbi–Anglong District in Assam, North Coastal Belt in Andhra Pradesh and Vidarbha in Maharashtra.
2. State level Strategy and Action Plans (LSAP's) for all Indian States and Union Territories.
3. Inter-state Eco-regional Strategy and Action Plans (IESAP's) for biological regions cutting across states (*i.e.* eastern ghats, western ghats, western coast and West Himalayas).
4. Separate detailed Thematic Strategy and Action Plan (TSAP's) for major topics related to biodiversity.
5. A National Strategy and Action Plan (NSAP) for the whole country will build on the four levels above.

West Bengal is unique as the only state in India to stretch from the sea shore to the loftiest peak of Himalayas. The hilly terrains with temperate and tropical forests, tidal creeks, swamps and mangroves, include the extensive sundarban make the biodiversity. As it is impossible to cover all the wetlands spanning over such an extensive area, some representative wetlands such as Bhomra beel, Ramnagar beel, Mathura beel, Subhas Sarobar, Mirik lake and East Kolkata Wetlands from different regions of West Bengal have been chosen as study areas and attempt has been made to compile some unforeseen facts and figures about them as well as developing sustainable participatory management plan of these different types of wetlands.

The threats as identified to conserve the biodiversity are:

1. Habitat loss,
2. Lack of biodiversity awareness especially in development initiatives.
3. Demographic pressure.

4. Preference to monoculture and indiscriminate use of agro-chemicals.
5. Unregulated marine fishing and prawn seed collection in different countries.

Today, wetlands represent the largest carbon pool with capacity of 770 GtC, overweighing the total carbon storage of farms, and temperate and rain forests. Besides, the carbon sequestration rate of wetlands is continuous and many wetlands have been carbon sinks for 4000–5000 years (Li *et al.*, 2006). The carbon sequestration potential of wetlands was 6.88 TgC, which accounts for 0.8 per cent of current annual C emission (Mauquoy *et al.*, 2002). It is necessary to maintain the carbon sequestration ability through wetland protection and restoration. Because wetlands are fragile ecosystems, it takes years for carbon sequestration ability of restored wetlands to reach the natural levels, such as peat land (Waddington *et al.*, 2001; Tuittila *et al.*, 1999), coastal marsh (Craft *et al.*, 2002) and forested wetlands (Cui *et al.*, 2005). Compared with wetland restoration, wetland protection is an efficient and economical approach to maintain carbon fixing function, most of which currently lack of effective protection.

Xiaonan *et al.*, 2008, investigated the carbon sequestration potential of lakes and swamps in China. They found significant differences among the carbon sequestration potential of various lakes, determined by natural conditions and human disturbance. In this study, swamps had a carbon sequestration potential of 4.90 TgC, much higher than lakes in China. Mangrove and coastal marsh have the highest carbon sediment rate among swamps. Carbon sequestration potential in returning farms to lakes and swamps was 30.26 and 0.22 $GgC.a^{-1}$, respectively. Under the ongoing national wetland conservation action plan in China, the carbon sequestration potential of wetland restoration was 6.57 $GgC.a^{-1}$. Protection and restoration measurements can improve carbon sequestration potential of wetlands.

Euliss Jr. *et al.* (2005), evaluated the potential of prairie wetlands in North America as carbon sinks. Agricultural conversion has resulted in the average loss of 10.1 Mg ha^{-1} of soil organic carbon on over 16 million ha of wetlands in this region. Wetland restoration has potential to sequester 378 Tg of organic carbon over a 10-year period. Wetlands can sequester over twice the organic carbon as no-till cropland on only about 17 per cent of the total land area in the region. We estimate that wetland restoration has potential to offset 2.4 per cent of the annual fossil CO_2 emission reported for North America in 1990.

Again according to Brevika *et al.*, 2003, wetlands have the potential to accumulate C at high rates over long time periods because they continuously accrete and bury organic-rich sediments, giving soils in wetlands a distinct advantage over many other environments in the sequestration of organic C. Given that wetlands are being lost worldwide, it is important to understand their C sequestration potential sediments in a southern California, USA wetlands complex were cored, and depositional environments were interpreted. Suitable materials were radiocarbon dated. Bulk density and organic C were grouped by depositional environments, and average mass of C per unit volume and C accumulation rates in each depositional environment were calculated. The total organic C sequestered and rates of sequestration in each depositional environment were in the following order from most (fastest) to least (slowest): lagoon, intertidal, salt marsh, freshwater marsh, aeolian. This study demonstrated that high levels of organic C are sequestered per unit volume of sediment (35.9F 3.2 kg m^{-3}), and the mean rate of C accumulation was high (0.033F 0.0029 kg C $m^{-2}year^{-1}$) over a long time period (5000 years). Results of this study strongly demonstrate the importance and necessary high priority for preserving and restoring wetlands both in the USA and internationally. However, despite their excellent potential to sequester C, significant losses of wetlands are occurring in the United States and elsewhere in the world.

Study Area

Four areas as delineated in Figure 8.1 in West Bengal in different locating representing different climate condition are chosen.These are

1. Mirik Lake is situated in the Kurseong sub division of Darjeeling district of West Bengal. On the western side close to Mirik flows Mechi River that demarcates the Indo-Nepal border. The total lake area is about 40 acres with approximate length of 1.025 Km and peripheral road about 3.5 km. With the construction of the lake in 1979,
2. Ramnagar beel is located in a village (84 Nebutala Mouza) under Ramanagar Borochupria-1 Gram Panchayat, Hanskhali Block of Nadia district, West Bengal. This wetland (92.92 acre) is an ox bow lake, which derives its origin from the river Ichhamati. The river Ichhamati in turn forms the boarder between India and Bangladesh.
3. Bhomra beel is situated under Kastadanga-1 and Kastadanga-2 Gram Panchayet, Haringhata Block of Nadia District, West Bengal. It is an oxbow lake covering 264 hectare, which derives its origin from the river Yamuna. The Bhomra Beel is located in between 22°59′N to 23°N and 88°38′E to 88°40′E.
4. Mathura Beel is located in the district of Nadia 24 Parganas (North). It is a meander cut off covering 184 hectare, which derives its origin from the river Yamuna.
5. The Subhas Sarobar (SS) (Latitude of 22°34′ to 22°34′30″N and Longitude 88°24′ to 88°24′30″E), under the administrative control of Kolkata Improvement Trust (KIT), represents the lung of some fringe areas of eastern Kolkata with massive environmental fillip.
6. The total area of East Kolkata Wetland is 12741.30 hectare comprising of 364 sewage fed fisheries, several other bodies, agricultural area, garbage disposal site, urban and rural settlement area. Out of this, total water area is about 5852.14 ha. Before 1930, the main source of water for the fisheries was the tidal Bidyadhari river. With the silting up of the river Bidyadhari, entire area became a vast derelict swamp with the elimination of tidal water in these areas.

Biodiversity Value of the Study Area

Biodiversity Value of Ramnagar Beel, Bhomra Beel and Mathura Beel

Human communities depend on freshwater biodiversity for a variety of resources. The various habitats provide a wide range of resources as well as services such as transport, supply of clean water, and energy. Fish is the most important resource and has a tremendous economic importance. Fish are also for fertilizer, medicine, the pet trade and form an integral part of the dynamic food chain. There are approximate 200 species of aquatic and semiaquatic plants in India. Most of the species have a very wide distribution all over the country. Site specific or endemic species are generally absent. Common plants belong to families Poaceae, Cyperaceae; Hydrocharitaceae, Alismaceae.

These beels are very much rich in biodiversity level. The different groups of animals recorded from the different wetlands in these three districts represent the different tropic levels of the food chain. Plants, being the primary producers, fix the solar energy into chemical energy by photosynthesis,

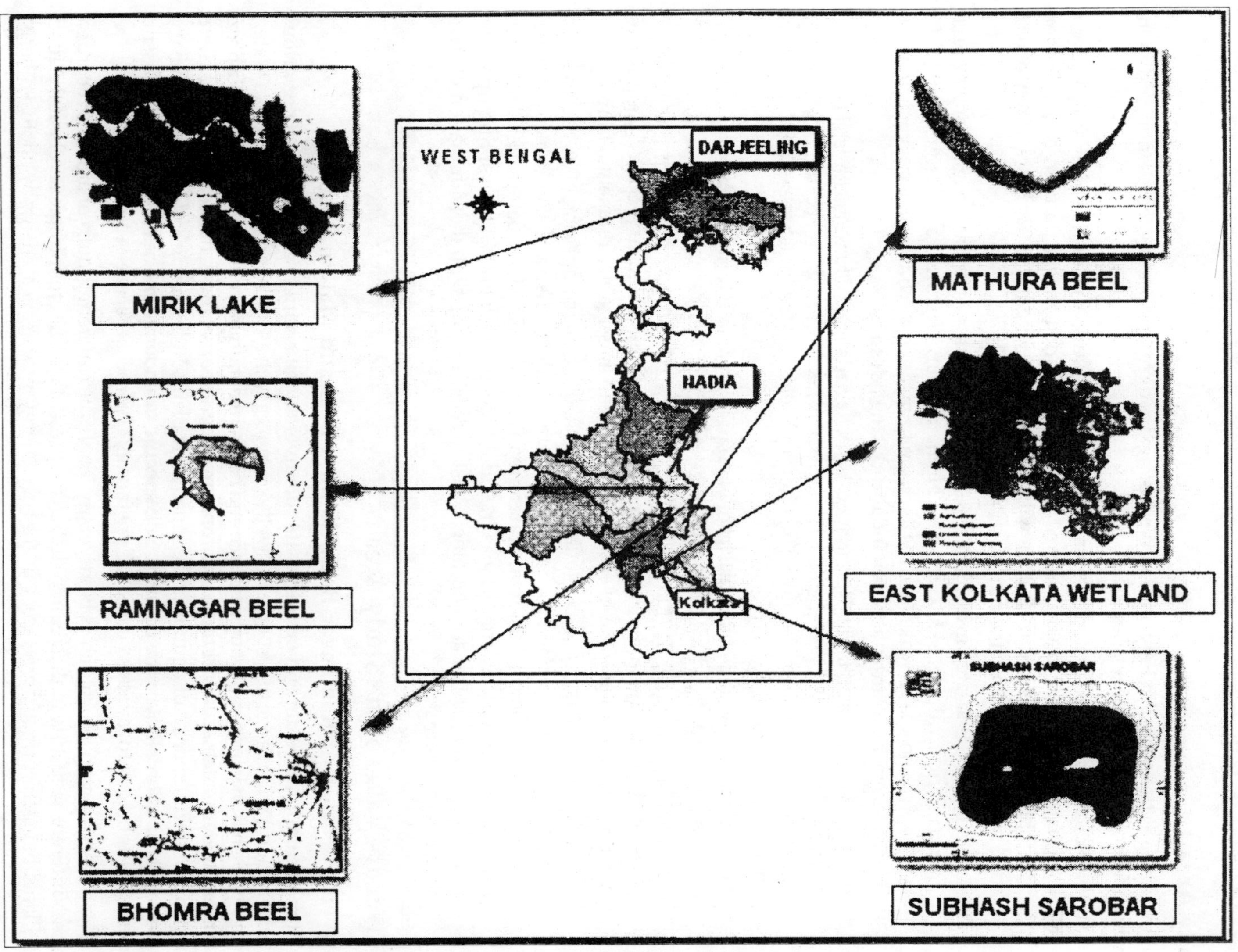

Figure 8.1: Location Map of Study Area

which is transferred to the different consumer levels successively. The zooplanktonic population like 'Daphnia, Moina, Ceriodaphnia, are the primary consumer groups, representing the first trophic level. They occupy both the littoral and limonitic zones. Among the aquatic insects, the Ephemeropterans and Coleopterans are the primary consumers. The ephemeropterans are periphytons, while the coleopterans comprise the neuston groups which are the surface dwellers. The odonates and hemipteran insects are carnivorous animals which represent the secondary and tertiary consumer groups of the food chain. They are the periphytons of the littoral zone. The Piscean populations are representatives of the neckton groups. They inhabit both the littoral and limnetic zones. Among them, Catla, species from the Puntius groups, the Notopterans are primary consumers, while Tilapia, Wallago Attu etc. represent the secondary consumer groups. The snails, earthworm, leach are the benthic populations occupying the primary consumer level in the food chain. A good number of birds are associated with these waterbodies. Among them some are residential, while others are migratory. They usually migrate during the winter. The present survey availed the opportunity of recording 60 species of the Avi-fauna of the Bhomra Beel, Ramnagar Beel and Mathura Beel.

Table 8.1: Birds Recorded in Bhomra, Mathura and Ramnagar Beel (Kundu *et al.*, 1997)

English Name	*Local Name*	*Scientific Name*
Little grebe	Pandubi	*Tachybaptus ruficollis*
Little cormorant	Choto pankauri	*P. niger*
Grey heron	Sada Kank	*Ardea cineria*
Large egret	Dhal bok	*Egretta alba*
Intermediate egret	Majhari bok	*E. intermediate*
Little egret	Chhoto bok	*E. garzetta*
Chestnuts bittern	Khayari bok	*Ixobrychus cinnamomes*
Open bill stork	Shamukh khol	*Anastomus oscitans*
Pin tail	Digh Hansh	*Anas acuta*
Common teal	Pati Hansh	*A. crecca*
Widgeon	Chhoto Lalshir	*A. penelope*
White eyed pilchard	Bhuti Hansh	*Athya nyroca*
Cotton teal	Bali Hansh	*Nettapus coromandelianus*
Ruddy Crake	Lal khairy	*Porzana fusca*
White breasted water hen	Dhaluk	*Amaurornis phoenicurus*
Moorhen	Jal Murgi	*Gallinula chloropus*
Common red shank	Chotobatan	*Tringa totanus*
Pintail snipe	Kada Khocha	*Gallinago stenura*
Black headed gull	Gang chill	*L. ridibundus*
Lesser pied kingfisher	Fatka machranga	*Ceryle rudis*
Common kingfisher	Choto machranga	*Alcedo hercules*
White breasted kingfisher	Sadabuk	*Halcyonpileata*
Stork billed kingfisher	Gurial	*Pelargopsis copensis*

Table 8.2: Aquatic Birds of Subhas Sarobar

Sl.No.	Scientific Name	Common Name	Occurrence	Migrature Nature
1	*Dendrocygna javanica*	Lesser whistling-teal	Common	Winter visitor
2.	*Alcedo atthis*	Common kingfisher	Common	Localized
3.	*Halcyon smyrnensis*	White-throated kingfisher	Common	Localized
4.	*Halcyon capensis*	Stork-billed kingfisher	Scarce	Localized
5.	*Amaurornis phoenicurus*	White-breasted waterhen	Common	Localized
6.	*Gallinule chloropus*	Common moorhen	Scarce	Winter visitor
7.	*Fulica atra*	Common coot	Rare	Winter visitor
8.	*Tringa hypoleucos*	Common sandpiper	Scarce	Winter visitor
9.	*Metopidius indicus*	Bronze-winged jacana	Common	Localized
10.	*Phalacrocorax niger*	Little cormorant	Common	Localized
11.	*Phalacrocorax fuscicollis*	Indian cormorant	Common	Localized
12.	*Phalacrocorax carbo*	Great cormorant	Scarce	Localized
13.	*Egretta garzetta*	Little egret	Scarce	Localized
14.	*Casmerodius albus*	Great egret	Common	Localized
15.	*Bubulcus ibis*	Cattle egret	Abundant	Localized
16.	*Ardeola grayil*	Indian pond-heron	Abundant	Localized
17.	*Ardea cinerea*	Grey heron	Scarce	Localized
18.	*Nycticorax nycticorax*	Black-crowned night-heron	Common	Localized

Table 8.3: Aquatic Macrophytes of Subhas Sarobar

Sl.No.	Scientific Name	Family	Common Name	Life Form
1.	*Alternanthera phylloxeroides*	Amaramthaceae	Alligator grass	Wetland herb
2.	*Aeschynomene aspera*	Fabaceae	Shoal	Wetland spongy herb
3.	*Alisma* sp.	Alismataceae	–	Emergent
4.	*Alternanthera philoxeroides*	Amaranthaceae	–	Wet land plants
5.	*Altrovanta vesicusa*	Cyperaceae	–	Free floating/submerge herb
6.	*Aponogeton* sp.	Aponogetonaceae	Ghechu	Submerged
7.	*Azolla pinnata*	Azollaceae	Azolla	Free floating
8.	*Carex* sp.	Cyperaceae	–	Wetland grass
9.	*Ceratophyllum* sp.	Ceratophyllaceae	–	Submerged/floating
10.	*Colocasia esculenta*	Araceae	Kachu	Root emergent
11.	*Commelina benghyalensis*	Commelinaceae	Kanchara	Wetland herb
12.	*Cyperus rotundus*	Cyperaceae	Mutha ghass	Wet land grass
13.	*Eclipta prostrate*	Compositae	–	Wet land herb
14.	*Eichhornia crassipes*	Hydrocharitaceae	Kochuri pana	Free floating
15.	*Enhydra fluctuans*	Compositae	Hingcha	Wet land herb

Contd...

Table 8.3–Contd...

Sl.No.	*Scientific Name*	*Family*	*Common Name*	*Life Form*
16.	*Hydrilla verticillata*	Hydrocharitaceae	Jhangi	Roote submerged
17.	*Hygrophila quadrivalvis*	Acanthaceae	Kule khara	Wetland herb
18.	*Hygrophila spinosa*	Acanthaceae	Kanta kule khara	Spiny wet land herb
19.	*Ipomoea aquatica*	Convolvulaceae	Kalmi shak	Free floating
20.	*Isoetes* sp.	Isoetaceae	–	Submerged/emergent
21.	*Lemna perpusilla*	Lemnaceae	–	Free floating
22.	*Lemna minor*	Lemnaceae	Common duck weed	Free floating
23.	*Ludwigia adscendens*	Onagraceae	Keshar dam	Spongy, wet land herb
24.	*Marsilea minuta*	Marsileaceae	Susnishak	Wet land plants
25	*Monochoria hastate*	Pontederiaceae	–	Wet land plant
26.	*Nelumbo nucifera*	Nelumboceae	Padma	Rooted with floating leaves

Table 8.4: Birds of Mirik Lake and its Surrounding (Ornithofauna)

Name	*Scientific Name*
Rufus turtle Dove	*Streptopelia oriertalis*
Small cuczoo	*Cuculus poliocephaalus*
Great Himalayan Barbet	*Megalaima virens*
Searlet Minivet	*Pericrocotus flammeus*
White Cheeked Bulbul	*Pyenonofus leueooggenys*
Black Bulbul	*Hypsipetes*
Black Drongo	*Dicrusus adssiimilis*
Green Maypie	*Cissa chinenisis*
Green Backed Tit	*Parus monficolus*
Himalayan Treepie	*Dendrocitta formosae*
White Browed Tit Babbler	*Alcippe vinipeefus*
Silvereared Mesia	*Argentauris*
Redhilled Leiothrix	*Leciothrix lutea*
Sptted Forktail	*Enicurur maculates*
Himalayan Whistling Thrush	*Myiophorus*
Common Maayna	*Acridotheres fristis*

Table 8.5: Free Floating Suspended Rooted Floating Plant of Subhas Sarobar

English Name	*Scientific Name*	*Family*
Water Hyacinth		
Simrayo (water cress)	*Nasturtium officinala*	Brassicaceae
Bojo	*Aquorous calamus*	Monocot

Biodiversity Values of East Kolkata Wetlands

There are about 100 plants species which have been recorded in and around the East Kolkata Wetlands. These include *Sagittaria montividensis, Cryptocoryne Ciliata, Cyperus* spp., *Acrostichum aureum, Ipomoea aquatica* etc.

Amongst the rare mammal's marsh mongoose and small Indian mongoose, palm civet and small Indian civet are significant in and around East Kolkata Wetland areas. About 20 mammals are reported from this region. Threatened reptiles like, Indian mud turtle is also reported occasionally from the adjacent locality. Presently more than 40 bird species comprising of local and migratory types are reported to visit these cluster of Wetlands. Among these, grebe, coot, cormorant, teals, egrets, jacanas, snipes, tern, eagles, sand pipers, gulls, rails and kingfishers are significant.

Table 8.6: Algal General that are Present in the Sewage Fed Ponds in EKW

Euglenophyta	*Chrysophyta*
Euglena	Navicula
Phacus	Cyclotella
Chlorophyta	Cynophyta
Chlamydomonas	Ocsillatoria
Chlorogonium	Anabaena
Endorina	
Pandorina	
Pyrobitrys	
Chlorella	
Carteria	
Rhodomonas	

Expected Contributions of Carbon Sequestration to these Wetlands

Besides the other carbon sinks in the area, these wetlands play an important role of carbon sequestration. These wetlands continue to sequester carbon from the air through photosynthesis by their aquatic plants. Carbon contained in the nearby trees and other vegetation in these wetlands generates carbon rich peats, organic soils and sediments which have been built up with a long period. The magnitude of carbon storage in these wetlands depends upon wetland type and size, vegetation, depth of wetland soils, groundwater levels, nutrient levels, pH and other factors.

The trees and other plants in these wetlands convert atmospheric CO_2 into biomass through photosynthesis and the living animals in these wetlands feed upon them which maintain a food chain in these wetlands. Photosynthesis and decomposition processes in peat lands of these wetlands are complex and fluctuate in a specific setting, depending upon so many factors. Peat lands also convert CO_2 to methane which is also a more active Green House Gas (GHG). Increased CO_2 in the atmosphere of the surrounding wetlands will result in increased plant growth and hence the potential for carbon sequestration will increase.

These wetlands also take part in conversion of CO_2 to carbonate minerals by a series of reactions with aqueous ions. The conversion of CO_2 to carbonate minerals via mineral trapping may contribute significantly to CO_2 sequestration through the long time period.

Action Plan

1. Quantification of carbon stocks in these wetlands.
2. Actual carbon sequestration by the aquatic flora of these wetlands.
3. Influence of climate change by carbon sequestration of these wetlands.
4. Estimation of other GHGs emission from these wetlands.
5. Verify models that couple hydrology, ecosystem processes and carbon sequestration.
6. Cost effective management approaches and technologies to mitigate loss of carbon from these wetlands.
7. To develop national strategies, plans, etc., for conservation and sustainable use of biodiversity.
8. To integrate, as far as possible and appropriate, the conservation and sustainable use of biological diversity into relevant sectoral or cross-sectoral plans
9. To identify and monitor the components of biodiversity and activities which have or might have significant adverse impacts.
10. To establish protected areas or areas where special measures are needed and to regulate or manage biological resources important to biodiversity.
11. To promote protection of ecosystems and natural habitats.
12. To promote environmentally sound and sustainable development in areas adjacent to protected areas;
13. To prevent introduction of species from outside a country which could threaten native ecosystems or species.
14. To preserve and maintain knowledge and practices of indigenous and local communities embodying traditional lifestyles.
15. To develop or maintain necessary legislation and other regulatory provisions for protection of threatened species and populations.
16. To cooperate in providing financial and other support for conservation of wetlands.
17. To integrate consideration of conservation and sustainable use into national decision making.
18. To adopt economically and socially sound measures that act as incentives for the conservation and sustainable use of components of biological diversity.
19. To establish programs for scientific and technical education and training in identification, conservation, sustainable use of biodiversity and promote research that contributes to conservation and protection biodiversity.
20. To have programs for public education and awareness.
21. Formulation of policies for the protection of wetlands, grasslands, and sacred groves significant from the point of view of biodiversity.
22. The country needs to urgently finalize a comprehensive biodiversity strategy and action plan that will provide an overall assessment of the current state of India's biodiversity and formulate concrete steps, mechanisms, and guidelines to develop institutional structures for its implementation.

References

Brevik, Eric C. and Homburg, Jeffrey A., 2003. A 5000 year record of carbon sequestration from a coastal lagoon and wetland complex, Southern California, USA. *Catena*, 57: 221–232

Craft, C., Stephen, B. and Campbell, C., 2002. Fifteen years of vegetation and soil development after brackish-water marsh creation. *Restoration Ecology*, 10: 248–258.

Cui, J.B., Li, C.S. and Trettin, C., 2005. Analyzing the ecosystem carbon and hydrologic characteristics of forested wetland using a biogeochemical process model. *Global Change Biology*, 11: 278–289.

Euliss, Ned H. Jr., Gleason, R.A., Olness, B., McDougal, R.L., Murkin H.R, Robarts, R.D., Bourbonniere, R.A. and Warner, B.G., 2005. North American prairie wetlands are important non-forested land-based carbon storage sites. *Science of the Total Environment*, 361: 179–188.

Kothari, A., 2000. *Planning for the Future: India's Biggest Environment and Development*.

Kundu, Nitai, Bhattacharya, Mohit and Mukherjee, Ambarish, 1997. *Managing Wetlands*. Institute of Wetlands Management and Ecological Design, Basumati Corporation Limited.

Mauquoy, D., Engelkes, T. and Groot, M.H.M., *et al.*, 2002. High-resolution records of late-olocene climate change and carbon accumulation in two north-west European ombrotrophic peat bogs. *Palaeogeography, Palaeoclimatology, Palaeoecology*, 186(3): 275–310.

Olson *et al.*, 1983. *Global Distribution of Fresh Water Wetlands for Use in STOCHEM*.

Tuittila, E.S., Komulainen, V.M. and Vasander, H., *et al.*, 1999. Restored cutaway peatland as a sink for atmospheric CO_2. *Oecologia*, 120: 563–574.

Vijayan *et al.*, 2004. *Inland Wetlands of India Conservation Priorities*. Salim Ali centre for Ornithology and Natural History.

Waddington, J.M. and Warner, K.D., 2001. Atmospheric CO_2 sequestration in restored mined peatlands. *Ecoscience*, 8: 359–368.

Xiaonan, Duan, Xiaoke, Wang, Lu, Fei and Zhiyun, Ouyang, 2008. Primary evaluation of carbon sequestration potential of wetlands in China. *Acta Ecologica Sinica*, 28(2), Online English edition of the Chinese language journal.

Chapter 9

Bioefficacy of Partially Purified Toxin from Entomopathogenic Fungi *Nomuraea rileyi* (F.) Samson Against Groundnut Sucking Pest *Aphis craccivora* (Homoptera : Aphipidae)

S. Karthick Raja Namasivayam* and K. Soumya
Department of Biotechnology and Bioinformatics, SRM University, Ramapuram Campus, Chennai – 89, Tamil Nadu

ABSTRACT

The present study undertaken to evaluate the partially purified toxin isolated from culture filtrate of entomopathogenic fungi *Nomuraea rileyi* (F.) Samson grown in liquid media. The chloroform and methanol extract of the culture filtrate was concentrated and the final residues were separated using column chromatography with suitable solvent system and the collected fractions was found to be protein and this protein was screened for pesticidal activity against *A. craccivora* under pot and microplot assay. Distinct effect could be observed in both cases. The toxin shows activity in both pot and micro plot assay. In pot assay the mortality was ranged from 55.21 to 100.0 per cent while 25.11 to 100.0 per cent in microplot assay at respective days.

Keywords: Nomuraea rileyi, Aphis craccivora.

* E-mail: skrn.microbiol@gmail.com

Introduction

Groundnut/Peanut (*Arachis hypogoea* Family–Leguminosae) it is the 13th most important food crop of the world. It is world 4th most important source of edible oil and 3rd most important source of vegetable protein. The groundnut which is also an important cash crop is an annual legume. Its seeds are rich in high quality–edible oil–50 per cent, easily protein–25 per cent and carbohydrates–20 per cent (Aher *et al.*, 2004). Among the various biotic and abiotic constraints in groundnut production, insect pests play a vital role. More than 360 species of insects and mites were reported to attack the groundnut crop in the field and the pods in storage in all over the world. In India during 1960's the major pests were aphids like. *Aphis craccivora* (sucking pest), *Aproaerema modicella* (leaf miners), *Spodoptera litura* (tobacco army worm), gram pod borer *Helicopvera armigera*. The loss due to these pests is in high magnitude all over the world particularly in India. The sucking pests complex of groundnut consists of Aphids, Leaf hopper and thrips which generally causes considerable yield loss in groundnut. *Aphis crassivora* Koch (Homoptea; Aphpidae) is one of the major sucking pest causing yield loss by sucking plant sap and transfer viral disease such as Peanut Rostee Virus (PSV). Though chemical pesticides are applied to control, increasing public sensitivity to environmental pollution and problems of pest resistance to chemical pesticides has provided the impetus for the development of alternate, benign strategies for pest control (Dhingra and Sarup, 1990; Alegbeja, 1997).

Biological control based on microorganism is now being extensively utilized because of their high efficacy towards the pests, non target effect and persistence etc. Among the different groups, Fungi known as entomopathogenic fungi is found to be used as biocontrol agents against a wide range pests throughout the world. *Nomuraea rileyi* (F.) Samson (Mitosporic fungi; Hypomycetes) a cosmopolitan entomopathogenic fungi is reported to infect a wide spectrum of insect pests of economic important crops. *N. rileyi* have been found to produce metabolites including enzymes like chitinase, proteinases and phospholipases and toxins and these factors play a vital role in insect pathogenesis. In the present study the partially purified toxin isolated from culture filtrate of *Nomuraea rileyi* (F.) Samson was evaluated against *Aphis craccivora* Koch in groundnut under pot and microplot assay.

Materials and Methods

Fungal Culture

Nomuraea rileyi–Mother culture was obtained from MTCC (Microbial type culture collection, Chandigarh), and sub-cultured on Potato Dextrose agar and the pure culture was maintained on PDA slant.

Preparation of Conidial Suspensions

The re-isolated conidia were sub cultured in SDYB (Sabouraud dextrose maltose yeast extract broth) medium. Fungal conidia were collected from 15 days old culture by scrapping off with a sterilized glass rod. A homogenous conidial suspension was prepared in sterile distilled water by adding a few drops of the wetting agent Tween 80 (0.01 per cent). The conidial concentration of the suspension was determined using an improved Neubauer haaemocytometer (Germany).

Extraction of Crude Compound

The spore suspension was inoculated into sabouraud dextrose yeast extract broth and incubated at 26°C with constant aerations for 12 days. After which the mycelia was separated from the broth by filtering with Whattman no. 1 filter paper and the filtrate was extracted with Chloroform and methanol and concentrated in rotary evaporator at 35°C. The final residues were separated using silica gel column with chloroform : methanol : ethyl acetate (20 : 1 : 1 ratio). The collected fractions were identified by TLC qualitatively and used for bioassay.

Bioassay on *Aphis crassivora* Koch

Pot Evalution with Whole Spore Suspension

Aphis craccivora a life stages were inoculated (40–45/plant) in groundnut plants were grown in pots. After 10 days of inoculation of aphids the treatments were carried out with knapsack sprayer, *N. rileyi* spore suspension (10^8 spores) was sprayed and subsequently the aphids population was observed for 2, 3, 5, 7 and 10 days after spraying of fungi. Five replicates were maintained.

Pot Evaluation with Toxin

A. craccivora life stages were inoculated (40–45/plant) in groundnut plants were grown in pots. After 15 days of inoculation, 100 mg of final fraction dissolved in 100 ml of sterile distilled water was sprayed on pest. infested plants. Observations were made daily and the mortality was recorded. Five replicates were maintained.

Microplot Evaluation

Microplot trail was carried on University campus, the plot size of 5 × 3 m leaving a gangway of one meter all around the plot has been maintained for this experiment (40–45/plant). After 10 days of inoculation of aphids, the treatments were carried out with hand sprayer, 100 mg of final fraction dissolved in 100 ml of sterile distilled water and sprayed on the Aphids infested plants. Five replicates were maintained including control and the mortality rate was recorded up to ten days.

Results and Discussion

Effect of Whole Spore Suspension (Pot Culture)

The spore formulation of *Nomureae rileyi* when sprayed against the *A. craccivora* has shown toxic effects to the pest. The mortality was noticed at second day and it ranged from 50 to 100 per cent at second and eighth day of application. The infected aphids turned into whitish green color due to germination of fungal spores (Table 9.1) The pathogenicity of *N. rileyi* against various insect pests were reported by many workers (Adrian *et al.*, 2005; Boucias *et al.*, 2002; Crisan, 1991).

Table 9.1: Effect of Whole Spore Suspension and Toxin of *Nomuraea rileyi* on *A. craccivora* Under Pot Culture

Timer (Days)	*Mortality (Per cent)*	
	Spores	*Toxin*
1	0.0	0.0
2	50.12*	55.21*
4	75.01	86.21
6	92.23	100.0
8	100.0*	100.0*
10	100.0*	100.0*
CD (P > 0.05)		

Mean values of five replications.

* Column carries the same letter is not statistically significant.

Table 9.2: Microplot Assay of Partially Purified Protein Component from Culture Filtrate on *A. craccivora*

Time (Days)	*Temperature (°C)*	*Relative Humidity (Per cent)*	*Mortality (Per cent)*	
			Toxin	*Spore*
1	40.3	60.1	0	0
2	39.4	64.2	25.11	20.2
4	37.1	69.4	76.42*	56.1
6	42.2	58.4	100.0*	76.1
8	39.3	70.1	100.0*	76.3
10	38.1	75.4	100.0*	79.4

Mean values of five replications.

*: Significant at 5 per cent level by DMRT.

Chromotographic Separation of Pesticidal Compound

The components of culture filtrate were separated using silica gel column chromatography. The fraction was obtained after time period of 3 hrs 20 mins at flow rate of 3 drops per minute.

The fraction collected was subjected to TLC with solvent system of butanol : acetic acid water (9 : 4 : 1).The sheet was sprayed with Ninhydrin and showed purple coloured band which confirmed the nature of compound as protein.

Effect of Partially Purified Protein Component from Culture Filtrate

As in spore suspension the toxin also shows distinct effect on the mortality of the pest, the partially purified compound reveals 55.11, 86.21 and 100 per cent mortality at second, forth and sixth days of application. Similarly the peptide was isolated from the fungi was active against *Anticarsia gemmatalis* 3rd instar larvae (Onofre *et al.*, 2002). Hamill (2005) isolated Beauvericin a new depsipeptide antibiotic from *Beauveria bassiana* and evaluated against *Artemia salina.* Similarly Brousseau (1996) isolated destruxin toxin from deuteromycetes entomopathogenic fungi *Metarhizium anisopliae* and tested against various insects.

Micro Plot Assay with the Toxin

In microplot assay 25.11, 76.42 and 100.0 per cent mortality was recorded As in whole spore suspension treatment, the mortality started with second day of application and complete mortality could be reached even at 8th day of application. But the maximum mortality with spores was only 79.4 per cent. It is clear from the above results that the environmental factor did not cause any distinct effect on the action the partially purified compound on the pest. The results clearly reveals the partially purified protein toxin shows distinct effect against *A. crassivora* in both *in vitro* and micro plot assay than the spores. Moreover the environmental factors may influence the fungal activity in the field. But in this study the toxin caused distinct mortality and environmental factors didn't cause any effect. So we can use this toxin with various environmental factors. Further study such as structure, characterization, formulation and evaluation against various pests is now progress and this highly useful to use this toxin as the major biocontrol strategy against *A. crassivora.*

Acknowledgement

We are grateful to the management and Director, SRM University Ramapuram Campus, Chennai for encouragement and support.

References

Adrian, T., Gillespie, Norman and Claydon, 2005. *National Products as a Source for New Agricultural Chemicals,* Organised by the Pesticides Group of the Society of Chemical Industry, held in London on 21–22 February.

Aher, S.K., Thite, S.V. and Punde, B.N., 2004. Aerobiology and epidemiology of certain diseases of groundnut. *Journal of Current Sciences,* 5(1): 229–232.

Alegbeja, M.D., 1997. Effect of intercropping groundnut with cereals on the incidence of groundnut rosette virus. *International Arachis Newsletter,* 5: 21–25.

Boucias, D.G., Brasdford, D.L. and Barfield, C.S., 2002. Susceptibility of the velvetbean caterpillar and soybean looper (Lepidoptera : Noctuidae) to *Nomurea rileyi*: Effects of pathotype, dosage, temperature, and host age. *J. Econ. Entomol.,* 14: 121–126.

Brousseau, C., Charpentier, G. and Belloncik, S., 1996. Susceptibility of spruce budworm, *Choristoneura fumiferana* Clemens, to Destruxins, cyclodepsipeptidic Mycotoxins of Metarhizium anisopliae. *J. Invert. Pathol.,* 19(4): 176–179.

Crisan, E.V., 1991. Mechanism responsible for release of toxin by *Metarhizium anisopliae* spores in mosquito larvae. *J. Invert. Pathol.,* 17: 260–264.

Dhingra, S. and Sarup, P., 1990. Development of techniques for defecting resistance in crop pest to insecticides. *Journal of Entomological Research,* 156–165.

Hamill, R.L., Higgens, C.E., Boaz, H.E. and Gorman, M., 1969. The structure of beauvericin: A new depsipeptide antibiotic toxic to *Artemia salina*. *Tet. Letters,* 49: 4255–4258.

Onofre, S.B., Gorzaler, R.R. and Barros, N.M., 2002. LC50 of the peptide produced by the entomopathogenic fungus *Nomuraea rileyi* (F.) Samson against third instar larvae of *Anticarsia gemutalis* (Lepidoptera, Noctuidae). *Brazilian Archieves of Biology and Technology,* 45: 269–275.

Padmaja, V., 2005. Role of entomopathogenic fungi in insect pest management. In: *Green Pesticides for Integrated Pest Management,* (Eds.) S. Ignachimuthu and S. Jayaraj. Narosa Publishing House, p. 324–327.

Chapter 10

Oxide Ferroelectric: Where Do We Start?

M.R. Panigrahi[1]*, S. Panigrahi[1], K.M. Purohit[2]

[1]Department of Physics, [2]Department of Chemistry, National Institute of Technology, Rourkela – 8, Orissa

ABSTRACT

Ferroelectric effect is an electrical phenomenon whereby certain materials may exhibit a spontaneous dipole moment, the direction of which can be switched between equivalent states by the application of an external electric field. The term ferroelectricity is used in analogy to ferromagnetism, in which a material exhibits a. permanent magnetic moment. Ferroelectricity is a phenomenon which was discovered in 1921. Ferroelectricity has also been called Seignette electricity, as Seignette or Rochelle Salt (RS) [$NaKC_4H_4O_6.4H_2O$] was the first material found to show ferroelectric properties such as a spontaneous polarization on cooling below the Curie point, ferroelectric domains and a ferroelectric hysteresis loop. A huge leap in the research on ferroelectric materials came in the 1950's, leading to the widespread use of barium titanate ($BaTiO_3$) based ceramics in capacitor applications and piezoelectric transducer devices.

A large class of ferroelectric crystals is made up of mixed oxides containing corner sharing octahedra of O^{2-} ions. Perovskite type compounds are corner sharing oxygen octahedral. Perovskite is a family name of a group of materials and the mineral name of calcium titanate ($CaTiO_3$) having a structure of the type ABO_3. Many piezoelectric (including ferroelectric) ceramics such as Barium Titanate (BaTiO3), Lead Titanate ($PbTiO_3$), Lead Zirconate Titanate (PZT), Lead Lanthanum Zirconate Titanate (PLZT), Lead Magnesium Niobate (PMN), Potassium Niobate ($KNbO_3$), Potassium Sodium Niobate ($K_xNa_{1-x}NbO_3$), and Potassium Tantalate Niobate

* E-mail: manash_123india@yahoo.co.in

(K [Ta_xNb_{1-x})O_3] have a perovskite type structure. In 1945, the ferroelectric behavior of $BaTiO_3$ was reported for the first time. This material is of stable perovskite type one of the most fundamental crystal lattice structure.

Keywords: *Perovskite, Transition temperature, Phase transition, Spontaneous polarization, Ferroelectric domain, Grain size.*

Introduction

$BaTiO_3$ was discovered during World War II in 1941 and 1944 in the United States, Russia, and Japan. At least in the USA, the research was accelerated because of the war. At that time, mica was used in most capacitors, but U-boats threatened the supplies of mica to the U.S.A. from South America. The initial reports were based on doping studies of TiO_2 with BaO, which produced ceramic materials with enhanced dielectric permittivities.

The mixed oxides were made by Thurnaurer and Deaderick at the American Lava Co. as early as 1941 (Thumaurer and Deaderick, 1941). The high permittivities were found by measurements made at the Erie Resistor Company, with dielectric constant exceeding 1000, ten times greater than any other ceramic known at that time, such as TiO_2 ($\varepsilon r = 110$).

Later, more precise studies were reported by Wainer and Solomon in the United States, Ogawa and Waku in Japan, and Wul and Goldman in Russia (Wainer and Solomon, 1942 and 1943; Ogawa and Waku 1944; Wul and Goldman, 1945).

The race to understand the nature of the dielectric anomaly in ceramic BaO–TiO_2 Ceramics continued, and in 1945 and 1946, von Hippel (USA), and Wul and Goldman (Russia) demonstrated ferroelectric switching in the ceramics (Hippel *et al.*, 1946; Wul and Goldman, 1946). The discovery of ferroelectricity in BaO–TiO_2 ceramics was extremely important, as it demonstrated for the first time that ferroelectricity could exist in simple oxide materials, and it was not always associated with hydrogen bonding. $BaTiO_3$ is a member of the perovskite family based on the mineral $CaTiO_3$, as predicted from fundamental crystal chemical principles described much earlier by Goldschmidt (1926). The first detailed description of the crystal structure of $BaTiO_3$ in the high temperature ferroelectric phase was proposed by Helen D. Megaw (1945) in the United Kingdom and confirmed soon after with the work of Miyake and Veda (1946). Figure 10.1 shows the basic structure of barium titanate with the titanium ion occupying the octahedrally coordinated site and the Ba ion in the twelve coordinated site in a high temperature cubic symmetry.

Barium titanate ($BaTiO_3$) has a paraelectric cubic phase above its Curie point of about 130°C. In the temperature range of 130°C to 0°C the ferroelectric tetragonal phase with a cia ratio of ~1.01 is stable. The spontaneous polarization is along one of the [001] directions in the original cubic structure. Between 0°C and –90°C, the ferroelectric orthorhombic phase is stable with the polarization along one of the [110] directions in the original cubic structure. On decreasing the temperature below –90°C the phase transition from the orthorhombic to ferroelectric rhombohedral phase leads to polarization along one of the [111] cubic directions.

The spontaneous polarization on cooling $BaTiO_3$ below the Curie point T_c is due to changes in the crystal structure. As shown in Figure 10.2 the paraelectric cubic phase is stable above 130°C with the center of positive charges (Ba^{2+} and Ti^{4+} ions) coinciding with the center of negative charge (O^{2-}). On

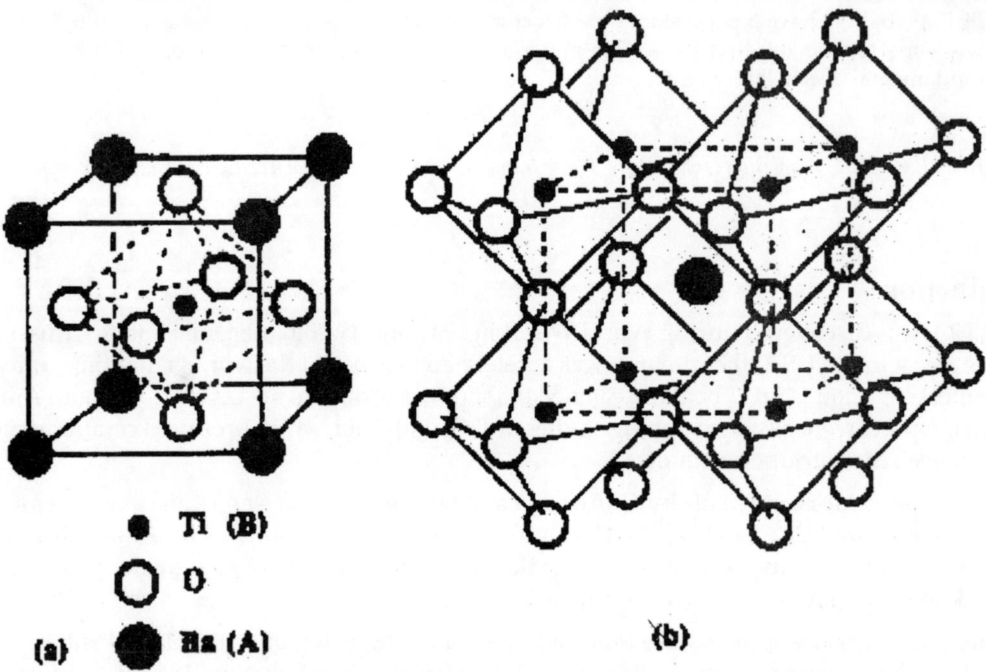

Figure 10.1: (a) A cubic ABO_3 ($BaTiO_3$) perovskite-type unit cell and (b) Three dimensional network of comer sharing octahedral of O^{2-} ions (Xu, 1991)

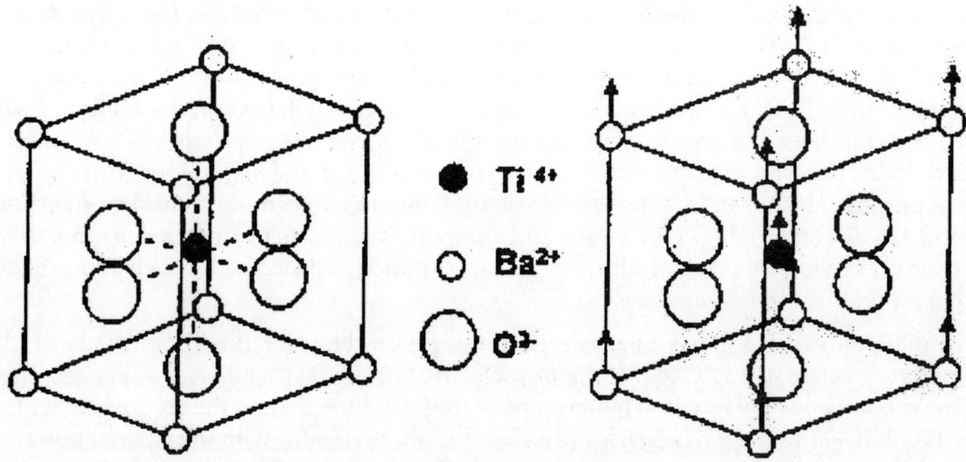

Figure 10.2: The Crystal Structure of $BaTiO_3$ (a) Above the Curie point the cell is cubic; (b) Below the Curie point the structure is tetragonal with Ba^{2+} and Ti^{4+} ions displaced relative to O^{2-} ions

cooling below the Curie point T_c, a tetragonal structure develops where the center of Ba^{2+} and Ti^{4+} ions are displaced relative to the O^{2-} ions, leading to the formation of electric dipoles. Spontaneous polarization developed is the net dipole moment produced per unit volume for the dipoles pointing in a given direction (Pepinsky *et al.*, 1960).

Phase Transition

The ferroelectric phase at room temperature is tetragonal with oxygen and titanium ions shifting to produce a spontaneous polarization. Later, Kay and Vousden (1949) showed that $BaTiO_3$ had a series of ferroelectric-ferroelectric transitions from tetragonal → orthorhombic, at 5°C and at –90°C a transition from orthorhombic to rhombohedral, Figure 10.2 (Kay and Vousden, 1949). These data indicated coexisting regions and hysteresis consistent with first order behavior in the transition that was earlier detected by Wul (1946) and also Harwood, Popper, and Rushman (1949) with humps in the specific heat data at the transition temperatures (Wul and Goldman, 1946; Harwood *et al.*, 1950). Precise refinements for tetragonal $BaTiO_3$ were performed later by Pepinsky and Shirane and compared with the tetragonal phase of $PbTiO_3$ (Shirane and Pepinsky, 1953).

Better understanding of the nature of the phase transitions came with the growth of $BaTiO_3$ crystals. The first crystals were produced in and around 1947; however, these were small. Later in 1954, Remeika produced larger crystals that enabled the ferroelectrics community to extend the fundamental understanding of the dielectric anisotropy, domain switching, and electro-optical and electromechanical properties of ferroelectric $BaTiO_3$ (Remeika, 1954).

In 1949 Merz produced some classical polarization and permittivity data to illustrate dielectric anisotropy in the ferroelectric phases of $BaTiO_3$. Later, Cross revisited the $BaTiO_3$ crystals for polarization data and dielectric anisotropy in the lower temperature ferroelectric phases, as the Merz data were plagued by multidomain effects (Merz, 1949; Cross, 1953).

With the increased availability of $BaTiO_3$ single crystals, understanding of the ferroelectric domain structures was improved through consideration of the low energy configurations and size effects in both the tetragonal and orthorhombic ferroelectric phases. These observations were made with optical microscopes, Kay (1948), Matthias and von Hippel (1948) and Blattner, Kanzig and Merz (1949).

It is now accepted that $BaTiO_3$ has elements of both displacive soft-mode phonons and order disorder behavior. (Comes *et al.*, 1968; Heimann and Ushioda, 1974; Harada *et al.*, 1971; Kneipkamp *et al.*, 1954). The debate on the true microscopic nature of the $BaTiO_3$ phase transition continues to challenge each new generation of ferroelectricians.

Dielectric Constant

The dielectric properties of $BaTiO_3$ are found to be dependent on the grain size (Kneipkamp *et al.*, 1954; Hennings, and Rosenstein, 1984; Bell, 1994). Figure 10.3 shows the variation of dielectric constant with temperature for $BaTiO_3$ ceramics with a fine (~ 1 μm) and coarse (~ 50 μm) grain size. Large grained $BaTiO_3$ (≥ 1 μm) shows an extremely high dielectric constant at the Curie point. This is because of the formation of multiple domains in a single grain, the motion of whose walls increases the dielectric constant at the Curie point. For a $BaTiO_3$ ceramic with fine grains (~ 1 μm), a single domain forms inside each grain. The movement of domain walls are restricted by the grain boundaries, thus leading to a low dielectric constant at the Curie point as compared to coarse grained $BaTiO_3$ (Swartz, 1990). The room temperature dielectric constant (ε_r) of coarse grained (≥ 10 μm) BT ceramics is found to be in the range of 1500-2000. On the other hand, fine grained (~1 μm) BT ceramics exhibit a room temperature dielectric constant between 3500-6000. The grain size effect on the dielectric constant at room temperature has been explained by the work of Buessem *et al.* (1966) and Arlt *et al.* (1966). Buessem and coworkers proposed that the internal stresses in fine grained $BaTiO_3$ must be

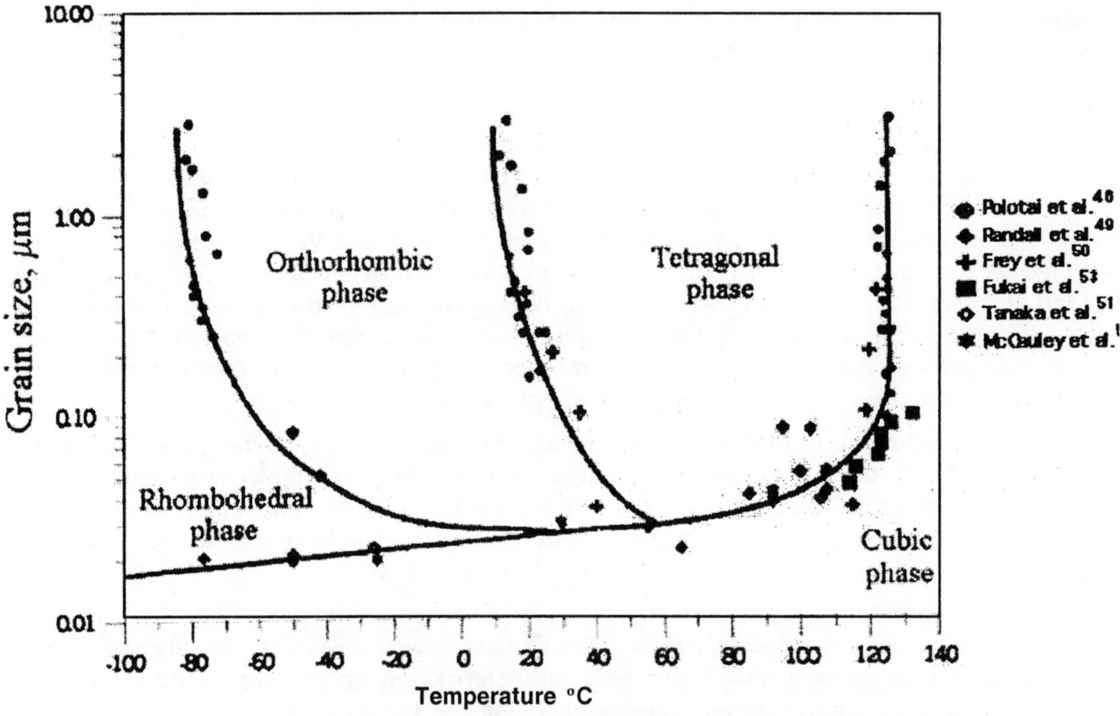

Figure 10.3: Grain Size Variations Affecting Transition Behavior

much greater than the coarse grained ceramic, thus leading to a higher permittivity at room temperature. Arlt studied the domain structures in BT ceramics and showed that the room temperature ε_r reached a peak value at a critical grain size of ~0.7 μm. He concluded that the enhanced dielectric constant was due to the increased 90° domain wall density. The mobility of the 90° domain walls in very fine grained BT is hindered and only less than 25 per cent of the ε_r was achieved.

As the BT ceramics have a very large room temperature dielectric constant, they are mainly used multilayer capacitor applications. The grain size control is very important for these applications.

PTCR Effect

Polycrystalline $BaTiO_3$ appropriately doped with an excess of donors and TiO_2 can lead to unusual non-linear resistors known as positive temperature coefficients resistors (PTCRs). This phenomenon involves an enhanced resistivity at the Curie transition temperature at 125°C.

A schematic representation of this PTCR phenomenon is shown in Figure 10.4. First discovered by Subari in 1959, the semiconductor properties are strongly influenced by the ferroelectric transition, with up to 6–7 orders of magnitude change in resistivity in the best $BaTiO_3$ PTCR thermistors. This phenomenon is only observed in polycrystalline $BaTiO_3$, with the spontaneous polarization of the ferroelectric domains modifying the band-bending and, therefore, the electronic transport across the grain boundary (Heywang, 1964).

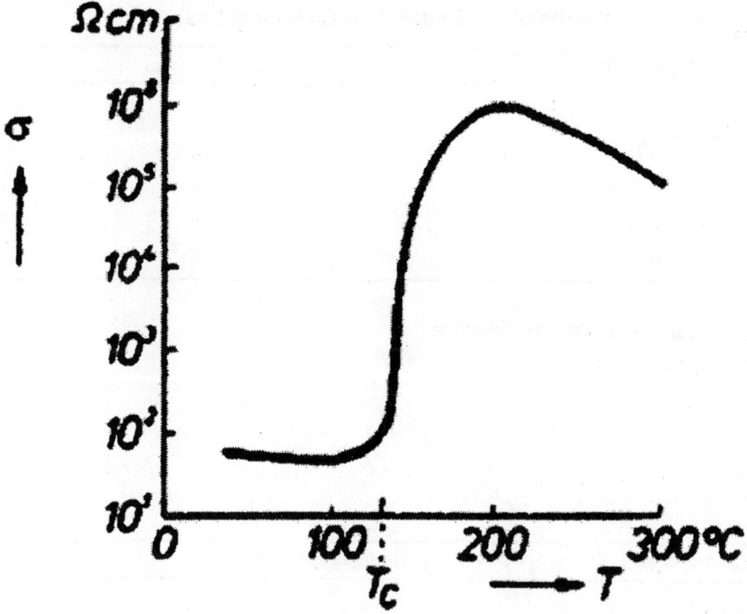

Figure 10.4: Conductivity vs. Temperature

Some Properties of Barium Titanate

Temperature Dependent Permittivity

Temp. in °C	*Permittivity, f = 1.5 mc/s*
20	400
60	500
100	600
115	2000
120	4900
140	2800
145	2000
180	1500

Frequency Dependent Permittivity

Frequency	*Permittivity*
1 KHZ	1500
100 KHZ	1490
107 HZ	1375
109 HZ	1000
1011 HZ	200

Temperature Dependent Dielectric Loss

Temp in °C	*Delectric Loss*
0	0.24
40	0.20
80	0.16
120	0.06
160	0.04

Physical and Optical Properties of Barium Titanate

Crystal structure: Tetragonal, P4MM

Lattice parameter: a = 3.992 Å, c = 4.036 Å

Melting point: 1625°C

Density: 6.02 gm/cc

Hardness: 5 Mohs

Electro-optic Coefficient		*Unclamped*	*Clamped*
In unit pm/V	r_{13}	9.5	8
	r_{33}	97	28
	r_{42}	1460	820

Application of $BaTiO_3$

Over the years, $BaTiO_3$ has been used in a wide variety of applications in various forms of crystals, bulk ceramics, multilayer, and thin films (Chazono and Kishi, 2001). The ability to be used in so many different applications is a consequence of the complex coupling and the various interactions associated with ferroelectric phenomena. $BaTiO_3$ nanoparticles can find use for future multilayer dielectrics (Chazono and Kishi, 2001; Polotai *et al.*, 2003; O'Brien *et al.*, 2001). These are fundamental and experimental studies to understand the size affect that control interfacial and surface phenomena in greater detail.

As Transducers

$BaTiO_3$ was the very first ceramic transducer. To establish piezoelectric activity from the $BaTiO_3$ ceramics, the domains had to be aligned to induce a macroscopic polarization. Once induced, the $BaTiO_3$ ceramics were employed for a variety of transducers, which control current, either electrical or mechanical energy, and the conversion of one into the other.

It was as early as September 1946 that an engineer (R.B. Gray) at the Erie Resistor Company, working on the D.C. degradation of the first $BaTiO_3$ capacitors, recognized that the "hum" was the result of an induced piezoelectric effect (Gray, 1949). Later, S. Roberts recognized the poling effect in his publication in 1947 (Roberts, 1947). Some of the early work was discussed by Hans Jaffe (1950). $BaTiO_3$ phonographic pickups were the first use of piezoelectric ceramics. The British Navy used some underwater sonar devices based on the piezoelectric response of $BaTiO_3$ ceramics.

As Photorefractive Crystals

$BaTiO_3$ is an excellent photorefractive crystal for self pump phase conjugation, four wave mixing and many other photorefractive processes. It has high beam coupling gain and short beam coupling rise time. The highest sppc reflectivity was reported by using $BaTiO_3$. It can be operated at visible and near infrared regions (Wadhawan).

References

Arlt, G., Hennings, D. and With, G. de, 1985. *J. Appl. Phys.*, 58: 4.

Bell, A.J. 1994. *Proc. Int. Symp. Appl. Ferroelectrics.*

Buessem, W.R., Cross, L.E. and Goswami, A.K., 1966. *J. Am. Ceram. Soc.*, 75: 2927.

Chazono, H. and Kishi, H., 2001. *Jap. J. Appl. Phys.*, 40(9B): 5624.

Comes, R., Lambert, M. and Guinier, A., 1968. *Comm. Solid State Phys.*, 6: 715.

Cross, L.E., 1953. *Phil. Mag.*, 44: 1161.

Forsbergh, P.W. Jr., 1949. *Phys. Rev.*, 76: 1187.

Goldschmidt, V.M., Shrifter Nofke Videnskaps-Akad and Oslo, I., 1926. *mat-Naturv.* Kl. No. 2, 8.

Gray, R.B., 1949. U.S. Patent No. 2, 486, 560, November 1; filed in 1946.

Harada, J., Axe, J.D. and Shirane, G., 1971. *Phys. Rev.*, B4: 155.

Harwood, M.G., Popper, P. and Rushman, D.F., 1950. *Nature*, 160: 58.

Heimann, D. and Ushioda, S., 1974. *Phys. Rev.*, B9: 2122.

Hennings, D. and Rosenstein, G., 1984. *J. Am. Ceram. Soc.*, 67: 249.

Heywang, W., 1964. *J. Am. Ceram. Soc.*, 47: 484.

Hippel, A. Yon, Breckenridge, R.G., Chesley, F.G. and Tisza, L., 1946. *Ind. Eng. Chem.*, 38: 1097.

Jaffe, H., 1950. *Ind. Eng. Chem.*, 42(2): 264–268.

Kay, H.F. and Vousden, P., 1949. *Phil. Mag.*, 40: 1019.

Kneipkamp, H., Heywang, W. and Angew, Z., 1954. *Phys.*, 6: 385.

Megaw, H.D., 1945. *Nature*, 155: 484.

Merz, W.J., 1949. *Phys. Rev.*, 76: 1221.

Miyake, S. and Ueda, R., 1946. *J. Phys. Soc. Jap.* 1: 32.

O'Brien, S., Brus, L. and Murray, C.B., 2001. *J. Am. Chem. Soc.*, 123(48): 12085.

Ogawa, T. and Waku, S. 1944. T. Ogawa, Busseiron, Kenkyu, No. 61.

Pepinsky, R., Okaya, Y. and Unterleitner, F., 1960. *Acta. Cryst.*, 13: 1071.

Polotai, A.V., Ragulya, A.V. and Randall, C.A., 2003. *Ferroelectrics*, 88: 93–102.

Remeika, J.P., 1954. *J. Am. Ceram. Soc.*, 76: 940.

Roberts, S., 1947. *Phys. Rev.*, 890–5.

Shirane, G. and Pepinsky, R., 1953. *Phys. Rev.*, 91: 812.

Subari, O., 1961. *J. Am. Ceram. Soc.*, 44: 54.

Swartz, S.L., 1990. *IEEE Trans. Electrical Insulation*, 25: 935.

Thumaurer, H. and Deaderick, J., 1941. U.S. Patent No. 2, 429, 588, October 21; filed.

Wadhawan, V.K. *Introduction to Ferroic Material.*

Wainer, E. and Solomon, A.N., 1942 and 1943. Titanium Alloy Manufacturing Co. Report Nos. 8 and 9.

Wul, B. and Goldman, J.M., 1945. *C.R. Acad. Sci. URSS*, 46: 139.

Wul, B. and Goldman, J.M., 1946. *C.R. Acad. Sci. URSS*, 51: 21.

Xu, Y., 1991. *Ferroelectric Materials and their Applications*. North Holland, Amsterdam.

Chapter 11

A Comparative Study of the Exhaust Emissions from a Low Heat Rejection Engine with Two Different Levels of Insulation with Pongamia Oil

M.V.S. Murali Krishna[1*], S. Naga Sarada[2], G. Sudha Rani[3], K. Kalyani Radha[2] and P.V.K. Murthy[4]

[1]Mechanical Engineering Department, Chaitanya Bharathi Institute of Technology, Gandipet, Hyderabad – 500 075

[2]Mechanical Engineering Department, J.N.T.D College of Engineering, Kukatpally, Hyderabad – 500 085

[3]Mechanical Engineering Department, Jaya Prakash Narayan College of Engineering, Dharmapur, Mahabubnagar – 509 001

[4]Gnayana Saraswathi College of Engineering and Technology, Dharmaran (B), Dichpally, Nizamabad – 503 230

ABSTRACT

In the context of fast depletion of fossil fuels and increase of pollution levels with conventional fuels, the search for alternate fuels has become pertinent. Alcohols and vegetable oils are probable candidates to replace diesel fuels as they are renewable in nature. However, use of alcohols in internal combustion engines is limited, as the properties of alcohol fuels are not compatible with diesel fuels. Hence much emphasis is given to the non-edible vegetable oils as alternate fuels in

* E-mail: mvsmk_in@yahoo.com

internal combustion engines. However, the drawbacks of low volatility and high viscosity associated with non-edible vegetable oils call for hot combustion chamber, provided by low heat rejection diesel engine. The investigations are carried out on low heat rejection diesel engines with varied air gap thickness with crude vegetable oils at different injection pressures. Pollution levels decreased with high degree of insulation with vegetable oil in comparison with conventional engine with pure diesel operation.

Introduction

The major pollutants emitted from diesel engine are smoke and oxides of nitrogen (NOx). Excessive breathing of smoke (Sharma, 1996; Fulekar, 1999) causes tuberculosis and may also lead to death. It also causes detrimental effects on animal and plant life besides environmental (Khopkar, 1993) disorders. Inhaling of oxides of nitrogen causes dizziness, vomiting, sensation, severe headache etc. Hence, globally, stringent regulations are made for permissible pollutants in the exhaust of engines.

It is well known fact (Heywood, 2004) that about 35 per cent of the energy supplied is lost through the coolant and the 35 per cent is wasted through friction and other losses, thus leaving only 30 per cent of energy utilization for useful purposes. In view of above, major thrust in engine research during the last two decades has been on development of low heat rejection engines. Several methods adopted for achieving low heat rejection to the coolant were using ceramic coatings (Krishnan *et al.*, 1980) on piston, liner and cylinder head, creating air gap in the piston (Ramamohan, 1995) and other components with low-thermal conductivity materials like supemi, mild steel etc. However, this method involved the complications of joining two different metals. A few researchers (Jabez Dhinagar *et al.*, 1993) used different crown materials with different thickness of air gap in between the crown and the body of the piston. As cetane number of the vegetable oils is closer to that of diesel fuel, attempts were made to induct vegetable oils in the forms of crude vegetable oil (Rehman *et al.*, 1995), emulsified form (Kiannejad *et al.*, 1993) and bio-diesel (Annapuma Devi *et al.*, 2006) in the conventional engine. However, due to high viscosity of the vegetable oil, these vegetable oils were used in low heat rejection (Bhaskar *et al.*, 1993; Murali Krishna, 2004) diesel engine. However, no systematic investigations are reported so far about the use of crude vegetable oils in low heat rejection diesel engines with varied air gap thickness and hence attempt is made in this direction and the exhaust emissions are reported and compared with that of pure diesel operation.

Experimental Programme

The low heat rejection (LHR-1) diesel engine contains a two-part piston–the top crown made of low thermal conductivity material, supemi-90 (an alloy of nickel) screwed to aluminum body of the piston, providing a 3mm-air gap in between the crown and the body of the piston. A supemi-90 insert is screwed to the top portion of the liner in such a manner that an air gap of 3-mm is maintained between the insert and the liner body. Investigations are also conducted on LHR-2 engine with same configuration with 3.2-mm air gap thickness. At 500°C the thermal conductivity of supemi-90 and air are 20.92 and 0.057 W/m-K respectively. Figure 11.1 shows the experimental set up. The conventional engine is a four-stroke, water cooled diesel engine with 3.68 kW at a rate speed of 1500 rpm having compression ratio of 16 : 1. The engine is connected to an electric dynamometer for measuring brake power of the engine. Burette method is used for finding fuel consumption of the engine. Air consumption of the engine is measured by air-box method. The naturally aspirated engine is provided with water-cooling system in which, inlet temperature of water is maintained at 60°C by adjusting the water flow rate. The injection pressures are varied from 190 bars to 270 bars (in steps of 40 bars) using nozzle-

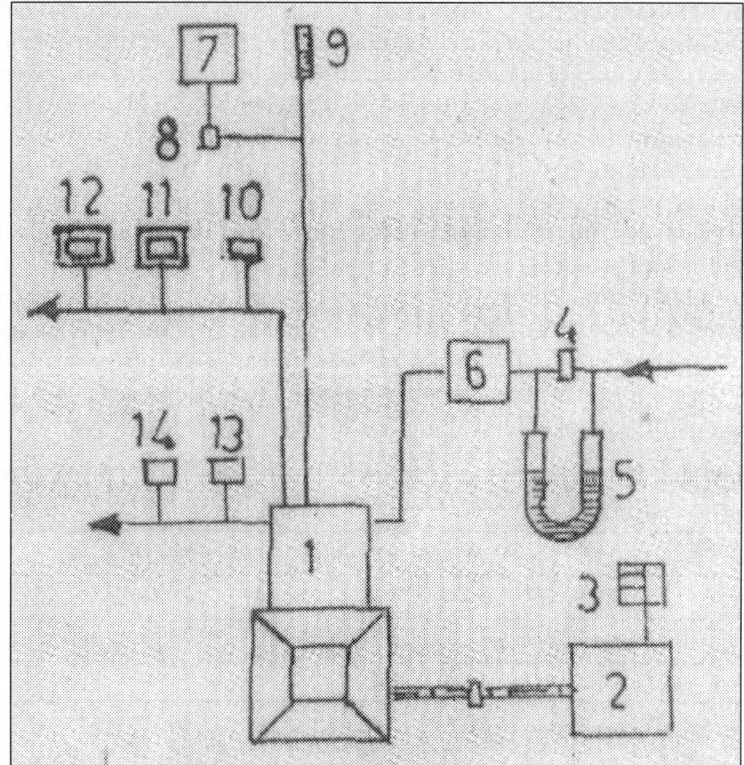

Figure 11.1: Experimental Set-up

1: Engine; 2: Electrical Dynamometer; 3: Load box; 4: Orifice meter; 5: V-tube water manometer; 6: Air box; 7: Vegetable oil tank; 8: Three-way valve; 9: Burette; 10: Exhaust gas temperature indicator; 11: A VL Smoke meter; 12: Netel Chromatograph NOx analyzer; 13: Outlet jacket water temperature indicator; and 14: Water flow meter.

testing device. The exhaust gas temperature (EGT) is measured with thermocouple made of iron and iron-constantan. Pollution levels of smoke and NOx are recorded by A VL smoke meter and Netel Chromatograph NOx analyzer respectively at the peak load operation of the engine. Pongamia oil substituted 100 per cent for diesel fuel is used as fuel in the investigations. The experiments are conducted on conventional engine (CE), LHR-1 engine and LHR-2 engine. Results are compared with pure diesel operation on conventional engine. The properties of crude vegetable oils are given in reference.

Results and Discussion

Smoke Levels

The data of variation of smoke levels at peak load with different versions of the engine at normal temperature of the vegetable oil and at different injection pressures is shown in Table 11.1. Drastic increase of smoke levels is observed at the peak load operation in both versions of the engine with vegetable oil operation compared with pure diesel operation on conventional engine. This is due to the higher magnitude of the ratio of C/H, (C-Number of carbon atoms, H-Number of hydrogen atoms in the composition of the fuel) when compared with pure diesel. The increase of smoke levels is also due to decrease of air-fuel ratios and volumetric efficiency with vegetable oil operation compared with pure diesel operation. Smoke levels are proportional to the density of the fuel. Since vegetable oils have higher density compared to diesel fuels, smoke levels are higher with vegetable oils. Due to higher molecular weight, crude vegetable oils have low volatility and because of their unsaturation, crude vegetable oils are inherently more reactive than diesel fuels, which results that they are more susceptible to oxidation and thermal polymerization reactions. However, LHR engines marginally decreased smoke levels due to efficient combustion and less amount of fuel accumulation on the hot combustion chamber walls of the LHR engines compared to the conventional engine. Increase of air gap thickness in LHR engine decreased smoke levels marginally with vegetable oil operation. This is due to decrease of ignition delay and combustion duration. Smoke levels decreased with increase of injection pressure, in both versions of the engine. This is due to improvement in the fuel spray characteristics at higher injection pressures and increase of air entrainment causing lower smoke levels.

Table 11.1: Data of Smoke Levels with Different Versions of the Engine at Peak Load Operation with Different Test Fuels

	Smoke Levels (Hartridge Smoke Units, HSU)					
Engine →	*Pure Diesel Operation*			*Crude Pongamia Oil Operation*		
	Injection Pressure (bar)			*Injection Pressure (bar)*		
Version →	*190*	*230*	*270*	*190*	*230*	*270*
CE	48	38	34	75	65	60
LHR-1	55	50	45	65	60	55
LHR-2	68	63	58	62	57	52

CE: Conventional engine; LHR-1: Air gap insulated piston and air gap insulated line with 3-mm air gap; LHR-2: Air gap insulated piston, air gap insulated liner with 3.2-mm air gap.

The data of variation of NOx levels at peak load with different versions of the engine at normal temperature of the vegetable oil and at different injection pressures is shown in Table 11.2.

Table 11.2: Data of NOx Levels with Different Versions of the Engine at Peak Load Operation with Different Test Fuels

	NOx Levels (ppm)					
Engine →	*Pure Diesel Operation*			*Crude Pongamia Oil Operation*		
	Injection Pressure (bar)			*Injection Pressure (bar)*		
Version →	*190*	*230*	*270*	*190*	*230*	*270*
CE	850	890	930	740	750	780
LHR-1	1300	1280	1260	1265	1235	1200
LHR-2	1400	1350	1300	1325	1300	1250

CE: Conventional engine; LHR-1: Air gap insulated piston and air gap insulated liner; LHR-2: Air gap insulated piston, air gap insulated liner and ceramic coated cylinder head.

NOx levels are lower in the conventional engine while they are higher in the LHR engines with crude vegetable oil at the peak load when in comparison with pure diesel operation. This is due to lower heat release rate because of high duration of combustion causing lower gas temperatures with the vegetable oil operation on conventional engine, which reduced NOx levels. Increase of combustion temperatures with the faster combustion and improved heat release rates in the LHR engines cause higher NOx levels. Increased injection pressures decreased NOx levels in LHR engines while they increased in conventional engine due to decrease of gas temperatures in LHR engines and increase of the same in the conventional engine.

Conclusions

Increase of air gap thickness decreased smoke levels and marginally increased NOx levels in LHR engine with vegetable oil operation. Increase of injection pressure deceased pollution levels in LHR engine. However, increase of air gap thickness increased pollution levels with pure diesel operation in LHR engine.

Acknowledgements

The authors are thankful to the authorities of Chaitanya Bharathi Institute of Technology, Hyderabad for providing facilities to carryout this work. The financial assistance provided by the All India Council for Technical Education, New Delhi is also gratefully acknowledged.

References

Annapurna Devi, N. *et al.*, 2006. Production and evaluation of bio-diesel from sunflower and niger seed oil with performance studies on a diesel engine. In: *Proc. International Congress on Renewable Energy*, p. 478-484.

Bhaskar, T. *et al.*, 1993. The effect of two ignition improving additives on the performance of pongamia oil in low heat rejection diesel engine. In: *Proc. IV International Conference on Small Engines and their Fuels*, Thailand, p. 14–19.

Fulekar, M.H., 1999. Chemical pollution: A threat to human life. *Indian J. Env. Prot.*, 1: 353–359.

Heywood, J.B., 2004. *Internal Combustion Engine Fundamentals*. Tata Mcgraw Hill Publishing, New Delhi.

Jabez Dhinagar, S. *et al.*, 1993. A comparative study of the performance of a low heat rejection engine with four different levels of insulation. In: *Proc. International Conference on Small Engines and Fuels*, Chang Mai, Thailand, p. 121–126.

Khopkar, S.M., 1993. *Environmental Pollution Analysis*. New Age International (P) Ltd., Publishers, New Delhi.

Kiannejad, F. *et al.*, 1993. Performance and emissions of a 1.5 litre single cylinder diesel engine with low cetane number vegetable oil fuel and emulsification with water. In: *Proc. IV International Conference on Small Engines and Fuels*, Chang Mai, Thailand, p. 32–39.

Krishnan, D.B. *et al.*, 1980. Performance of an Al-Si graphite particle composite piston in a diesel engine. *Transactions of Wear*, 2: 205–215.

Murali Krishna, M.V.S *et al.*, 2004. Investigations on low heat rejection diesel engine with alternate fuels. *Ph.D Thesis*, J.N.T. University, Hyderabad.

Rama Mohan, K., 1995. Performance evaluation of an air gap insulated piston engine. *Ph.D. Thesis*, Kakatiya University, Warangal.

Rehman, A. and Singhai, K.C., 1995. Vegetable oils as alternate fuels for diesel engine. In: *Proc. of IV Asian Pacific International Symposium on Combustion and Energy Utilization*, Hong Kong, p. 924–928.

Sharma, B.K., 1996. *Engineering Chemistry*. Pragati Prakashan (P) Ltd., Meerut.

Chapter 12

CETP Combined Sludge Application (Textile Dye and Tannery) for its Interaction and Effect on Soil and Groundnut Crop

K.E. Malathy*, S. Avudainayagam and V. Davamani

Department of Environmental Sciences,
Tamil Nadu Agricultural University, Coimbatore – 641 003

ABSTRACT

The Common Effluent Treatment Plant (CETP) tannery and textile/dye sludge are rich in soluble salts (4.5 dSm^{-1} and 4.23 dSm^{-1}) respectively. The textile/dye sludge contains high amount of aquaregia extractable iron (55,091 mg kg^{-1}) and the tannery sludge contains high amount of aquaregia extractable chromium (9,531 mg kg^{-1}). These two industrial solid wastes are combined together to minimize the deleterious effect of one other through chemical interactions and at the same time increasing the beneficial effect for crop production. The combined application of tannery and dye sludge significantly changed the concentration of water soluble iron. However, the changes did not show a definite trend within treatments. The concentration of water soluble iron increased with increasing incubation days for both soils. The water soluble chromium concentration was higher in red soil for normal dose (4.6 t ha^{-1}) in tannery sludge and high in textile/ dye sludge mixed treatment in initial period. The chromium concentration significantly increased with increasing incubation days in both the soils. The results indicate that the behaviour of water soluble iron and chromium are independent and no meaningful interaction effect was

* E-mail: malathy_ke@rediffmail.com

observed all through the incubation days. This experiment was carried out in two different soils *viz.*, black and red soil using groundnut as test crop. The content of chromium was high in kernels in tannery sludge applied soils than combined sludge applied soils.

Keywords: *Groundnut, Tannery and Textile sludge, Chromium and Iron.*

Introduction

Textile and tannery industries are the important industries in India. Dye effluent and sludge are the byproducts from textile industries, similarly tannery effluent and sludge are generated from tanning industries. Solid waste from these industries containing large amount of essential nutrients in unavailable form. Agricultural utilization of sludge is one way of recycling, which depends on nutritive status and ameliorative potential. The nutrients encourage plant growth, the excess nutrients and toxic constituents affect the growth and leads to accumulation of toxic constituents.

Sludge application also had a marked effect on plant and produce yield. In general, there was an appreciable increase in both shoot and produce yield uptake of total chromium (Cr) concentration of 750 mg kg^{-1}. However, the Cr uptake by the plants was significant in all the test crops. Out of the total plant Cr uptake was 30–40 per cent accumulated in shoots and 10–15 per cent in fruits (Sara Parwin Banu *et al.*, 2000).

The studies are lacking on utilizing the combination of industrial solid wastes to minimize the deleterious effect of one another through chemical interactions and at the same time increasing the beneficial effect for crop production. In the present study the potentiality of CETP dye and tannery sludge as source of nutrients and their interaction effect to minimize the uptake of toxic trace metals was evaluated with the following objectives.

1. To assess the interaction effect of iron and chromium in minimizing the availability of chromium in soils for uptake by crop.
2. To assess the accumulation of chromium in various parts of the test crop.

Materials and Methods

Samples of textile and tannery sludge ware collected from CETP, Perundurai Erode and Ranipet, Vellore district, respectively. Samples were shade dried and processed for the analysis of various parameters. The preserved samples was taken for further analysis such as pH, EC, water soluble Fe and Cr were measured by (1 : 5) sludge water suspension and then kept for shaking for 2 hrs in end to end shaker (Avudainayagam *et al.*, 2001). Similarly for exchangeable extracts of Fe and Cr was carried out by (1 : 10) mixture of sludge and Magnesium nitrate, then kept for shaking for 4 hrs in end to end shaker. The filtrate was collected and stored for further estimation in AAS (Krishnamoorthy, 1995). Two types of soil was taken for pot culture experiment *viz.*, red (Typic haplustalfs–Sommayanur series) and black soil (Vertic ustropept–Periyanaikenpalayam series) respectively. For plant analysis the roots, shoots, kernels of groundnut plant was oven dried, powdered and kept for digestion using triacid ($HNO_3 : H_2SO_4 : HClO_4$@ 9 : 2 : 1) (Piper *et al.*, 1996).

Groundnut crop was raised as test crop, since it is an oilseed crop and also needs both Ca and S in high quantities, which were rich in the CETP sludge. Seven kilograms of soil was taken in the each pot. The seeds were sown, initially three plants were allowed and then two plants were maintained in each pot. Calculated quantity of sludge and recommended dosage (400 kg/ha) of gypsum were applied on 45th day after sowing.

Treatments Details

T_1: Control (soil alone)
T_2: Soil + dye sludge (0.4 t ha^{-1})
T_3: Soil + dye sludge (0.4 t ha^{-1}) + tannery sludge (2.3t ha^{-1})
T_4: Soil + dye sludge (0.4 t ha^{-1}) + tannery sludge (4.6 t ha^{-1})
T_5: Soil + dye sludge (0.4 t ha^{-1}) + tannery sludge (6.9 t ha^{-1})
T_6: Soil + tannery sludge (4.6 t ha^{-1}) + dye sludge (0.2 t ha^{-1})
T_7: Soil + tannery sludge (4.6 t ha^{-1}) + dye sludge (0.5 t ha^{-1})
T_8: Soil + tannery sludge (4.6 t ha^{-1}) + dye sludge (0.6 t ha^{-1})
T_9: Soil + tannery sludge alone (4.6 t ha^{-1})

Results and Discussion

Water Soluble Iron

From the results it shows that, in black soil the concentration of water soluble iron in control (T_1) was 0.10 mg kg^{-1} which has decreased over the days. In dye sludge treatment (T_2) the concentration of water soluble iron was 0.19 mg kg^{-1} and increased to 1.16 mg kg^{-1} on 90th day of incubation. In tannery sludge treatment water soluble iron was 0.42 mg kg^{-1} and increased to 2.36 mg kg^{-1} over the days. But in the combined sludge treatment (T_3–T_5) the concentration of water soluble iron ranged from 0.25–0.39 mg kg^{-1}, which was increased over the days (Table 12.1).

Table 12.1: Concentration of Water Soluble Iron (mg kg^{-1})

Treatment	*Incubation Period*								
	30 Days			*60 Days*			*90 Days*		
	Black Soil	*Red Soil*	*Mean*	*Black Soil*	*Red Soil*	*Mean*	*Black Soil*	*Red Soil*	*Mean*
T_1	0.10	0.42	0.26	0.11	BDL	0.057	BDL	BDL	BDL
T_2	0.19	0.56	0.38	0.60	0.1	0.800	1.16	1.23	1.20
T_3	0.35	0.34	0.34	0.61	0.61	0.61	0.55	0.90	0.72
T_4	0.39	0.38	0.39	0.36	0.47	0.42	1.05	1.06	1.05
T_5	0.25	0.49	0.37	0.87	1.43	1.15	1.17	0.98	1.08
T_6	0.32	0.56	0.44	0.61	1.61	1.11	1.76	1.77	1.76
T_7	0.37	1.15	0.76	0.71	1.18	0.95	1.68	1.45	1.56
T_8	0.32	1.52	0.92	0.822	1.57	1.19	1.66	1.71	1.68
T_9	0.42	1.59	1.00	1.17	1.37	1.27	2.36	1.42	1.89
Mean	0.305	0.782	0.543	0.65	1.03	0.842	1.26	1.17	1.22

In red soil the concentration of water soluble iron in control (T_1) was 0.42 mg kg^{-1} which has decreased over the days. In textile/dye sludge treatment (T_2) the concentration of water soluble iron was 0.56 mg kg^{-1} and increased to 1.23 mg kg^{-1} on 90th day of incubation. In tannery sludge treatment water soluble iron was 1.59 mg kg^{-1} and decreased to 1.42 mg kg^{-1} over the days. But in the combined sludge treatment (T_3–T_8) the concentration of water soluble iron ranged from 0.25–0.39 mg kg^{-1}, which was increased over the days.

However, the studies are less for the combination of two iron rich sludges. So the water soluble iron may be increased over the days. But in normal soil system reported by Troug (1948) that as the pH increased more than 6.5, soluble forms of ferrous ion tended to become oxidized to ferric which as insoluble under neutral and alkaline conditions.

Water Soluble Chromium

In black soil the concentration of water soluble chromium in control (T_1) was in below detectable levels. In textile/dye sludge treatment (T_2) the concentration of water soluble chromium was 0.12 mg kg^{-1} and increased to 0.28 mg kg^{-1} on 90th day of incubation. In tannery sludge treatment water soluble chromium as 0.23 mg kg^{-1} and increased to 0.31 mg kg^{-1} over the days. But in the combined sludge treatment (T_3–T_8) the concentration of water soluble chromium ranged from 0.07–0.17mg kg^{-1}, which was increased over the days (Table 12.2).

Table 12.2: Concentration on Water Soluble Chromium (mg kg^{-1})

Treatment	Incubation Period								
	30 Days			60 Days			90 Days		
	Black Soil	Red Soil	Mean	Black Soil	Red Soil	Mean	Black Soil	Red Soil	Mean
T_1	BDL	BDL	BDL	BDL	BDL	BDL	BDL	BDL	BDL
T_2	0.12	0.20	0.16	0.23	0.24	0.24	0.28	0.36	0.32
T_3	0.12	0.12	0.12	0.11	0.18	0.14	0.44	0.33	0.38
T_4	0.07	0.09	0.08	0.21	0.21	0.21	0.32	0.33	0.32
T_5	0.17	0.28	0.23	0.23	0.19	0.21	0.28	0.42	0.35
T_6	0.12	0.32	0.22	0.35	0.35	0.35	0.26	0.44	0.34
T_7	0.14	0.23	0.19	0.33	0.29	0.31	0.31	0.36	0.31
T_8	0.16	0.31	0.23	0.32	0.34	0.33	0.29	0.33	0.39
T_9	0.23	0.27	0.25	0.47	0.28	0.37	0.31	0.47	0.39
Mean	0.128	0.206	0.16	0.253	0.234	0.24	0.279	0.27	0.310

In red soil the concentration of water soluble chromium in control (T_1) was in below detectable levels. In textile/dye sludge treatment (T_2) the concentration of water soluble chromium was 0.20 mg kg^{-1} and increased to 0.27 mg kg^{-1} on 90th day of incubation. In tannery sludge treatment water soluble chromium was 0.27 mg kg^{-1} and increased to 0.47 mg kg^{-1} over the days. But in the combined sludge treatment (T_3–T_8) the concentration of water soluble iron ranged from 0.25–0.39 mg kg^{-1}, which was increased over the days.

Cary *et al.* (1977b) observed extremely low concentration of Cr (VI) in the soil solution by applying Cr compounds in pot experiments and also reported that reduction of Cr (VI) to Cr (III) is more rapid in acid soils. Also there will be less or no adsorption of Cr (VI) under alkaline soils. Hence organic matter could be the other factor that reduced Cr (VI) to Cr (III) or Cr (III) might have been formed as mixture of hydrated oxide of Cr and Fe after reduction of Cr (VI).

Exchangeable Iron

In black soil the concentration of exchangeable iron in control (T_1) was 0.29 mg kg^{-1} which has decreased over the days. In textile/dye sludge treatment (T_2) the concentration of exchangeable iron was 0.34 mg kg^{-1} and increased to 5.77 mg kg^{-1} on 90th day of incubation. In tannery sludge treatment exchangeable iron was 3.20 mg kg^{-1} and increased to 8.87 mg kg^{-1} over the days. But in the combined sludge treatment (T_3–T_8) the concentration of exchangeable iron ranged from 0.90–7.29 mg kg^{-1}, which was increased over the days (Figure 12.1)

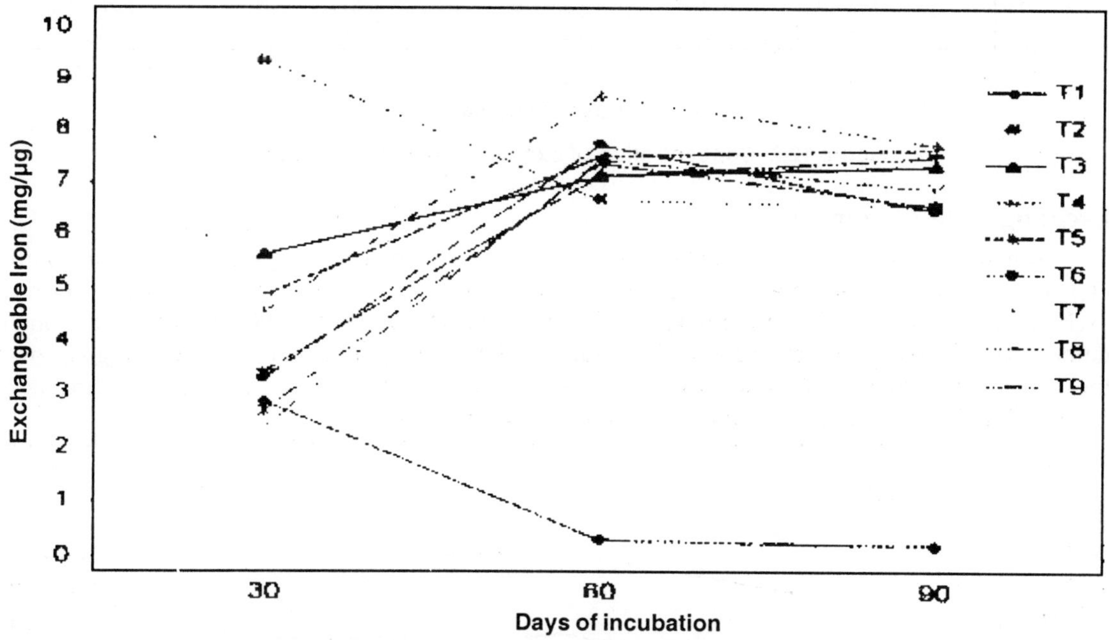

Figure 12.1: Concentration of Exchangeable Iron in Red Soil

In red soil the concentration of exchangeable iron in control (T_1) was 2.88 mg kg^{-1} which has decreased over the days. In textile/dye sludge treatment (T_2) the concentration of exchangeable iron was 9.34 mg kg^{-1} and decreased to 0.27 mg kg^{-1} on 90th day of incubation. In tannery sludge treatment the exchangeable iron was 3.43 mg kg^{-1} and increased to 7.52 mg kg^{-1} over the days. But in the combined sludge treatment (T_3–T_8) the concentration of exchangeable iron ranged from 2.39–5.62 mg kg^{-1}, which was increased over the days (Figure 12.2)

Dahiya and Singh (1979) reported that the exchangeable forms of iron were higher in normal soil, which decreased with the increase in salinity and alkalinity and it also showed a gradual decrease with the increase in time of incubation in all the soils.

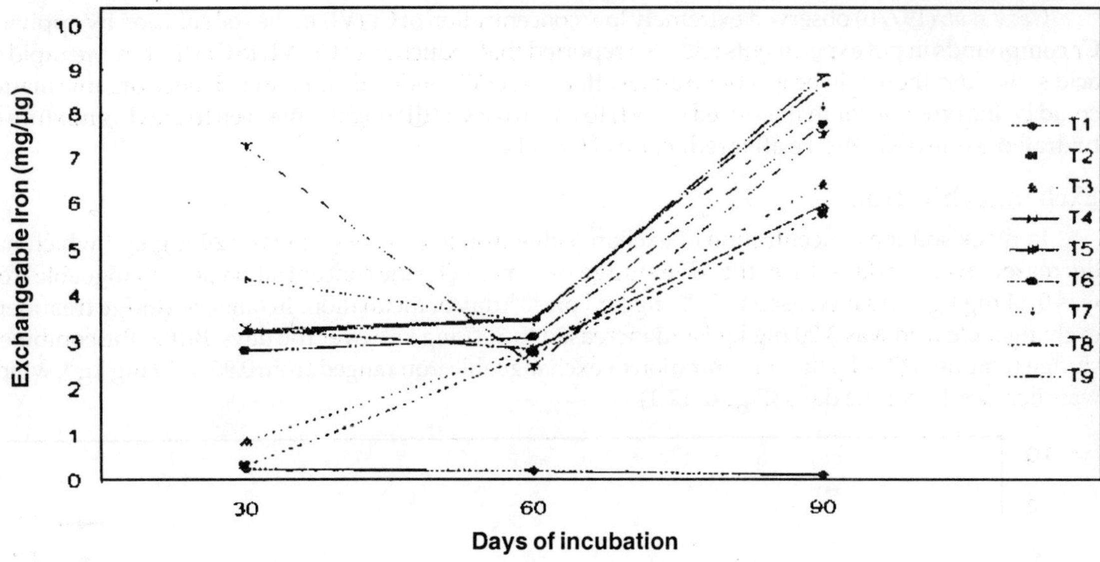

Figure 12.2: Concentration of Exchangeable Iron in Black Soil

Exchangeable Chromium

From the results it shows that, in black soil the concentration of exchangeable chromium control (T_1) was in below detectable levels. In textile/dye sludge treatment (T_2) the concentration of exchangeable chromium was 0.44 mg kg^{-1} and increased to 0.53 mg kg^{-1} on 90th day of incubation. In tannery sludge treatment exchangeable chromium was 0.83 mg kg^{-1} and increased to 0.96 mg kg^{-1} over the days. But in the combined sludge treatment (T_3–T_8) the concentration of exchangeable chromium ranged from 0.18–0.57 mg kg^{-1}, which was increased over the days (Figure 12.3).

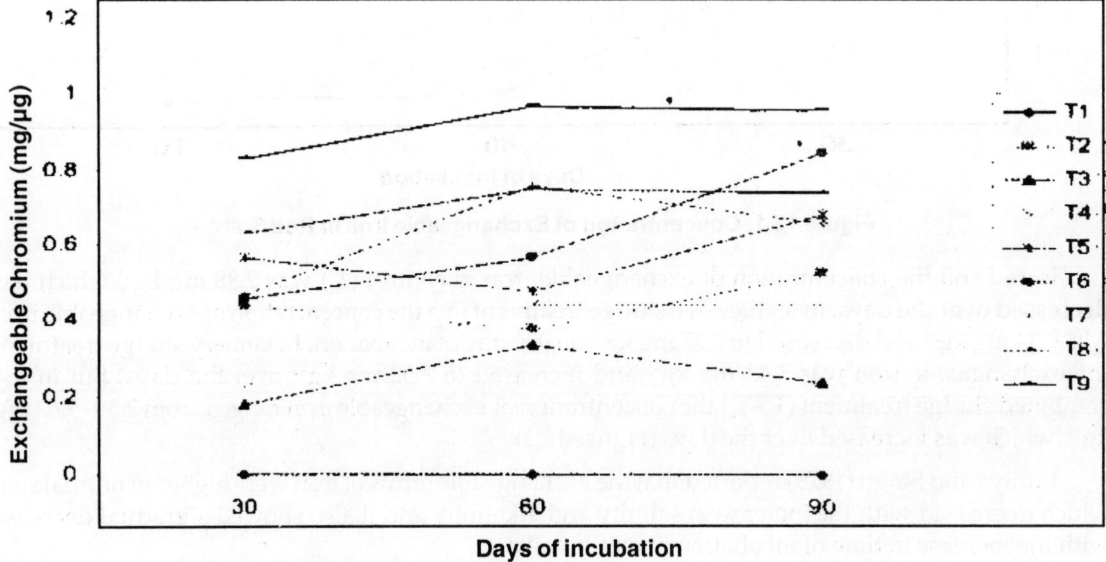

Figure 12.3: Concentration of Exchangeable Chromium in Black Soil

In red soil the concentration of in control (T_1) was absent. In dye sludge treatment exchangeable chromium (T_2) the concentration of exchangeable chromium was 0.38 mg kg^{-1} and decreased to 0.26 mg kg^{-1} on 90th day of incubation. In tannery sludge treatment exchangeable chromium was 0.94 mg kg^{-1} and increased to 0.97 mg kg^{-1} over the days. But in the combined sludge treatment, (T_3–T_8) the concentration of exchangeable chromium ranged from 0.66–0.91 mg kg^{-1}, which was increased over the days (Figure 12.4).

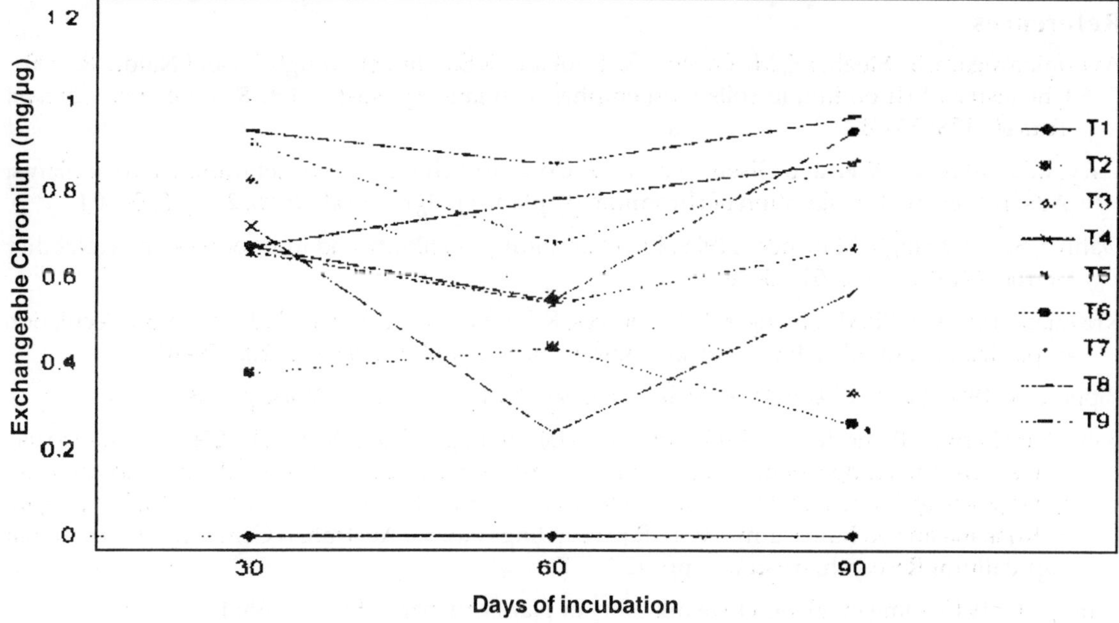

Figure 12.4: Concentration of Exchangeable Chromium in Red Soil

The concentration of chromium in tannery sludge treatment may be due to the interaction effect of iron with chromium, which facilities the precipitation.

Content of Chromium in Plants Parts (mg kg^{-1})

It was noted that concentration of chromium was high in root than aerial parts and, in root the accumulation was recorded high in the treatment that received tannery sludge application T_9 (6.75 mg kg^{-1}), shoot T_7 (8.75 mg kg^{-1}) and in kernel concentration was recorded high in T_9 (3.20 mg kg^{-1}) in black soil. Similarly in red soil root accumulation more in T_9 (8.75 mg kg^{-1}), shoot accumulation in T_6 (5.70 mg kg^{-1}) and in kernel was T_9 (3.20 mg kg^{-1}) in combined sludge treatment.

Sara Parwin Banu *et al.* (2000) reported that the amount of Cr taken by plants varied considerably both in plants species and rates of sludge application. Increase in sludge application rate led to an appreciable increase in Cr uptake by plants. The increased uptake may be attributed to increased Cr in the soil solution and to the alkaline pH of the soil at the higher applications of sludge.

Summary

From results of this study it was confirmed that the behavoiur of water soluble iron and chromium are independent and no meaningful interaction effect was observed. Both the concentration of exchangeable iron and exchangeable chromium was significant with each other. The concentration of exchangeable iron was higher than exchangeable chromium and this was more pronounced in both the soils. The presence of high concentration of iron is obvious because of the application of iron rich combined sludge. The combined application of CETP tannery sludge and dye sludge favored groundnut crop growth and yield due to the presence of major nutrients and calcium. However, the uptake of trace metals chromium in the edible parts was of concern and requires further for the studies.

References

Avudainayagam, S., Megharaj, M., Owens, G., Kookana, R.S., Chittleborough, D. and Naidu, R., 2001. Chemistry of chromium in soils with emphasis on tannery waste sites. *Rev. Environ. Contain. Toxicol.*, 178: 53–91.

Cary, E.E., Allaway, W.H. and Olsen, O.E., 1977. Control of chromium concentration in food plants: Adsorption and translocation of chromium by plants. *J. Agric. Food Chem.*, 25(2): 300–304.

Dahiya, S.S. and Singh, Mahendra, 1979. Effect of salinity, alkalinity and iron sources on availability of iron. *Plant and Soil*, 51: 13–18.

Krishnamurthy, G.S.RP.M., Huang, K.C.J., Vanrees, K.C.J. Kozak and Rostad, H.P.W., 1995. Speculation of particulate bound cadmium of soils and its bioavailability. *Analyst*, 120: 659–665.

Piper, C.S., 1966. *Soil and Plant Analysis*. Interscience Publications, New York, p. 368.

Banu, Sara Parwin, Ramesh, P.T., Ramasamy, K., Mahimairajah, S. and Naidu, R., 2000. Is it safe to use tannery chrome sludge for growing vegetables results from a glass house study. In: *Towards Better Management of Soils Contaminated with Tannery Waste*, (Eds.) R. Naidu, I.R. Willet, S. Mahimairajah, R. Kookana and K. Ramasamy. ACIAR proceedings No. 88. Australian Centre for International Agricultural Research, Australia, pp. 127–137.

Trough, E., 1948. Lime in relation to availability of plants nutrients. *Soi. Sci.*, 65: 1–7.

Chapter 13

Base Composition Analysis of DNA Using HPLC and Short 3μ-cartridge Column

Nupur Mathur
Environmental Toxicology Unit, Department of Zoology,
University of Rajasthan, Jaipur

ABSTRACT

In the present study a reverse phase Column 18 (3μ particle size) was used to identify plant DNA nucleobases. All the four major bases and a few important minor bases like 5-methyl Cytosine could be well resolved within a few minutes. The precision and accuracy of this HPLC method can help in studying the tissue specific and developmental stage specific differences in the levels of 5-methyl Cytosine and 7-methyl Guanine in DNA of various plants that normally are rich in Guanine and Cytosine. Furthermore, it can be used to study the relation between malignancy and overall methylation status of DNA.

Keywords: *High Pressure Liquid Chromatography (HPLC), Plant DNA nucleobases, 5-methyl Cytosine.*

Introduction

Besides the four major bases, the DNA of most organisms contains one or more minor bases. The most common of these, in higher plants, is 5-methyl Cytosine, though N6-methyl adenine and 7-methyl Guanine are also found. In vertebrates 3-6 per cent of DNA Cytosine is methylated but this value decreases going down the evolutionary scale so that in many insects and single celled eukaryotes there is no detectable 5-methyl Cytosine. In contrast, plants may have up to 30 per cent of their DNA Cytosine methylated. Considerable importance has been attached to the Cytosine methylation of plant DNA, probably owing to its implications in regulation of replication, gene expression, differentiation, etc. (Vanyushin and Kirnos, 1988 and Vanyushin, 2006). Methylated cytosine in DNA appears to be an important modulator of the expression of some genes (Catania and Fairweather, 1991).

The relationship between levels of DNA methylation and gene activity has been known for some time. As genetic research advances, many different approaches have been employed to determine the overall level of DNA methylation in a genome or to reveal the methylation state of particular nucleotide residues, starting from semi quantitative methods up to new and powerful techniques (Havli and Trbusek, 2002). Many of the early procedures developed gave only somewhat limited information about methylation patterns, for example, the total level of 5-methyl cytosine in the genome or the frequency of methylation of cytosines within certain restriction sites. However, in the last few years, there has been an explosion of interest in DNA methylation, and with it, many new and powerful techniques have been developed to facilitate its study (Oakeley, 1999).

Several chromatographic approaches (Bird and Taggert, 1980 and Christman, 1982) have been used to separate, identify and quantify, major and minor methylated nucleobases including thin layer, gas and paper chromatography. The first attempts to separate the minor methylated bases employed either anion exchange or cation exchange resins. However, the time required achieving this separation was at least two hours (Essigmann, *et al.*, 1977).

High Pressure Liquid Chromatography (HPLC) on the other hand displays considerable advantages for separation of bases on account of the elution speed, reproducibility and resolving capacity. High performance liquid chromatography (HPLC) is unquestionably the most widely used analytical separation technique. The reasons for its popularity are its suitability for separating substances whose chemical properties range from nonpolar to ionic, the availability of a multitude of detection modes such as UV absorbance, refractive index, fluorescence, electrochemical, radioactivity, and mass spectrometric detection, as well as its ready adaptability to accurate quantitative measurements.

While in conventional Column chromatography the mobile phase passes through the stationary phase by gravity or low pressure, in HPLC the mobile phase is pumped through the column under high pressure. Unlike normal partition chromatography, in reverse phase chromatography, the mobile phase is more polar than the stationary phase.

The concept of performing fast chromatography using columns packed with very small particles is not new. A.J.P. Martin and R.L. Synge postulated "the smallest HETP (Height Equivalent of Theoretical Plate) should be obtained by using very small particles and a high pressure difference across the length of the column" (Martin and Synge, 1941). In the present study a reverse phase Column 18 (3μ particle size) column was used to identify plant DNA nucleobases.

Materials and Methods

Apparatus

A Perkin-Elmer 410 L C pump equipped with an L C 235 Diode array detector working at 190nm–340nm and L C-100 computing integrator were used for routine HPLC analysis. A reverse phase C18 (311 particle size) column (33 mm × 4.6 mm) was used.

Preparation of Standards

All the standards were prepared in 0.01 N HCl with the final concentration as 1 μg/ml. Initially 3M KOH was used to dissolve the bases. pH of the bases was adjusted to 5.5. 20 μl of this stock was further diluted to 1ml and this was used as working solution. The buffer used was 2.3mM $NH_4H_2PO_4$, pH 3.5. All the solutions used were filtered and sonicated before use.

DNA Isolation and Sample Preparation

DNA was isolated from the leaves of sugarcane (*Saccharum officinarum*) according to Dellaporta method (Dellaporta *et al.*, 1983). Purified DNA (RNA free) was hydrolyzed in sealed capillary tubes with 90 per cent Formic acid at 175°C for 25 min. Formic acid was then evaporated under vacuum and the dry residue was dissolved in 0.01 N HCl.

Chromatographic Conditions

All the runs were made at room temperature (25°C). Samples were loaded on the column equilibrated at 1.2 ml/min. with a mixture of 99 per cent of 2.3 mM Ammonium dihydrogen phosphate buffer, pH 3.5 and 1 per cent water. Chromatographic runs were performed isocratically at the starting conditions for 1.8min. Then a linear gradient in 0.3 minutes resulting in mixture of 20 per cent of 2.3 mM $NH_4H_2PO_4$ buffer, pH 3.5 and 80 per cent water at F.R. of 2.0 ml/minute.

Results and Discussion

In eukaryotic cells nuclear DNA is subjected to enzymatic methylation resulting in formation of 5-methylcytosine residues. In plants and animals, this DNA methylation is species-, tissue-, and organelle-specific. DNA methylation controls all genetic processes in the cell (replication, transcription, DNA repair, recombination, gene transposition) and it is a mechanism of cell differentiation, gene discrimination, and silencing (Vanyushin, 2004). Most of the methods used for separating nucleobases use longer reverse phase columns (250 mm × 4 mm) with larger particle size, about 5–10 μ. Total separation has been obtained in about 18–20 min or more in all these methods (Gehrka *et al.*, 1984 and Wietstock, 1995).

This is one of the first methods that use a small cartridge column (33 mm × 4.6 mm) with a particle size of 3μ. The advantage being that the same numbers of plates per column are generated by reducing particle size and column length. For comparison refer to Table 13.1. The buffer composition was 2.3 mM Ammonium dihydrogen phosphate buffer, containing 6 per cent (v/v) methanol, pH 3.5. Individual bases were then loaded to characterize their elution time.

Table 13.1: Performance Levels of 10μ, 5μ and 3μ Columns

Particle Diameter (μ)	*Typical/Lengths (cm)*	*Plates/Column Length*
10	25–30	10,000–12,000
5	12.5–25	10,000–20,000
3	3.3–15	4,950–22,500

Guanine, first standard base to be loaded, gave a retention time of 1.10 min., peak purity index of 1.1 and Abs. Max. of 246 nm (Figure 13.1). Next a mixture of Guanine and Cytosine was loaded. Cytosine gave a retention time of 0.93min., peak purity index of 2.5 and Abs. Max. of 274nm (Figure

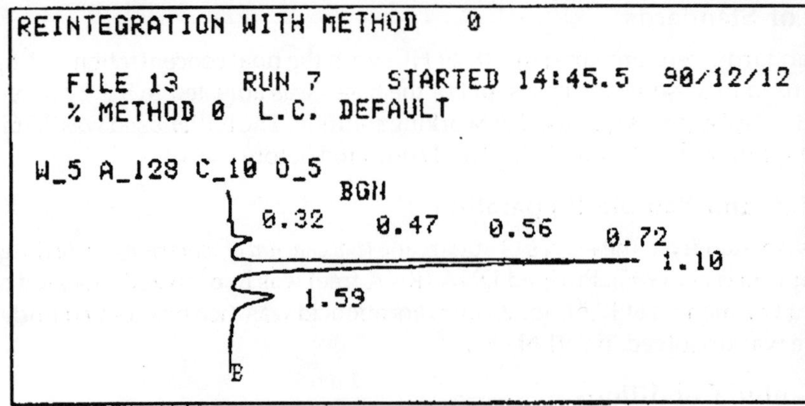

Figure 13.1: Guanine Peak, Retention Time of 1.10 min., Peak Purity Index of 1.1 and Abs. of 246 nm

13.2) while Guanine eluted at 1.06 min., purity index of 3.6 (Figure 13.2). Peaks of Guanine and Cytosine were not totally resolved due to peak merging and possible change in elution profile of the mixture. Adenine gave retention time of 2.12 min., purity index of 1.9 and Abs. Max. of 260 nm (Figure 13.3) while Thymine gave retention time of 1.3 min., purity index of 1.1 and Abs. Max. of 265nm (Figure 13.4).

A mixture of all the four bases was then loaded. The merging of Cytosine (purity index of 1.3) and Guanine (purity index of 3.4) persisted probably due to fronting of Cytosine peak and tailing of Guanine peak. However Adenine and Thymine peaks were well separated (Figure 13.5).

On increasing the pH of the above buffer, Cytosine and Guanine peaks merged further. Hence, a lower pH of 3.0 was used. At this pH Cytosine and Guanine peaks were completely resolved (Figure 13.6). Cytosine eluted in 1.13 min. at a purity index of 1.3.

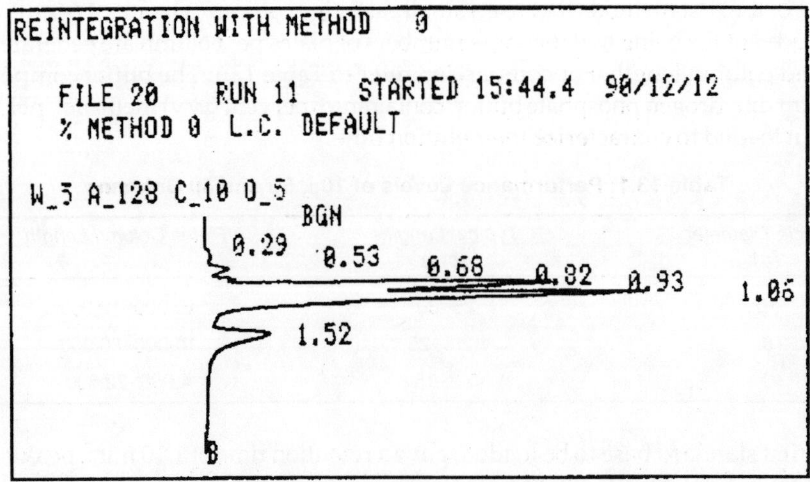

Figure 13.2: Guanine Eluted at 1.06 min., Purity Index of 3.6 and Cytosine, Retention Time of 0.93 min., Peak Purity Index of 2.5 and Abs. Max. of 274 nm

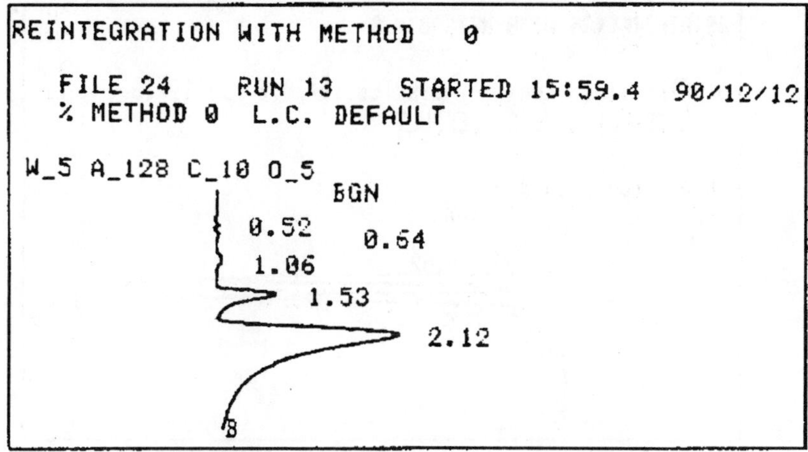

Figure 13.3: Adenine Peak, Retention Time of 2.12 min., Purity Index of 1.9 and Abs. Max. of 260 nm

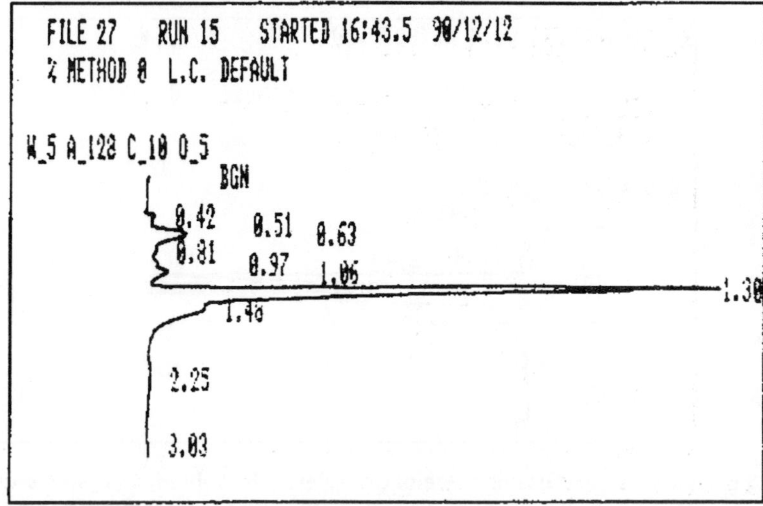

Figure 13.4: Thymine Peak, Retention Time of 1.3 min., Purity Index of 1.1 and Abs. Max. of 265 nm

Next 5-methyl Cytosine was loaded, which had a retention time of 1.23 min., peak purity index of 1.0 and Abs. Max. of 284 nm. However on loading a mixture of Cytosine, Guanine and 5-methyl Cytosine, Guanine and 5-methyl Cytosine peaks merged, most likely due to 5-methyl Cytosine peak tailing (purity index of 14) (Figure 13.7).

Thus as changing the pH did not help in separation of all four major and one minor base, original 2.3 mM Ammonium dihydrogen phosphate buffer, pH 3.5, was used and methanol concentration was decreased from 6 per cent (v/v) to 2 per cent (v/v). This elution buffer resolved Cytosine (purity index

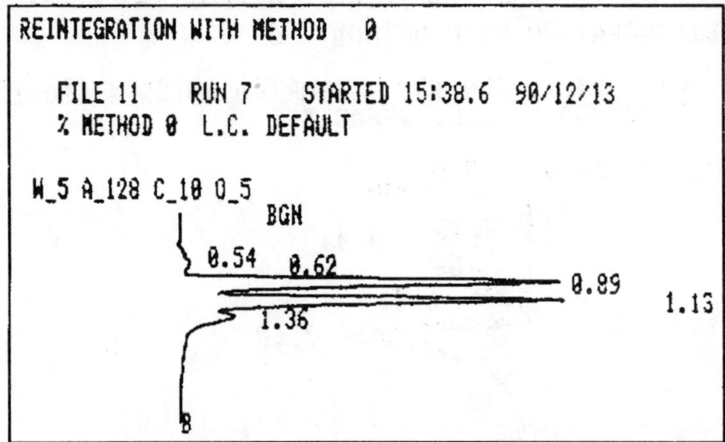

Figure 13.5: A Mixture of all the Four Bases Loaded. The merging of Cytosine (purity index of 1.3) and Guanine (purity index of 3.4), Adenine and Thymine peaks well separated; buffer composition: 2.3 mM Ammonium dihydrogen phosphate buffer, containing 6 per cent (v/v) methanol, pH 3.5.

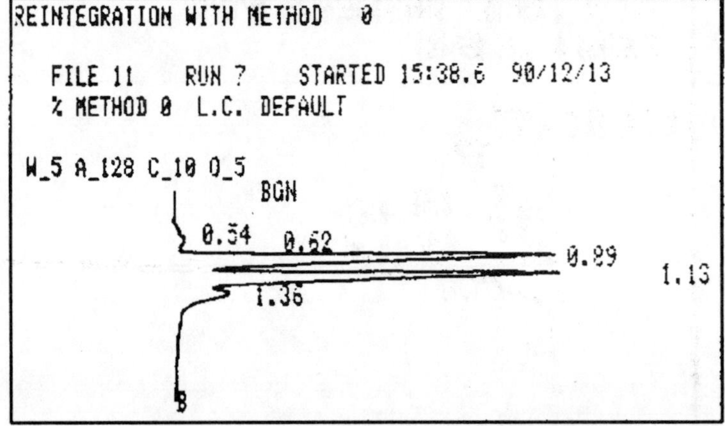

Figure 13.6: Cytosine and Guanine Peaks Completely Resolved; Cytosine eluted in 1.13 min. at a purity index of 1.3; buffer composition: 2.3 mM Ammonium dihydrogen phosphate buffer, containing 6 per cent (v/v) methanol, pH 3.0.

of 1.1) and Guanine (purity index of 1.4) completely. Thymine (purity index of 4.6) and 5-methyl Cytosine (purity index of 5.9) which were completely merging at pH 3.0 were now partially resolved (Figure 13.8). Further resolution of Thymine and 5-methyl Cytosine was obtained by using buffer having no methanol at all (Figure 13.9). Thymine eluted at 2.25 min with purity index of 1.1 and Abs. Max. of 265 nm and 5-methyl Cytosine at 2.63 min. with purity index of 1.3 and Abs. Max. of 281 nm.

Hence using 2.3 mM Ammonium dihydrogen phosphate buffer (no methanol), pH 3.5 at F.R. = 0.8 ml/min., all four major and one minor base could be well resolved. To reduce the elution time of Adenine linear gradient was used with higher flow rates of 1.2 ml/min. for 1.8 min. and then 2.0 ml/

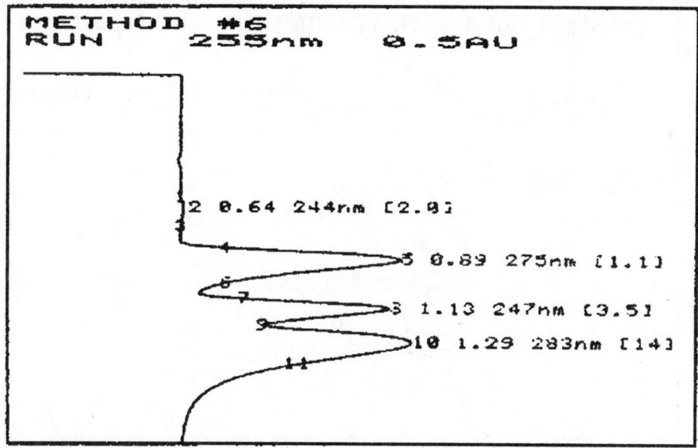

Figure 13.7: 5-methyl Cytosine Peak, Retention Time 1.23 min., Peak Purity Index of 1.0 and Abs. Max. of 284 nm

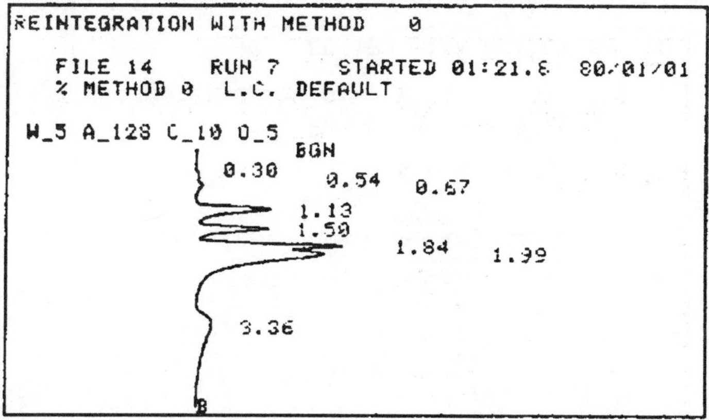

Figure 13.8: Cytosine (Purity Index of 1.1) and Guanine (Purity Index of 1.4) Completely Resolved. Thymine (purity index of 4.6) and 5-methyl Cytosine (purity index of 5.9) partially resolved; buffer composition: 2.3 mM Ammonium dihydrogen phosphate buffer, pH 3.5, methanol concentration 2 per cent (v/v).

min. till the end of the run. Under these conditions, Adenine elutes out at 2.66 min. with purity index of 1.2 (Figure 13.10). Under similar conditions, 7-methyl Guanine eluted out at 2.91 min. with purity index of 1.5 and Abs. Max. of 249 nm (Figure 13.11).

Thus separation of all the four standard major and two minor bases were obtained within 3 min. All these bases were characterized by not only their retention times but also by their Abs. Max. in UV range, as shown in Table 13.2.

This method was then used to analyze the genomic sugarcane DNA (hydrolyzed) sample (Figure 13.11). All the five bases Adenine, Guanine, Cytosine, Thymine and 5-methyl Cytosine could be identified. 7-methyl Guanine peak's retention time was same as in case of standard and it eluted just after Adenine but its Abs. Max. was 283 nm. Hence, it mayor may not be 7-methyl Guanine since the

```
REINTEGRATION WITH METHOD   0

   FILE 6     RUN 4     STARTED 15:33.0   90/12/19
   % METHOD 0  L.C. DEFAULT

 W_5 A_128 C_10 O_5
                   BGN
              0.64
              0.97
                        1.24
                     1.81
                             2.24
               2.61
```

Figure 13.9: Further resolution of Thymine eluted at 2.25min with purity index of 1.1 and Abs. Max. of 265nm and 5-methyl Cytosine at 2.63min. with purity index of 1.3 and Abs. Max. of 281nm; buffer composition: 2.3 mM Ammonium dihydrogen phosphate buffer, pH 3.5, no methanol.

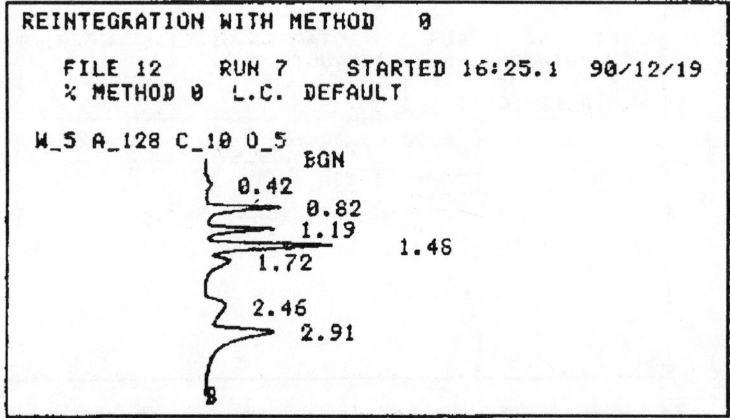

Figure 13.10: All Four Major and One Minor Base Well Resolved; Buffer Composition: 2.3 mM Ammonium dihydrogen phosphate buffer (no methanol), pH 3.5 at F.R. = 0.8 ml/min.

Table 13.2: Characterization of Nucleobases

Base	*Retention Time (Minutes)*	*Absorption Maxima (UV Range) (nm)*
Cytosine	0.83	274
Guanine	1.2	246
Thymine	1.48	265
5-methyl Cytosine	1.73	281
Adenine	2.41	260
7-methyl Guanine	2.91	249

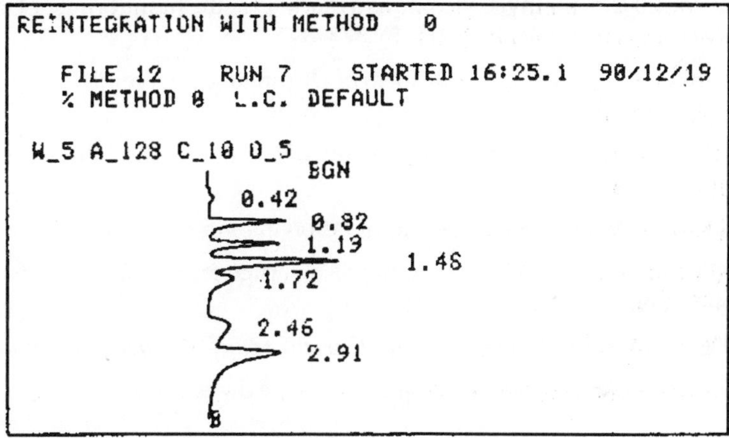

Figure 13.11: 7-methyl Guanine Peak at 2.91 min. with Purity Index of 1.5 and Abs. Max. of 249 nm; buffer composition: 2.3 mM Ammonium dihydrogen phosphate buffer (no methanol), pH 3.5 at F.R. = 0.8 ml/min.

bases are known to change Abs. Max. with change in pH value. An additional undetermined compound (X) eluted just before Cytosine, at OAmin., having max. Abs. at 211 nm. It could either be a normal constituent of DNA or result from degradation in the course of DNA extraction, purification or hydrolysis.

The precision and accuracy of this HPLC method can help in studying the tissue specific and developmental stage specific differences in the levels of 5-methyl Cytosine and 7-methyl Guanine in DNA of various plants that normally are rich in Guanine and Cytosine. Furthermore, it can be used to study the relation between malignancy and overall methylation status of DNA.

References

Bird, A.P. and Taggert, M.H., 1980. Variable patterns of total DNA and rDNA methylation in animals. *Nucleic Acids Research,* 8(7): 1485–1497.

Catania, J. and Fairweather, D.S., 1991. DNA methylation and cellular ageing. *Mutation Research/ DNAging,* 256(2–6): 283–293.

Christman, J.K., 1982. Separation of major and minor deoxyribonudeoside monophosphates by reverse-phase high-performance liquid chromatography: A simple method applicable to quantitation of methylated nucleotides in DNA. *Anal Biochem.,* 119(1): 38–48.

Dellaporta, S.L., Wood J., Hicks, J.B., 1983. A plant DNA minipreparation: Version II. *Plant Molecular Biology Reporter,* 1: 19–21.

Essigmann, J.M., Busby, W.F. Jr. and Wogan, G.N., 1977. Quantitative determination of transfer RNA base composition using high pressure liquid chromatography. *Anal. Biochem.,* 81: 384–394.

Gehrka, C.W., McCune, R.A., Gama-sosa, M.A., Ehrlich, M. and Kuo, K.C., 1984. Quantitative reversed-phase high performance liquid chromatography of major and modified nucleosides in DNA. *J. Chromatogr.,* 301(1): 199–219.

Havli, J. and Trbuiek, M., 2002. 5-methyl Cytosine as a marker for the monitoring of DNA methylation. *Abstract Journal of Chromatography B*, 781(1–2): 373–392.

Martin, A.J.P. and Synge, R.L.M., 1941. A new form of chromatogram employing two liquid phases. *Biochem. J.*, 35: 1358–1368.

Oakeley, E.J., 1999. DNA methylation analysis a review of current methodologies. *Pharmacology and Therapeutics*, 84(3): 389–400.

Vanyushin B.F. and Kirnos, M.D., 1988. DNA methylation in plants. *Gene* 74: 117.

Vanyushin, B.F., 2005. Enzymatic DNA methylation is an epigenetic control for genetic functions of the cell. 70(5): 488–499.

Vanyushin, B.F., 2006. DNA methylation in plants. *Current Topics Microbiol. Immunol.*, 301: 67–122.

Wietstock, S.M., 1995. DNA composition analysis by nuclease digestion and HPLC. *J. Chem. Educ.*, 72: 950–952.

Chapter 14

Vibrational Spectral Studies of Doped Cu^{2+} L-Arginine Phosphate Crystal

A. Ramalingam, D. Karunanithy, V. Karunakaran and C. Arunagiri

Department of Physics, Government Arts College,
Ariyalur – 621 713, T.N.

ABSTRACT

The vibrational (Infrared) Spectra of Cu^{2+}–LAP crystal have reported in the region 400–4000cm^{-1}. Cu^{2+}–LAP crystal compound play an using KBr pellet. The observed fundamental bands were assigned assuming the molecules under point group. These bands have been correlated with various modes of vibrations reported in the literature. The lattice lines are fairly strong compared with the others and show pronounced intensity variations with the orientation of the crystal. The high-frequency lines are assigned to the vibrations of the molecular groups.

Introduction

The vibrational assignment of infrared spectrum of Cu^{2+}–LAP crystal have been extensively studied in present investigation. In the present work a detailed study of the polarized infrared absorption spectra of Cu^{2+}–LAP crystal and essentially complete vibrational assignments of frequencies of the said molecule. Recent spectral investigation on Cu^{2+}–LAP crystals find interesting application in agriculture and pharmacy (Fanivan, 1979; Weinman, 1979). The studied of the vibrational assignments in light of the previous work.

Experimental

Freshly crushed powder of the pure and metal (Cu^{2+}) doped LAP crystals were mixed with KEr in the ratio 1 : 10 and palletized using a hydraulic press and then subjected to FTIR studies. FTIR of these compound were used recorded on its purity was confirmed by BRUKER IFS 66V model FTIR spectrophotometer. Elemental analysis and melting point determination.

Results and Discussion

The molecular structure of the compound Cu^{2+}-LAP crystal is shown in Figure 14.1. FTIR spectrum in KEr and plyethylene pellets is shown in Figure 14.2. The observed fundamental vibrational frequencies and their proposed assignments of Cu^{2+}-LAP crystal is given in Table 14.1.

Table 14.1: FTIR Assignments of CU-LAP Crystal

Observed FTIR Frequency (in cm^{-1})	*Intensity of the Peaks*	*Assignments*
3453	w	V_{as} (N–H)
3335	w	V_s (N–H)
3169	W	V_{as} (NH_3^+)
1690	W	V (C–N)
1653	Vs	(N–H) bending
1569	s	V_{as} (N–H)
1528	s	V_{as} (NH_3^+)
1455	s	V_{as} (O–H)
1410	S	V_s (COO^-)
1373	ms	β(C–C–H)
1334	w	$\gamma_w(CH_2)$
1322	W	$\gamma_w(CH_2)$
1299	W	V (P = 0)
1176	W	$\beta_{as}(NH_3^+)$
1133	w	$\gamma_w(NH_2)$
1047	W	$\gamma_1(PO_4)$
963	S	γ(P–O)
879	mS	v(C–C)
785	S	$\beta_{as}(NH_2)$
762	s	$\beta_s(COO^-)$
700	s	$v_s(NH_2)$
645	s	$v_s(OH)$
551	w.	$\beta_s(COO^-)$
531	w	$v_s(PO_4)$

v: Stretching; v_{as}: Asymmetric stretching; v_w: Very weak; v_s: Symmetric stretching; γ_w: Wagging; sb: Strong broad; β: In-plane-bending; β_s: Scissoring; m: Medium; VVW: Very very Weak; Vs: Out-of-plane bending; V_{sb}: very strong broad; β_{as}: Rocking; s: Strong; MS: Medium strong; VS: very strong.

$$
\begin{array}{ccc}
H & & H \\
|| & & || \\
H_2N-C-COO^- & -X- & {}^-OOC-C-NH_2 \\
|| & & || \\
H-C-H & & H-C-H \\
|| & & || \\
H-C-H & & H-C-H \\
|| & & || \\
H-C-H & & H-C-H \\
|| & & || \\
H_2N-C-NH_2 & & H_2N-C-NH_2 \\
|| & & || \\
NH & & NH
\end{array}
$$

Figure 14.1: Structure of Cu^{2+}-LAP Crystal

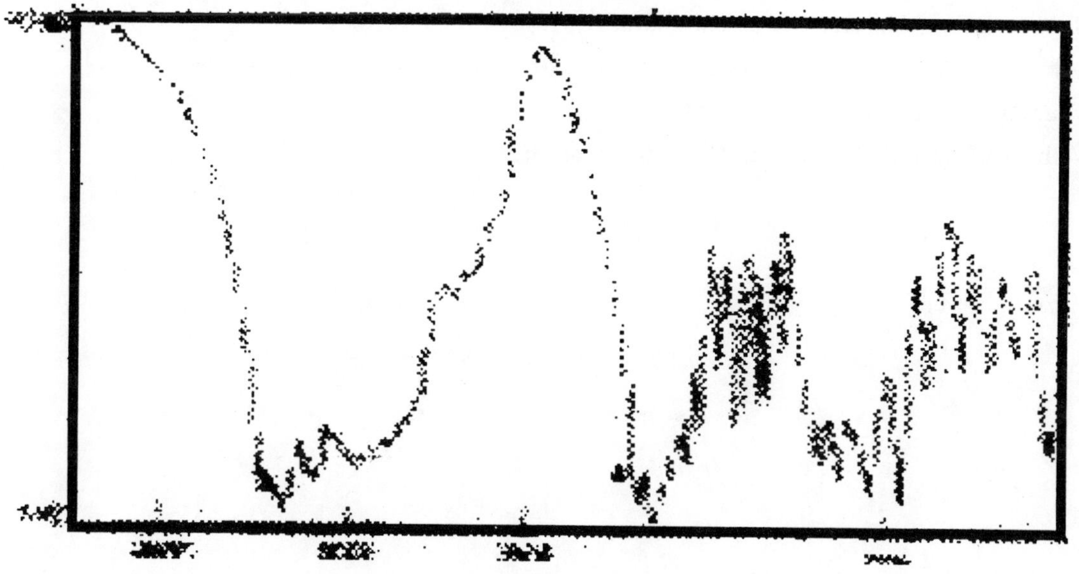

Figure 14.2: FTIR Spectrum Cu^{2+}-LAP Crystal

Group Vibration

N–H Group

The N–H stretching vibrations of Amine compounds occur in the region 3000–3500 cm^{-1} (Bellany, 1959). The FTIR band at 3453 and 3335 cm^{-1} for Cu^{2+}-LAP crystal is assigned to N–H symmetric, asymmetric modes.

Amino Group

Amino group stretching vibrations of amino compounds occur in the region 3500–2600 cm^{-1} (Mooney, 1964). The FTIR bands at 3169 and 1528, 1176 cm^{-1} for Cu^{2+}-LAP crystal is assigned to Amino symmetric, asymmetric and rocking modes.

C–N Group

The frequencies observed at 1690 cm^{-1} in Cu^{2+}-LAP crystal is assigned to C–N stretching modes. The assignments of C–N in stretching vibration (Seshadri *et al.*, 2007). The symmetric mode of COO^- group has produced its characteristic peaks around 1400 cm^{-1} (Gupta *et al.*, 1988).

In the formation of phosphate ion, P–O, P = O stretching are observed at 1299 and 963 cm^{-7}. Move over, rocking mode, symmetric deformation and asymmetric bending were found from the FTIR spectrum of Cu^{2+}-LAP Crystal. From this FTIR spectrum of Cu^{2+}-LAP crystal is assigned only one C–C stretching group region at 879 cm^{-1}. This finds support from the work of various investigations.

References

Bellany, L.J., 1959. *The Infrared Spectra of Complex Molecules.* John Wiley, New York.

Fanivan, J.A., 1979. *Spectrochim, Acta,* 38A: 1257.

Gupta, S.P., Sharma, S.D., Jetly, U.K., Yadav, B.S. and Seema, K.M., 1988. *Journal of Indian Council of Chemists,* 4(3).

Mooney, E.E., 1964. *Spectrochim Acta,* 20: 1021.

Seshadri, S., Arunagiri, C. and Ramanathan, R., 2007. *Acta Ciencia Indica,* 23(1): 65.

Weinman, C.J., 1979. *J. Flairine Chem.*

Wiebengs, E.H and Strits, D.W., 1950. *Acta Cryst.,* 3: 265.

Chapter 15

Role of VAM in Uptake of Heavy Metals

K.E. Malathy*, S. Avudainayagam and V. Davamani
Department of Environmental Sciences,
Tamil Nadu Agricultural University, Coimbatore – 641 003

ABSTRACT

In this study, the effects of VAM on the growth of maize and the uptake of some heavy metals by the plant were studied. Maize roots were inoculated with VAM fungus and sludge, and the rate of infection on the roots was determined. Heavy metals contents increased in soil, root, stem, and leaves, depending on the increased doses of Cr, Fe, Cd, Zn, and Cu, being applied to both inoculated and non-inoculated treatments. The inoculation of VAM along with sludge decreased the translocation of chromium from root to aerial portion through chelation of the metal with the cells. The decrease in the translocation of chromium from root to aerial parts was observed only in black soil.

Keywords: *VAM, Tannery and Textile sludge, Chromium.*

Introduction

Farm soils irrigated with heavy metals rich industrial water increases their heavy metals contents. Heavy metals probably will be strongly toxic to crops, inhibit the formation of mycorrhiza and affect the nutrition of the host plant.

Chao and Wang (1991) stated that the content and concentration phosphate in corn above ground portion significantly reduced at higher concentration of heavy metals. Zinc, copper and chromium

---

* E-mail: malathy_ke@rediffmail.com

were much inhibits the phosphate content in the plant tissues than other heavy metals. The heavy metals in nutrient solution increased the concentration and content of heavy metals in roots were two to seventeen times higher than in ground portion.

Loth and Hofner (1995) reported that the heavy metal uptake of mycchorizal oat plants (*Avena sativa* L.) was evaluated in two soils differing in heavy metals accumulation. The effect of the fungal isolates on the uptake of the immobile metals Zn and Cu differed between the two soils. The mycorrhizal colonies increased heavy metals uptake up to 37 per cent. In the highly contaminated soil mycorrhizal infection lead to a higher uptake (max 59 per cent) in root but to a reduced translocation to the aerial pants architecture by increasing the specific root length (km per pot). This increased absorbing surface of the roots was a major, but not the cause for the differences in heavy metals uptake.

Ricken and Hofner (1996) reported that in pot experiment with a sewage sludge treated soil, the influence of arbuscular mycorrhizxal fungi (AMF) on plant growth and on the uptake of heavy metals by alfalfa (*Medicago sativa* L.) and Oat (*Avena sativa* L.) was investigated. Alfalfa showed an increase of biomass with mycorrhizal infection only to a small extent. In oat AMF inoculation increased the growth of both root and shoot by up to 70 per cent and 55 per cent respectively. The root length was significantly decreased in mycorrhizal alfalfa plants (up to 38 per cent). The translocation of heavy metals into the shoot of mycorrhizal alfalfa was significantly increased. Mycorrhizal infection of oat led to higher concentrations of Zn, Cd, and Ni the root but to less Zn in the shoot. The translocation of heavy metals to the oat shoot was clearly decreased by mycorrhizal colonization.

At the present situation disposal of sludge is a greater problem. Since textile and tannery sludge contains high amount of iron and chromium these two sludges could be combined in order to reduce toxicity of Cr. Studies are also lacking in these aspects, so combination of two sludges can reduces deleterious effects.

Materials and Methods

In this study, the effect of VAM on the growth of maize and the uptake of some heavy metals by the plant were studied. Maize roots were inoculated with VAM fungus and sludge, and the rate of infection on the roots was determined. Heavy metals contents increased in soil, root, stem, and leaves, depending on the increased doses of Cr, Fe, Cd, Zn, and Cu, being applied to both inoculated and non-inoculated treatments.

Pot Culture Experiment

The crop maize was tested for uptake of heavy metals, which were rich in the CETP sludge. Two kilograms of soil was taken in the each pot. The seeds were sown; two plants were maintained in each pot. Calculated quantity of tannery and dye sludge were mixed and applied. VAM also inoculated separately.

Test crop	:	Maize
Variety	:	CO_2
Date of sowing	:	12.05.2004
Design	:	FCRD
Replications	:	Three
Soil type	:	Red and black

Treatment Details

T_1: Control

T_2: Soil + tannery sludge + 100g VAM

T_3: Soil + tannery sludge alone

Result and Discussion

Biometric Observation

Plant Weight (Table 15.1)

Maize crop was raised in both black and red soil. A significant change has been observed in T_3 (5.65 g pot^{-1} of plant weight) which was grown in black soil. Among the treatments considerable change has been due notified to sludge application. But in red soil, no significant change has been observed.

Table 15.1: Effect of VAM on Whole Plant Weight (g pot^{-1})

Treatments	*Observation on 45th day*		
	Black Soil	*Red Soil*	*Mean*
T_1	3.05	2.00	2.52
T_2	4.35	1.70	3.02
T_3	5.65	2.50	4.0
Mean	4.35	2.066	3.20
		SEd	*CD (0.05)*
	S	0.094	0.230
	T	0.115	0.282
	S×T	0.163	0.391

S: Soil; T: Treatment; S×T: Soil treatment interaction.

Root Weight (Table 15.2)

In black soil, compared to control T_1 root weight has been increased due to VAM application, which was observed in T_2 (2.358 g pot^{-1}). The same significant change has been notified in red soil also.

Root Length (Table 15.3)

With the application of sludge in black soil root length has been increased, in T_3 (26.758 cm) compared to T_1 (24.75 cm) and T_2 (75.20 cm of root length). But in case of red soil, root length has been increased in T_2 (26.75 cm) than T_1 (18.50 cm) and T_3 (17.15 cm).

Shoot Length (Table 15.4)

From the analyzed data it was confirmed that the shoot length has been more in control (13.40 cm) than T_2 (11.75 cm). So there was decreasing trend in the shoot length due to sludge application in black soil. A similar change has been observed in the red soil also. T_1 (15.5 cm) has recorded more shoot length values compared to T_2 (8.75 cm) and T_3 (8.6 cm). Here, also the considerably decreasing trend has been noticed.

Table 15.2: Effect of VAM on Root Weight (g pot^{-1})

Treatments	*Observation on 45th day*		
	Black Soil	*Red Soil*	*Mean*
T_1	1.45	0.75	1.10
T_2	2.35	0.94	1.64
T_3	2.10	1.27	1.68
Mean	1.96	0.99	1.47
		SEd	*CD (0.05)*
	S	0.078	0.192
	T	0.096	0.235
	S×T	0.136	0.332

S: Soil; T: Treatment; S×T: Soil treatment interaction.

Table 15.3: Effect of VAM Inoculation on Root Length (cm)

Treatments	*Observation on 45th day*		
	Black Soil	*Red Soil*	*Mean*
T_1	24.25	18.50	21.37
T_2	26.75	26.85	25.98
T_3	25.20	17.15	21.95
Mean	25.40	20.80	23.10
		SEd	*CD (0.05)*
	S	0.333	0.815
	T	0.408	0.998
	S×T	0.577	1.412

S: Soil; T: Treatment; S×T: Soil treatment interaction.

Content of Chromium (mg kg^{-1}) in Plants (Table 15.5)

Due to VAM application, increased content of chromium was observed in T_2 (0.034 mg kg^{-1}) than other treatments, this shows that content of chromium has a significant trend over the treatments in both the soils.

Conclusion

The inoculation of VAM along with sludge decreased the translocation of chromium from root to aerial portion through chelation of the metal with the cells. The decrease in the translocation of chromium from root to aerial parts was observed only in black soil. Due to the inoculation of VAM and sludge application, the content of chromium was increased in root than shoots in both the soils. The content chromium of was increased in root parts due to VAM infection, and compare to root, shoot part have less accumulation. Because translocation of metal has been stopped to aerial portion due to VAM infection. According to the studies conducted by Rieken and Hofner (1996) reported that the translocation of heavy metals to the oat shoot was clearly decreased by mycorrhizal colonization. This may be based on the ability of fungal tissues to complex heavy metals at the cell walls, thus excluding metals from the shoot.

Table 15.4: Effect of VAM Inoculation on Shoot Length (cm)

Treatments	*Observation on 45th day*		
	Black Soil	*Red Soil*	*Mean*
T_1	11.75	8.75	10.25
T_2	13.40	15.55	14.97
T_3	12.75	8.65	10.70
Mean	12.63	10.98	11.80
		SEd	*CD (0.05)*
	S	0.206	0.504
	T	0.252	0.617
	S×T	0.357	0.873

S: Soil; T: Treatment; S×T: Soil treatment interaction.

Table 15.5. Effect of VAM Inoculation on Chromium Content (mg kg^{-1})

Treatment	*Root*		*Shoot*	
	Black Soil	*Red Soil*	*Black Soil*	*Red Soil*
T_1	BDL	BDL	BDL	BDL
T_2	0.034	0.004	0.020	0.040
T_3	0.002	0.001	0.053	0.033
Mean	0.01	0.001	0.024	0.024
		SEd	*CD(0.05)*	
	S	0.006	0.001	
	T	0.007	0.001	
	S×T	0.001	0.002	

S: Soil; T: Treatment; S×T: Soil treatment interaction.

References

Chao, C.C. and Wang, Y.P., 1991. Effects of heavy metals on the infection of vesicular arbuscular mycorrhizae and the growth of maize. *Journal of the Agricultural Association of China*, 152(1): 34–45.

Loth, F.G. and Hofner, W., 1995. Influence of VAM on heavy metal uptake of oat (*Avena sativa* L.) from soils differing in heavy metal contamination. *Zeitschrift Fur Pflanzenernahrung and Bodenkunde,* 158: 339–345.

Rieken, B. and Hofner, W., 1996. Effect of arbuscular mycorrohizal fungi (AMF) on heavy metal tolerance of alfalfa (*Hedicago sativa* L.) and oat (*Avena sativa* L.) on a sewage sludge treated soil. *Zeitschrift Fur Pflanzenernahrung and Bodenkunde,* 159(2): 189–194.

Chapter 16

Estimation of Firing Temperature of Archaeological Pottery Shreds Recently Excavated in Salamankuppam, Tamil Nadu, India

G. Velraj and K. Janaki
Department of Physics, Periyar University, Salem – 636 011, Tamil Nadu, India

ABSTRACT

Clays and related raw materials have been used by mankind since earliest times. The different thermal reactions of clay minerals on temperatures ranging from 500–1100°C are important for formation of potteries as well as for reconstruction of their production conditions. Clay minerals as main raw material for production of potteries show some characteristic transformation effects in the course of their thermal decomposition. The property of thermal transformation in clay minerals during firing and process of firing provide a means to estimate the firing temperatures of the artifacts. Hence, an attempt has been made in the present work to estimate the firing temperature of the archaeological potteries that have been collected at five different depths from the archaeologically important site Salamankuppam, near famous Mahabalipuram in Chennai using Fourier Transform Infrared (FT-IR) spectroscopic technique.

Keywords: FT-IR, Firing temperature, Archaeological potteries.

Introduction

The transformation of clay into pottery was an important step in the development of artisans. The technology of production of archaeological materials reveals the level of the technical background of the people and can be used as an index of ancient civilization and their interactions (Maniatis *et al.*, 1982; Venkatachalapathy *et al.*, 2002). Clays undergoes several structural and compositional changes when exposed to increasing temperature. Up to 200°C adsorbed water evaporates. When the temperature increases to 450–600°C dehydroxylation and the loss of structurally water occur. The above transformation has been widely studied by Fourier-Transform infrared (FT-IR) spectroscopy for a range of temperature up to 1400°C. It is intended that these results will help in revealing the potential of FT-IR spectroscopy as a tool in analyzing earthenware pottery shreds and their methods of production in Tamil Nadu (Francesco Berna *et al.*, 2007). To learn more about the technological conditions such as firing temperature under which potteries were produced, potteries collected horn Salamankuppam site has been investigated by using FT-IR spectroscopy.

Materials and Methods

The samples collected from Salamankuppam site was re-fired using the muffle furnace in the laboratory to temperatures at 550, 650, 750 and 850°C. The refired samples were powdered using Agate Mortar and were palletized by mixing with spectra grade KBr at the ratio of 1 : 20 by weight. The FT-IR spectra were recorded for the powdered samples in the explored range of frequencies 4000–400 cm^{-1} using Nicolet Avatar360 FT-IR Spectrometer (Francesco Berna *et al.*, 2007; Benedetto, 2005).

Results and Discussions

Five samples at different depths have been collected from the recently excavated site Salamankuppam near famous Mahabalipuram in Chennai, Tamil Nadu. The samples are named as SMK1, SMK2, SM3, SMK4 and SMK5 in the order of increasing depth. The FT-IR spectra of the refired samples are depicted in Figures 16.1–16.5. The firing temperature of the potteries can be determined with the help of the study of thermal transformation of the clay minerals in the pottery samples. The bands at about 3440 cm^{-1} and 1640 cm^{-1} which are attributed to stretching and bending of absorbed water molecules present in the samples. The water molecules are absorbed by the potteries during burial under the earth. These bands are sensitive to temperature effect. After refiring the samples to 250°C these bands diminished in intensity and this reflects the evaporation of absorbed water molecules (Venkatachalapathy *et al.*, 2002). These two bands are present in all the samples taken from Salamankuppam site. As stated by (Venkatachalapathy *et al.*, 2002), the spectrum of the refired samples shows the band at 3440 cm^{-1} which gets diminished in intensity as temperature increases and the band at 1640 cm^{-1} diminishes and disappears at 850°C.

Francesco Berna *et al.* (2007) have studied that when pottery samples heated to 500°C, the Kaolinite absorption bands at 3695 cm^{-1}, 3650 cm^{-1}, 3620 cm^{-1} disappears due to collapse of the structure due to the loss of hydroxyl groups. Maniatis *et al.*, (1982) have reported that the persistence of a weak band around 3600 cm^{-1} indicates the presence of the iron hydroxyls still in the clay body and points to incomplete dehyroxylation. This band persists up to 800°C. Among the samples taken from Salamankuppam site the sample SMK shows a very weak band at 3626 cm^{-1} and indicates incomplete dehydroxylation and the remaining samples SMK2 to SMK5 do not show any band at 3620 cm^{-1} and hence dehydroxylation is completed in all these samples. So the sample SMK1 may fired below 800°C and SMK2–SMK5 might have been fired above 800°C. The bands at 1100 cm^{-1} and 915 cm^{-1} due to

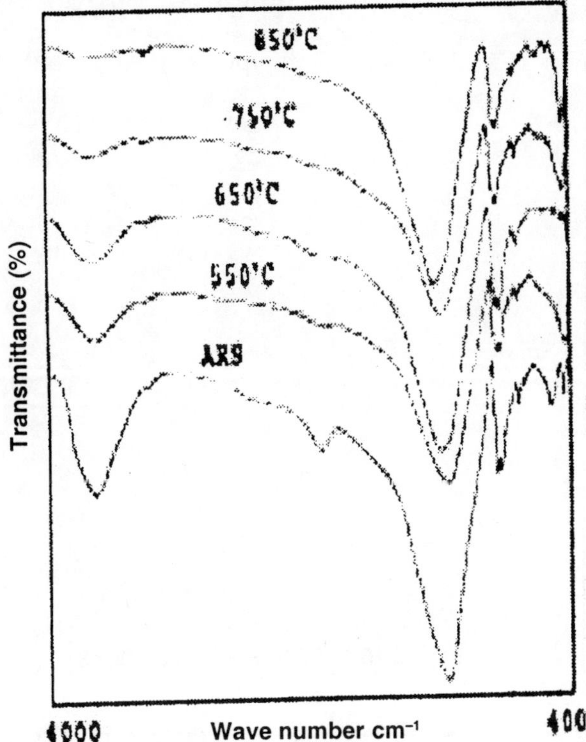

Figure 16.1: FT-IR Spectra of SMK1

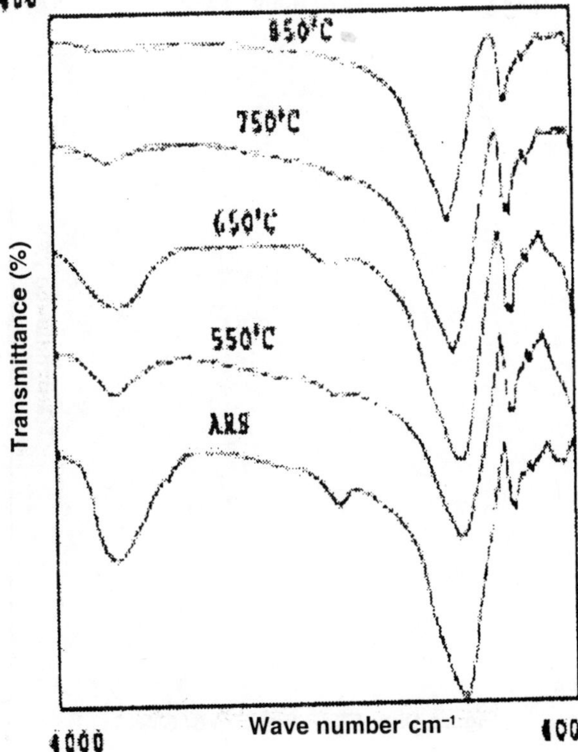

Figure 16.2: FT-IR Spectra of SMK2

850°C
750°C
650°C
550°C
ARS
Transmittance (%)
4000
Wave number cm⁻¹
400

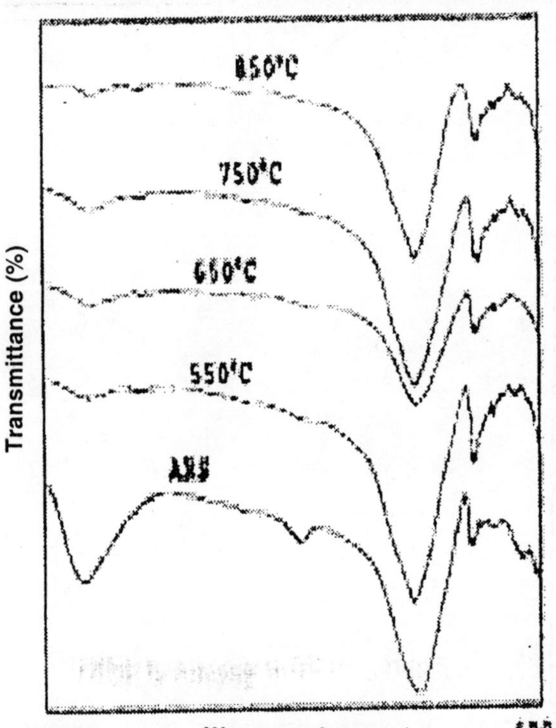

Figure 16.3: FT-IR Spectra of SMK3

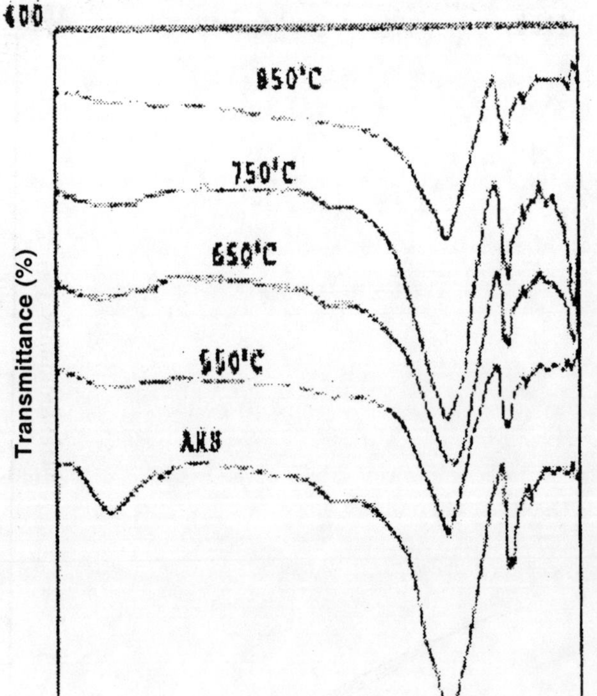

850°C
750°C
650°C
550°C
ARS
Transmittance (%)
4000
Wave number cm⁻¹
400

Figure 16.4: FT-IR Spectra of SMK4

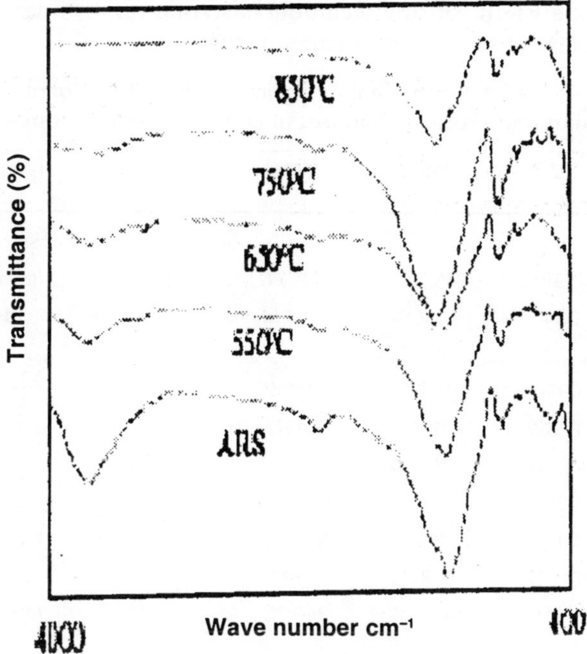

Figure 16.5: FT-IR Spectra of SMK5

Al(OH) vibrations in the octahedral sheet structure. This band begins to disappear with increasing temperatures and at 500°C the band at 915 cm^{-1} disappears completely (Venkatachalapathy *et al.*, 2002). The FT-IR spectra at the received state (ARS) itself do not show any band at 915 cm^{-1}. Hence, all the samples taken from the Salamankuppam site have been refired to a temperature above 500°C. This can be predicted in the samples at received state spectra itself so all the samples were refired above 500°C *i.e.*, 550, 650, 750 and 850°C. The weak shoulder band appearing at 875 cm^{-1} in the IR spectra is due to iron rich clay minerals. The band has been attributed to Fe Al(OH) group mineral present in the clay. It further indicates the persistence of the octahedral sheet structure and the presence of Al-OH in the clay mineral and its incomplete dehydroxylation (Mendeluvici *et al.*, 1979). This band does not appear in any of the sample taken from Salamankuppam site and hence indicates disappearance of octahedral sheet structure (Mendeluvici *et al.*, 1979).

Calcite is particularly reflective and can serve as a useful diagnostic mineral to estimate the maximum firing temperature (Eiland, 2000). Legodi and Waal (2007) have stated that the presence or calcite confirms the processing temperature below 800°C. The sample SMK1 shows the strong band at 1427 cm^{-1} in ARS itself hence may be tired below 800°C and this band persists up to 650°C, hence may be fired above 650°C *i.e.*, in the range 650–750°C. The remaining samples do not show the Calcite band in the received state itself and hence fired above 800°C. These results are tabulated in Table 16.2.

Conclusion

From the above discussions it is clear that the sample SMK1 is fired in the range 650–750°C and remaining samples SMK2–SMK5 taken from the Salamankuppam site have been fired above 850°C. The artisans belonging to this site knew the art of controlling the temperature of firing to produce good quality potteries and from the FT-IR refiring technique also one can estimate the range of firing temperature achieved by the artisans at the time of manufacture.

Table 16.1: Tentative Vibrational Assignments for FT-IR Spectra of Salamankuppam Pottery Samples at Different Temperatures

Sample Code	*ARS*	*550°C*	*650°C*	*750°C*	*850°C*	*Tentative Vibrational Assignment*
SMK1	3626VW	–	–	–	–	O-HSlr.crystalline hydroxyl
	3442VW	3429VW	3435VW	3433VW	3422VW	Inner O-H group of
	1637VS	1634S	1635S	1627 M	–	adsorbed water
	1427S	1430S	1427S	–	–	Calcite
SMK2	3416M	3433W	3439VW	3443W	l402W	Inner O-H group of
	1624W	1633W	1634S	1617M	–	adsorbed water
SMK3	3408VS	3423W	3412VW	3412W	3429VW	Inner O-H group of
	1635VS	1625M	1616VS	1616VS	–	adsorbed water
SMK4	1435VS	3413S	3430W	3444W	3382S	Inner O-H group of
	1634M	1615W	1638VW	1634W	–	adsorbed water
SMK5	3440VS	3431S	3428 M	344 7S	3406W	Inner O-H group of
	1634W	1638W	1636VW	1633VW	–	adsorbed water

Table 16.2: Estimated Firing Temperature of the Pottery Shred Collected from Salamankuppam site

Sample Code	*Dehydroxylation of Hydroxyl Band*	*Octahedral Sheet Structure*	*Estimated Firing Temperature*
SMK1	Incomplete	Disappeared	<850°C
SMK2	Completed	Disappeared	>850°C
SMK3	Completed	Disappeared	>850°C
SMK4	Completed	Disappeared	>850°C
SMK5	Completed	Disappeared	>850°C

Acknowledgements

The authors are highly thankful to Mr. S. Mathivanan, Conservation Assistant, Archaeological Survey of India (ASI), Salem branch for the timely help rendered by him to collect the samples. The authors also thank Mr. S Srinivasan, Prof. and Head, Department of Chemistry, Government College of Engineering, Salem for his permission to use the Muffle Furnace to refire the samples and staff in charge who helped to record the FT-IR spectra in CSIL lab, Annamalai University, Annamalainagar.

References

Benedetto, G.E., DeFabbri, B., Gualtieri, S., Sabbatini, L. and Zamgonin, P.G., 2005. FT-IR chemometric tools as aids for data reduction and classification of pre-roman ceramics. *Journal of Cultural Heritage*, 6: 205–211.

Eiland, Murray L. and Williams, Quentin, 2000. Infrared spectroscopy of ceramics from Tell Brak, Syria. *Journal of Archaeological Science*, 27: 993–1006.

Francesco Berna, Adi Behar, Ruth Shahack-Gross, John Berg, Elisabetta Boaretto, Ayelet Gilboa, Ilan Sharon, Suriel Shalev, Sana Shilstein, Naama Yahalom-Mack, Jeffrey Zorn and Steve Weiner, 2007. Sediments exposed to high temperatures: Reconstructing pyrotechnological processes in Late Bronze and Iron Age strata at Tel Dor (Israel). *Journal of Archaeological Science*, 34: 358–373.

Legodi, M.A.D. de.Waal, 2007. Raman spectroscopic study of ancient South African domestic clay pottery. *Spectrochimica Acta*, 66(A): 135–142.

Maniatis, Y., Simopoulos, A. and Kostikas, A., 1982. In: *Archaeological Ceramics*, (Eds.) J.S. Olin and A.D. Frenklin. Smithsonian Institution Press, Washington, Sc., pp 97–108.

Mendelovici, E., Yariv, S.H and Villalba, 1979. Iron-bearing kaolinite in Venezuelan laterites: Infrared spectroscopy and chemical dissolution evidence. *Clay Minerals*, 14: 323–331.

Venkatachalapathy, R., Sridharan, T., Dhanapandian, S. and Monoharan, C., 2002. Determination of firing temperature of ancient potteries by means of infrared and Mossbauer studies. *Spectroscopy Letters*, 35(6): 769–779.

Chapter 17

Hydrogeochemistry and Quality of Chittar-Uppodai River Water, Tambaraparani River Basin, Tirunelveli District, Tamil Nadu: Role of Geo-environment

T. Jeyavel Raja Kumar[1]*, A. Balasubramanian[2], R.S. Kumar[1] and K. Manokaran[1]

[1]Department of Earth Sciences, Annamalai University, Annamalainagar – 608 002

[2]University of Pondicherry, Pondicherry, India

ABSTRACT

In the present study about eleven water samples were collected from the river during post-monsoon season (January 2005). The analysis has been carried out for pH, Electrical Conductivity (EC), major cations such as Ca, Mg, Na, K and anions namely HCO_3, Cl, SO_4, NO_3 and Fe. Besides, other parameters like Total Dissolved Solids (TDS), Permeability Index (PI), Corrosivity Ratio (CR) and Sodium Absorption Ratio (SAR) were calculated. The pH level varied from 7.1 to 8.2 with an average value of 7.6. The EC, TDS and Total Hardness values were varied from 45 µ mohs/cm to 2910 µ mohs/cm, 32 mg/l to 1862 mg/l and 10 mg/l to 525mg/l respectively. The average major cations such as Ca, Mg, Na, K and anions HCO_3, Cl, SO_4, NO_3 are 104.7 mg/l, 25.1 mg/l, 99.18 mg/l, 2.0 mg/l and 220 mg/l, 377.6 mg/l, 170.5 mg/l, 0.28 mg/l respectively. The Sodium Absorption Ratio (SAR) varied from 0.17 to 3.60. The value of Permeability Index and

* E-mail: tjeyavel@rediffmail.com

Corrosivity Ratio are varied from 0.36 to 2.12 and 0.71 to 12.85. To understand the water quality and its usage altitude, the data has been interpretations with US. Salinity Laboratory diagram and Gibb's plot. It is observed that the catchment area water samples were fell in C_1S_1 and C_2S_1 categories which are classified as good class. Rest of the samples was fallen in C_3S_1 and C_4S_1 categories which are classified as satisfactory class and salt tolerant crops class. The Gibb's interpretation indicated that the water samples show the precipitation zone for catchment area water samples and remaining were found in the border zone of rock dominance and evaporation zone. It is observed that apart from weathering process agriculture practice, fertilizer application, livestock waste, soil erosion and anthropogenic activities have contributed for contaminating the river water.

Keywords: *Chittar-Uppodai River, Hydrogeochemistry, Quality, Tirunelveli District.*

Introduction

The present days' advancement and development of the world is directly linked with the harnessing and utilization of natural resources of the river basins. India is gifted with a river system comprising more than 20 major rivers with several tributaries. Many of these rivers are perennial and some of these are seasonal (Raghunath, 1983). India faces challenges in the water sector in the form of declining per capita availability of water, over-exploitation of groundwater resources leading to lowering of groundwater table in certain pockets, deterioration in the water quality, cost and time over-runs in the completion of irrigation and multi-purpose projects and poor maintenance of the existing systems (Todd, 1980). The quality of the water is getting higher priority, due to the increasing trend of contamination level. The primary source for contamination of water is the atmospheric contact and the other important sources are nature of soil, rock-types, live stock excrements, industrial effluents and anthropogenic activities. Waste disposal from cattle yards running through open drains and waste disposal from villages in the margins contributes accountable pollution elements to the stream (Babar and Kaplay, 1999; Babar, 2000; Chavan *et al.*, 2002; Babar *et al.*, 2003). In fact, socio-economic progress is impossible without adequate supplies of good quality water. In the initial period of water quality deterioration on damage may be low but it is not controlled at the right time, this water will not be suitable for any purposes (Karanth, 1987). Hence, it is essential that the study on quality of the water is important not only for determining the level of concentration but also to learn about the contaminating sources of elements to rivers. In addition, information on river chemistry is essential to assess water quality for domestic, agricultural and industrial usage. Therefore, the management and suitable development of the river basins have prime importance. In this present attempt, the Chittar-Uppodai river water chemistry and quality has been discussed.

Study Area

The study area of Chittar-Uppodai Water shed is a sub-basin of Tambaraparani river originates from the Western Ghats. The area is situated in between the East longitudes of 77°10′ to 77°50′ and the North latitudes of 8°52′ to 9°10′ of the Survey of India topo sheets 58G/8,12,13,16 and 58H1,5,9. The groundwater is main source for both domestic and agricultural purposes. The chamockite and biotite gneiss are the major rock types in the study area. Since, the river is a seasonal water flow for hardly 2–3 months and lingering periods being without water flow. The areas adjacent to Western Ghats are of dry to moist sub humid climatic terrains (Ram Mohan, 1984). The average maximum temperature recorded during April and May is about 39°C. The average minimum Temperature is about 24°C recorded usually during the months of November and December. The average rainfall is about 918 mm and the depth to water table during post monsoon varies from 1 m to 10m below ground level.

Materials and Methods

In this present study about 11 water samples were collected from the river during post-monsoon season (January 2005) as shown in the Figure 17.1. The samples were collected in a clean polythene bottles as prescribed by Hem (1985), Brown *et al.* (1970) and APHA (1985) and the analysis has carried out in the Tamil Nadu Water and Drainage Board Laboratory, Manjaneerkayal, Tuticorin. The study has been carried out for pH, Electrical Conductivity (EC), major cations such as Ca, Mg, Na, K and anions namely HCO_3, Cl, SO_4, NO_3 and Fe. Besides, other parameters like Total Dissolved Solids (TDS), Permeability Index (PI), Corrosivity Ratio (CR) and Sodium Absorption Ratio (SAR) were calculated and given in Table 17.1. EC values are expressed in μ mohs/cm at 25°C and except pH other parameters are expressed in mg/l.

Results and Discussion

In the present study area, the ionic concentration of Hydrogen level varied from 7.1 to 8.2 with an average value of 7.6. It is observed that low value found in Courtallam location where the water starts to runoff from the western ghats. The water sample location Vadakarai is also originated from western ghats but it shows with the value of 7.4.The value gradually increases towards its downward runoff and increased to 7.9 at Gangaikondan location. In Kayathar location of Uppodai river shows pH value with 7.7 and after merged with the Chittar river the pH level increased further to 8.2 at Sevalaperi location.

The EC value observed 45 μ mohs/cm as minimum and 2910 μ mohs/cm as maximum level with an average of 3225 μ mohs/cm. The low value observed in Courtallam location and found 640 μ mohs/cm in Vadakarai location. The level has increased to 940 μ mohs/cm from 640 μ mohs/cm, when it mixed with the Chittar river. In Uppodai it observed with the value of 2760 μ mohs/cm. After merging with the main Chittar river water the level has come down to 1270 μ mohs/cm. But the high level has noticed in Seliyanallur location with the value of 2910 μ mohs/cm. The TDS and Total Hardness values are varied from 32 mg/l to 1862 mg/l and 10 mg/l to 525mg/l respectively. The TDS value found low in Courtallam and high in Gangaikondan location. In Vadakarai location it found with 448 mg/l and in Uppodai with 1932 mg/l, after mixing with the Chittar it gained further from its traveling path and reached the maximum level of 1862 mg/l in Gangaikondan. Similarly, the TH also observed low in Courtallam location and in Vadakarai it shows 186 mg/l. The maximum level noticed in Seliyanallur location as 525 mg/l. But the level observed in Kayathar location is 525 mg/l and after merged with Chittar water its value decreased to 253 mg/l in Sevalaperi location.

The Ca concentration is varying from 2 mg/l to 178 mg/l with an average value of 104.70 mg/l. In Courtallam the level is very less and in Vadakarai it is 47 mg/l. The maximum level has found in Seliyanallur and Gangaikondan locations where as its level observed as 137 mg/l in Kayathar location. After merging of these two river waters the concentration level decreased to 57 mg/l. The Magnesium concentration is varying from 1 mg/l to 35 mg/l with an average value of 25.1 mg/l. With the less contribution such as 1 mg/l and 16 mg/l have found in the Courtallam and Vadakarai locations but the concentration is 39 mg/l in Kayathar location. The higher level of Mg encountered in VKPudur location and decreased to 27 mg/l in Sevalaperi location.

The Sodium concentration is varying from 3 mg/l to 189 mg/l with an average value of 99.18 mg/l. Similarly, the less concentration from both Courtallam and Vadakarai locations it has

Table 17.1: Hydrogeochemical Data of the Study Area

Sl.No.	Parameters	pH	EC	TDS	TH	Ca	Mg	Na	K	HCO_3	Cl	SO_4	NO_3	F	SAR	PI	CR
			μ mohs/cm	mg/l													
1.	Courtallam	7.1	45	32	10	2	1	3	0	17	5	0	0.06	0.2	0.43	2.12	0.71
2.	Nainanagaram	7.4	1840	1288	408	110	32	89	1.9	356	3.7	186	0.18	0.6	1.92	0.52	0.72
3.	Vadakarai	7.4	640	448	186	47	16	9	1.8	212	38	58	0.64	1.2	0.28	0.55	0.80
4.	Kilpavour	7.6	940	658	313	81	27	7	1.6	263	147	29	0.2	0.8	0.17	0.36	1.50
5.	Agaram	7.6	1720	1204	414	109	34	110	3.6	267	302	207	0.06	1.4	2.35	0.52	3.78
6.	VK Pudur	7.4	1965	1375	505	149	35	95	3.1	339	398	124	0.06	1.4	1.81	0.44	3.32
7.	Ukkirankottai	7.8	2600	1820	505	145	34	154	3.4	216	631	264	0.52	2.4	2.98	0.51	8.66
8.	Seliyanallur	7.7	2910	2037	525	178	19	189	2.3	195	774	252	0.6	2.4	3.60	0.53	11.3
9.	Gangaikondan	7.9	2668	1862	424	178	19	189	2.3	157	689	252	0.08	2.4	3.37	0.52	12.8
10.	Kayathar	7.7	2760	1932	505	137	39	174	3.5	170	631	372	0.64	1.4	3.60	0.52	12.0
11.	Sevalaperi	8.2	1270	889	253	57	27	67	4.6	233	228	103	0.08	0.4	1.83	0.60	2.96

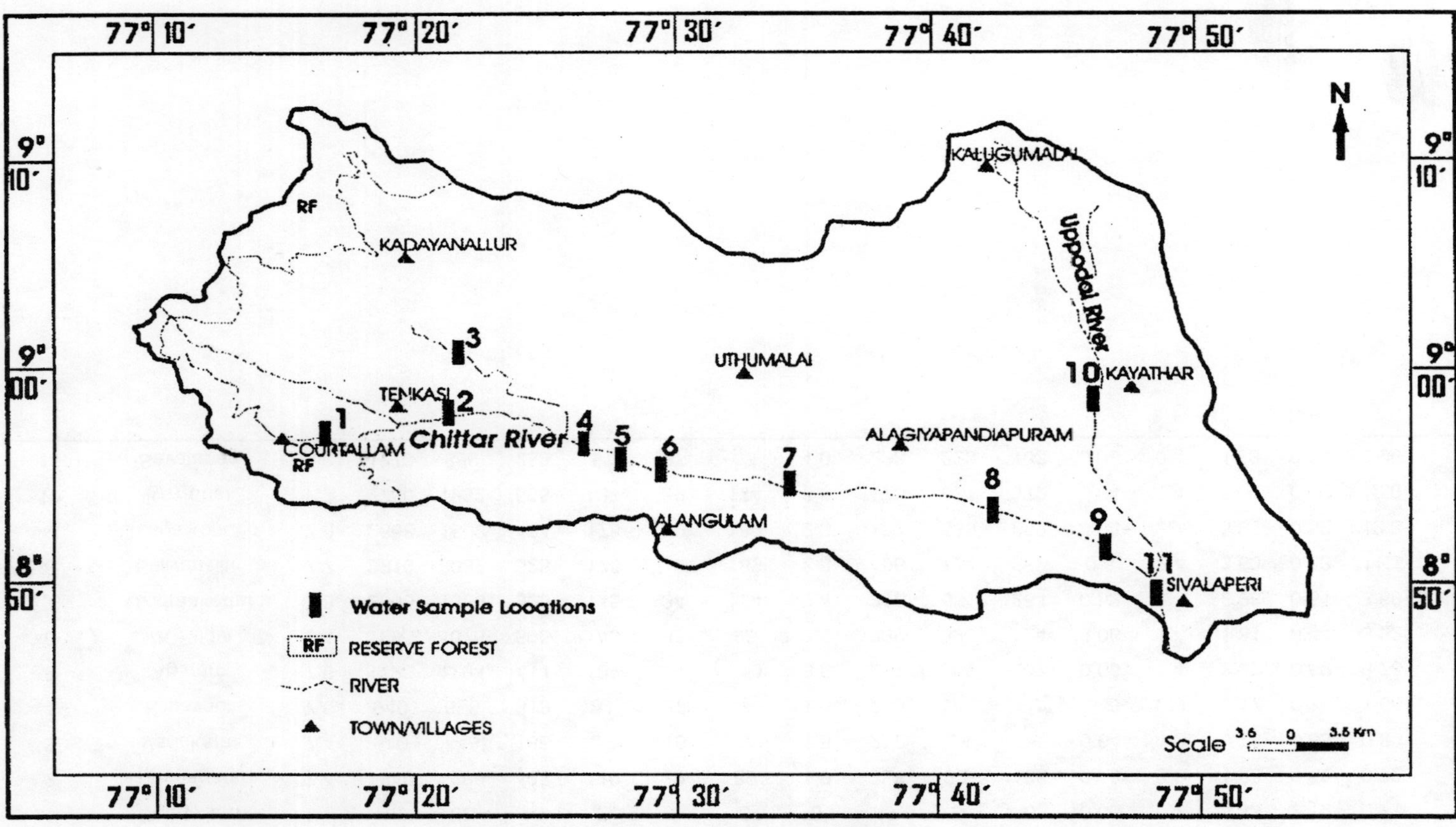

Figure 17.1: Map of Water Sample Location

reached a maximum level of 189 mg/l in Agaram location. Even the concentration of 174 mg/l from the Kayathar location the level decreased to 67 mg/l in Sevalaperi location after merging the two river waters. The Potassium concentration is varying from none to 3.6 mg/l with an average value of 2.0 mg/l. The concentration observed low in the origin point of both Courtallam and Vadakarai locations. The level observed in Kayathar is 3.5 mg/l and after merged with Chittar river water the level has increased further to the maximum of 4.6 mg/l in Sevalaperi location.

The Cl concentration is varying from 5 mg/l to 774 mg/l with an average value of 377.36 mg/l. The less amount of concentration was observed in both Courtallam and Vadakarai locations and it observed 621 mg/l in Kayathar location. The level increased gradually from the origin point and reached maximum level at Seliyanallur location and decreased to 228 mg/l in Sevalaperi location. The Sulphate concentration is varying from none to 264 mg/l with an average value of 170.54 mg/l. The Bicarbonate concentration varies from 17 mg/l to 339 mg/l with an average value of 220 mg/l. The concentration of both SO_4 and HCO_3 are observed with less concentration in the origin point and found higher concentration in Ukkirankottai and VKPuram respectively. Even with the significant amount of concentration from Kayathar location the level has decreased in the end point of Sevalaperi location. The Nitrate and Fluoride concentrations vary from 0.06 mg/l to 0.64 mg/l and 0.2 mg/l to 2.4 mg/1 respectively. Similarly, the Sodium Absorption Ratio (SAR) level varies from 0.17 to 3.60 mg/l.

The study result shows that the water samples near the catchment area having a smaller amount concentration of chemical elements. Whereas in the discharging area has higher amount of concentration with drastic variation. In Seliyanallur and Ukkirankottai locations have shown higher level for different chemical elements. In these areas, the river depth also comparatively deep and water passes through the cutting of the rock beds. Besides, a storage check dam has been constructed in Seliyanallur location and stored continuously for a long time in certain quantity of water leads to change its chemical composition due to fertilizers, aquatic weeds, climatic changes and weathering process. Similarly, the Uppodai water input also plays a substantial level to increase the overall concentration variation in the Chittar river water. For supporting this view it is observed that in Kayathar location it has higher amount of chloride and potassium concentrations. The important other factors such as geographic variation, soil and rock types, rate of runoff, agricultural practice and cattle excrements etc also could be involved in the present hydrogeochemical changes of the river water.

Permeability Index

Recent studies have indicated that soil permeability is influenced by long term use of irrigation water containing Sodium and Bicarbonate contents. Doneen (1964) have evolved a criterion for assessing the suitability of water for irrigation based on permeability index. The permeability index value varies from 0.34 to 2.12 with the average value of 0.65.The index range has been classified into three categories as 1: Excellent to Good (0.5); 2: Good to Injurious (0.5 to 2.0) and 3: Injurious to Unsatisfactory (>2). The present study result shows that about 73 per cent water samples (8) are coming under the category–I, whereas 18 per cent are under second category and 1 per cent under third category.

Corrosivity Ratio

Corrosivity ratio is defined as the corrosive tendency of groundwater due to the proportional occurrence of alkaline earths and saline salts along with sulphides in the groundwater. As along the river bed is an intensive agriculture practice area, it is essential to discern whether the water can be

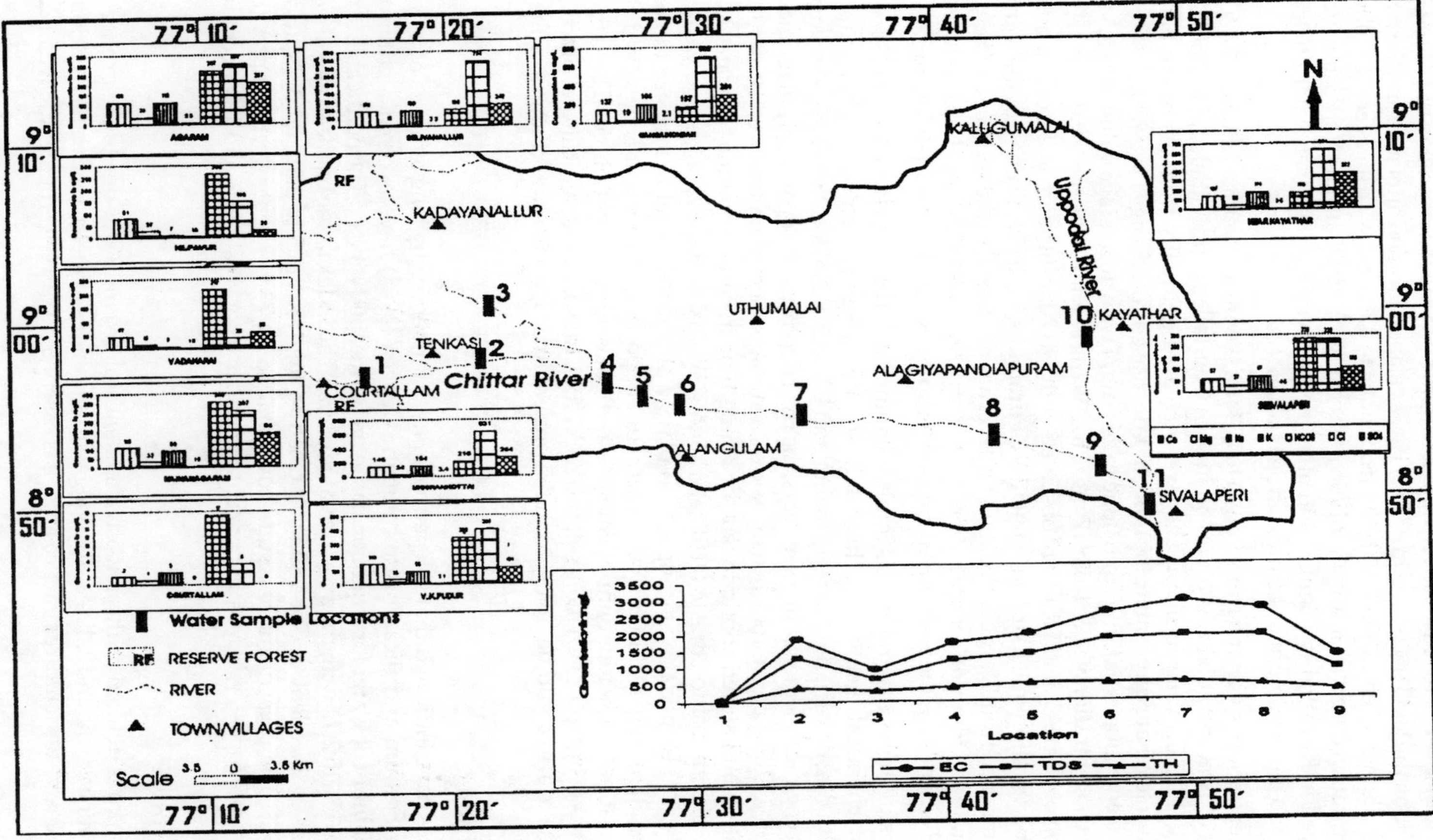

Figure 17.2: Concentration Level Along the River Course

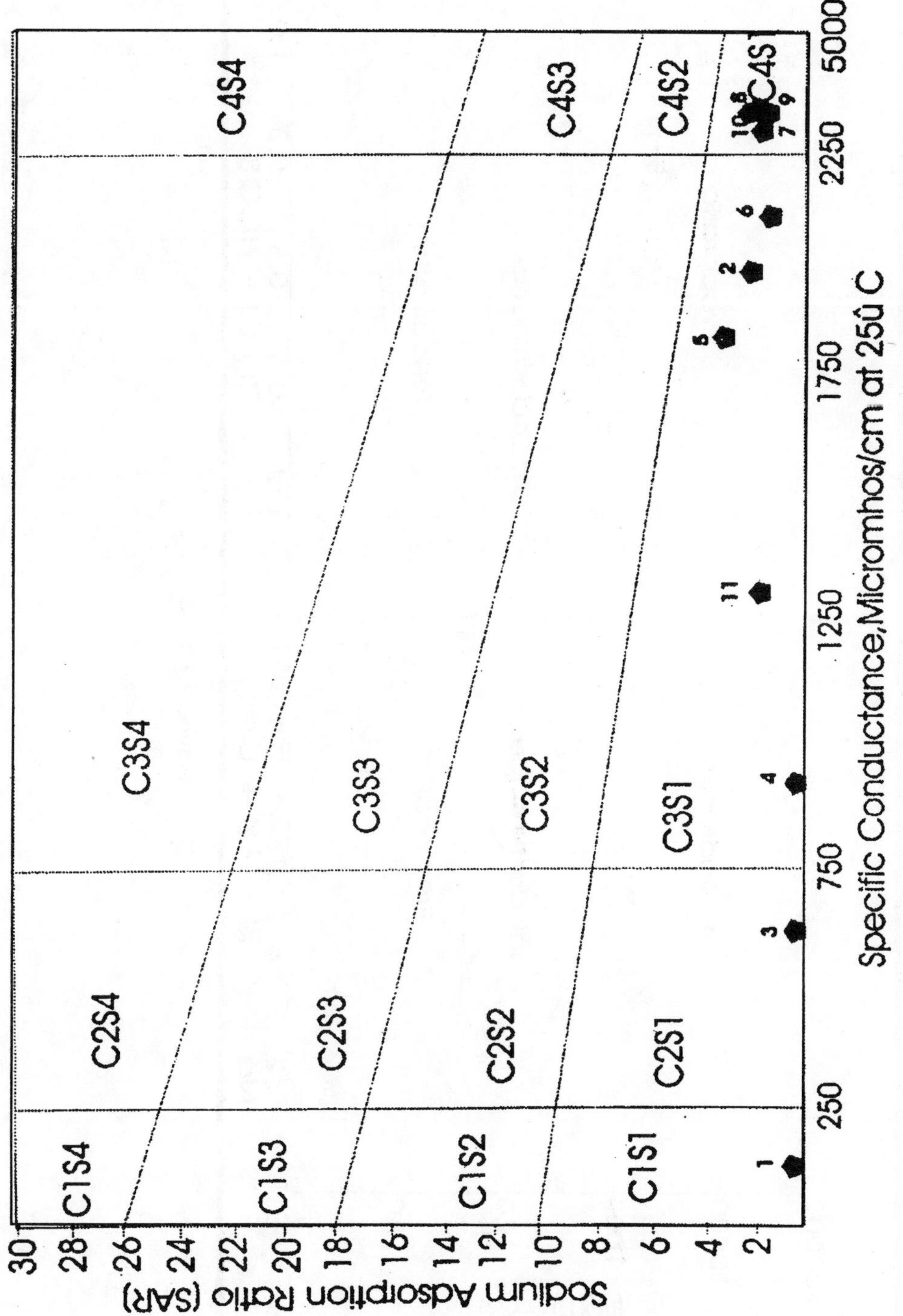

Figure 17.3: USSL Diagram

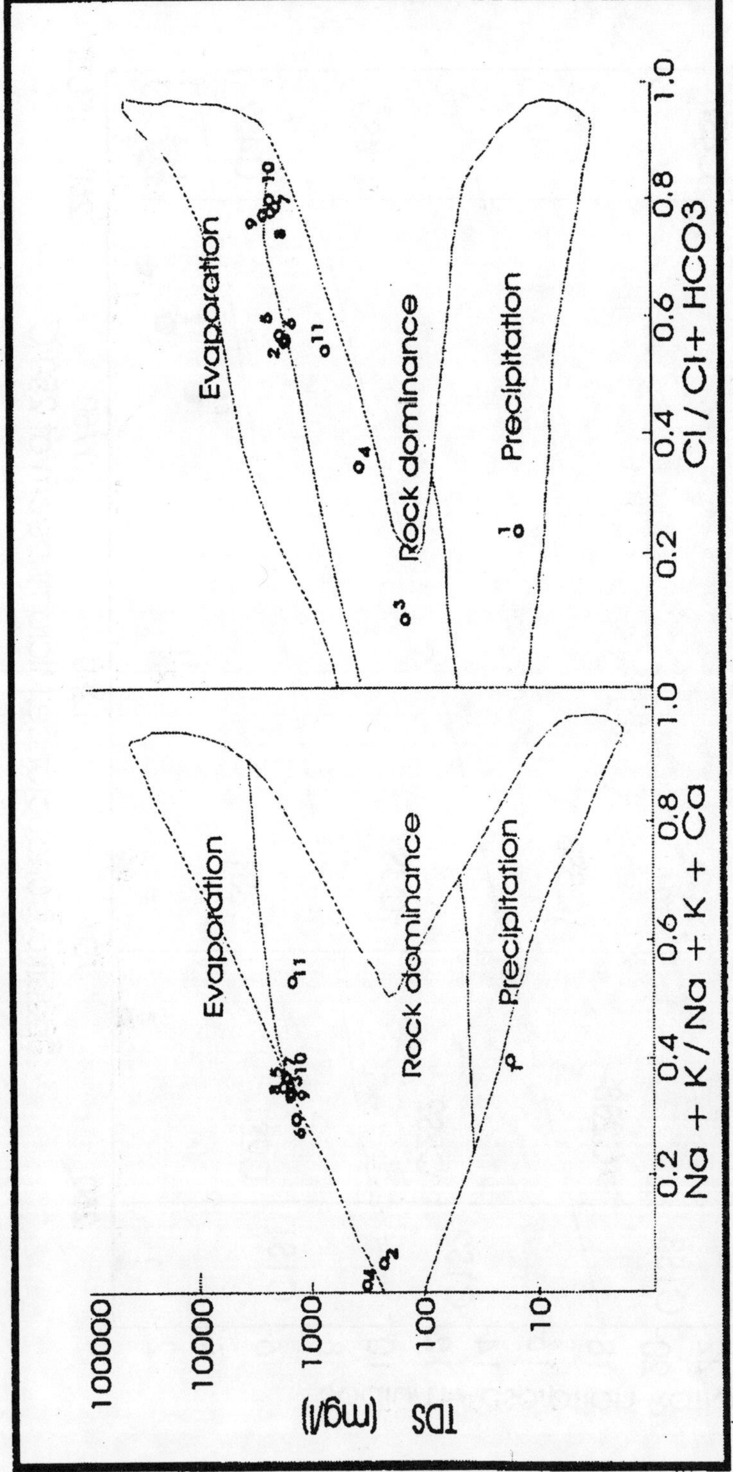

Figure 17.4: Gibb's Plot

transported in metallic pipes or not. The index values were calculated using the methods proposed by Ryzner (1944). The study area river water samples corrosivity value is varied from 0.71 to 12.85. According to Balasubramanian (1986) the corrosivity value less than one is considered to be safe for transport of water in any pipe. It is observed that the value is low (< 1.0) in the catchment areas which is upto Kilpavur location, whereas the value gradually increased up to 12.85 in consecutive sample locations due to water stagnation in low lying areas (VK Pudur and Ukkirankottai), rock water interaction in check dam zones (Seliyanallur and Kayathar) and fertilizers discharge.

USSL Diagram

In order to understand the quality and source for the present chemical elements concentration, an attempt has been made with U.S. Salinity Laboratory (1954) and Gibb's plot (1970) interpretations.

The USSL (1954) diagram has been used to understand the alkali hazard of the groundwater samples for any study area, because this interpretation is very much useful for judging the quality of groundwater for the use of agricultural purposes (Kumar *et al.*, 2005). The USSL diagram prepared for the groundwater samples of the study area is shown in the Figure 17.3, where the sodium adsorption ratio is plotted against specific conduction. It is observed that the catchment area water samples were fell in C_1S_1 and C_2S_1 categories which are classified as good class. Rest of the samples was fallen in C_3S_1 and C_4S_1 categories which are classified as satisfactory class and salt tolerant crops class. However, according to the SAR classification all the water samples are coming under excellent water class (SAR<10).

Gibb's Diagram

Gibb's (1970) proposed a diagram to understand the relationship of the chemical components of waters from their respective aquifer lithologies. The Gibb's ratio–I, Cl/Cl +HCO_3 for anions and ratio–II, Na+K/Na+K+Ca for cations of the samples plotted separately against respective values of total dissolved solids as shown in Figure 17.4. The interpretation has indicated that the water samples show the precipitation zone for catchment area water samples and remaining were found in the border zone of rock dominance and evaporation zone. Hence, it indicates that the source for chemical concentration of water samples in the downward area not only from rock water interaction but also the agricultural practice, climate change, oxidation and reduction process etc were involved.

Conclusion

From the present study, it is observed that the water of Chittar River gets its concentration from runoff path and the level of concentration increases gradually towards the down locations. In general, the river water flow will be less during post-monsoon season and hence, the possibilities of locking and stagnating of water in deeper and low lying area for a long period would be high. This condition would lead to weaken and disintegrate the lithology and change the water chemistry gradually. During the post-monsoon period the disintegrated and weathered fragments of source rock will dissolve and convert the chemistry of water, where it stored in check dams and stagnating places. As the present study water samples have shown higher level in Ukkirankottai and Seliyanallur locations, the storing of water in a place could be the key source. Apart from this source agriculture practice, fertilizer application, livestock waste, soil erosion and anthropogenic deeds are also contributed certain extent. However, the permeability index is within the acceptable range in most of the water sample locations whereas the corrosivity ratio is high at many sample locations. Hence, the usage of metal pipes in these areas has to be avoided to transport the river water for different usages such as drinking, domestic and irrigational. The present study revealed to understand the quality and geochemistry of the Chittar-Uppodai water for various usages. This study realized the importance of basin management and attention for monitoring water quality periodically. Because the surface water will lead to groundwater quality changes over a long period.

References

American Public Health Association, American Water Works Association, Water Pollution Control Federation, 1985. *Standard Methods for the Examination of Water and Wastewater*, 16th Edition, Washington D.C.

Babar, Md., 2005. Hydrogeochemistry of Pingalgad Nala stream in Parbhani District, Maharashtra. *Poll. Res.*, 24(3): 665–668.

Babar, Md. and Kaplay, R.D. 1999. Groundwater quality around Pingalgad Nala, Parbhani district. *J. Ecology Env. and Conservation*, 5(2): 141–142.

Babar, Md., Muley, R.B. and Joshi, Rohini K., 2003. Water pollution in relation to sewage and waste disposal in natural streams and man made lakes in Parbhani city (Maharashtra), India. In: *Proceeding Volume of National Seminar on Geo-environmental Hazards: A Future Perspective* held at Faizpur during October 6–7, pp. 194–197.

Babar, Md., 2000. Management of groundwater resources through watershed developments in Parbhani district of Maharashtra. In: *Resource Management and Contours of Development Reflections through Macro-Micro Narratives*, (Ed.) B.C. Barik. Rawat Publications, Jaipur, pp. 350–364.

Balasubramanian, 1986. Hydrogeological investigations in the Tambaraparani River basin, Tamil Nadu, Unpublished *Ph.D. Thesis*, University of Mysore, 348p.

Brown, E., Skougstad, M.W. and Fishman, M.J., 1970. In: *Techniques of Water Resources Investigation of US Geological Survey Book 5*, US Government Printing Office, Washington D.C.

Chavan, S.P., Joshi, P.K., Pawar, S.B. and Babar, Md., 2002. Qualitative analysis of planktons in wells around Pingalgad Nala, Parbhani district, Maharashtra, India. In: *Biotechnology in Agriculture, Industry and Environment–Proc. Vol. of International Conf. of SAARC Countries*, (Ed.) A.M. Desshmukh, pp. 336–342.

Doneen, L.D., 1964. Notes on water quality in agriculture. Published as Water Science and Engineering, Univ., California, Davis.

Gibbs, R.J., 1970. Mechanism controlling world's water chemistry. *Science*, 170: 1088–1090.

Hem, J.D., 1985. *Study and Interpretation of the Chemical Characteristics of Natural Water*. U.S. Geol. Survey, Water supply Paper No. 2254: 264.

Karanth, K.R., 1987. *Groundwater: Assessment, Development and Management*. Tata McGraw Hills Publications Company Limited, New Delhi, 720 p.

Raghunath, H.M., 1983. *Groundwater*. Wiley eastern Ltd, New Delhi, p. 459.

Kumar, Rakesh, Singh, R.D. and Sharma, K.D., 2005. Water resources of India. *Current Science*, 89(5): 794–811.

Ram Mohan, H.S., 1984. A climatological assessment of the water resources of Tamil Nadu. *Ind. Jour. Power and River Valley Develop.*, pp. 58–63.

Ryzner, 1944. A new index for determining amount of calcium carbonate scale formed by a water. *J. Amer. W.W. Assoc.*, 36: 472–486.

Todd, D.K., 1980. *Groundwater Hydrology, 2nd Edn.* John Wiley and Sons, New York, p. 535.

Hazarika, U.M., Goswami, D.C., Bora, P.K. and Patra, S.C., 2006. Flood hazard In Jiadhal River Basin, Dhenaji District, Assam. *J. of Applied Hydrology,* 19(1 and 2): 59–67.

U.S. Salinity Laboratory, 1954. *Diagnosis and Improvement of Saline and Alkali Soils.* U.S. Dept. Agriculture Handbook–60, Washington, D.C., 160 p.

Chapter 18

Decolourisation and Biodegradation of Sulphonated Azo Dyes by Fungi to Clean Dye Contaminated Soil Environments

E. Laxminarayana[1], M. Thirumala Chary[1],
Nikhat Kousar[2] and M.A. Singara Charya[2]
[1]Department of Chemistry, Kakatiya Institute of Technology and Science, Warangal – 506 015
[2]Department of Microbiology, Kakatiya University, Warangal – 506 009

ABSTRACT

Frequently encountered thirteen fungi isolated from dye contaminated soils were assessed for their decolourisation and biodegradation of three sulphonated azo dyes. The fungal organisms showed 41–80 per cent decolourisation of scarlet red and 80–90 per cent of fast greenish blue and 48–89 per cent of brilliant violet. For the decolourisation the responsible enzymes, K-amylase, protease and catalase were estimated. *A. niger* showed maximum secretion of K-amylase in 14 days of incubation. The protease production was ranged from 25–38 U/ml by majority of the fungi tribe *T. notatum* was found to be efficient in production of catalase. This study is the fungal based treatment system for the reclamation of the dye contaminated soils.

Keywords: Decolourisation, Biodegradation, Sulphonated azo dyes, Fungi.

Introduction

One of the very pressing environmental problems of textile industry is the removal of colour from effluents prior to discharge to local sewage treatment. Brightly coloured water-soluble dyes are problematic, as they pass unaffected through conventional sewage treatment systems. To depollute the dye wastewater, the physico-chemical methods, adsorption, chemical precipitation, flocculation, photolysis, chemical oxidation and reduction, electro-chemical treatment and ion-pair extraction have proved to be costly and less effective (Zollinger 1987; Cripps *et al.*, 1990; Ollikka *et al.*, 1993; Sheshadri *et al.*, 1994; Sani and Benerjee, 1999). In the recent past, considerable efforts have been made to develop suitable treatment systems for these coloured effluents (Nicola *et al.*, 1998). Banat *et al.* (1996) recorded that various bacteria and fungi are effective in many cases and adsorption of dyes to the microbial cell surface is the primary mechanism for the decolourisation. It was observed that enzymes, which are involved in the degradation of the substrates also, participate in the decolourisation of dyes (Chivukula *et al.*, 1995; Vyas and Molitorid, 1995). It has been suggested that only biotechnological approach can offer complete degradation of the dye, with a co-reduction in BOD and COD (Paul, 1995; Hobson and Wales, 1998; Reemtsma and Lakobs, 2001). Hence, in this study certain textile dyes were studied for their removal from textile mill effluents by using the fungal biomass, metabolites and enzymes.

Materials and Methods

Thirteen mostly occurred fungal species in the soils contaminated with textile effluent was selected and experimented (Table 18.1) Asthana and Hawker's broth (5 g glucose, 3.5 g KNO_3, 1.75 g KH_2PO_4, 0.75 g $MgSO_4.7H_2O$ and 1 L distilled water) was prepared and added with 1 per cent dye solution before autoclaving. The sterilized flasks were inoculated with fungi and incubated at room temperature and after 7, 14, and 21 days of incubation the growth, enzyme production and decolourisation of dye was arrayed.

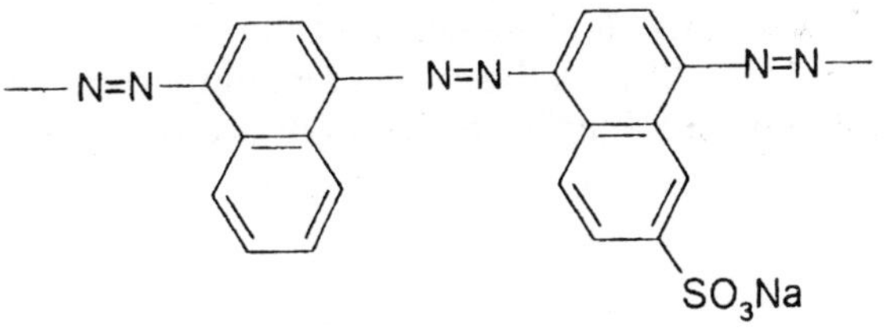

N=N
N=N
N=N
SO_3Na

Ink Blue-69cd

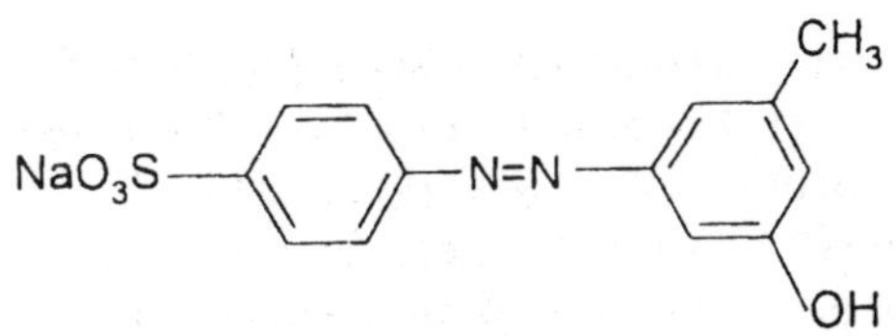

CH_3
NaO_3S
N=N
OH

Direct rawlin green-81.

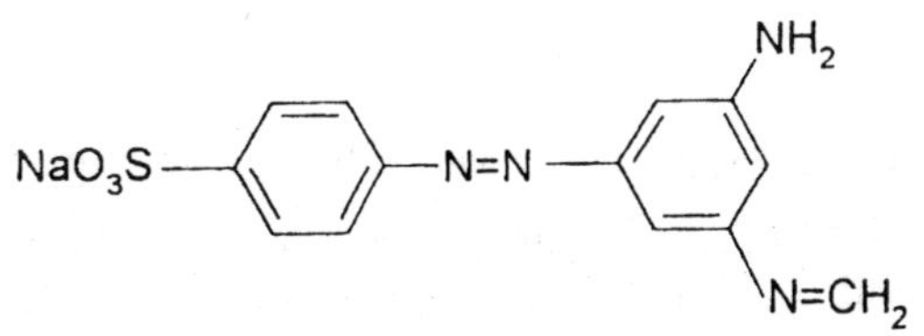

NH_2
NaO_3S
N=N
$N=CH_2$

Fast direct pink-38

The growth was measured and indicated as +1-meager, +2 and +3-moderate and +4-high. In some cases of mycelial biomass was weighed on pre weighed filter papers, and the final and the initial weights were measured to get the actual weight of fungal biomass. The influence of metal ions (Cu^{+2}, Fe^{+2}, Ni^{+2}, Mn^{+2} and Zn^{+2}) was assayed on the growth, pH and enzyme production of amylase, protease, catalase, peroxidase and laccase after appropriate incubations. 25 ml of modified Asthana and Hawker's media was taken into the conical flasks and added with 20 ppm of respective metal ions and 0.025 M of veratryl alcohol inoculated with seven days old fungal species, after sterilizing the media period decolourisation percentage and the enzymes were assayed from the centrifuged culture extracts.

Decolourisation Assay

Decolourising activity was expressed in terms of per cent decolourisation as described by Yatome *et al.* (1993). The decrease in the absorbance was monitored at Ar_{660}, A_{570}, A_{530} nm for ink blue, direct rawlin green, fast direct pink, scarlet red, direct blue, brilliant violet, crystal violet and malachite green respectively. Decolourisation activity was calculated according to the following formula:

$$D\% = 100 \times \frac{(A_{ini} - A_{fin})}{A_{ini}}$$

D: Decolourisation

A_{ini}: Initial absorbance

A_{fin}: Final absorbance of dye, after incubation time.

Two types of controls were used: (*a*) Uninoculated sterile control with dye, and (*b*) inoculated control without dye. Control (*a*) indicated the effect of medium components in decolurisation and control (*b*) showed the inhibition of enhancing effects of dyes on enzyme production.

Enzyme Estimation

K-amylase

Activity of K-amylase in the fungal culture filtrates was estimated as suggested by Ross *et al.* (1973).

1 ml of culture extract was taken into a conical flask and added with 5 ml buffered starch substrate followed by 3 ml of 0.1 M Citrate buffer (pH 4.6) and 1 ml of CaO_2. The contents were incubated at 40°C for 10 minutes. After incubation, 2 ml of 1N HCl was added and thoroughly shaken for 1015 minutes. From this solution 0.2 ml reaction mixture was taken and added with 40 ml of distilled water followed by 0.5 ml of 1N HCl and 0.1 ml of 10 per cent iodine solution. The developed blue colour was measured with spectrophotometer at 620 nm. Blank was prepared with out enzyme and amylase content was measured and expressed in terms of decrease in optical density at 620 nm/g as U/ml of culture filtrate.

Protease

Protease enzyme was estimated as procedure suggested by Nanniperi *et al.* (1980).

One ml of culture extract, 1 ml of casein solution were and incubated for 1 hour at 37°C. This reaction mixture was added with 2 ml of 0.4 M TCA and incubated for 20 minutes at 37°C. The contents were filtered through Whatmann No. 1 filter paper. 1 ml of filtrate was taken and added with 5 ml of 0.4 M Na_2CO_3 and 1 ml of diluted FC reagent. The contents were incubated at 37°C for 1 hour and the developed blue colour was read at 660 nm. Blank was prepared without cultured extract and activity was expressed in μ/ml.

Results and Discussion

The decolourisation of three sulphonated azo dyes by thirteen fungi after 7, 14, and 21 days was analysed and presented in Figures 181.1–18.3.

From Figure 18.1 it was evident that different fungal organisms showed 41 to 80 per cent decolourisation of scarlet red with the maximum on 21 days of incubation. Aniger and *T. viride* proved to be more efficient in decolourising the scarlet red (80 per cent). Other fungi, *C. globosum, D. macropus, D. rostrata, T. herbarum* and *H. fuscostra* showed moderate decolourisation that ranged between 50 to 65 per cent.

The influence of thirteen fungi on the decolourisation of fast greenish blue was presented in Figure 18.2. The range of activity on decolourisation was up to 90 per cent with *T. viride*, in 21 days of incubation. Decolourisation was up to *C. lunata, M. mucedo* and *P. notatum* fluctuated between 80 to 90 per cent in 21 days of incubation. In general, the maximum colour removal was witnessed in 21 days of incubation in almost all the organisms. The moderate activity of decolourisation was with *A. alternata, C. globosum, D. macropus, T. herbarum* and *H. fuscoatra.*

The impact of fungal organisms on the decolourisation of brilliant violet was presented in Figure 18.3. The range of decrease of the dye with these organisms ranged between 48 to 89 per cent. All the fungi showed decolourisation above 70 per cent in 21 days. Moderate activities were recorded in *D. rostrata, T. herbarum* and *H. fuscostra*. No significant reaction could be observed in the enzyme production and dye decolourisation.

Enzyme Production during Decolourisation

Enzymes, K-amylase, protease and catalase estimated during the decolourisation by different fungal isolates was given in Tables 18.1–18.3. The production of K-amylase in the decolourisation of scarlet red (Table 18.1) at three stages varied from 1 to 18 (μ/ml) in different organisms. *A. niger* showed maximum secretion of this enzyme in 14 days of incubation, while its absence was observed in 21 days in *C. lunata, F. oxysporum* and *D. macropus.* In general, 14 days of incubation showed maximum protease production in all the fungi and ranged from 25 to 38 (μ/ml). Except, *L. actaea*, all other fungi showed decreased trend of protease production in 21 days of incubation. Catalase was secreted in abundant quantities in 21 days of incubation by all the fungi. The activity of this enzyme in fungal isolates varied 0.2 to 3.2 (μ/ml). *P. notatum* was found to be efficient in the production of catalase. The least quantity of catalase was estimated in *T. viride* both 7th and 14th day of incubation. The pH was acidic with fungal cultures of *A. niger, M. mucedo* and *P. notatum* and neutral in the remaining organisms. The dry weight of the fungal biomass ranged from 220 to 540 (mg) with maximum in 21 days of incubation.

It is clear from Table 18.2 that K-amylase production varied considerably with different fungal isolates in the decolourisation of the fast greenish blue. The range of activity was between 1 to 8 (μ/ml). The initial incubation of *A. alternata, C. lunata, F. oxysporum, L. actaea* and *T. viride* failed to secrete this enzyme. Minimum and maximum protease was 4 and 41 (μ/ml) for *A. niger* and *C. lunata* respectively after 7 days of incubation. Catalase production was maximum (5.0 μ/ml) in *A. niger* and minimum in (1.3 μ/ml) in *C. lunata*. All the remaining organisms showed moderate activity. The pH was acidic in *A. niger, M. mucedo, P. notatum* and *T. viride* and neutral in all other organisms. The range of biomass varied among different fungal organisms was in between 220 to 840 mg.

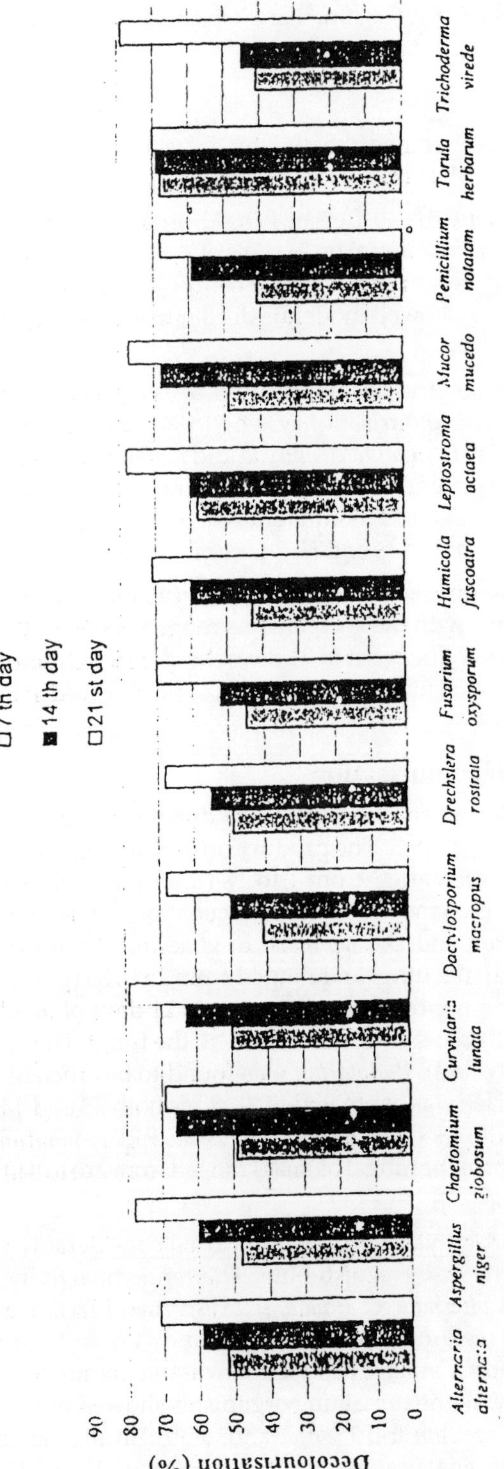

Figure 18.1: Decolourisation of Scarlet Direct Red-23, by Thirteen Fungal Organisms after 7, 14 and 21 Days of Incubation

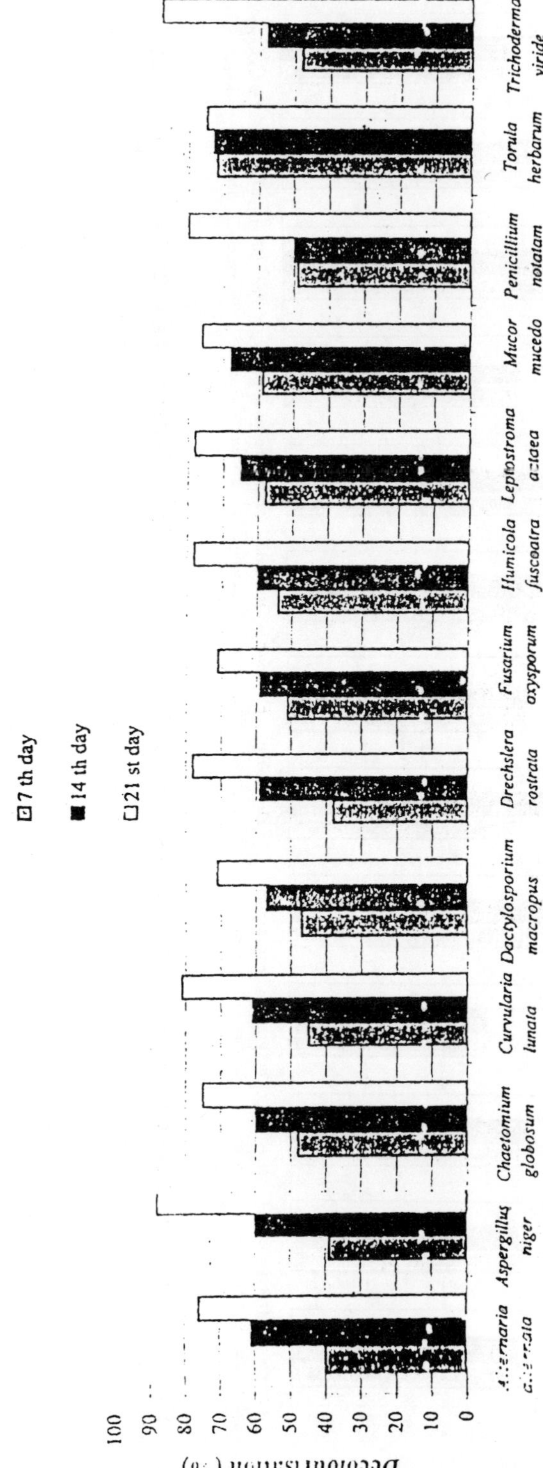

Figure 18.2: Decolourisation of Fast Greenish Blue-86, by Thirteen Fungal Organisms after 7, 14 and 21 Days of Incubation

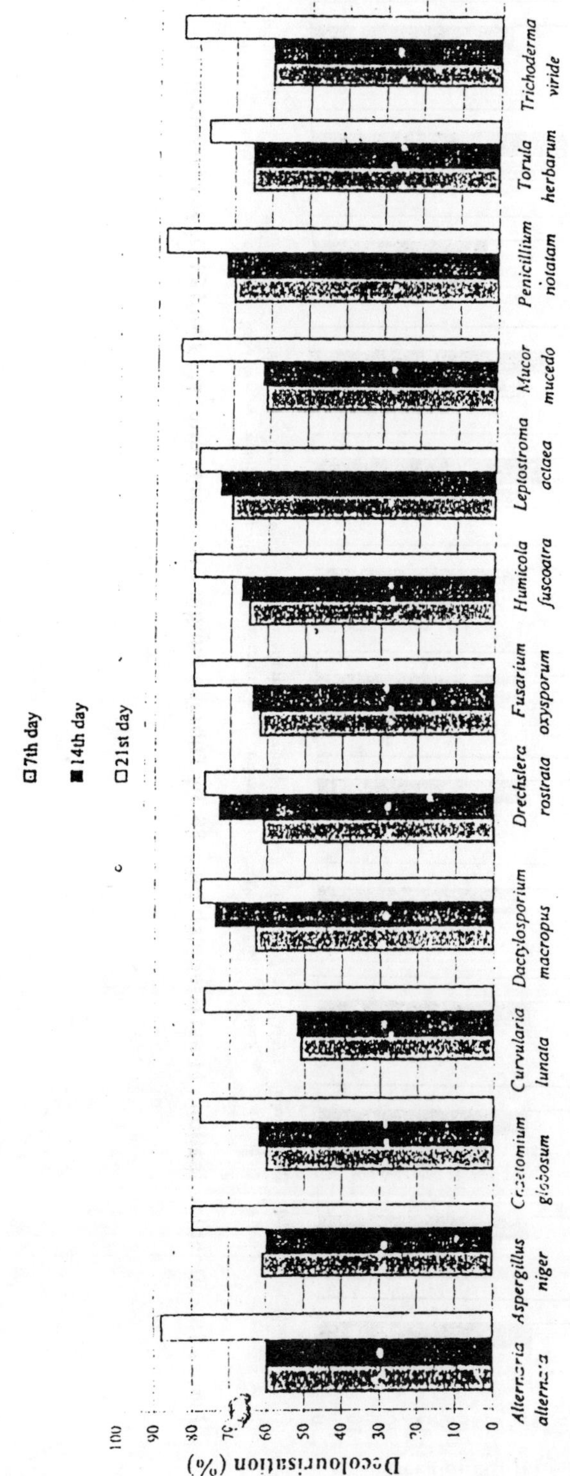

Figure 18.3: Decolourisation of Brilliant Direct Violet-9, by Thirteen Fungal Organisms after 7, 14 and 21 Days of Incubation

Table 18.1: pH, Dry Weight and Enzyme Production by Thirteen Fungi during the Decolourisation of Scarlet Direct Red-23

Name of the Fungi	*Days of Incubation*	*pH*	*Dry Weight (mg)*	*α-amylase (μ/ml)*	*Protease (μ/ml)*	*Catalase (μ/ml)*
Alternaria alternata	7	6.5	320	4	13	1.2
	14	6.5	430	2	25	1.2
	21	6.5	520	3	14	1.3
Aspergillus niger	7	3.5	240	3	8	1.2
	14	3.5	280	18	25	1.4
	21	3.5	480	12	8	1.5
Chaetomium globosum	7	6.5	280	2	13	0.9
	14	6.5	270	1	28	0.5
	21	6.0	280	4	18	0.6
Curvularia lunata	7	7.5	340	1	13	0.8
	14	7.0	320	2	38	0.9
	21	7.0	520	2	25	1.2
Dactylosporium macropus	7	6.5	240	1	14	0.8
	14	6.5	280	1	38	1.2
	21	6.0	330	0	24	1.7
Drechslera rostrata	7	6.0	320	0	15	4.0
	14	7.5	320	1	32	0.5
	21	6.2	500	2	31	0.8
Fusarium oxysporum	7	7.5	270	1	14	1.1
	14	7.5	320	2	35	1.2
	21	7.5	480	2	38	1.5
Humicola fuscoatra	7	7.5	250	2	00	1.0
	14	7.5	250	2	38	1.8
	21	7.5	310	1	28	1.8
Leptostroma actaea	7	7.5	270	8	18	1.2
	14	7.5	320	11	34	1.2
	21	7.5	480	2	38	1.5
Mucor mucedo	7	4.0	240	4	25	2.6
	14	4.0	340	2	28	7.5
	21	4.0	480	1	35	1.8
Penicillium notatum	7	3.5	220	8	12	0.5
	14	4.0	380	8	28	0.5
	21	4.0	520	2	19	3.2
Torula herbarum	7	7.5	240	1	16	1.1
	14	7.5	250	2	34	1.4
	21	7.5	310	1	28	1.8
Trichoderma viride	7	6.0	280	1	18	0.2
	14	6.0	330	2	35	0.2
	21	6.0	480	1	22	3.1

Table 18.2: pH, Dry Weight and Enzyme Production by Thirteen Fungi during the Decolourisation of Fast Greenish Direct Blue-86

Name of the Fungi	*Days of Incubation*	*pH*	*Dry Weight (mg)*	*α-amylase (μ/ml)*	*Protease (μ/ml)*	*Catalase (μ/ml)*
Alternaria alternata	7	6.5	310	0	00	1.4
	14	6.5	600	2	06	3.6
	21	6.5	840	1	08	4.0
Aspergillus niger	7	3.5	260	3	04	1.4
	14	3.5	480	2	04	5.0
	21	4.0	530	1	04	3.8
Chaetomium globosum	7	6.5	280	2	00	1.4
	14	6.5	540	1	39	4.7
	21	6.5	560	0	03	4.2
Curvularia lunata	7	7.5	320	0	00	1.4
	14	7.5	490	1	41	4.2
	21	7.2	580	2	04	4.2
Dactylosporium macropus	7	6.5	270	1	00	1.6
	14	6.5	500	2	38	3.0
	21	6.4	480	3	38	4.0
Drechslera rostrata	7	6.0	320	1	00	1.6
	14	6.2	550	2	39	3.5
	21	6.0	520	3	39	4.2
Fusarium oxysporum	7	7.5	260	1	00	1.3
	14	7.5	520	2	32	3.8
	21	7.5	700	3	03	4.8
Humicola fuscoatra	7	7.5	240	2	09	2.3
	14	7.0	440	3	35	3.3
	21	7.0	700	1	35	4.3
Leptostroma actaea	7	7.5	280	1	00	1.3
	14	7.5	390	2	08	3.2
	21	7.5	570	3	08	4.3
Mucor mucedo	7	4.0	250	1	00	1.3
	14	4.0	440	2	12	3.9
	21	4.5	640	1	20	4.1
Penicillium notatum	7	4.0	220	1	00	2.3
	14	4.0	480	1	33	3.0
	21	4.0	590	0	35	4.3
Torula herbarum	7	7.5	340	2	08	2.4
	14	7.5	440	4	34	3.2
	21	7.5	530	8	34	4.3
Trichoderma viride	7	4.5	280	1	00	1.4
	14	4.5	440	0	34	3.8
	21	5.0	480	1	34	4.4

Table 18.3: pH, Dry Weight and Enzyme Production by Thirteen Fungi during the Decolourisation of Brilliant Direct Violet-9

Name of the Fungi	Days of Incubation	pH	Dry Weight (mg)	α-amylase (μ/ml)	Protease (μ/ml)	Catalase (μ/ml)
Alternaria alternata	7	6.5	240	2	04	0.5
	14	6.5	380	4	08	0.6
	21	6.5	430	1	17	0.7
Aspergillus niger	7	3.5	240	2	04	0.5
	14	3.5	430	1	06	0.6
	21	3.5	520	4	18	0.7
Chaetomium globosum	7	6.5	260	1	04	0.2
	14	6.5	290	2	04	0.2
	21	6.5	430	2	14	0.4
Curvularia lunata	7	7.5	240	2	00	0.4
	14	7.0	310	3	18	0.5
	21	7.0	500	1	00	0.6
Dactylosporium macropus	7	6.5	240	2	05	6.1
	14	6.5	430	3	08	0.8
	21	6.5	380	0	17	0.9
Drechslera rostrata	7	6.0	270	3	09	0.2
	14	6.5	350	1	04	0.8
	21	6.0	430	0	13	0.3
Fusarium oxysporum	7	7.5	280	1	15	0.5
	14	7.0	310	3	16	0.9
	21	7.0	440	1	18	0.9
Humicola fuscoatra	7	7.5	230	2	13	0.4
	14	7.5	320	4	09	1.3
	21	7.5	440	6	19	1.3
Leptostroma actaea	7	7.5	260	1	11	0.8
	14	7.0	440	.0	12	1.2
	21	7.0	520	1	14	1.4
Mucor mucedo	7	4.0	310	1	12	0.4
	14	4.5	360	1	11	0.4
	21	4.0	480	3	19	0.9
Penicillium notatum	7	4.0	300	1	14	0.2
	14	4.0	350	·2	15	1.2
	21	4.0	420	4	15	0.5
Torula herbarum	7	7.5	280	1	13	0.2
	14	7.5	290	2	08	1.3
	21	7.5	390	1	14	1.3
Trichoderma viride	7	4.5	290	1	08	0.2
	14	4.0	240	2	09	1.3
	21	4.0	440	0	14	0.3

The decolourisation of brilliant violet supported for varied production ranges of fungal enzymes (Table 18.3). The K-amylase activity was 1 to 6 (μ/ml). *M. mucedo, D. rostrata*, and *T. herbarum* faded to secrete K-amylase at all incubation times. The minimum and the maximum production of protease were between 4 to 19 (μ/ml) for all the fungi. The catalase production was maximum (6.1 μ/ml) after 7 days of incubation in *D. macropus* and minimum in *C. globosum* (0.2 μ/ml) and *H. fuscoatra, T. viride, L. aclaea* and *T. herbarum* showed moderate catalase activity (1.3 μ/ml). The pH was acidic in *A. niger, M. mucedo,* and *T. viride,* while all other organisms showed neutral pH. The dry weight of the organisms varied between 240 to 520 mg.

The mineralization of the sulphonated azo dyes by bacterial consortium (Mechsner and Wurhamann, 1982; Diniz *et al.*, 2002) and Oxidation of different sulphonated azo dyes by certain brown-rot fungi also on the characteristics of important degradative enzymes involved in the process was investigated by Dey *et al.* (1994). Various researchers (Meyer, 1981; Glenn and Gold, 1983; Brown and Humberger, 1987; Bumpus and Brock, 1988; Mills *et al.*, 1989; Spadaro *et al.*, 1992; Bollag, 1992 and Kim and Shoda, 1999) have established the importance of white-rot fungi in decolourisation of dyes, and they noted that: (*i*) lignin degrading culture of *Phanerochaele chrysoporium* capable of mineralizing a variety of toxic dyes, and (*ii*) the rate at which *P. chrysosporium* mineralized aromatic rings of dyes is dependent on the nature of ring substitutes, Novel procedures for the decolourisation of different dyes by various wood rot fungi have been developed (Borchert and Libra 2001; Nyanhango *et al.*, 2002; Libra *et al.*, 2003) while the role of common soil fungi in the dye-amended soils is yet not well understood.

The present study is thus an effort to develop a fungal based treatment system for the cleaning of dye industry effluents and for bio remediation of dye contaminated soils using fungal-based primary and secondary metabolites. The extraction, purification and development of commercial products from efficient strains of these fungal isolates is yet another line of study to be given utmost importance.

References

Banat, I.M., Nigam, P., Singh, D. and Merchant, 1996. Microbial decolourisation of textile dye-containing effluents: A review. *Biores. Technol.*, 58: 399–404.

Bollag, J.M., 1992. Enzymes catalyzing oxidative coupling reactions of pollutants. Metal Ions. Biol. Sys., 28: 205–217.

Borchert, M. and Libra, J.A., 2001. Decolourisation of reactive dyes by the white-rot fungus *Trametes versicolor* in sequencing bath reactors. *Biotechnol. Bioeng.*, 75: 313–321.

Brown, D. and Hamburger, B., 1987. The degradation of dye stuffs–III: Investigations of their ultimate degradability. *Chemosphere*, 16: 1539–1553.

Bumpus, J.A. and Brock, B.J., 1988. Biodegradation of crystal violet by white rot fungus *Phanerochaete chrywsporium. Appl. Environ. Microbiol.*, 54: 1143–1150.

Chivukula, M. and Renganathan, V., 1995. Phenolic azo dye oxidation by laccase from *Pyricularia oryzae. Appl. Environ. Microbiol.*, 61: 4374–4377.

Cripps, C., Bumps, J.A. and Aust, D., 1990. Biodegradation of azo and heterocyclic dyes by *Phanerochaete chrysosporium. Appl. Environ. Microbiol.*, 56: 1114–1118.

Dey, S., Maiti, T.K. and Bhattacharya, B.C., 1994. Production of some extra cellular enzymes by a lignin peroxidase producing brown rot fungus, *Polypoyous osteiformis* and its comparative abilities for lignin degradation and dye degradation. *Appl. Environ. Microbiol.*, 60: 4216–4218.

Glenn, J.K. and Gold, M.H., 1983. Decolourisation of several polymeric dyes by the lignin degrading basidimycetis *Phanerochae techlysosporium*. *Appl. Environ. Microbiol.*, 43: 1741–1747.

Hobson, D.K. and Wales, D.S., 1998. Green days. *JSDC*, 42: 42–44.

Kim, S.J. and Shoda, M., 1999b. Purification and characterization of a novel peroxidase from *Geotrichum candidum* involved in decolourisation of dyes. *Appl. Environ. Microbiol.*, 65: 1029–1035.

Libra, J.A., Brochert, M. and Banet, S., 2003. Competition strategies for the decolourisation of a textile reactive dye with the white rot fungi *Trametes versicolor* under non-sterile conditions. *Biotechnol. Bioeng.*, 82: 736–744.

Mechsner, K. and Wurhamann, K., 1982. Cell permeability as a rate limiting factor in the microbial reduction of sulphonated azo dyes. *Eur. J. Appl. Microbiol. Biotechnol.*, 15: 1195–1202.

Meyer, U., 1981. Biodegradation of synthetic organic colourants. *FEMS. Symp.*, 12: 371–385.

Milis, C.D., Caim, D., Stankovich, M.T. and Tien, M., 1989. Oxidation-reduction potentials and ionization states of extra cellular peroxidases from lignin degrading fungus *Phanerochaete chrysosporium. Biochemistry*, 28: 8484–8489.

Nicola, W., Jim, G. and Nelson, G., 1998. The biotechnology approach to colour from textile effluent. *JSDC*, 114: 38–41.

Nyanhongo, G.S., Gomes, J., Gubitz, G.M., Zvaurya, R.R., Read, J. and Steiner, W., 2002. Decolourisation of textile dyes by laccases from a newly isolated stain of *Trametes modesta. Water Res.*, 36: 1449–1456.

Ollikka, P., Alhonmoki, K., Leppanen, Y.M., Glumoff, T., Raijola, T. and Suominen, L., 1993. Decolourisation of azo triphenyl methane, heterocyclic and polymeric dyes by lignin peroxides isozymes from phenerocheate chrysosporium. *Appl. Environ. Microbiol.*, 59: 4010–4016.

Paul, F.H., 1995. The impact of biotechnology on the textile industry. *Textile Magazine*, 3: 6–10.

Reemtsma, T. and Jacobs, J., 2001. Concentrated chemical and microbial degradation of sulphophthalimides formed from sulphophthalocyanine dyes by white rot fungi. *Environ. and Technol.*, 35: 4655–4659.

Sani, R.K. and Benarjee, U.C., 1998. Comparison of static and shake culture in the decolourisation of textiles dyes and dye effluents by *Phanerochaete chrysosporium*. *Folia Microbiol.*, 43: 85–88.

Seshadri, S., Bishop, P.L. and Agha, A.M., 1994. Anaerobic/aerobic treatment of selected azo dyes in wastewater. *Waste. Mang.*, 14: 127–137.

Spadaro, J.T. and Renganathan, V., 1994. Peroxidase-catalysed oxidation of azo-dyes: Mechanism of disperse yellow degradation. *Arch. Biochem. Biophy.*, 312: 301–307.

Vyas, B.R.M. and Molitoris, H.P., 1995. Involvement of extra cellular H_2O_2 dependent ligninolytic activity of the white rot fungus *Pleuritus ostreatus* in the decolourisation of remazol brilliant blue R. *Appl. Environ. Microbiol.*, 61: 3919–3927.

Zolliner, H., 1987. *Colour Chemistry: Synthesis, Properties and Applications of Organic Dyes and Pigments*. VCH Publication, New York.

Chapter 19

Designing a Real-Time Mobile GIS

D. Vijayalakshmi[1] and I.V. Murali Krishna[2]
[1]Professor, Mahatma Gandhi Institute of Technology, Hyderabad
E-mail: psridhar_babu_2k@yahoo.com
[2]Professor, Centre for Spatial Information Technology, JNTU, Hyderabad
E-mail: ivm@ieee.org

ABSTRACT

Recent developments in mobile technologies have enabled GIS information to be taken into the field as digital maps on compact, powerful mobile computers, providing field access to enterprise geographic information. This enables organizations to add real-time (and near real-time) information to their enterprise database and applications, speeding up analysis, display, and decision making by using up-to-date, more accurate spatial data.

Introduction

Mobile GIS

Mobile computing is creating fundamental changes by adding the ability to take GIS with you into the field and interact directly with the world around you. Mobile GIS is composed of a number of technologies.

1. GIS
2. Mobile hardware in the form of lightweight devices and ruggedized field PCs
3. Global positioning systems (GPS)
4. Wireless communications for Internet GIS access

Traditionally, the process of field data collection and editing has been time consuming and error prone. Geographic data has traveled into the field in the form of paper maps. Field edits were performed using sketches and notes on paper maps and clipboards. Once back in the office, these field edits were deciphered and manually entered into the GIS database. The result has been that GIS data has often not been as up-to-date or accurate as it should have been. Consequently, GIS analysis and decisions have been delayed.

Mobile GIS Application Compare with Wireless GIS Application

Mobile GIS and Wireless GIS is very similar to wire-line Web-GIS, There are three main components common to both:

1. The client
2. The server (combination of web, map and data servers)
3. The network

The main differences are the client devices and the network type (mobile uses wireless communication service providers). The server functions and structures are similar to Internet GIS. Mobile applications are a bit like 'thin client' applications of Internet GIS.

There is an additional element in mobile, which is the location.

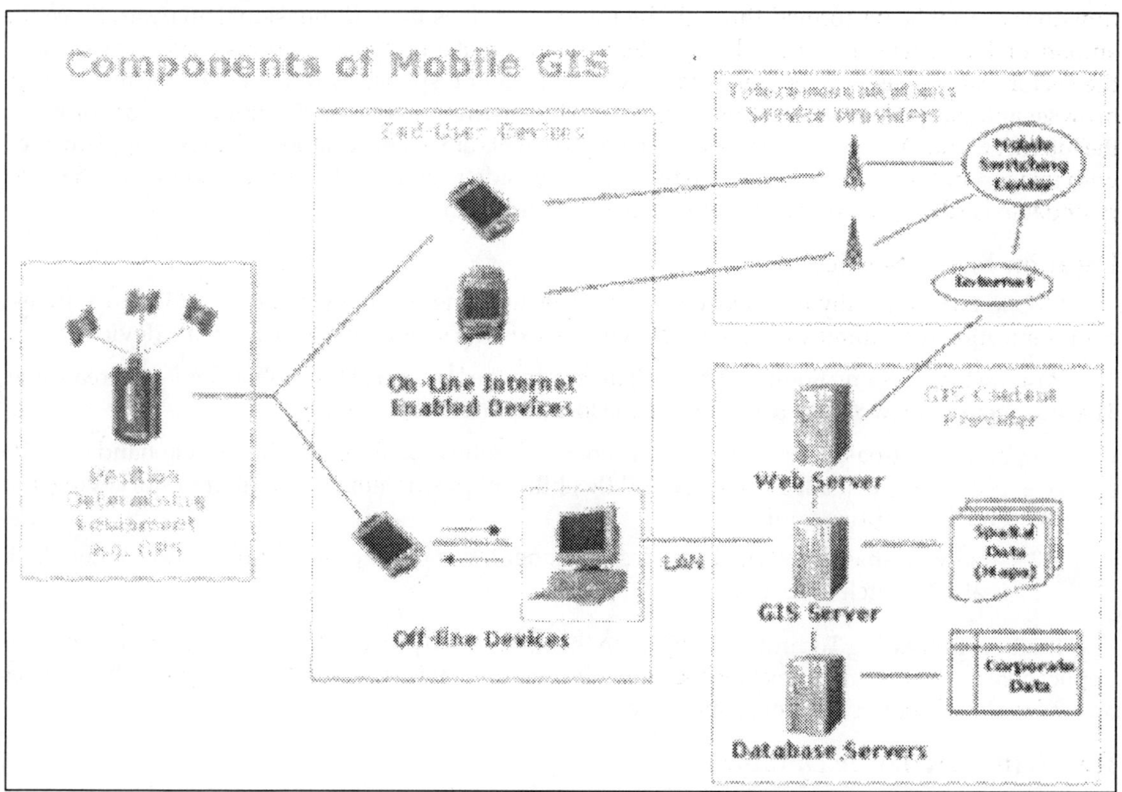

Figure 19.1: Components of Mobile GIS

Designing a Real-Time Mobile GIS

The Mobile GIS applications can be divided into two main groups: system construction and system utilization. The former one concentrates on creating the GIS, such as data collection, data checking, data updating among other aspects. The system utilization aspect concentrates on the use of an existing system, for example, location searching, path finding, and information seeking. In this paper, the utilization issue is the focus.

A wireless web GIS application extends the use of Internet GIS. Through the web content provider's service offering and by pointing to a web-browser on a mobile device and open to a web site, the mobile users can perform various functions. This kind of application uses all the advantages of Internet, mobile computing and GIS. This also allows users to interact with GIS data and maps on the Web without owning mobile GIS software and data. Mobile GIS applications mainly serve the public and the business to customer (B2C) services.

Limitations of Wireless Web GIS Application

As wireless web GIS application combines the wireless, Internet and GIS technology, it has all the limitations of the above three technologies. In the second generation of mobile systems, the low bandwidth and low reliability are the major obstacles in application development. These obstacles constrain the applications design. Moreover, the small screen and low resolution of mobile devices degrade the visibility and create a non user-friendly interface, and certain sophisticated and enhanced functions cannot be performed through the Internet, such as three-dimensional analysis. Since the amount of data sent over web is only a quarter to one-third of the amount of corresponding raster data, the vector display has a far greater productivity and faster response. Unfortunately, the mobile web browser and plug-ins do not support vector data. Furthermore, there is no common vector-format in the Internet world. As a result, the mobile web page contents are mainly composed by text, supplemented with images, but of low resolution. Lastly, the server side performs most processing works. Hence, poor database design has downgraded the overall performance.

Location-based Service

Mobile GIS is usually coupled with Global Positioning System satellites (GPS) and wireless communication to facilitate exchanges between the existing spatial server and mobile devices.

Position Determining Equipment (PDE) determine the location of the mobile devices in real-time. There are generally two ways to determine the location of a mobile device:

1. A handset-based system that relies on GPS, which is placed at the mobile handset. GPS provides unequalled accuracy and flexibility of positioning for navigation, surveying and GIS data capture, and
2. A network-based system that relies on triangulation of the cellular information, where PDE is at the switch centres

Position-Processing Technology (PPT) can then be used to process, track, manage and help other applications to query and retrieve the location information sent from a PDE. Thus, mobile GIS becomes an essential component to wireless mobility.

Real-Time Dynamic Database

A spatial temporal data set is defined as a collection of spatial objects including their whole lifetime (history). Each member in the spatial temporal data set can be treated as a snapshot of a particular spatial object in a particular moment. The notation is:

$$ST = \{s_i(t):g(s_i(t)) \quad A, I = 1,2, \quad ,n;0<t<\square\} \tag{1}$$

where $s_i(t)$ refers to a spatial object 'i' at time t. Traditionally, a change of attribute can be defined as the difference between the two elements in the same spatial temporal data set that corresponds to the same object. For example, there is only one building in zone A at time t1. In time t2, however, there are three buildings in the same zone. In a mobile environment, we should focus on the location of the user. At different locations, a mobile user may encounter different objects in different locations. The user also gets a different picture at the same location at different times. The changes of a particular object itself are meaningless unless the 'entire' scene is considered. The spatial temporal change should be described as the environmental change according to the point of view of the mobile user at his current position and time. In Real-Time mobile GIS applications, we distinguish between attribute, spatial and temporal changes. Attribute change comprises the inclusion of only the application-related attributes that are retained. Temporal change includes the alteration of the same spatial attributes at a different time, and parts of the object are discarded if they are invalid for the application. Spatial change is the environmental change while the mobile user is at a different position. After applying the new spatial temporal data set in the original database, a new subset is created and the contents will continue to be updated as the location changes. We term this as a "Real-Time dynamic database". Real-Time dynamic databases consist of the "value-based" relationships where typically, the relationship is specified at a retrieval time and the locations of related records are discovered during retrieval. Therefore, the content and number of records are altered from time and time.

Figure 19.2 shows the spatial temporal data set. For experimentation, an assumption is made that a mobile service is requested to find the coffee shop at a particular location. In stage 1, it shows the structure of the original GIS database. Each row corresponds to one spatial object. The attribute

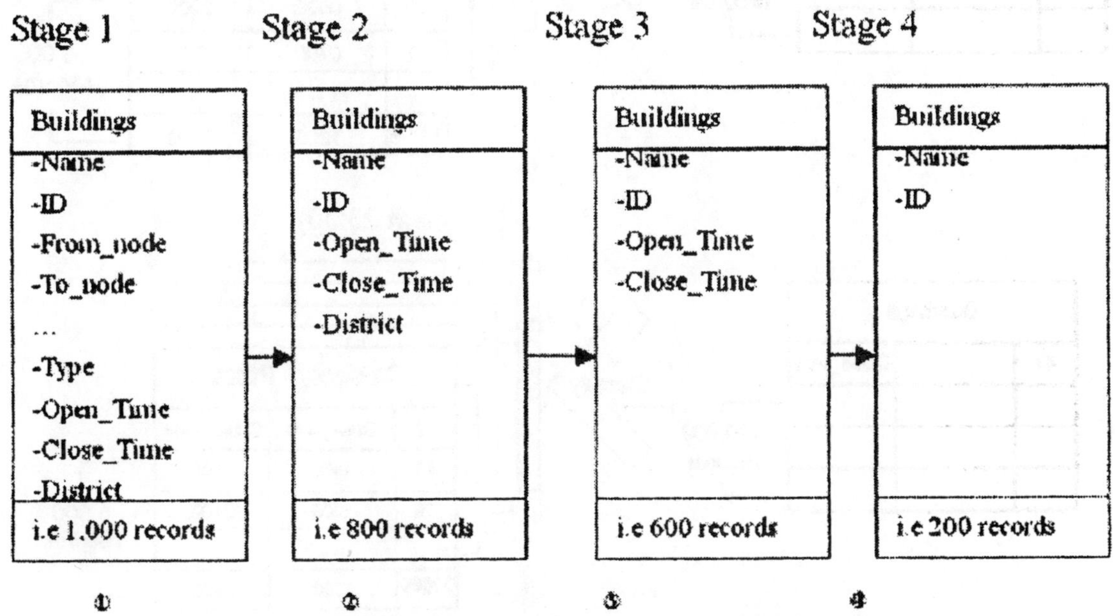

Figure 19.2: Spatial-Temporal Set Concept in Spatial Data

1: Original DB; 2: DB with Attribute Constraint; 3: DB with Spatial Constraint; 4: DB with Temporal Constraint

constraints are added in stage 2. The relevant fields such as "ID", "Open_time", "Close_time", "District" and "Name" are preserved. "ID" acts as the unique key while "Open_time" and "Close_time" indicate the working hours. "District" is the location identification and "Name" is the label. A spatial constraint is added in stage 3, where all objects within the mobile user's scopes are extracted. For example, the spatial data are kept in the same district of where the user is located. In the final stage, a temporal constraint is implemented and the meaningless information is discarded. For example, even if the buildings are in the area of interest, but do not have the same office hour temporal properties, they can be ignored. As a result, a subset is generated.

Physically, the contents in each Real-Time dynamic database are extracted from the original database and stored separately. However, the sum of records in all Real-Time dynamic databases may not be equal to the number of records in the master database.

Figures 19.3a and b illustrate that the number of records varies from time to time. For example, there are 100,000 records in the attribute table and 30,000 are categorized as fast food shops. At time 17:00, there are 30,000 records. However, at 23:30, only 5,000 records are stored because they are still within working hours.

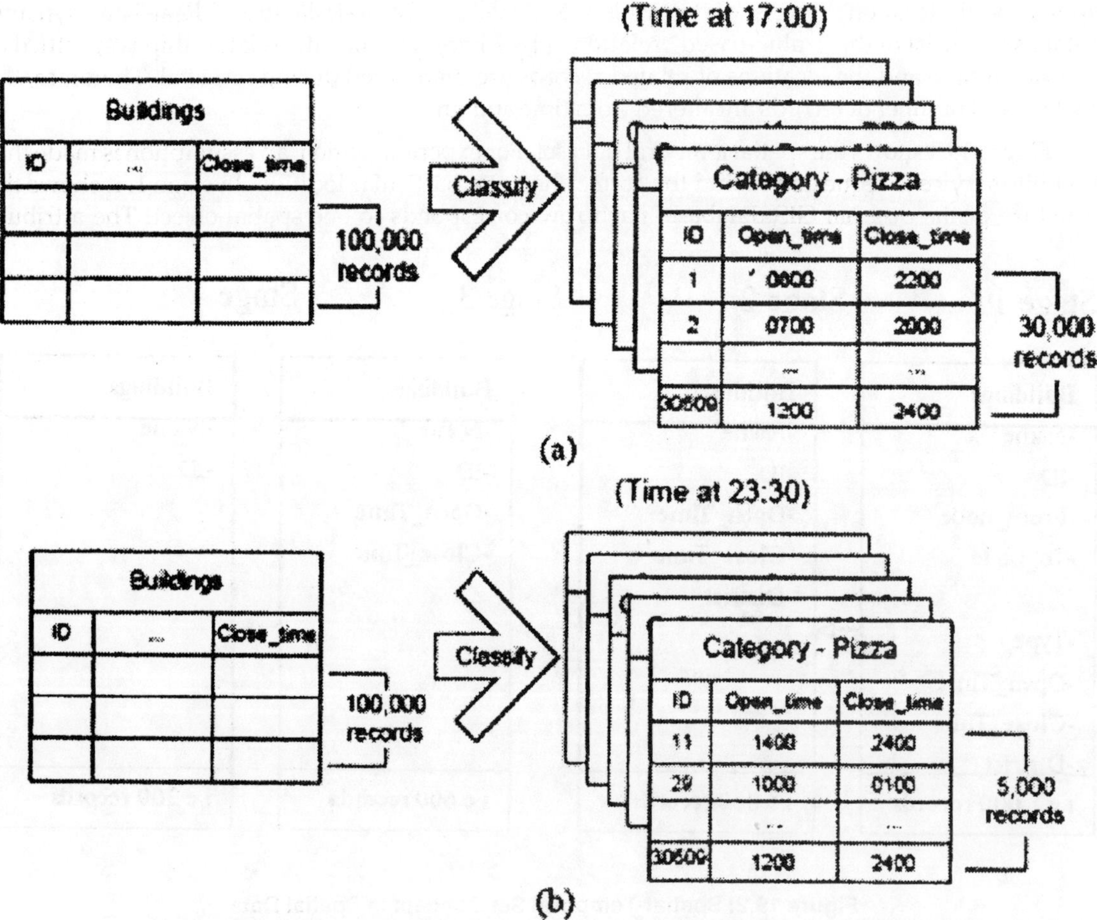

Figure 19.3(*a*) **Number of records at time 17:00,** (*b*) **Number of records at time 23:30**

Figure 19.4 shows the attribute value changes in a Real-Time dynamic database. A particular value, *i.e.* cost value, is varied from time to time. While there is environmental change, some contents are also changed to reflect the truth. A typical example is the optimum path measurement. Optimum path means the best route from the source to destination path is calculated from all objective factors, such as the traffic condition, the cost to pass, the penalty for the turns, and so on.

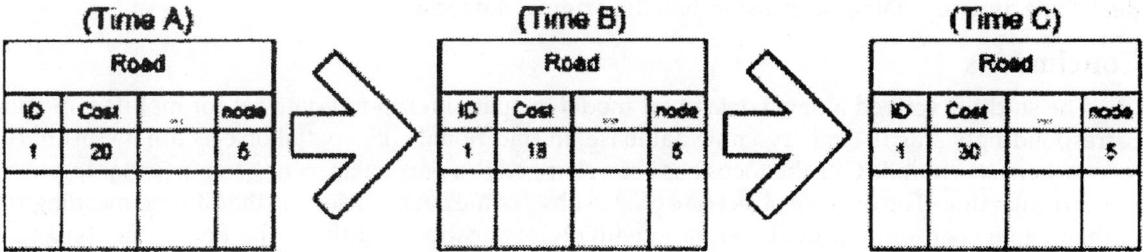

Figure 19.4: Changes In Attribute Value

Performance Evaluation

The model discussed in this study has been simplified with the following assumptions:

1. Time is one-dimensional and linearly ordered;
2. Relational databases are used to develop the GIS model;
3. Connection to the server and data retrieval via an active TCP/IP connection using the Hyper Test Transfer Protocol (HTTP) utilized as data transfer protocol; the Mobile user provides an initial position.

A simplified location-based service has demonstrated how the Real-Time dynamic database works. A mobile user chooses a district, a current location, pass and destination points, and maximum walking distance. Based on the above criteria, the optimum path can be found and displayed on the mobile device. We characterize the performance of this function in the following way. The average response time from the mobile client is measured as the time spent (in seconds) from the moment the query is issued to the moment the results of the query are generated.

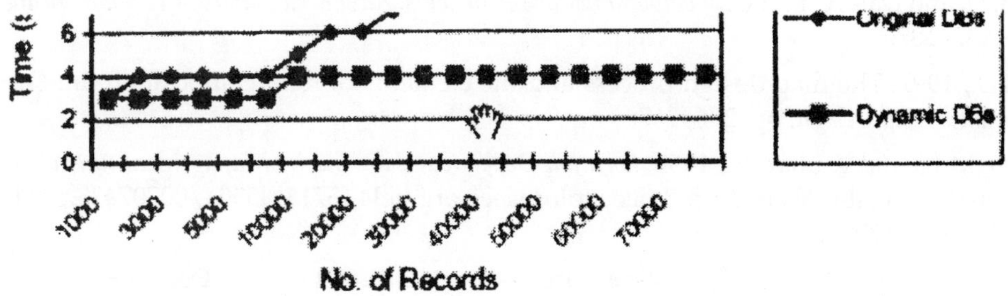

Figure 19.5: Average Response Time

Analysis of the Results

The result is depicted in Figure 19.5, which is plotted as a two-dimensional graph to illustrate the results of average response time. The line with square nodes corresponds to the time measurement from the dynamic database, while the line with diamond nodes represents the time measurement from the original database. We observe that the time saved increases as the database size expands. For the number of records, over 60,000, the time required in the original database is three times more than that in the Real-Time dynamic database. It is noted, however, that this is related to the actual time of the query, since the database volume is varied as time changes. The response time is slightly different to the result shown in Figure 19.5. Nevertheless, we can easily discover that the time response on the Real-Time dynamic database is faster than the original database.

Conclusions

The study described a new conceptual model- a spatial temporal data set for mobile GIS and correspondingly a Real-Time dynamic database. In traditional GIS, spatial objects are the primary focus whereas in mobile GIS, the focus has been changed to a particular scene of a particular location at a particular time. The relevant objects are only a subset of the overall objects in the GIS and meaningful to the mobile user for a particular situation–involving location and time. The Real-Time dynamic database is automatically updated to mirror the current world. The application and test were developed using a wireless web GIS environment. We compared the difference between a traditional database with the proposed Real-Time dynamic database in terms of response. From the results we can see that the processing time of the conventional database is proportionally increased with the data size in the database while the response time can be dramatically reduced in the proposed Real-Time dynamic database. This is even more effective for a database with a large data volume. In the Real-Time dynamic database, the response time is minimized and query accuracy is improved. These two advantages will greatly benefit mobile users in their applications.

References

Aberdeen Group, Inc., 2000. *The Benefits of Mobilising Enterprise Applications with Handheld Devices*, August.

Chan, Y., Si, A. and Leong, H.V., 1998. Cache management for mobile databases: Design and evaluation. In: *Data Engineering, 1998, Proceedings, 14th International Conference*, pp. 54–63.

Coates, P., 2000. Mapping and the mobile future. *GI News*, June.

CONNECT Computer Corp., 2000. *What is Thin Client Computing*? [online] Available from: http://www.connectcomputer.com/thin–client.asp [Accessed 13 December 2001].

Darwen, H. and Date, C.J., 2000. *Temporal Database Systems: Advanced Database Technology and Design,*, pp. 137–187.

Dent, B.D., 1997. The map design process and the elements of map composition. In: *Cartography: Thematic Map Design*, pp. 236–267.

Dunham, M.H. and Kumar, V., 1998. *Location Dependent data and its Management in Mobile Database* [online]. Available ITom: http://ieeexplore.ieee.org/iel4/5718/15304/00707433.pdf [Accessed 15 December 2001].

Francica, J., 2000. Location-based services wireless meets GIS. *GEO World*, December.

Hanttula, D., 2001. *The Future of the Pocket PC*. SemperAptus.com, January.

Hecht, L., 2000. Stand up for Standards. *GEO Europe*, September.

Iyengar, S.S. and Pangburn, B.E., 1998. Real-time computing: Meeting the client's demand for data access via the internet. In: *IEEE Workshop on Application–Specific Software Engineering Technology, ASSET–98 Proceedings*, pp. 14–19.

Ka-wai, Kwan and Wen-zhong, Shi. *A Study of Database in Mobile GIS.*

Koeppel, I., 2000. *What are Location Services?–from a GIS Perspective* [online]. Available from: http://www.jlocationservices.com/company/esri/What%20are%20%Location%20Services.html. [Accessed 8 July 2001].

Laurini, R., 2001. Real-time spatio-Temporal database. *Transactions in GIS*, 5(2).

Lauzac, S.W. and Chrysanthis, P.K., 2001. *Programming Views for Mobile Database Clients* [online]. Available from: http://ieeexplore.ieee.org/iel4/5718/15304/00707432.pdf [Accessed 17 December 2001].

Lowe, J.W., 2001. Which where will win? Dueling wireless location solution. *Geospatial Solutions*, March.

Meshulam, Y., 2001. says report *HK Economics Times*, 10 October.

Mirghafori, N. and Fontaine, A., 1995. A design for file access in a mobile environment. In: *Proceedings Workshop on Mobile Computing Systems and Applications*, pp. 57–62

Open Location Services Initiative, 2001. *Open Location Services: Vision and Objectives, Corporate Roles and Work Program Summary* [online]. [Accessed 16 December 2001].

Palm Inc., 2000. *Wireless Enterprise Application for Mobile Information Management* [online]. Available from: http://www.palm.com/pdfs/wireless–ent_apps.pdf [Accessed 15 December 2001].

Chapter 20

Characterization of Potteries from Melchittamur by a Combined FT-IR and Chemical Analysis

G. Velraj and K. Janaki
Department of Physics, Periyar University, Salem – 636 011, Tamil Nadu

ABSTRACT

As ceramics is a quite stable material, pottery is abundant in the archaeological context. For archaeological purposes, primary, secondary and accessory minerals are good enough to characterize the pottery material and to determine the clay and its sources. From this knowledge, information about the skill of the artisans in ancient societies can be established. Clay minerals, chemical composition of potteries and the atmosphere prevailed by the artisans belonging to the archaeological site Melchittamur were identified in the present study using Fourier Transform Infrared Spectroscopy (FT-IR) and Chemical analysis.

Keywords: *FT-IR, Chemical analysis, Clay minerals, Firing atmosphere, Archaeological potteries.*

Introduction

During the last two decades, architectural heritage preservation has reached a rising interest for Scientists, Architects, Engineers and Archaeologists. This subject being an interdisciplinary research area (Paula Lopez-Aree *et al.* (2003). The study of mineralogical phases of archaeological ceramics may be very helpful in unraveling the history of an ancient shred, particularly by means that investigate the process of its production. Phase analysis can potentially provide much information on the technology of the production of ceramics, which possess apart from provenance and dating, a very relevant and important archaeological question that need to be solved. Revealing the technical skill of ancient potters has been the subject of much research, since it is one of the most important issues for gaining a deep insight of bygone culture (Angela Zoppi *et al.* 2000). In the present work, an attempt has been made to identify the clay minerals composition and atmosphere prevailed to the archaeological potteries collected from the recently excavated site Melchittamur using Fourier Transform Infrared spectroscopy (FT-IR) and Chemical analysis. This site was excavated by State Department of Archaeology, Chennai.

Materials and Methods

The Infrared spectra of the pottery samples were recorded in the mid region 4000-450 cm^{-1} using Avatar 360 FT-IR spectrophotometer with 1 cm^{-1} resolution in its 100 scan mode using KBr pellet technique (Palanivel and Velraj, 2007). Titration method was utilized to find the chemical composition of potteries (Jeffery and Hutchison, 1986).

Results and Discussion

Fourier Transform Infrared Spectroscopy (FT-IR) have been extensively utilised to characterize archaeological potteries. The usefulness of infrared spectroscopy in characterising archaeological ceramics lies primarily in its ability to provide a finger print spectral pattern that can be readily associated with the mineralogic constituents of samples. Hence an attempt has been made to find the clay mineral constituents of potteries from Melchittamur site. The samples were collected at five different depths and are named as MCR1, MCR2, MCR3, MCR4 and MCR5 in increasing order of depth. Their respective Fourier transform infrared spectra are presented in the Figure 20.1. Except lower most spectrum other spectra were depicted vertically. The absorption observed in the spectra with their relative intensities and the tentative assignments are given in the Table 20.1. Russell (1987) has suggested that a band at 3626 cm^{-1} is due to O-H stretching of hydroxyl group. The samples MCR1 and MCR5 show a weak band at 3625 and 3626 cm^{-1} respectively may be assigned to O-H stretching of hydroxyl group.

A broad band of absorption around 3440 cm^{-1} is indicative of hydroxyl group present in the sample. The broad absorption band around 3440 cm^{-1} is always accompanied by a band around 1640 cm^{-1} suggests the presence of H-O-H bending of adsorbed water molecules (Palanivel and Velraj, 2007; Duraisamy *et al.*, 1991; Russell, 1987). The strong and weak bands in the region 3401–3441 cm^{-1} in all the samples MCR1-MCR5 have been assigned to O-H adsorbed water (Duraisamy *et al.*, 1991). In addition, the medium bands in the region 1632–1637 cm^{-1} in the samples MCR1–MCR5 are also due to the presence of water (Duraisamy *et al.*, 1991; Lara Maritan *et al.*, 2005; Colombini *et al.*, 2005; Roberts *et al.*, 2007; Claret *et al.*, 2003) have observed that the two week peaks that can be observed at 2852 cm^{-1} and 2922 cm^{-1} are due to C-H stretching mode and reveal the presence of some organic contribution. The present study shows a very weak band around 2923 cm^{-1} and 2852 cm^{-1} in all the Melchittamur samples which may be due to C-H stretching (Roberts *et al.*, 2007). The strong bands at 1427 cm^{-1} and 1421 cm^{-1} in MCR1 and MCR5 respectively are analogous to the accessory mineral Calcite.

Benedetto *et al.* (2002). As suggested by Benedetto *et al.* (2002) a very strong band at 1050 cm^{-1} in MCR1 may be assigned to Microcline.

The presence of very strong band around 1040 cm^{-1} in MCR2 and MCR4 is due to Si-O stretching of clay minerals (Palanivel and Velraj, 2007). A very strong band at 1080 cm^{-1} in MCR3 and MCR5

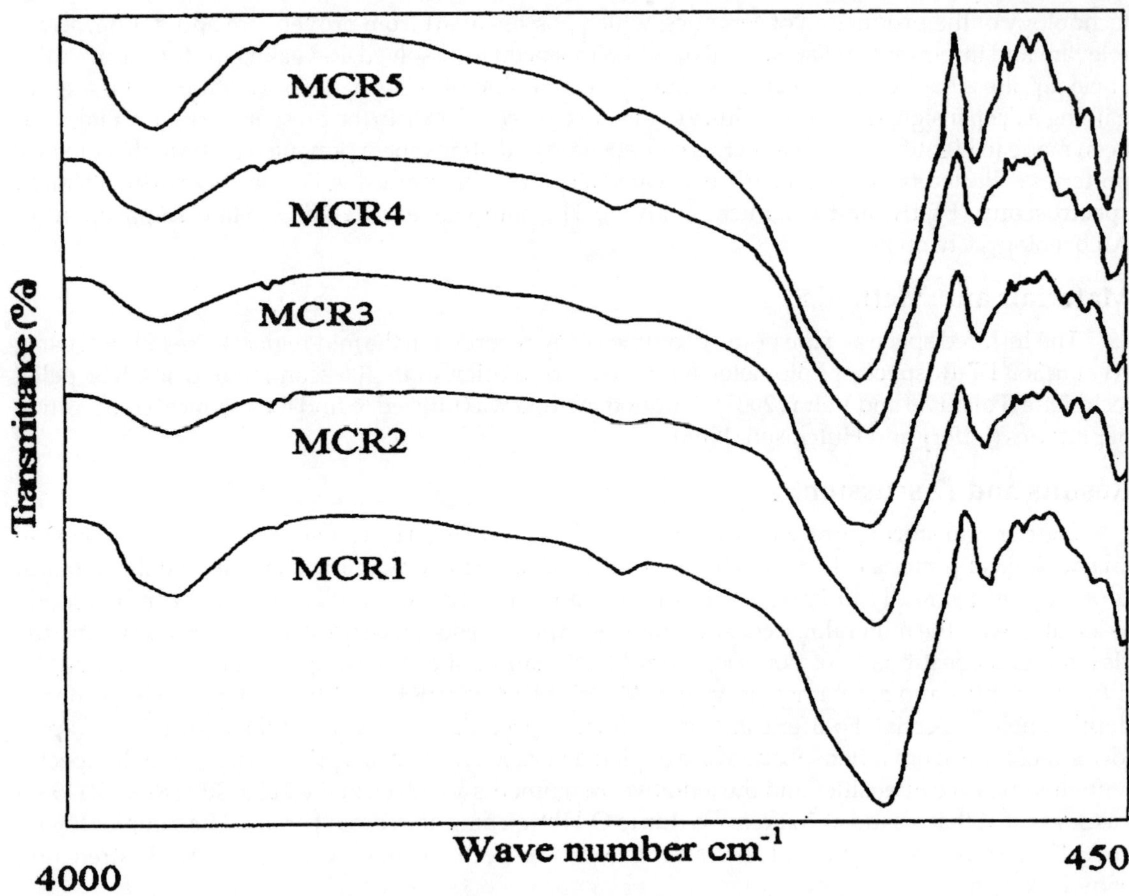

Figure 20.1: FT-IR Spectra of Melchittamur Sample

reveals the presence of Quartz (Barilaro *et al.*, 2005). In addition, the bands around 779 cm^{-1} and 694 cm^{-1} in all the Melchittanmr samples indicate the presence of Si-O Quartz (Barilaro *et al.* (2005). The band around 644 cm^{-1} which are present in MCR1–5 is analogous to Manganese Oxide (Barilaro *et al.*, 2005). Considering the iron rich compounds, Magnetite (Akyuz *et al.*, 2007) is present in MCR2, MCR3 and MCR4 which are evident from the medium absorption bands around 580 cm^{-1}. The medium band at 536 cm^{-1} in MCR4 is due to Hematite (Legodi and Waal, 2007).

The strong and very strong bands in the region 458–469 cm^{-1} in all the samples belonging to the archaeological site Melchittamur have been assigned to Hematite (Legocli and Waal, 2007). The ratio of extinction coefficient due magnetite and hematite was also calculated as suggested by (Ramasamy and Venkatachalapathy, 1991). This ratio is present only for the samples MCR2, MCR3 and MCR4. The ratio of extinction coefficient is very small (0.0134, 0.0119 and 0.0082 respectively) and the visual color of the samples are Red slipped ware. Hence, as suggested by (De Benedetto *et al.*, 2002) the atmosphere prevailed by the artisans is oxidizing atmosphere. Regarding the remaining samples of Melchittamur site, the magnetite band is not detected, they contains only hematite as shown in Table

20.1 and also the colour of the samples is Red and Red slipped ware and hence these two also may be fired in the oxidizing atmosphere (De Benedetto *et al.*, 2002). The results are tabulated in Table 20.3.

Table 20.1: FT-IR Vibrational Frequencies and Tentative Assignments of the Pottery Samples Excavated from Melchittamur (MCR) Site

MCR1	*MCR2*	*MCR3*	*MCR4*	*MCR5*	*Tentative Vibrational Assignment*
	Frequency in (cm^{-1}) with Relative Intensity				
3625 W	–	–	–	3626 W	O–H Str. Hydroxyl group
3433 S	3401 W	3430 W	3430 W	3441 S	O–H Str. adsorbed water
2923 VW	2923 VW	2918 VW	2923 VW	2923 VW	C–H Stretching
2858 VW	2852 VW	2852 VW	2847 VW	2858 VW	C–H Stretching
1633 M	1633 M	1634 M	1637 M	1632 M	O–H Str. adsorbed water
1427 S	–	–	–	1421 S	Calcite
1050 VS	–	–	–	–	Microcline
–	1043 VS	–	1042 VS	–	Si–O Str. clay minerals
–	–	1080 VS		1080 VS	Si–O–Quartz
778 VS	778 M	778 M	777 M	778 VS	Si–O–Quartz
692 M	692 W	692 VW	692 VW	692 S	Si–O–Quartz
645 M	645 M	644 VW	643 VW	634 S	MnO
–	582 M	582 M	580 M	–	Magnetite–FeO– Fe_3O_4
–	–	–	536 M	–	Hematite–FeO–Fe_3O_4
465 VS	458 VS	460 S	462 S	469 VS	Hematite–FeO–Fe_3O_4

VS: Very strong; S: Strong; M: Medium; W: Weak; VW: Very Weak; bd: Broad; sp: Sharp.

The samples collected from the Melchittamur site have also been analyzed using Chemical analysis. From the Table 20.2, it is obvious that Silica is present above 55 per cent. Alumina (Al_2O_3) is present above 12 per cent in all the samples collected from this site. The flux component K_2O is also more and varies as 3.17–5.42 per cent in all the samples. The flux component TiO_2 is beyond the detecting limit in MCR5 and is present around 1 per cent in all the remaining samples taken from the

Melchittamur site. The percentage of MgO varies around 2 per cent. The percentage of Calcium oxide is less than 6 per cent in the all the samples and hence the type of clays in this site belongs to Non-Calcareous type (Maniatis and Tite, 1981) and the variation CaO with five samples from Melchittamur site are depicted in Figure 20.2. As the total flux content is more than 9 per cent in MCR1–MCR5, the clay type are low refractory in nature (Maniatis and Tite, 1981). The chemical analysis reveals that the samples are rich in Hematite than Magnetite. And also the samples are Red and Red slipped ware in color, hence the atmosphere prevailed by the artisans at the time of manufacture is open or oxidizing atmosphere (De Benedetto *et al.*, 2002). These results coincide very well with that of FT-IR result. The variations of Hematite and Magnetite in all the five different pottery samples were shown in Figure 20.3. The results are tabulated in Table 20.3.

Table 20.2: Chemical Composition of Archaeological Pottery Shreds Excavated at the Site Melchittamur

Oxides	*MCR1*	*MCR2*	*MCR3*	*MCR4*	*MCR5*
SiO_2	59.72	50.99	55.76	62.60	55.86
Al_2P_3	11.65	17.99	12.53	12.078	13.06
K_2O	5.18	4.15	3.17	5.42	3.68
Fe_2O_3	5.02	5.82	5.06	5.25	4.80
CaO	4.50	4.22	3.08	3.10	2.94
MgO	2.71	2.84	2.10	2.23	2.78
TiO_2	1.37	1.21	1.19	0.74	BDL
Fe_3O_4	4.46	4.82	4.57	4.69	4.31

Table 20.3: Determination of Firing Atmosphere Using Chemical Analysis and FT-IR

Sample Code	*Colour*	*Clay Type*	*$K_{Mag/Hem}$ from FT-IR Spectra*	*Magnetite/ Hematite from Chemical Analysis*	*Atmosphere Prevailed*
MCR1	Red slipped	NC	–	0.888	Oxidizing
MCR2	Red slipped	NC	0.0134	0.828	Oxidizing
MCR3	Red slipped	NC	0.0119	0.903	Oxidizing
MCR4	Red slipped	NC	0.0082	0.893	Oxidizing
MCR5	Red ware	NC	–	0.897	Oxidizing

Conclusion

The archaeological pottery samples collected from Melchittamur site was analyzed by using FTIR and Chemical analysis for its chemical composition, clay minerals identification and firing atmosphere. From the FT-IR analysis it has been found that the secondary mineral Quartz is present in all the samples belonging to Melchittamur site. The various accessory minerals present are Calcite, Manganese oxide, Microcline, Hematite and Magnetite. Among these minerals, Hematite alone is present all Melchittamur samples and the remaining are present in only some of the samples taken from Melchittamur site. The Chemical analysis shows the presence of SiO_2, Al_2O_3, K_2O, $TiOz_2$, MgO,

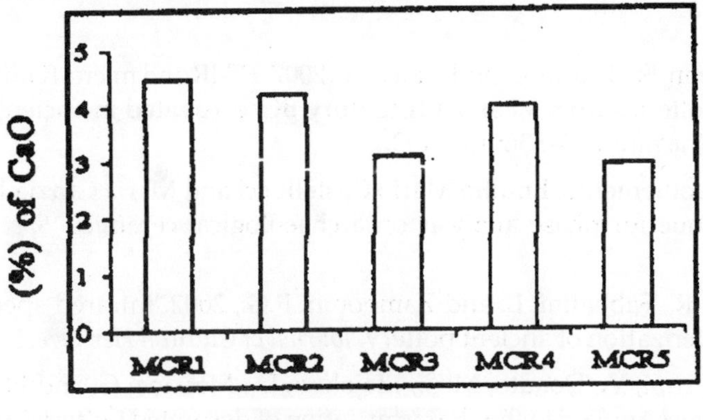

Figure 20.2: Variation of CaO

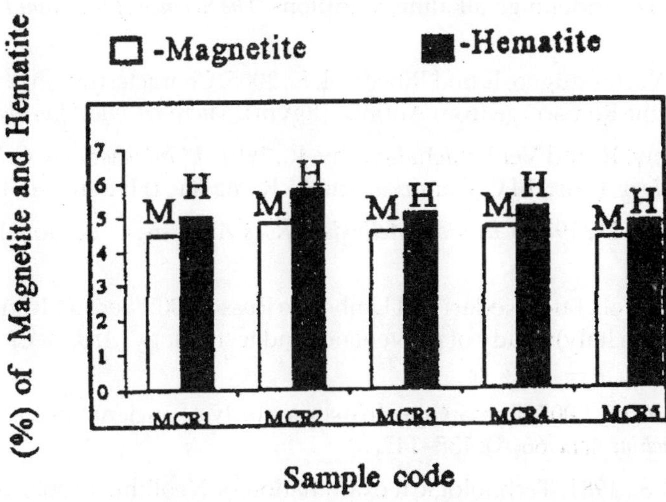

Figure 20.3: Variation of Magnetite and Hematite

CaO, Fe_2O_3 and Fe_3O_4 in different percentage in all the samples. The visual color identified is Red ware. The FT- IR and Chemical analysis shows that all the samples have been fired in oxidizing atmosphere. From the above analysis, one can infer that the artisans of Melchittamur site used potteries of various chemical compositions (potteries rich in Quartz, Alumina and Iron oxides) and they have used different minerals for making the potteries and they are well known about the art of firing the utensils in open air or oxidizing atmosphere

Acknowledgements

The authors acknowledge with thanks to The Head, Department of Ancient History and Archaeology, University of Madras, for his timely help and guidance for the collection of Archaeological pottery samples. The authors are also thankful to Mr. S. Srinivasan, Department of Chemistry, Annamalai University for his timely help to take FT-IR spectra in Centralized Instrumentation Service Laboratory (CISL), Annamalai University, Annamalainagar.

References

Akyuz, S., Akyuz, T., Basaran, S., Bolacal, C. and Gulec, A., 2007. FT-IR and micro-Raman Spectroscopic study of decorated potteries from VI and VII century BC, excavated in ancient Ainos, Turkey. *Journal of Molecular Structure*, 834–836: 150–153.

Angela Zoppi, Cristiana Lofuemento, Emillio Mario Castellucci and Maria Grazia Migllorini, 2000. Micro Raman technique for phase analysis on archaeological ceramics. *Spectroscopy Europe*, 14/5.

Benedetto, G.E., Laviano, R., Sabbatini, L. and Zambonin, P.G., 2002. Infrared spectroscopy in the mineralogical characterization of ancient pottery. *Journal of Cultural Heritage*, 3: 177–186.

Barilaro, D., Barone, G., Crupi, V., Donato, M.G., Majolino, D., Messina, G. and Ponterio, R., 2005. Spectroscopic techniques applied to the characterisation of decorated potteries from Caltagirone (Sicily, Italy). *Journal of Molecular Structure*, 744–747: 827– 831.

Claret, F., Schafer, T., Bauer, A. and Buckau, G., 2003. Generation of humic and fulvic acid from Callovo Oxfordian clay under high alkaline conditions. *The Science of the Total Environment*, 317: 189–200.

Colombini, M.P., Giachi, G., Modugno, F. and Ribechini, E., 2005. Characterisation of organic residues in pottery vessels of the Roman age from Antinoe (Egypt). *Micro-chemical Journal*, 79: 83–90.

Duraisamy, D., Ramasamy, K. and Venkatachalapathy, R., 1991. *1st International Colloq. on the Role of Chemistry in Archaeology*, (Eds.) M.C. Ganorkar and N. Rama Rao (Hyderabad, India), pp. 47–54.

Jeffery, P.G and Hutchison, D., 1986. *Chemical Methods of Rock Analysis*, 4th edition. Pergamon Press, pp. 48–150.

Lara Maritan, Claudio Mazzoli, Luca Nodari and Umbeerto Russo, 2005. Second Iron age grey pottery from Este (northeastern Italy): Study of provenance and technology. *Applied Clay Science*, 29: 31–44.

Legodi, M.A. and D. de Waal, 2007. Raman spectroscopic study of ancient South African domestic clay pottery. *Spectrochim Acta*, 66(A): 135–142.

Maniatis, Y. and Tite, M.S., 1981. Technological examination of Neolithic-bronze age pottery from central and southeast Europe and from the near east. *Journal of Archaeological Science*, 8: 59–76.

Palanivel, R. and Velraj, G., 2007. FT-IR and FT-Raman spectroscopic studies of fired clay artifacts recently excavated in Tamil Nadu, India. *Indian Journal of Pure and Applied Physics*, 45: 501–508

Paula Lopez-Aree, Javier Garcia, Mercedes Garcia and Joaquin Obis, 2003. Bricks in historical buildings of Toledo city: Characterization and restoration. *Materials Characterization*, 50: 59–68.

Ramasamy, K. and Venkatachalapathy, R., 1991. Infrared spectroscopic study of Neyveli lignite samples from mine cut I and II. *Spectroscopy Letters*, 24(6): 759–778.

Roberts, S., Sofaer, J. and Kiss, V., 2007. Characterization and textural analysis of middle Bronze Age Transdanubian inlaid wares of the encrusted pottery culture, Hungary: A preliminary study. *Journal of Archaeological Science*, pp. 1–9.

Russell, J.D., 1987. *A Handbook of Determinative Methods in Clay Mineralogy*, (Eds.) M.J. Wilson. Blackie and Son Ltd., pp. 163.

Chapter 21

On Some Improved Product Type Estimator

Manish Sharma[1], S.E.H. Rizvi[1] and Sharad Bhatnagar[2]
[1]Division of Agricultural Economics and Statistics,
Faculty of Agriculture, SKUAST-J, Chatha – 180 009, Jammu
[2]Professor (Statistics), CCSHAU, Hisar, Haryana, India

ABSTRACT

In the present chapter a product type estimator of population mean has been worked out from the general class of biased product type estimators proposed by Sharma *et al.* (2007). Its large sample properties were compared with the conventional product estimator as well as the estimators proposed by Robson (1957), Singh (1989) and Dubey (1993). The proposed estimator showed an improvement over these estimators according to mean squared error criterion.

Keywords: *Product type estimator, Conventional estimator, Population mean, Relative bias, Relative mean squared error.*

Introduction

Considering the auxiliary characteristic is negatively correlated with the characteristic under study, Robson (1957) proposed the biased product estimator and also developed an unbiased estimator. Later on, Singh (1989) and Dubey (1993) considered almost unbiased product estimators. Sharma *et al.* (2007) proposed a general class of product type estimator and worked out its efficiency. In the present investigation a product type estimator has been proposed and its large sample properties has been compared with the conventional product as well as product estimators proposed by Robson (1957),

Singh (1989) and Dubey (1993). Based on mean squared criterion, an improvement has been observed over the above mentioned estimators.

Estimators and their Properties

Based on a simple random sample of size n drawn from a finite population of size N, we denote by $\bar{x}$ and $\bar{y}$ the sample means which are the unbiased estimators of their respective parameters *viz.* $\bar{X}$ and $\bar{Y}$, the population means of the auxiliary variable and the variable under study, respectively.

For estimating $\bar{Y}$, Robson (1957) proposed the following product estimator

$$\bar{y}_P = \frac{\bar{y}\bar{x}}{\bar{X}} \tag{2.1}$$

whose relative bias and relative mean squared error up to O(n^{-2}) are as

$$RB(\bar{y}_P) = \frac{C_{11}}{n} \tag{2.2}$$

$$RM(\bar{y}_P) = \frac{1}{n}\left[C_{20} + C_{02} + 2C_{11}\right] + \frac{1}{n^2}\left[C_{20}C_{02} + 2C_{11}^2 + 2(C_{21} + C_{12})\right]$$

where,

$$S_{xy} = \frac{1}{N-1}\sum_{i=1}^{N}(x_i - \bar{X})(y_i - \bar{Y})$$

and

$$C_{ab} = \frac{1}{N-1}\sum_{i=1}^{N}\left(\frac{x_i - \bar{X}}{\bar{X}}\right)^a\left(\frac{y_i - \bar{Y}}{\bar{Y}}\right)^b$$

for non-negative integer a and b.

As the product estimator given by (2.1) is biased one, Robson (1957) developed the unbiased product estimator obtained by subtracting the unbiased estimate of the bias of $\bar{y}_P$ in (2.1) as follows

$$t^* = \frac{\bar{y}\bar{x}}{\bar{X}} - \frac{1}{n}\frac{s_{xy}}{\bar{X}} \tag{2.4}$$

where,

$$s_{xy} = \frac{1}{N-1}\sum_{i=1}^{n}(x_i - \bar{x})(y_i - \bar{y})$$

The relative variance, up to order O (n^{-2}), of t* is given by

$$RV(t^*) = RM(\bar{y}_P) - \frac{1}{n^2}\left[C_{11}^2 + 2(C_{21} + C_{12})\right] \tag{2.5}$$

From (2.5), it is observed 'that the unbiased estimator t* has smaller variance than the relative mean squared error of $\bar{y}_P$ provided

$$C_{11}^2 + 2(C_{21} + C_{12}) > 0 \quad 2.6$$

The condition (2.6) holds good for all symmetric bivariate distributions with $C_{21} = C_{12} = 0$.

Singh (1989) proposed an almost unbiased product estimator of $\bar{Y}$ as follows

$$t_1^* = \frac{\bar{y}\bar{x}}{\bar{X}}\left[1 + \frac{s_{xy}}{n\bar{x}\bar{y}}\right]^{-1} \quad 2.7$$

The relative bias and relative mean squared error of the estimator t_1^* upto O (n^{-2}) are given by

$$RB(t_1^*) = \frac{C_{11}^?}{n^2} \quad 2.8$$

$$RM(t_1^*) = RV(t^*) \quad 2.9$$

So, the estimator t_1^* is unbiased up to O (n^{-1}) and has relative mean squared error equal to relative variance of the estimator t*. It performs better than estimator $\bar{y}_P$ when the condition (2.6) holds.

Further Dubey (1993) proposed an almost unbiased product estimator as

$$t_2^* = \frac{\bar{y}\bar{x}}{\bar{X}} - \frac{s_{xy}}{n\bar{x}} \quad 2.10$$

whose relative bias and relative mean squared error upto order O(n^{-1}) and O(n^{-2}), respectively, are given by

$$RB(t_2^*) = -\frac{1}{n^2}(C_{20}C_{11} - C_{21}) \quad 2.11$$

$$RM(t_2^*) = RM(\bar{Y}_P) + \frac{1}{n^2}C_{11}(2C_{20} + C_{11}) + \frac{2}{n^2}(C_{21} + C_{12}) \quad 2.12$$

$$= RV(t^*) + \frac{2}{n^2}C_{11} * (C_{20} + C_{11})$$

From (2.11) and (2.12), it is revealed that the estimator t_2^* is unbiased up to order O(n^{-1}). Further, the estimator t_2^* has smaller relative mean squared error than conventional product estimator as well as the estimator proposed by Robson (1957) for bivariate normal population if

$$0 < \theta' < -\rho \quad 2.13$$

Here, ρ is the coefficient of correlation between the variable under study and the auxiliary variable, and

$$\theta = \left[\frac{C_{02}}{C_{20}}\right]^{1/2} \quad 2.14$$

Proposed Estimator and its Properties

Sharma *et al.* (2007) proposed a general class of biased product estimator as

$$\bar{y}_{P(q)} = \bar{y}\left[\frac{\bar{x}}{\bar{X}} + \frac{q}{n}\frac{s_x^2}{\bar{X}^2}\right] \quad 3.1$$

By setting q = 0 in (3.1), the estimator $\bar{y}_{P(q)}$ yields the conventional product estimator.

The relative bias and relative mean squared error of the estimator $\bar{y}_{P(q)}$ up to order $O(n^{-1})$ and $O(n^{-1})$, respectively, are given as

$$RB\left(\bar{y}_{P(q)}\right) = RB\left(\bar{y}_P\right) + \frac{qC_{20}}{n} \quad 3.2$$

$$RM\left(\bar{y}_{P(q)}\right) = RM\left(\bar{y}_P\right) + \frac{q}{n^2}\left[4C_{11}C_{20} + 2C_{02}C_{20} + qC_{20}^2\right] + \frac{2q}{n^2}\left[C_{30} + C_{21}\right] \quad 3.3$$

From (3.2), it is observed that the estimator $\bar{y}_{P(q)}$ has smaller bias than $\bar{y}_{P(q)}$ for all negative values of q.

From (3.3), it is found that the estimator $\bar{y}_{P(q)}$ is more efficient than $\bar{y}_{P(q)}$ under bivariate normal population if

$$-2\theta(\theta+2\rho) < \theta < 0 \text{ and } \theta > -2\rho \quad 3.4$$

Setting q = –1 in (3.1), we hereby a propose biased product estimator as

$$\bar{y}_{P(-1)} = \bar{y}\left[\frac{\bar{x}}{\bar{X}} - \frac{1}{n}\frac{s_x^2}{\bar{X}^2}\right] \quad 3.5$$

The relative bias and relative mean squared error of the estimator $\bar{y}_{P(-1)}$ up to order $O(n^{-1})$ and $O(n^{-2})$, respectively, are as

$$RB\left(\bar{y}_{P(-1)}\right) = RB\left(\bar{y}_P\right) - \frac{C_{20}}{n} \quad 3.6$$

$$RM\left(\bar{y}_{P(-1)}\right) = RM\left(\bar{y}_P\right) + \frac{1}{n^2}\left[-4C_{11}C_{20} - 2C_{02}C_{20} + C_{20}^2\right] - \frac{2}{n^2}\left[C_{30} + C_{21}\right] \quad 3.7$$

From (3.6), it is clear that, the estimator $\bar{y}_{P(-1)}$ has smaller bias than $\bar{y}_P$.

From (3.7), it is evident that the estimator $\bar{y}_{P(-1)}$ is more efficient than $\bar{y}_P$ under bivariate normal population, for the range of parameter θ as

$$0.71 < \theta < 2.23 \tag{3.8}$$

Now, if we compare the estimators $\bar{y}_{P(-1)}$ with t*, considering the mean squared error, we have

$$RM\left(\bar{y}_{P(-1)}\right) - RV(t^*) = \frac{1}{n^2}\left[C_{11}^2 - 4C_{11}C_{20} - 2C_{02}C_{20} + C_{20}^2\right] \tag{3.9}$$

From (3.9) it can be inferred that the estimator $\bar{y}_{P(-1)}$ has smaller relative mean squared error than t* provided

$$1 < \theta < 4.237 \tag{3.10}$$

Now, comparing the relative mean squared errors of the estimators t_2^* and $\bar{y}_{P(-1)}$, from (2.12) and (3.7) under bivariate normal population, we have

$$RM\left(\bar{y}_{P(-1)}\right) - RM\left(t_2^*\right) = \frac{1}{n^2}\left[-C_{11}^2 - 6C_{11}C_{20} - 2C_{02}C_{20} + C_{20}^2\right] \tag{3.11}$$

From (3.11), we conclude that the estimator $\bar{y}_{P(-1)}$ is more efficient than t_2^*

if $$0.578 < \theta < 2.155 \tag{3.12}$$

Thus, the proposed estimator $\bar{y}_{P(-1)}$ performs better than the estimators $\bar{y}_P$, t*, t_1^* and t_2^*

if $$1 < \theta < 2.155 \tag{3.13}$$

References

Dubey, V., 1993. An almost unbiased product estimator. *Jour. Indian Soc. Agril. Statist.*, 45(2): 226–229.

Robson, D.S., 1957. Application of multivariate polykas to the theory of unbiased ratio type estimators. *Jour. Amer. Statist. Assoc.*, 52: 511–522.

Sharma, M., Bhatnagar, S. and Grover, D., 2007. Improved product type estimator. In: *Proceedings of the National Seminar on Statistics and its Applications* on 19th to 20th January, in the Department of Mathematics, SLIET, Longowal, Punjab, India, p. 28-30.

Singh, H.P., 1989. A class of unbiased product estimators of population means. *Jour. Indian Soc. Agril. Statist.*, 41(1): 113–118.

Chapter 22

Ferroelectric Ceramics: Revisited

M.R. Panigrahi[1]*, S. Panigrahi[1], K.M. Purohit[2]
[1]Department of Physics, [2]Department of Chemistry,
National Institute of Technology, Rourkela – 8, Orissa

ABSTRACT

Ceramic materials and single crystals showing ferroelectric behavior are being used in many applications in electronics and optics. A large number of applications of ferroelectric ceramics also exploit properties that are an indirect consequence of ferroelectricity, such as dielectric, piezoelectric, pyroelectric and electro-optic properties. This review introduces the basic principles and characteristics of ferroelectric materials which show ferroelectric behavior. Various applications arising from ferroelectricity and related phenomena in ceramics and thin films have been listed at the end.

Keywords: *Ferroelectricity, Piezoelectricity, Pyroelectricity, Dielectrics, Ceramics, Single crystals.*

Introduction

Ferroelectric effect is an electrical phenomenon whereby certain materials may exhibit a spontaneous dipole moment, the direction of which can be switched between equivalent states by the application of an external electric field. The term ferroelectricity is used in analogy to ferromagnetism, in which a material exhibits a permanent magnetic moment.

* E-mail: manash_123inda@yahoo.co.in

Ferroelectricity is a phenomenon which was discovered in 1922. The name refers to certain magnetic analogies, though it is somewhat misleading as it has no connection with iron (ferrum) at all. Ferroelectricity has also been called Seignette electricity, as Seignette or Rochelle Salt (RS) was the first material found to show ferroelectric properties such as a spontaneous polarization on cooling below the Curie point, ferroelectric domains and a ferroelectric hysteresis loop. A huge leap in the research on ferroelectric materials came in the 1950's, leading to the widespread use of barium titanate ($BaTiO_3$) based ceramics in capacitor applications and piezoelectric transducer devices. Since then, many other ferroelectric ceramics including lead titanate ($PbTiO_3$), lead zirconate titanate (PZT), lead lanthanum zirconate titanate (PLZT), and relaxor ferroelectrics like lead magnesium niobate (PMN) have been developed and utilized for a variety of applications. With the development of ceramic processing and thin film technology, many new applications have emerged. The biggest use of ferroelectric ceramics have been in the areas such as dielectric ceramics for capacitor applications, ferroelectric thin films for non volatile memories, piezo-electric materials for medical ultrasound imaging and actuators, and electro-optic materials for data storage and displays.

In this article, an effort is made to introduce the basic principles governing ferroelectricity and list the various materials which exhibit these properties. The processing of ferroelectric ceramics in general, with a few examples is described. Finally, a few important applications of ferroelectric materials are briefly discussed.

General Properties of Ferroelectrics

Crystal Symmetry

The lattice structure described by the Bravais unit cell of the crystal governs the crystal symmetry. Though there are thousands of crystals in nature, they all can be grouped together into 230 microscopic symmetry types or space groups based on the symmetry elements (Nye, 1990; Newnham, 1975). Most of the crystals possess symmetry elements in addition to the repetitions expressed by the crystal lattice. The operation of any single symmetry element of the group leaves the pattern of symmetry unchanged. In studying the physical properties of crystals, only the orientations of the symmetry elements and not their relative positions are important. Hence, if only the orientations of the symmetry elements are taken into account, then the macroscopic symmetry elements in crystals reduce to a center of symmetry, mirror plane, 1-, 2-, 3-, 4- or 6- fold rotation axes and 1-, 2-, 3-, 4- or 6- fold inversion axes. A combination of these symmetry elements gives us the macroscopic symmetry also called as point groups. It can be shown by the inspection of the 230 space groups that there are just 32 point groups. As shown below, the seven crystal systems can be divided into these point groups according to the point group symmetry they possess.

We have,

2 space group in a Triclinic system: 1, 1

13 space group in a Monoclinic system: 2, m, 2/m

59 space group in an Orthorhombic system: 222, mm2, mmm

25 space group in a Rhombohedral system: 3, 3, 32, 3m, 3m

27 space group in a Hexagonal system: 6, 6, 6/m, 622, 6mm, 6m2, (6/m)mm

68 space group in a Tetragonal system: 4, 4, 4/m, 422, 4mm, 42m, (4/m)mm

36 space group in a Cubic system: 23, m3, 432, 43m, m3m

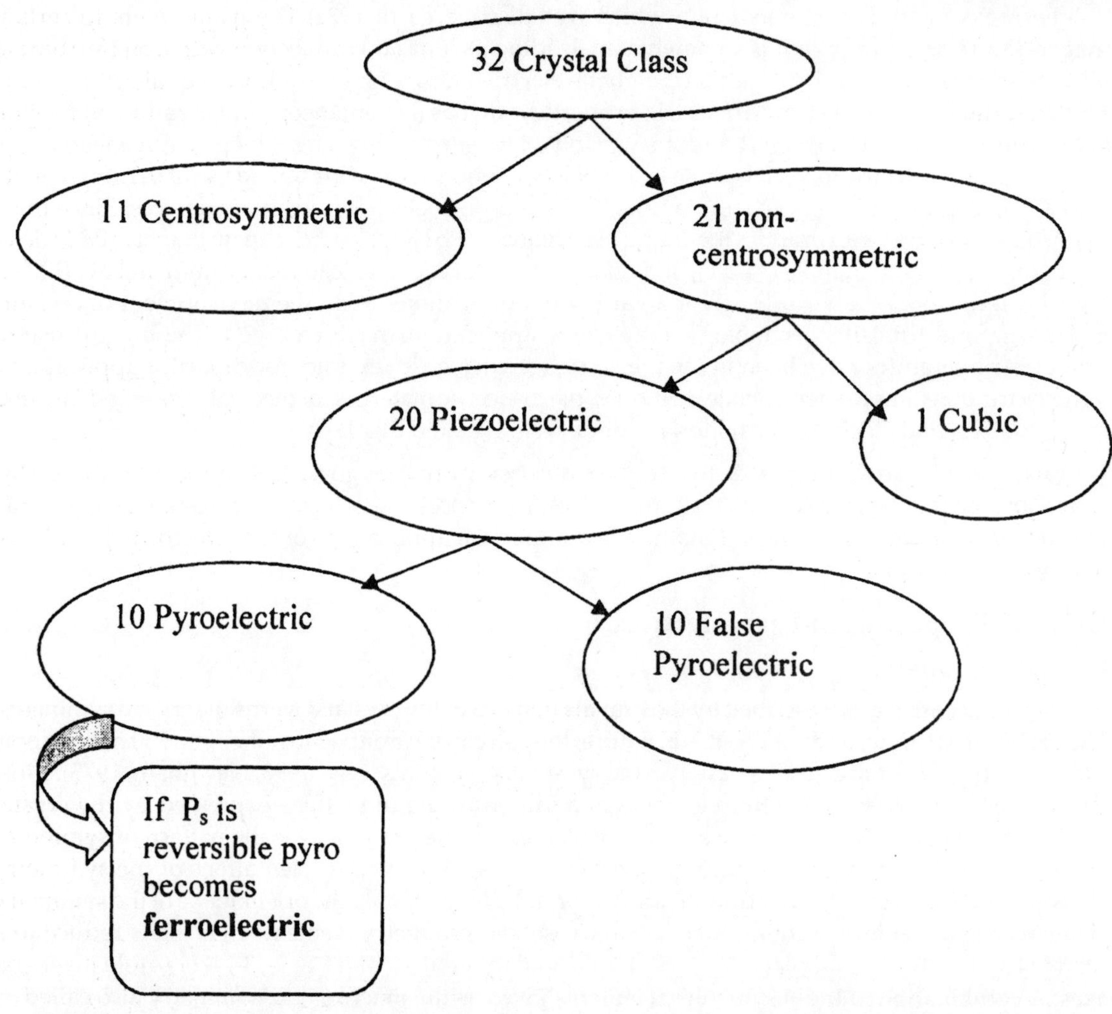

Figure 22.1: Classification of Crystal Class

The thirty-two point groups can be further classified into (*a*) crystals having a center of symmetry and (*b*) crystals which do not possess a center of symmetry as shown in Figure 22.1. Crystals with a center of symmetry include the 11 point groups labeled centro symmetric. These point groups do not show polarity. The remaining 21 point groups do not have a center of symmetry (*i.e.* non-centrosymmetric). A crystal having no center of symmetry possesses one or more crystallographically unique directional axes. All noncentrosymmetric point groups, except the 432 point group, show piezoelectric effect along unique directional axes. Piezoelectricity is the ability of certain crystalline materials to develop an electrical charge proportional to a mechanical stress.

Piezoelectric Crystal Classes: 1, 2, m, 222, mm2, 4, –4; 422, 4mm, –42m, 3, 32, 3m, 6, –6, 622, 6mm, 62m, 23, –43m

Pyroelectric Crystal Classes: 1,2, m, mm2, 3, 3m, 4, 4mm, 6, 6mm

Spontaneous Polarization and Pyroelectric Effect

The spontaneous polarization is given by the value of the dipole moment per unit volume or by the value of the charge per unit area on the surface perpendicular to the axis of spontaneous polarization. The axis of spontaneous polarization is usually along a given crystal axis. Although a crystal with polar axes (20 non-centro symmetric point groups) shows the piezoelectric effect, it is not necessary for it to have a spontaneous polarization vector. It could be due to the canceling of the electric moments along the different polar axes to give a zero net polarization. Only crystals with a unique polar axis (10 out of 21 non-centro symmetric point groups) show a spontaneous polarization vector Ps along this axis. The value of the spontaneous polarization depends on the temperature. This is called the pyroelectric effect which was first discovered in tourmaline by Teophrast in 314 B.C. and so named by Brewster in 1824 (Lang, 1974). The pyroelectric effect can be described in terms of the pyroelectric coefficient λ. A small change in the temperature ΔT, in a crystal, in a gradual manner, leads to a change in the spontaneous polarization vector ΔP_s given by,

$$\Delta P_s = \lambda \Delta T \tag{1}$$

Figure 22.2 shows the variation of the spontaneous polarization P_s with temperature for a $BaTiO_3$ ferroelectric crystal. An increase in the temperature leads to a decrease in the spontaneous polarization $BaTiO_3$ has a negative pyroelectric coefficient. The polarization suddenly falls to zero on heating the crystal above the Curie point (Merz, 1953).

Ferroelectric Domains and Hysteresis Loop

As described above, pyroelectric crystals show a spontaneous polarization P_s in a certain temperature range. If the magnitude and direction of P_s can be reversed by an external electric field, then such crystals are said to show ferroelectric behavior. Hence, all single crystals and successfully poled ceramics which show ferroelectric behavior are pyroelectric, but not vice-versa, For example tourmaline shows pyroelectricity but is not ferroelectric.

Ferroelectric crystals possess regions with uniform polarization called ferroelectric domains. Within a domain, all the electric dipoles are aligned in the same direction. There may be many domains in a crystal separated by interfaces called domain walls. A ferroelectric single crystal, when grown, has multiple ferroelectric domains. A single domain can be obtained by domain wall motion made possible by the application of an appropriate electric field. A very strong field could lead to the reversal of the polarization in the domain, known as domain switching (Batthais and Hippel, 1948; Hippel, 1950).

The main difference between pyroelectric and ferroelectric materials is that the direction of the spontaneous polarization in ferroelectrics can be switched by an applied electric field. The polarization reversal can be observed by measuring the ferroelectric hysteresis as shown in Figure 22.3. As the electric field strength is increased, the domains start to align in the positive direction giving rise to a

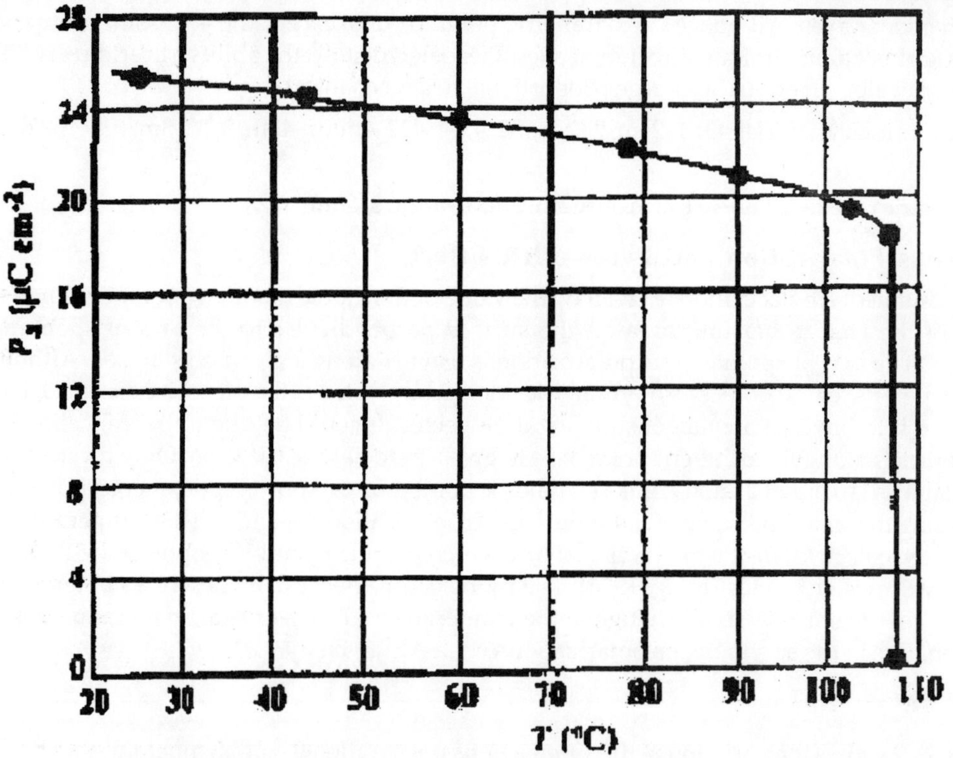

Figure 22.2: The Temperature Dependence of Spontaneous Polarization P_s for $BaTiO_3$ Ferroelectric Crystal

rapid increase in the polarization (OB). At very high field levels, the polarization reaches a saturation value (P_{sat}). The polarization does not fall to zero when the external field is removed. At zero external field, some of the domains remain aligned in the positive direction, hence the crystal will show a remnant polarization P_r. The crystal cannot be completely depolarized until a field of magnitude OF is applied in the negative direction. The external field needed to reduce the polarization to zero is called the coercive field strength E_c. If the field is increased to a more negative value, the direction of polarization flips and hence a hysteresis loop is obtained. The value of the spontaneous polarization P_s (OE) is obtained by extrapolating the curve onto the polarization axes (CE).

Curie Point and Phase Transitions

All ferroelectric materials have a transition temperature called the Curie point (T_c). At a temperature $T > T_c$ the crystal does not exhibit ferroelectricity, while for $T < T_c$ it is ferroelectric. On decreasing the temperature through the Curie point, a ferroelectric crystal undergoes a phase transition from a non-ferroelectric phase to a ferroelectric phase. If there are more than one ferroelectric phases, the temperature at which the crystal transforms from one ferroelectric phase to another is called the transition temperature. Early research work on ferroelectric transitions has been summarized by Nettleton (1970, 1971). Figure 22.4 shows the variation of the relative permittivity ε_r with temperature as a $BaTiO_3$ crystal is cooled from its paraelectric cubic phase to the ferroelectric tetragonal, orthorhombic, and

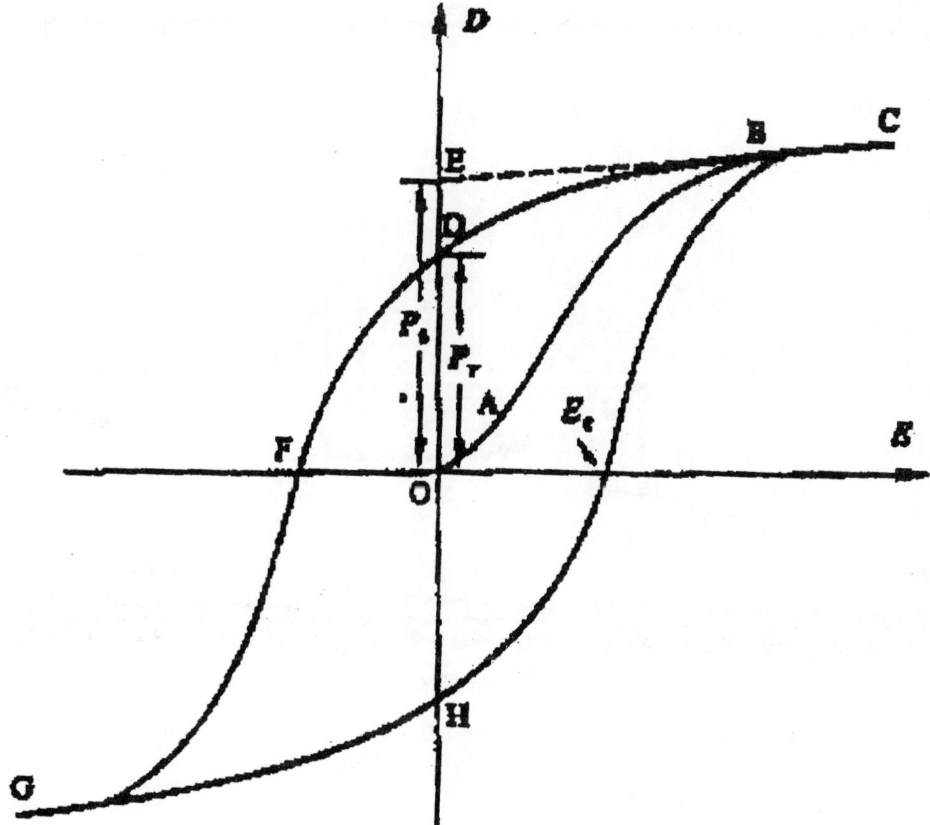

Figure 22.3: A Polarization vs. Electric Field (P-E) Hysteresis Loop for a Typical Ferroelectric Crystal (Xu, 1991)

rhombohedral phases. Near the Curie point or transition temperatures, thermodynamic properties including dielectric, elastic, optical, and thermal constants show an anomalous behavior. This is due to a distortion in the crystal as the phase structure changes. The temperature dependence of the dielectric constant above the Curie point ($T > T_c$) in ferroelectric crystals is governed by the Curie-Weiss law:

$$\varepsilon = \varepsilon_0 + C/(T - T_o) \quad (2)$$

where,

ε is the pennittivity of the material, ε_0 is the permittivity of vacuum, C is the Curie constant and T_o is the Curie temperature. The Curie Temperature T_o is different from the Curie point T_c. T_o is a formula constant obtained by extrapolation, while T_c is the actual temperature where the crystal structure changes. For first order transitions $T_o < T_c$ while for second order phase transitions $T_o = T_c$ (Jaffe *et al.*, 1971).

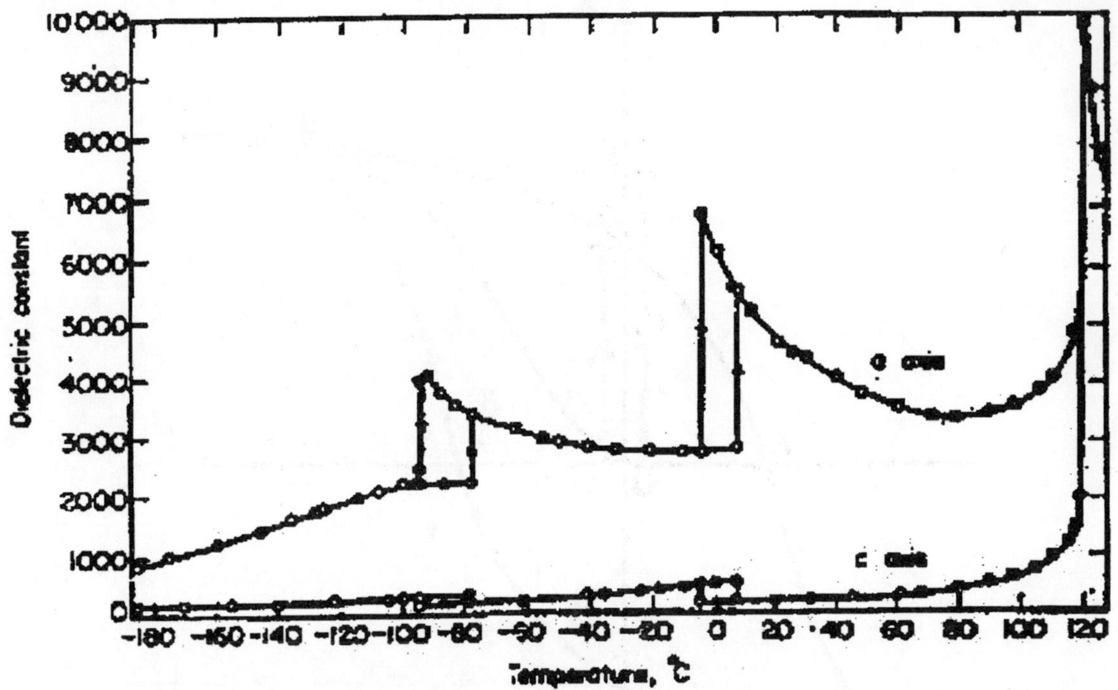

Figure 22.4: Variation of Dielectric Constants (a and c axis) with Temperature for $BaTiO_3$ (Merz, 1949)

Processing of Ferroelectric Ceramics

The final electromechanical properties of ferroelectric ceramic components greatly depend upon the processing conditions of the ceramic. Each step of processing has to be carefully monitored and controlled to get the best product.

The raw materials (metal oxides or metal carbonates) are first weighed according to the stoichiometric formula of the ferroelectric ceramic desired. The raw materials should be of high purity. The particle size of the powders must be in the submicron range for the solid phase reactions to occur by atomic diffusion.

The powders are then mixed either mechanically or chemically. Mechanical mixing is usually done by either ball milling or attrition milling for a short time. Chemical mixing on the other hand is more homogeneous as it is done by precipitating the precursors in the same container.

During the calcination step the solid phase reaction takes place between the constituents giving the ferroelectric phase. Proper calcination at the right temperature is necessary to obtain the best electrical and mechanical properties (Matsuo and Sasaki, 1965; Buckner and Wilcox, 1972).

After calcining, the lumps are ground by milling. The green bodies should have a certain minimum density before they can be sintered. The desired shape and a minimum green density can be provided by various techniques including powder compaction, slipcasting, and extrusion. The choice of the

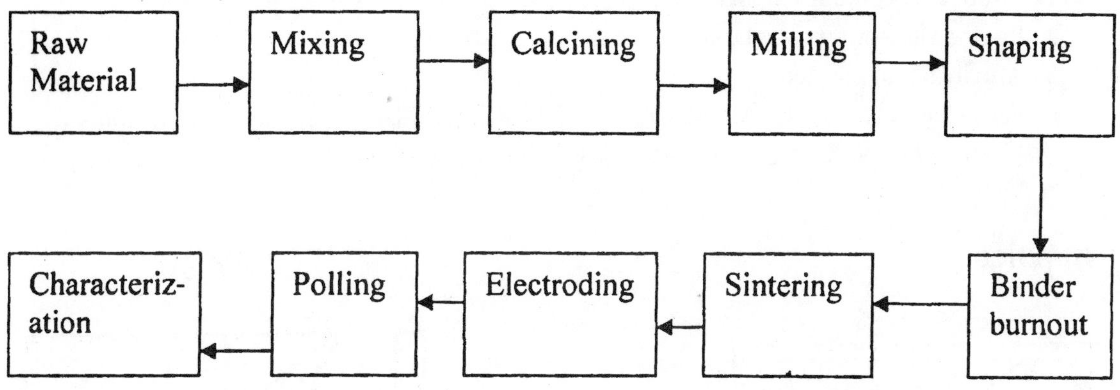

Figure 22.5: Flowchart for the Processing of Ferroelectric Ceramics

method depends on the type of powder used, particle size distribution, state of agglomeration, desired shape, and thickness of the particle.

After shaping, the green bodies are heated very slowly to between 500–600°C in order to remove any binder present. The binder burnout rate should be ≤ 1–2°C/min in order to allow the gases to come out slowly without forming cracks and blisters in the ceramic part. After the binder burnout is over, the samples are taken to a higher temperature for sintering to take place. The sintering temperature and time should be optimum for proper densification to occur without abnormal grain growth. The sintering of oxide ceramics must be carried out in an oxidizing atmosphere or in air. For lead containing piezoelectric ceramics (PZT, $PbTiO_3$, PLZT, etc.) lead loss occurs at temperatures above 800°C. In order to reduce the lead loss during sintering, the samples are kept in a sealed crucible with a saturated PbO vapor in it.

Ferroelectric Domain

The dipoles within a single domain have the same orientation. In ferroelectric ceramics with fine grain sizes (<1 µm) each grain is a single domain with the domain wall at the grain boundary. If the grain size is larger (> 1 µm) then there could be multiple domains in a single grain. As schematically shown in Figure 22.6(*a*), when the ferroelectric ceramic is cooled after sintering, it does not show any. piezoelectricity because of the random orientations of the ferroelectric domains in the ceramic. Piezoelectric behavior can be induced in a ferroelectric ceramic by a process called "poling". In this process a direct current (dc) electric field with strength larger than the coercive field strength is applied to the ferroelectric ceramic at a high temperature, but below the Curie point. On the application of the external dc field the spontaneous polarization within each grain gets orientated towards the direction of the applied field, as shown in Figure 22.6(*b*). This leads to a net polarization in the poling direction. All the domains in a ceramic can never get fully aligned along the poling axis because the orientations of the polarization is restricted by the symmetry.

Applications

Ferroelectric materials have been applied to a large field of applications because of their excellent dielectric, piezoelectric and ferroelectric properties. Some of the important applications are:

1. Multi layer ceramics capacitors.
2. Engine electronic control unit.
3. Anti lock brake system.
4. Ferroelectric memories: As DRAM, SRAM, EEROM, FERAM, in the microwave domain.
5. Electro-Optic application: Ferroelectric thin film wave guides, ferroelectric thin film optical memory display.

Un poled

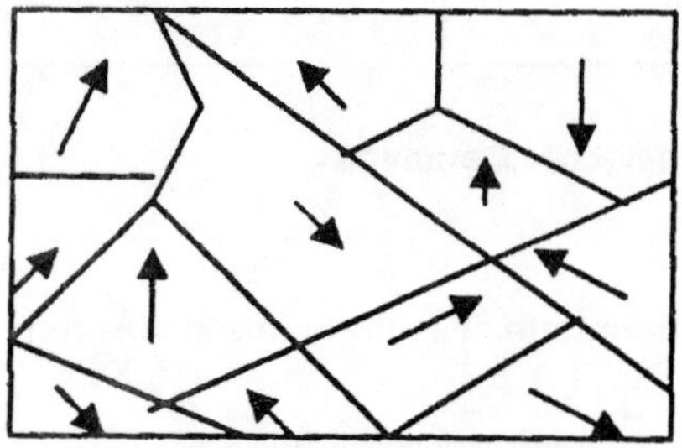

Poled

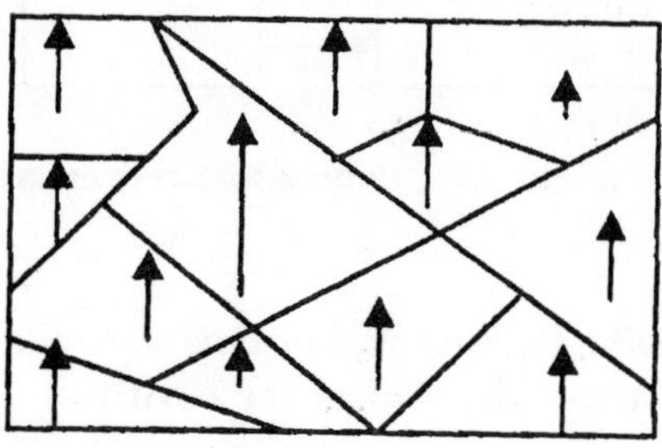

Figure 22.6: Schematic of the Poling Process in Piezoelectric Ceramics
(*a*) In the absence of electric field the domains have random orientation of polarization;
(*b*) The polarization within the domains aligns in the direction of the applied field.

Applications of Piezoelectric Ceramics

1. Medical ultra sound application.
2. Sensors and Actuators
3. Gas Ignitors.
4. Displacement transducers.
5. Accelerometers: Accelerometers is a device which gives an electrical out put proportional to the acceleration.
6. Piezoelectric transformer: Low voltage to high voltage transformer can be done by using a piezoelectric plate.

Summary

Ceramic materials and single crystals showing ferroelectric behavior are being used in many applications in electronics and optics. In this article, an effort has been made to introduce the basic principles governing ferroelectricity and some of their properties. The processing of ferroelectric ceramics has also been described. Finally, a number of applications of ferroelectric ceramics exploiting properties that are an indirect consequence of ferroelectricity, such as dielectrics, piezoelectricity, and pyroelectricity, are described.

References

Batthais, B.T. and Hippel A. von, 1948. *Phys. Rev.*, 73: 1378.

Buckner, D.A. and Wilcox, P.O., 1972. *Am. Ceram. Soc. Bull.*, 51: 218.

Burfoot, J.C., 1967. *Ferroelectrics*. Van Nostrand, New York.

Cady, W.G., 1962. *Piezoelectricity*. Dover Publications, New York.

Deri, M., 1969. *Ferroelectric Ceramics*. Gordon and Breach, New York.

Hippel A. von., 1950. *Rev. Modem Phys.*, 22: 221.

Jaffe, B., Cook, Jr. W.R. and Jaffe, H., 1971. *Piezoelectric Ceramics*, Academic Press, London.

Janas, V.F. and Safari, A., 1995. *J. Am. Ceram. Soc.*, 78: 2945.

Lang, S.B., 1974. *Source Book of Pyroelectricity*. Gordon and Breach, New York.

Matsuo, Y. and Sasaki, H., 1965. *J. Am. Ceram. Soc.*, 48: 289.

Megaw, R.D., 1957. *Ferroelectricity in Crystals*. Methuen, London.

Merz, W.J., 1949. *Phys. Rev.*, 76: 1221.

Merz, W.J., 1953. *Phys. Rev.*, 91: 513.

Mitsui, T., Tatsuzaki, I. and Nakamura, E., 1976. *An Introduction to the Physics of Ferroelectrics*. Gordon and Breach, London.

Nettleton, R.E., 1970. *Ferroelectrics*, p. 1, 3, 87, 93, 111, 121, 127, 207, 221.

Nettleton, R.E., 1971. *Ferroelectrics*, p. 2, 5, 77, 93.

Newnham, R.E., 1975. *Structure–Property Relations*. Springer Verlag, New York.

Nye, J.F., 1990. *Physical Properties of Crystals*. Clarendon Press, Oxford.

Xu, Y., 1991. *Ferroelectric Materials and their Applications*, North Holland, Amsterdam.

Chapter 23

Qualitative Analysis of Small-Signal Amplifier Circuits Configured by Coupling of BJTs

Satyendra Nath Tiwari, Kranti Kimesh Dubey, Jitendra Singh and Sachchida Nand Shukla*

Department of Physics and Electronics, Dr. R.M.L. Avadh University, Faizabad – 224 001

ABSTRACT

CE small signal RC coupled amplifier is considered as basic building block for electronic circuits. Imposing minor modifications in CE amplifier circuit, many significant amplifiers can be developed. We made the qualitative analysis of few small signal amplifier circuits developed by replacing the transistor of CE amplifier by coupled transistors in form of Darlington pair and two or more transistors in series or parallel. The dependence of voltage gain and bandwidth on various biasing components is exhaustively studied.

Keywords: Small signal amplifiers, Basic electronics circuits

Introduction

Most of the analog, digital or mixed analog-digital electronic systems generate weak output signals in milivolts or microvolts range, which can not be reliably processed to perform any useful

* Corresponding Author: E-mail: sachida_shukla@yahoo.co.in

function. The bipolar junction transistor (BJTs) can act as good amplifier when operated in linear region and popularly proved trustworthy to amplify weak signals of milivolts or microvolts range (Rebert, 2002). The performance of an amplifier can be determined in terms of frequency response, voltage gain, current gain or power gain and significantly depends on active and passive components used in the circuit for biasing purpose. The most popular BJTs amplifier recognized in audible frequency range is the high-fidelity Common Emitter RC coupled amplifier that produces enlarged output at the same frequency of input (Rebert, 2002; Bell, 2002). We have selected CE RC-coupled amplifier circuits for the present studies, due to its the wide and popular use of in electronics. Similarly, we selected another useful amplifier circuit that can be obtained by coupling two transistors in form of a Darlington pair. ACC/CE Darlington pair amplifier circuit possesses higher current gain with higher input resistance and lower output resistance than that of a single stage Emitter Follower (Grob, 1986). We also thought that just as resistors are connected in parallel or series to meet different current and voltage needs, transistors can also be connected in parallel or series for a required application (Grob, 1986; Zherebstov, 1988). Proceeded with this philosophy, we further selected single stage CE amplifier with two identical transistors in series or parallel.

In fact, an excessive of books and research papers are available to explore the model behaviour, characteristics and amplifier action of BJTs with or without analysis through simulation software (Rebert, 2002; Rashid, 2004). It is also a fact that only a little effort has been carried out to discuss the dependence of maximum voltage gain on passive biasing components of common emitter BJT amplifier when configured with one transistor, Darlington pair, two transistors in series and two transistors in parallel (Grob, 1986; Zherebstov, 1986). So, we dedicated the present studies to explore the dependence of voltage gain and bandwidth on various biasing components for selected amplifier circuits.

Experimental Setup and Observations

For the present studies, we have selected four different amplifier configurations and exhaustively studied their voltage gain performance with respect to R_L, R_E, R_C, V_{CC}, C_E, number of transistors in series or parallel and variation of bandwidth with supply voltage V_{cc}. The respective amplifier circuits are shown in Figures 23.1–23.4. Figure 23.1 is the conventional small signal CE amplifier. Figure 23.2 shows the CC/CE Darlington pair amplifier circuit while Figures 23.3 and 23.4 shows the CE amplifier circuits with two identical BJTs in series and parallel respectively. The bipolar junction transistor used here in for the present studies is a NPN transistor Q2N2222 (β=255.9) as active element while R_1=47KΩ, R_2=5KΩ, R_S=500Ω, R_C=10KΩ, R_E=2KΩ, R_L=10KΩ, C_1=1μF, C_2=1μF and CE= 10μF as passive biasing elements. All the circuits are biased with + 15V d.c. supply voltage V_{CC}. The observations are made with 1mV AC signal source at 1KHz frequency through PSpice simulation software (Rashid, 2004) (Student version 9.2).

Results and Discussion

Figure 23.5 shows how the maximum gain changes with increasing number of transistors in series and parallel for circuits of Figures 23.3 and 23.4. It is clear that the maximum gain and bandwidth (not shown in figure) reduces rapidly with increasing number of transistors in series (for amplifier of Figure 23.1 bandwidth is 1199 KHz and for amplifier of Figure 23.3 with 6 transistors in series it is 558 KHz). On the other hand, maximum voltage gain increases only a little with increasing number of transistors in parallel. It seems here that the series connection equally divide V_{CC} across the identical transistors in series due to which 'Q1' acts as voltage dropping resistor for collector of 'Q2' (circuit of Figure 23.3) that ultimately causes the reduction in voltage gain (Zherebstov, 1988). Similarly, the

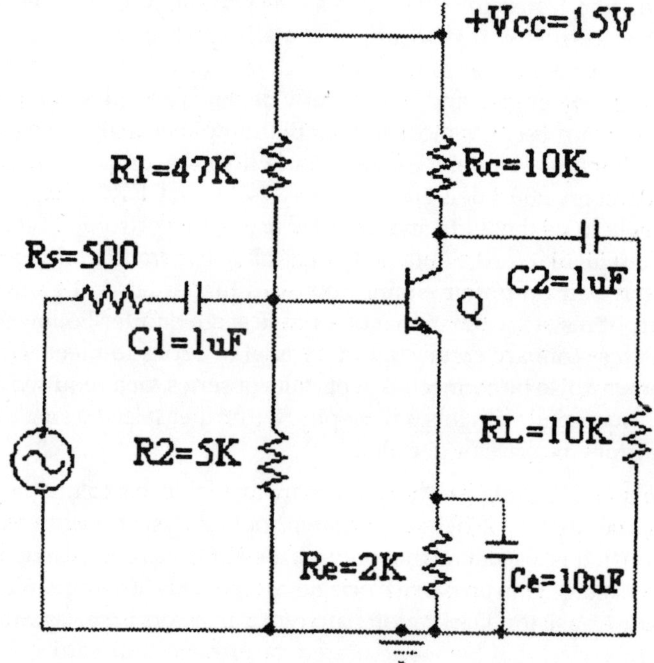

Figure 23.1

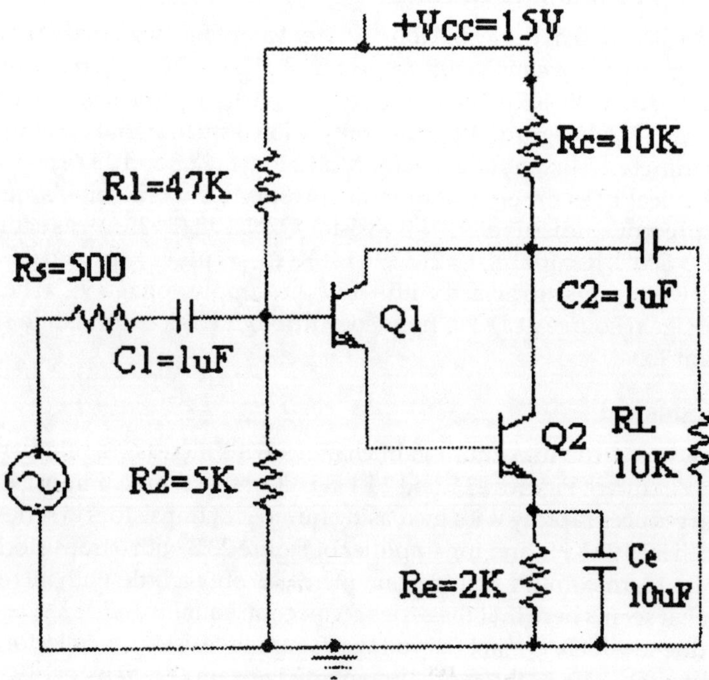

Figure 23.2

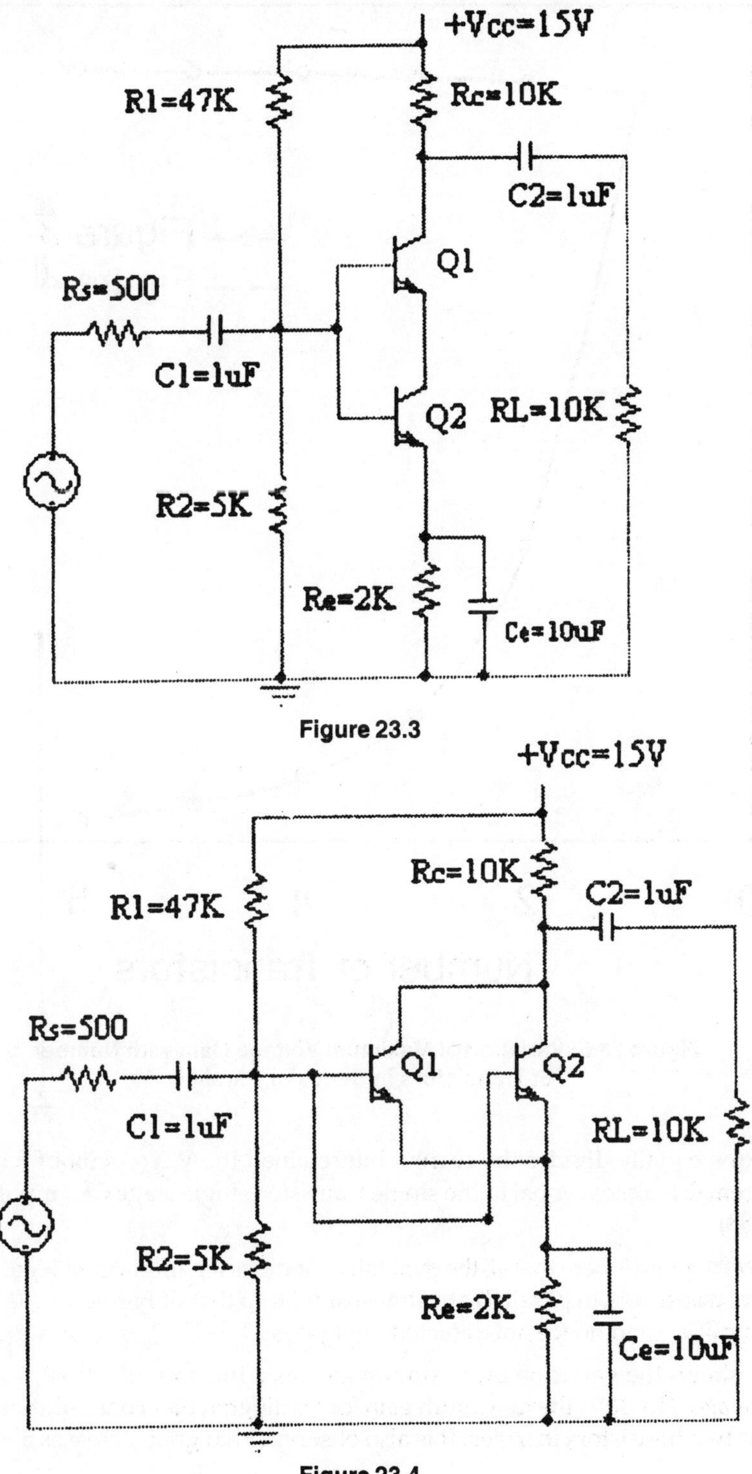

Figure 23.3

Figure 23.4

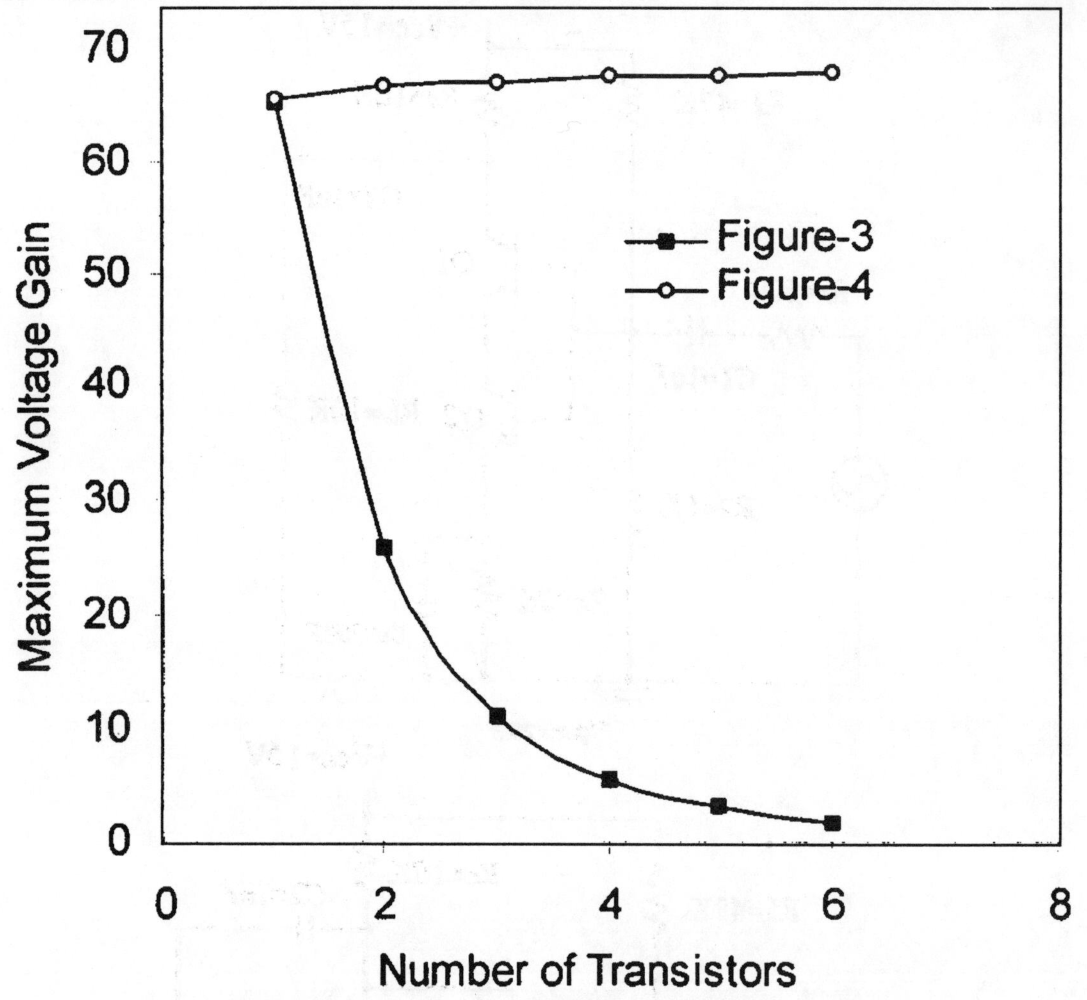

Figure 23.5: Variation of Maximum Voltage Gain with Number of Transistors in Series or Parallel

parallel connection equally divides the current but retained the V_{CC} (circuit of Figure 23.4), due to which the gain remains almost equal to the single transistor-single stage CE amplifier of Figure 23.1 (Zherebstov, 1988).

One noteworthy point is here that all the qualitative findings for amplifier of Figure 23.4 (configured with two or more transistors in parallel) are almost similar to that of Figure 23.1 (single stage-single transistor CE amplifier) and hence not depicted in Figures 23.6–23.11 as separate graphs.

Figure 23.6 shows the variation of maximum gain as a function of DC supply voltage V_{CC}. At lower supply voltages (10–18V) the maximum gain for Darlington pair configuration is observed to be less than that for two transistors in series. It is also observed that gain increases almost linearly with

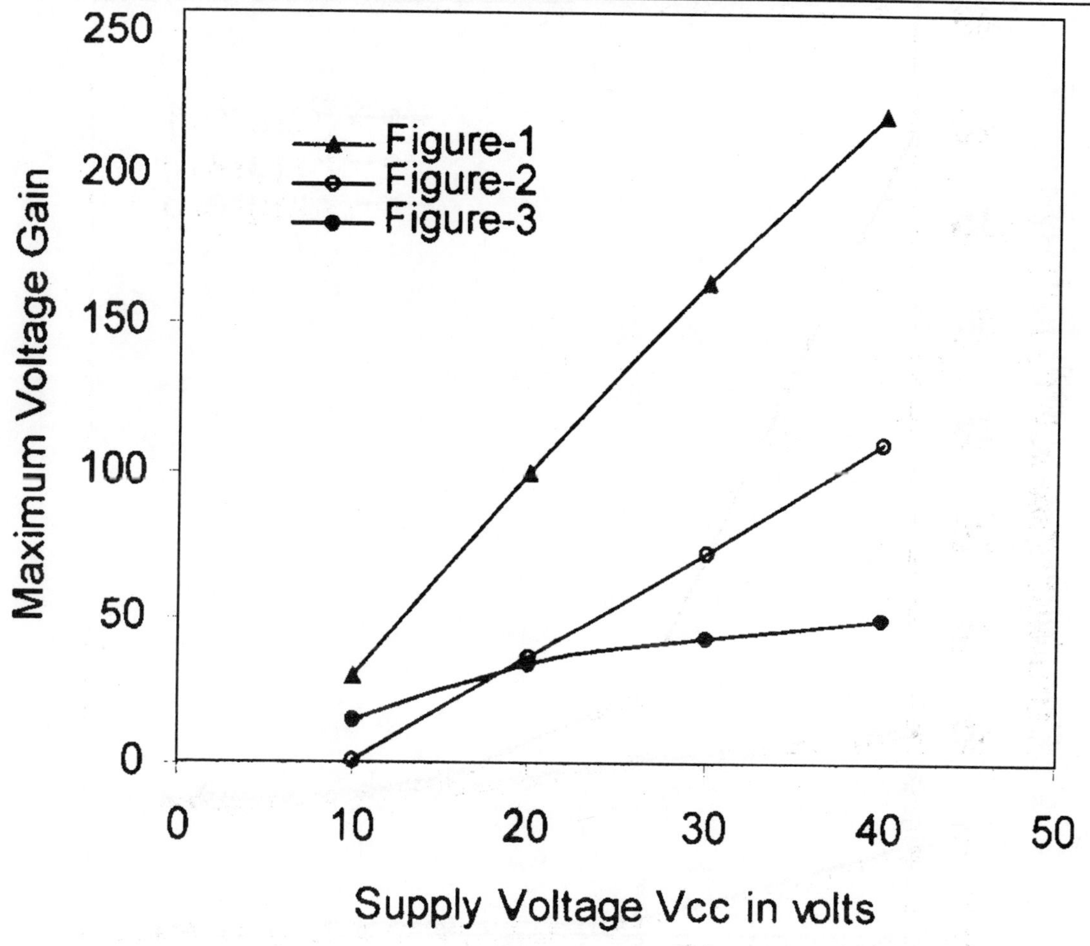

Figure 23.6: Variation of Maximum Voltage Gain with Supply Voltage V_{CC}

increasing supply voltage. Which is well in accordance of the usual behaviour of amplifiers (Rebert, 2002; Bell, 2002).

The Figures 23.7 and 23.8 shows the variation of bandwidth with load resistance R_L and supply voltage V_{CC} respectively. It is clear that the bandwidth decreases rapidly for rising values of R_L and V_{CC} in lower region of R_L and V_{CC} and at higher values of R_L and V_{CC} it acquires almost sustained level. It is also found that above to the 35 volts (Figure 23.8) the bandwidth for amplifier of Figure 23.3 is observed maximum than that of Figures 23.1 and 23.2.

Variation of maximum voltage gain with load resistance R_L is plotted in Figure 23.9. We observed that gain value rises up continuously at lower values of R_L up to 100KΩ while for higher R_L values it approaches towards a saturation level. It is also clear that the basic nature of the variation of maximum voltage gain with R_L for all the amplifier circuits are similar and only the performance values are getting changed. It is worth mentioning here that the maximum gain for amplifier of Figure 23.3 is

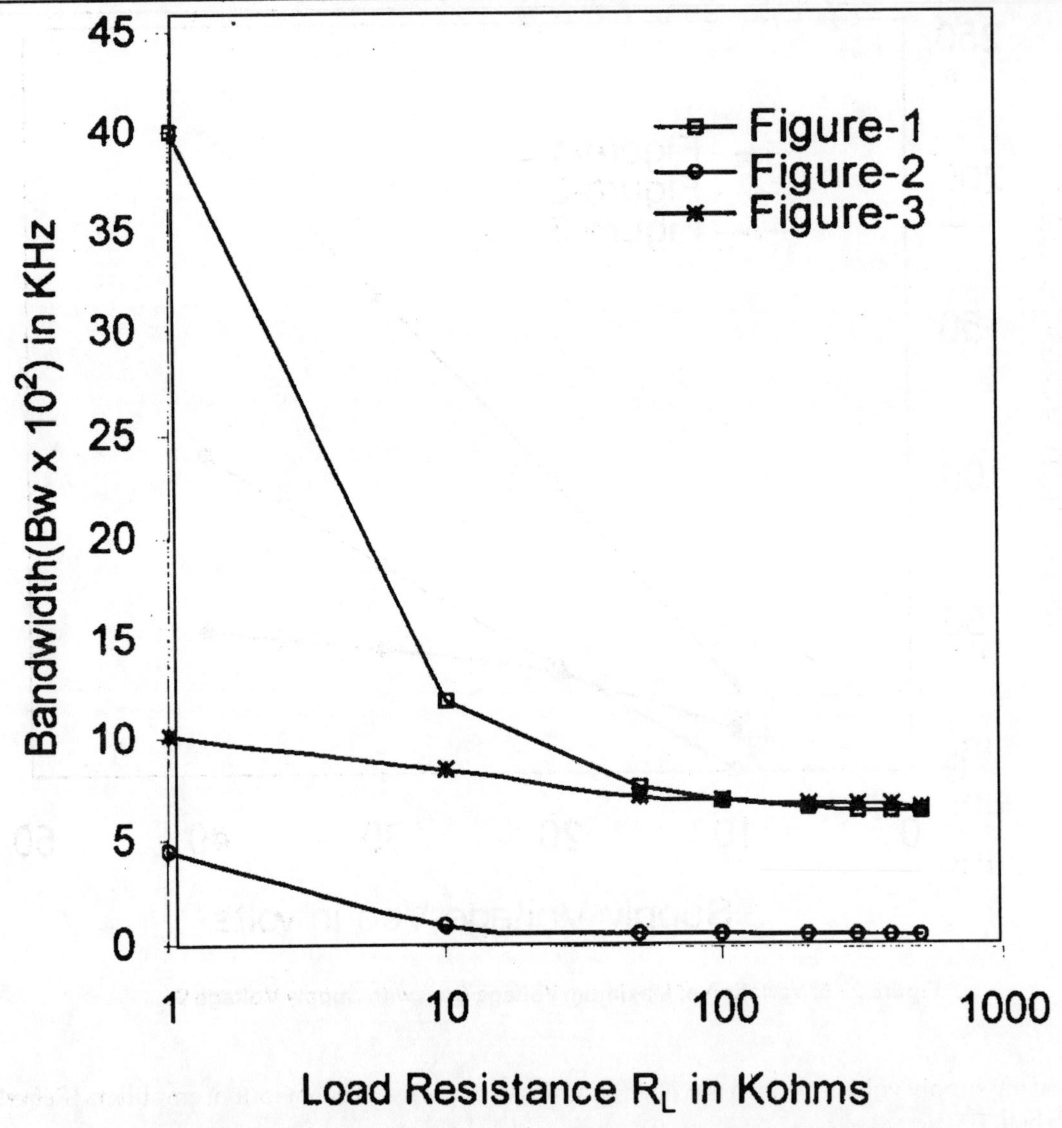

Figure 23.7: Variation Bandwidth with Load Resistance R_L (on Log Axis)

always higher than amplifier of Figure 23.2 at every value of load resistance. We also observed that in saturation region the voltage gain for a certain value of load resistance R_L increases with collector resistance R_C (not shown in figure) for all amplifiers. Taking the CE current gain β, transconductance g_m, collector resistance R_C, load resistance R_L and input series resistance R_S into account, the voltage gain of (CE) BJT amplifier can be expressed as:

$$A_V \sim \frac{\beta g_m (R_C \| R_L)}{\beta + g_m R_S}$$

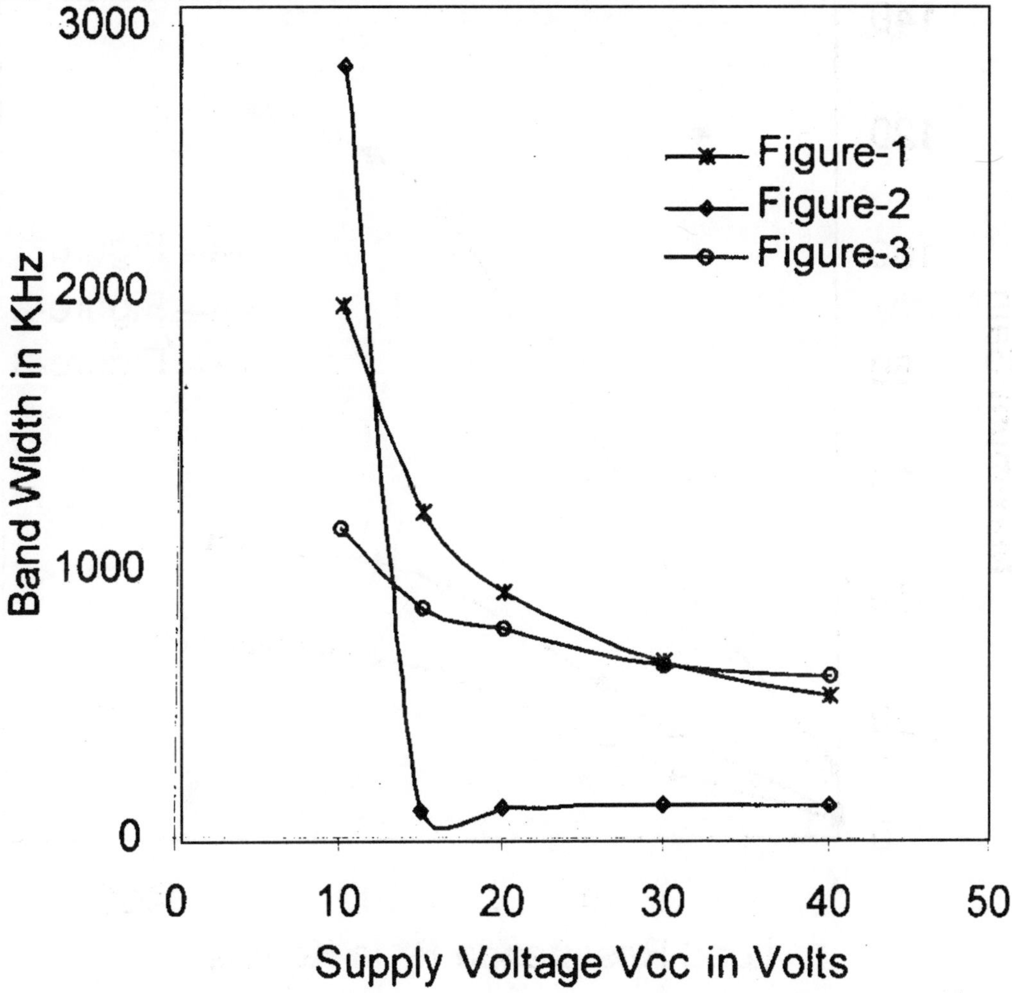

Figure 23.8: Variation of Bandwidth with Supply Volatge V_{CC}

From this equation it is clear that for a certain BJT, β and g_m are constant and hence the voltage gain A_V of common emitter amplifier for a certain value of R_C and R_S increases with R_L. The experimental nature of the curve is consistent with theoretical prediction (expression for A_V) that, voltage gain A_V should tends to saturate at large value of R_L (Motayed *et al.*, 2001; Sayed, 2002).

Variation of maximum voltage gain with collector resistance R_C is depicted in Figure 23.10. It is clearly indicated in Figure 23.10 that the maximum voltage gain increases with increasing collector resistance to a certain value of R_C for all cases and then falls down. For amplifier of Figure 23.1 it rises up to 25KΩ value of R_C to a level of 100 and ceased at 50KΩ while for circuits with two transistors it rises up to 50KΩ to a maximum value of 40 and there after start decreasing and ceased at 100KΩ. Thus the better gain performance for all amplifiers is always obtained for R_C <50KΩ.

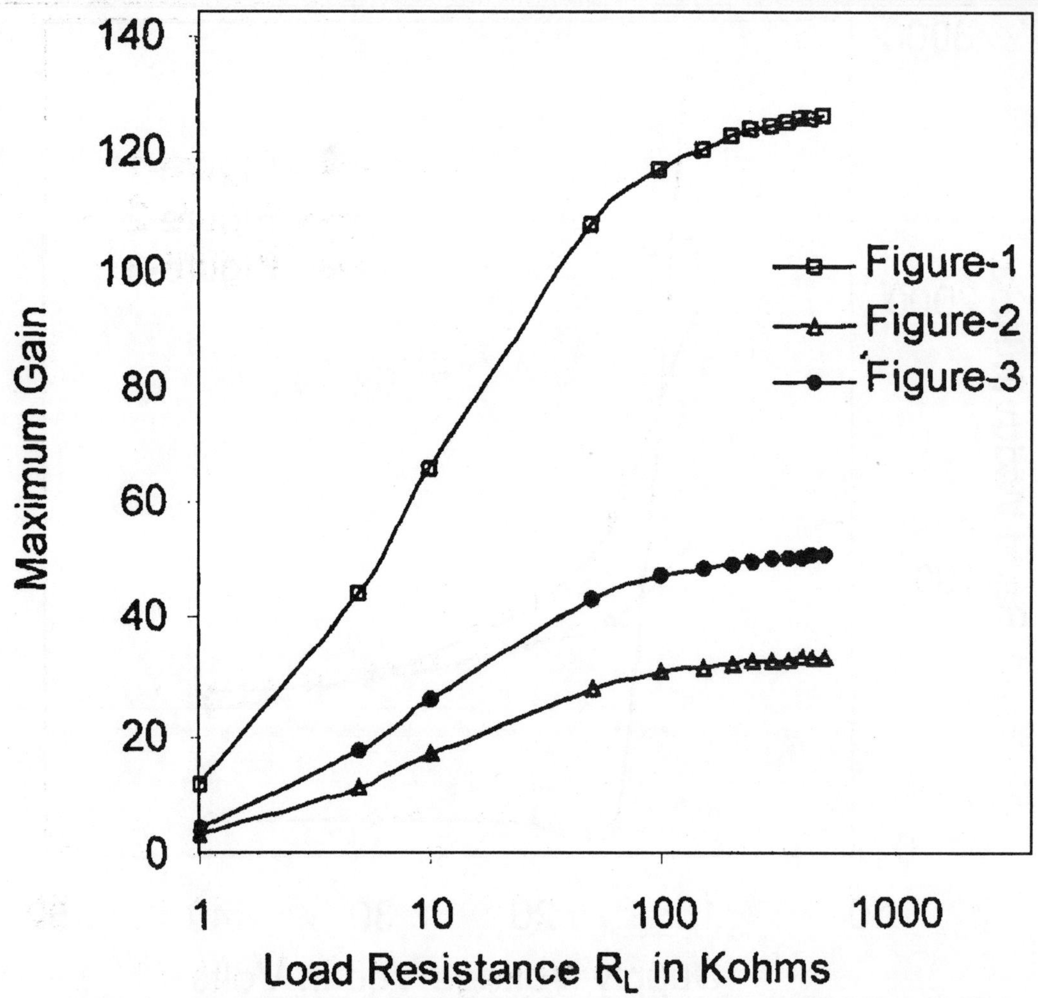

Figure 23.9: Variation of Maximum Voltage Gain with Load Resistance R_L (on Log axis)

In fact the voltage drop across the collector emitter junction can be mentioned as $V_{CE} = V_{CC} - I_C$, where I_C is the d.c. Collector current defined as $I_C = V_{CC}/R_C$. Thus when collector resistance R_C is increased, V_{CE} decreased, with this Q-Point change from its original position (Motayed *et al.*, 2001; Sayed, 2002). This led I_C to decrease the trans-conductance g_m, and hence voltage gain A_V. In our investigation of the effect of Collector resistance R_C and load resistance R_L on voltage gain clearly demonstrates that neither very large R_C nor very large R_L can provide the highest voltage gain of common emitter (CE) BJT amplifiers.

It is observed from Figure 23.11 that voltage gain has maxima at R_E=1KΩ and R_E=0.5KΩ amplifiers of Figures 23.1 and 23.2 respectively. On the other hand the voltage gain decreases slowly for amplifier of Figure 23.3 and has a maximum value at 0.01KΩ, which is lower than that of amplifiers of Figures

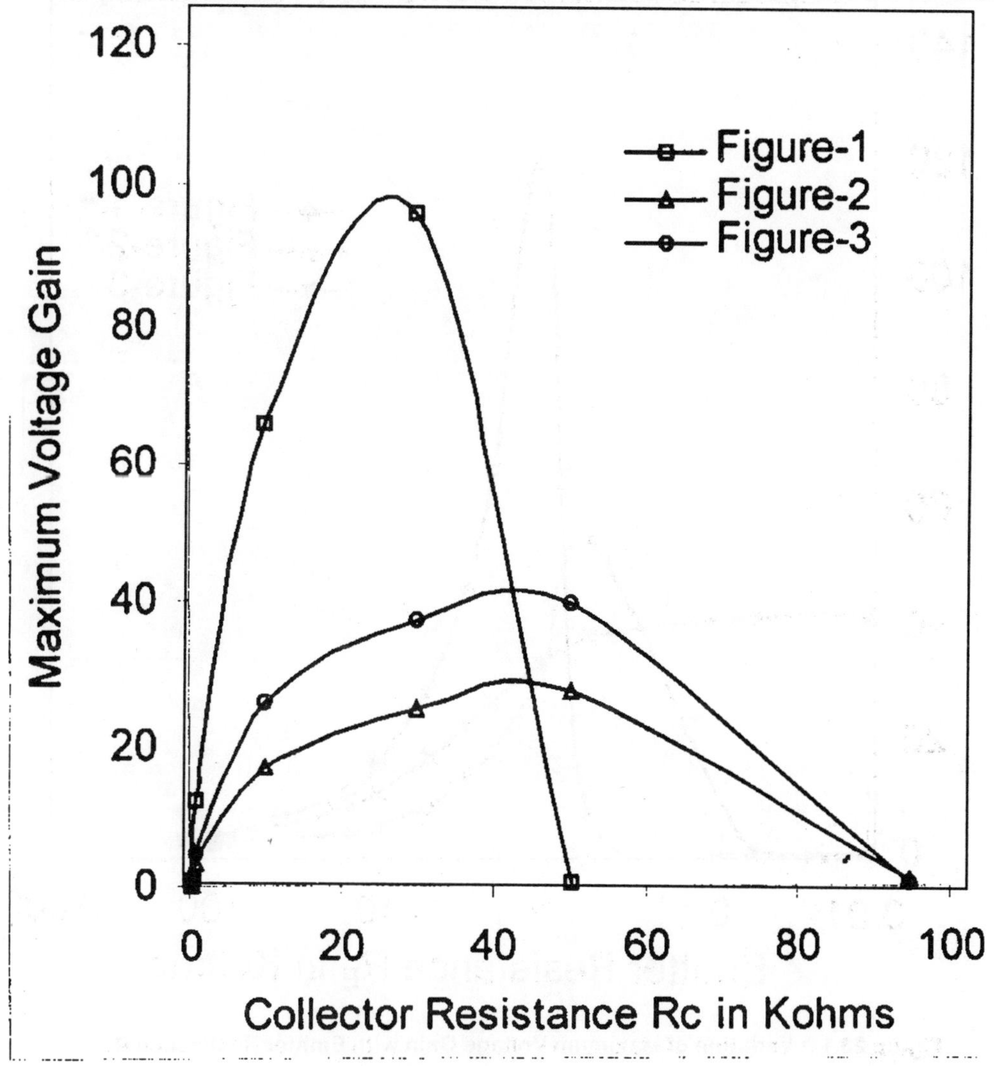

Figure 23.10: Variation of Maximum Voltage Gain with Collector Resistance R_C

23.1 and 23.2. Thus, the optimum value of RE should be less than 1KΩ for all the amplifiers to produce better gain performance.

We have also verified the variation of bandwidth as a function of C_E for all the amplifiers. We observed that on changing the value of C_E from 1uF to 500uF, the change in bandwidth is not very significant and voltage gain remains almost unaffected (not shown in figure). We observed that bandwidth changes from 1190 KHz to 1210 KHz for amplifier of Figures 23.1 and 23.4 while for amplifier of Figure 23.2 it changes from 102.06 to 102.12 KHz and for amplifier of Figure 23.3 it varies from 846 to 848 KHz for aforesaid values of C_E.

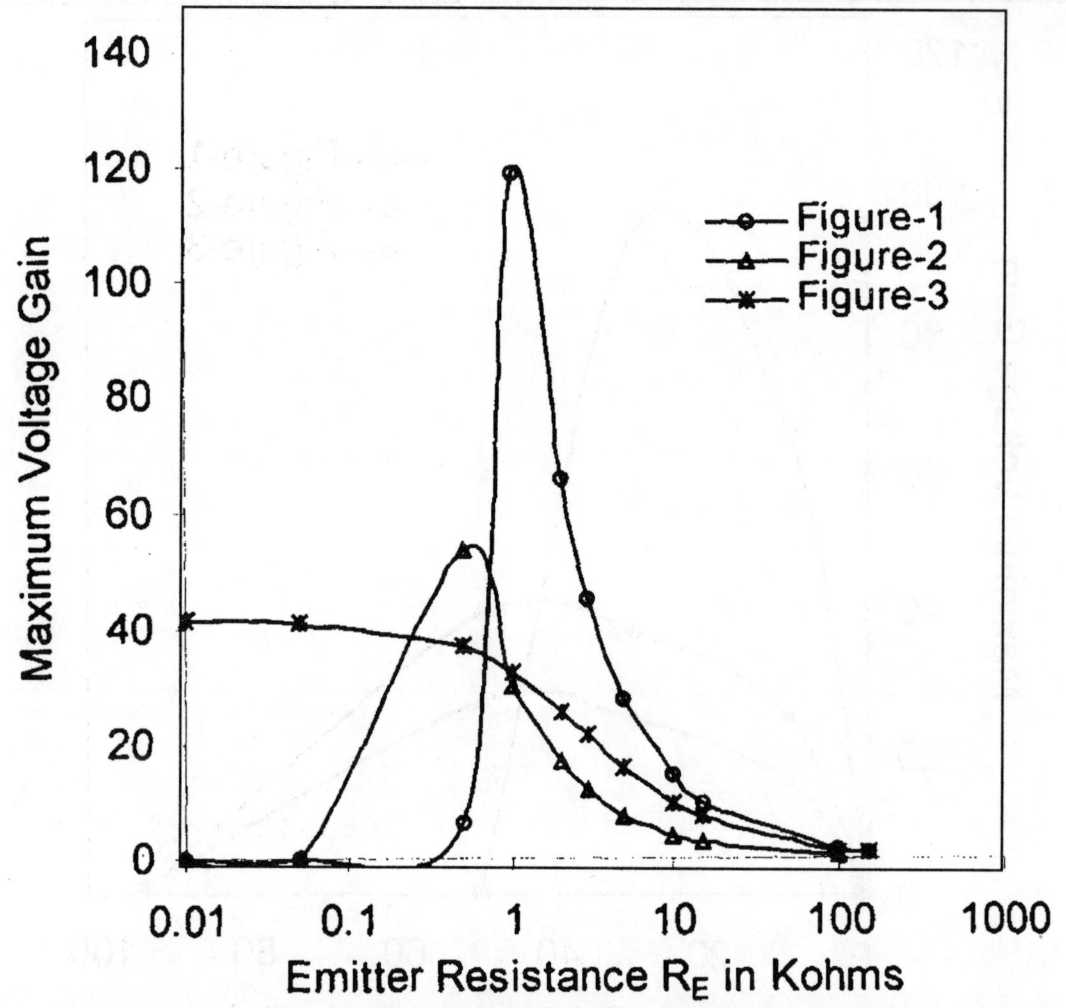

Figure 23.11: Variation of Maximum Voltage Gain with Emitter Resistance R_E

Conclusions

1. Increasing number of transistors in series in CE amplifier provides lowering of gain as well as bandwidth.
2. Maximum voltage gain increases and bandwidth decreases (for all the amplifiers) for rising values of V_{CC}.
3. Optimum values of biasing resistance R_C, R_E and R_L for better gain performance should always be less then 50KΩ.

References

Bell, David A., 2002. *Electronic Devices and Circuit*, 3rd Edn. Prentice Hall of India, p. 687–717.

Grob, Bernard, 1986. *Electronic Circuits and Applications*. McGraw Hill, p. 47–48.

Motayed, Browne T. E., Onuorah A.I. and Mohammad S.N., 2001. *Solid State Electronics*, 45: 325–333.

Rashid M.H., 2004. *Introduction to PSpice Using Or CAD for Circuits and Electronics*. Pearson Education, p. 210.

Rebert L. Boylestad and Louis Nashelsky, 2002. *Electronic Devices and Circuit Theory*. Pearson Education Asia, p. 389–396, 461–483, 627–633.

Sayed El Ahl A.M.H., Fahmi M.M.E. and Mohammad S.N., 2002. *Solid State Electronics*, 46: 593–595.

Zherebstov, I., 1988. *Basic Electronics*. Mir Publishers, Moscow, p. 46–48.

Chapter 24

Development of New Iterative Methods for Performing Binary Square Rooting Rapidly by Utilizing Multiplication

Subhash Kumar Sharma[1], L.K. Singh[1] and B.S. Rai[2]
[1]Department of Physics and Electronics, Dr. RML Avadh University, Faizabad
[2]Department of Electronics and Telecommunication, M.M.M.E. College, Gorakhpur

ABSTRACT

In this article, approaching has to be made, the development of new iterative methods of performing Binary square rooting rapidly which utilize multiplication. Each algorithm is considered for convergence rate, efficiency, and implementation. The most typical and efficient algorithm which utilizes multiplication, N algorithm, is introduced for the purpose of comparison with the new algorithms. The effect and importance of the initial approximation is considered and it is attractive to develop an iterative procedure to compute square root using multiplication.

Keywords: *Initial approximation, Newton-Raphson iteration scheme, Normalization technique, Pipe lining, Quadratic convergence, Square root.*

Introduction

The increasing use of computers in scientifically oriented real time applications, such as process control aerospace navigation and guidance, etc. has placed new demands on computational speed. Therefore, the frequently used software arithmetic operations such as division, square root, and other possible functions using a polynomial evaluation have to be performed in the hardware. In this chapter approaching has to be made concerned with generating and implementations the square root function, using new techniques that make use of fast multipliers found in large fourth generation scientific machines (Anderson *et al.*, 1967; CDC, 1965). Most general purpose scientific machines such as the CDC600, SDS930 use programmed subroutines to evaluate square root. Although an order of magnitude increase in speed is possible in this case, the cost of the additional hardware required can be prohibitive because of the complex sequencing that is needed in the square root generation. It is felt that some where between these extreme lies a host of computational techniques that reduce execution time by mailing use of the inherent parallelism of the algorithms. Hopefully they could also decrease the cost of additional equipments by maximum utilization of already existing hardware.

Classification of Square Rooting Methods

Direct Methods

These methods are based on the binary equivalent of the well known paper-and pencil methods to evaluate square root with decimal numbers. Among the several variations of this type of algorithm, the most efficient is the one that makes use of the non-restoring technique. The computation can be quite rapid, nearly as fast as division. It is limited in the same way as division, however, in that it must be processed one bit at a time and thus is relatively slow for a large operand.

Algorithms Based on Normalization Technique

There are two kinds of normalizations (Barnes *et al.*, 1968), one kind is continued-sum normalization and the other kind is continued-product normalization. Both of these normalizations are usually performed bit-by-bit and they become much faster by using redundant recoding since, the probability for the recoding factor to be 0 becomes 2/3 (Barnes *et al.*, 1968; Metze, 1965). In the case of square rooting by bit-by-bit normalization, continued-sum normalization. It takes slightly more than M/3 additions, where M is the length of the mantissa. This can be very fast for machines with small word lengths, but tends to become slow for those with larger word lengths due to the property of linear convergence. On the other hand, continued product normalization can also be performed by multiple bits. In division, the very efficient algorithm (Goldschmidt's division algorithm) belongs to this category (Andeson, 1967).

Algorithms Based on Newton-Raphson Iterative Formula

This class of algorithms contains the variations of the original Newton-Raphson iterative formula:

$$B_{k+1}=½(B_k+N/B_k) \tag{1}$$

where, $\lim_{k\to\infty} R_k=1/\sqrt{N}$

Which uses multiplication instead of division.

Among these types of algorithms the most efficient is the one that uses the following iterative formula and is called the N algorithm for convergence.

$$R_{k+1}=R_k/2(3-NR^2_k) \tag{2}$$

where, $\lim_{k\to\infty} R_k=1/\sqrt{N}$

After final iteration, $\sqrt{N}$ is obtained by one more multiplication $N^* 1/\sqrt{N} = \sqrt{N}$. This has a property of quadratic convergence which is important as far as speed is concerned, especially for relatively large word lengths.

We also present two new algorithms, R and F. R uses the following iterative formula:

$$R_{k+1} = R_k(2-B_k{}^*R_k)$$

where, $\lim_{k\to\infty} R_k = 1/\sqrt{N}$

$$B_{k+1} = ½(B_k+N^*R_{k+1}) \quad (3)$$

where, $\lim_{k\to\infty} B_k = 1/\sqrt{N}$

R is not as fast as N and R has the property of convergence whose rate is between linear and quadratic. But, R has the interesting capability of performing division by the same hardware and control circuitry. F uses the following iterative formula:

$$B_{k+1} = ½(B_k+N/B_k) = B_k + ½B_k(N-B^2{}_k) = B_k + f_p(B_k)(N-B^2{}_k)$$

where, $\lim_{k\to\infty} B_k = 1/\sqrt{N}$

and $f_p(x)$ means one-half reciprocal of x rounded to p bits and is achieved by the rounded-reciprocal generator. So F is slight variation of the original Newton-Raphson formula using rounded reciprocal instead of exact reciprocal. This has the property of linear convergence so that it can be slow for large word lengths, but may be better suited for smaller ones than N due to the simple form of its iterative formula.

Algorithm N

Properties of N

Since, this algorithm is well known, iterative formula:

$$R_{k+1} = R_k(3/2-N/2^*R^2{}_k) = R_k/2(3-N^*R^2{}_k) \quad (4)$$

where, $\lim_{k\to\infty} R_k = 1/\sqrt{N}$ and $N^*\lim_{k\to\infty} R_k = \sqrt{N}$

Now defining the relative error as

$$E_k = (R_k-1/\sqrt{N})/(1/\sqrt{N})$$

It can be shown that relative error function of R_k is as follows:

$$E_k = -½E^2{}_k(E_k+3) \quad (5)$$

E_k is independent of operand N. also the maximum relative errors can be obtained by the following iterative formula:

$$E_{mk+1} = -1/2\, E^2{}_{mk}\,(3-E_{mk}) \quad (6)$$

where, E_{mk} is the maximum of E_k and is a slightly slower rate of convergence, which is a very important property as far as speed is concerned.

The test of convergence rate was accomplished by calculating relative error after each iteration, using the iterative formula (5) under various initial approximations. The value $\{-\log_2(E_{mk})-1\}$ corresponds to the number of significant bits obtained after each iteration k. Actually the test confirms the fact that N has quadratic and monotonous convergence, and therefore, this algorithm converges much faster than the other two. It is obvious that the efficiency of N depends to some extent upon the initial approximation like the others. Since it usually takes more time or more hardware to increase the accuracy of initial approximations, optimization must be considered on this point.

Implementation

Two kinds of implementations are possible throughou, it is assumed that division is performed by Goldschmidt's division algorithm. Let us define α, basic hardware, as composed of two lookup tables, one multiplier, one adder, one complement or and one shifter. Then, implementation A requires {α–adder + 3's complementor}, while implementation B requires α plus one multiplier. On the other hand, A takes almost three multiplications for each iteration, while B requires almost two multiplications plus one addition for each iteration.

Algorithm F

Properties of F

In some of the fast computers an iterative multiplication is used to perform division. The performance depends to some extent upon the accuracy of initial approximation. This is usually obtained by use of a lookup table. Algorithm F assumes the availability of such a lookup table. Iterative formula:

$$B_{k+1} = B_k + f_p(B_k)(N-B^2_k) \tag{7}$$

where, $f_p(x)$ is one-half the reciprocal of x rounded to p bits, and $\lim_{k \to \infty} B_k = \sqrt{N}$

The basic idea of F is to use a rounded reciprocal instead of an exact reciproca1. But it still convergeses linearly, not quadratically.

Implementation

One possible implementation is Defining a as in the previous section, it takes exactly a basic hardware. And requires almost two multiplication plus two additions for each iteration.

Algorithm R

Properties of R

This algorithm uses the Newton-Raphson iteration for the reciprocal to replace division for each iterative approximations (one for obtaining reciprocal and the other for obtaining √N) equally for each iteration.

Iterative Formula

$$R_{k+1} = R_k(2-B_k{*}R_k)$$

where, $\lim_{k \to \infty} R_k = 1/\sqrt{N}$

$$B_{k+1} = \tfrac{1}{2}(B_k + N^*R_{k+1}) \tag{8}$$

where, $\lim_{k \to \infty} B_k = 1/\sqrt{N}$

This algorithm does not possess the property of monotonous convergence. The iterative formula for the relative error of Bk is as follows:

$$E_{k+1} = -\tfrac{1}{2}(E^2_{k-1}{}^*E_k + E^2_{k-1} - 4E_{k-1}E^2k - 6E_{k-1}E_k + 8E^2_k + 4E^3_k) \tag{9}$$

Here, E_k is independent of operand N also the test of convergence rate was accomplished by calculating the relative error after each iteration, using formula (9) for various initial approximations. R_0 and B_0 were assumed to have the same accuracy of initial approximation in these examples. Actually, it does not show the property of consistent convergence, and the convergence rate is between linear and quadratic. One feature of R is that division can be performed using no additional hardware and control circuitry because the Newton-Raphson iteration for the reciprocal is included *i.e.*

$$R_{k+1} = R_k(2 - D^*R_k) \tag{10}$$

where, $\lim_{k \to \infty} R_k = 1/D$

Implementation

One possible implementation takes exactly a basic hardware and three multiplication and one addition for each iteration.

Comparison of Algorithms R, F, and N

The required time per iteration and required hardware for the implementation of Algorithms R, F, and N. Algorithm N in implementation B takes the least time per iteration, but it requires one more multiplier. If N is implemented as A, *i.e.* if N uses no more hardware than the other two, then F takes the least time per iteration the number of iterations for various initial approximations (4.0116, 6.0589 and 8.3803 corresponding to 3,5-, and 7-bit precise approximation). The necessary iterations to get a result of at least M bits of precision versus initial approximation. In any case, N turns out to take the least number of iterations.

1. If N is implemented in B *i.e.*, If N is implemented by using one more multiplier than the other two, then it is obvious from previous discussion that N is always faster than the other two.
2. Therefore, if N is implemented by using no more hardware than the other two, then F can be faster than N in some cases especially with word lengths, although N is generally faster than the other two. For example, consider the case of 32 bit operand with 3 bit precise initial approximation. N takes four iterations. In short, R is always slower than N, but it has the feature that division can be performed by using no additional hardware and control circuitry is faster for relatively small word lengths and profitable in those cases where the rounded-reciprocal generator is used for division. N is generally faster than R and F and can be made more computers with large word lengths due to its property of quadratic convergence.

Algorithm G

Basic Principle

This algorithm is analogous to Goldschmidt's algorithm for division that is, it is a kind of normalization technique whose normalizing factor is the form of multiple bits.

The basic idea can be expressed by the following mathematical expressions. First, the approximation to $1/\sqrt{x}$ is attempted.

$$1/x = (1/x)*(r'_0/r'_0)*(r'_1/r'_1)*(r'_2/r'_2)*\ldots(r'_n/'_n)$$

$$= \left(\prod_{i=0}^{n} r'_i\right)/\left(x \prod_{i=0}^{n} r'_i\right)$$

Here, if $\left(x \prod_{i=0}^{n} r'_i\right) \cong 1$,

Then, $1/x \cong \left(\prod_{i=0}^{n} r'_i\right)$

Now taking r'_j in form of r'^2_i

$$1/x = 1/x*(r^2_0/r^2_0)*(r^2_1/r^2_1)\ldots(r^2_n/r^2_n)$$

$$= \prod_{i=0}^{n} r^2_i \times \prod_{i=0}^{n} r^2_i$$

Here if $x \prod_{i=0}^{n} r^2_i \cong 1$

Then $1/x \cong \prod_{i=0}^{n} r^2_i$

So, $1/\sqrt{x} \cong \prod_{i=0}^{n} r_i$

and $\sqrt{x} = x*1/\sqrt{x} \cong x*\prod_{i=0}^{n} r_i$

Now x_k and R_k can be defined as follows:

$$x_k = x*\prod_{i=0}^{k-1} r^2_{k-1} \quad X_0 = x$$

$$(\sqrt{x}) = R_k = x^*\prod_{i=0}^{k-1} r_{k-1} \quad R_0=x$$

Properties of G

It is important to choose the appropriate normalizing factor r^2_i since it affects the convergence rate and suitability for implementation.

It is known that, in case of division, the normalizing factor d_k is selected as 1+x when the parity normalized denominator x_k is (Anderson, 1967)

$$x_k = X^*\prod_{i=0}^{k-1} d_j = 1 - x$$

Implementation

To increase the efficiency of this algorithm several technique can be applied for implementation (Anderson, 1967) (*e.g.* IBM 360/91). Which are as follows:

Truncated Normalizing Factor

Now let us make use of the following properties to save multiplication time. Denoting $[x_k]_T$ as the value of x_k which is truncated up to T bits.

$$[x_k] = x_k - \varepsilon = 1 - x - \varepsilon$$

$$2 - [X_k]_T = 1 + x + \varepsilon$$

shifting the fractional part right by one bit,

$[r_k]_t = 1 + (1/2)x + (1/2)\varepsilon$ where $[r_k]_t$, means the modified r_k since $[r_k]_t$ is obtained from $[x_k]_t$ rather than x_k,

$$[r_k]_t^2 = (1 + ½x + ½\varepsilon)^2$$

$$= [1 + x + 1/4x^2 + \varepsilon(1 + ½x)]$$

$$x_{k+1} = x_k^*[r_k]_t^2$$

$$= (1-x)[1+x+¼x^2+\varepsilon(1+½x)]$$

$$= 1-¾x^2-¾x^3+¼\varepsilon(1-x)+\varepsilon(1-½x-½X^2)$$

$$\leq 1 - ¾x^2 - ¾x^3 + \varepsilon$$

Therefore, it is more advantageous to use the truncated normalizing factor during the first several iterations.

Initialization

It is more efficient to use the first normalizing factor r^2_0 read from the lookup table so that x_1 may possess several leading 1's or 2's, since it can lessen the necessary iterations. Here r_0 and r^2_0 must both be generated. If we select Goldschmidt's algorithm for division and Algorithm G for square rooting, then r^2_0 can be used both for division and square root. Also the size of the table must be increased to cover the range [¼, ½] in addition to [½, 1].

Pipe Lining and a Typical Model of Implementation

$R_{k+1} = R_k * r_k$ and $R_k^2 = r_k * r_k$ can be executed with the same hardware by pipe lining, since multiplier r_k is the same for both R_{k+1} and r_k^2, and the computation can be improved by using this fact. This will be more apparent by examining an implementation of the algorithm for the case of a 64 bit word, which is almost the same as the one used for division. Next, the formats of the intermediate results and their multiplier should be noted that r_4^2 and x_5 need not be obtained.

Evaluation of G Compared to N

Now let us evaluate the efficiency of algorithm G by comparing it with algorithm N, since N is generally the most efficient among the algorithms using iterative multiplications, except in some cases. Algorithm G and N both have quadratic convergence.For each iteration N requires three multiplications by using the same hardware and two multiplications plus one addition by using one more multiplier, while G requires two multiplications. Moreover, one of them is an abbreviated multiplication due to the truncated normalizing factor and G does not need any adder. Therefore, it can be seen that G is Faster than N and takes no more hardware than N, although it is believed that circuitry for G would be a little more complicated than N. Also it is obvious that G becomes very efficient in cases where Goldschmidt's algorithm is used for division.

Conclusion

Three new algorithms for square rooting are presented and compared with the well known Algorithm N. Algorithms R, F, and N belong to the same group, which are the variations original Newton-Raphson iterative formula using multiplications instead of division. Algorithm G is based on multiple-bit normalization. Algorithm R is not as fast as N, but it has the feature that division can be performed by using no additional hardware and control circuitry. Algorithm F is fast enough, compared to N, for relatively small word lengths and profitable in those cases where the rounded-reciprocal generator is used for division. Algorithm N is generally faster than R and F. However, it is obvious that algorithm G is faster than N even in the case where N uses one more multiplier. Both algorithms G and N are suitable form large word lengths due to quadratic convergence. Here, it could be mentioned that the mathematical functions of both algorithms N and G are essentially the same, although algorithm G is derived on the basis of the normalization technique for hardware implementation, while algorithm N is derived purely on the mathematical basis of the Newton iteration function. So in some sense algorithm G is an improved analog of algorithm N for hardware implementation.

References

Anderson, S.F., Earlie, J.G., Goldschmidt, R.E. and Powers, D.M., 1967. The IBM SYSTEMI360 MODEL 91: Floating point execution unit. *IBMJ.*, January.

Barnes, G.H., Brown, R.M., Kato, M., Kuck, D.J., Slotnick, D.L. and Strokes, R.A., 1968. The ILLIACIV computer. *IEEE Trans. Comput.*, C-17: 746–757.

CDC 6600 computer system/Library Function Manual. Publ. 60114500, Rev. A., 1965.

Metze, G., 1965. Minimal square rooting. *IEEE Trans. Electron. Comput.*, EC 14: 181–185.

Chapter 25

Trap Depth Evaluation of Nd, Eu and Sm Doped CaO Phosphors

Ashish Verma*, Shishir Jain, Swati Mishra and Rajendra

Department of Physics and Electronics,

Dr. Harisingh Gour Vishwavidyala, Sagar – 470 003, M.P.

ABSTRACT

In the present study the trap depth is calculating using formula like Randall and Wilkin's, Grosswiener's, Halperin and Braner's which is corrected by Chen's discussed. The activation energy is also calculated using Chen's formula for general order kinetics and discussed. The activation energy is also calculated using Chirstodoulides formulae.

In the present study the glow curves are go under thermal cleaning to calculate the activation energy in case of CaO phosphor activated with Nd and Eu. The TL glow curve of CaO : Sm phosphor is complex in nature so it was solved differently.

Keywords: Activation energy, Order of Kinetics, Glow curve.

Introduction

Luminescence can be delayed by an amount, governed by the mean lifetime τ spent by an electron in the trap and given by Arrhenius equation.

$$\tau = S^{-1} \exp (E/KT)$$

* E-mail: vermaashish31@.rediffmail.com

And the probability per unit time of the release of an electron from the trap, which is given by the equation

$$P = \tau^{-1} = S \exp(-E/KT)$$

Here the E is trap depth, K is Boltzmann constant and S is attempting to escape frequency.

It is observed that if the trap depth E is greater than the temperature of irradiation, then only one electron will remain trapped for a long period of time and there will exist a substantial population of trapped electrons after the removal of irradiation.

In the present study the trap depth is evaluated using different formulae like Randall and Wilkin's, Grosswiener formula corrected by Chen, Halperin and Branar formula corrected by Chen, Chen's Formula for general order Kinetics (Randall and Wilkins, 1945; Grosswiener, 1953; Chen, 1969; Halperin and Braner, 1960). Activation energy is also calculated using the width of the Thermoluminescence peak given by C. Christodoulides (Christodoulides, 1985).

Experimental Procedure

Preparation of Phosphor Sample

AR grade reagent $CaCO_3$ was fired with 99 per cent pure oxide of Nd, Eu and Sm in a tabular furnace at 900±20°C for 2 hours. Fast cooling was preferred for better emission intensity. In this was three series of Nd, Eu and Sm were prepared for 10^{-1} to 10^{-6} per cent of activator concentration.

Measurement

The present arrangement involves the measurement of Themoluminescence of phosphor. X-ray are used as exciting source agent which one produced from the copper target of a Radon X-ray machine model no. 11704 operated at 32 KV and 10 mA. The intensity of emission was recorded using GCA/mc pherson photo multiplier Model No. EU-701-30. The output of the P.M. tube was connected with the D.C. micro voltmeter model no. PP9004.

Thermal Cleaning of TL Curves

Thermal cleaning is a convenient method to separate glow peak, which do not have interference from other trap groups. In the case of Nd and Eu samples the peaks were distinct and it was possible to separate the glow peaks experimentally. The sample was heated to about 382 K, which was the maximum of the first peak, and the heating was discontinued. After the temperature returned room temperature, the heating was restarted and the second peak obtained. After that the second peak was subtracted from the total curve to obtain the first peak as an isolated one. The cleaning curves are shown in Figure 25.1 and 25.2.

Results and Discussion

Evaluation of Trap Depth

CaO : Eu

In this case the activation energy first calculated by Randall and Wilkins formulae assuming frequency factor $S = 10^9 s^{-1}$. The value comes out to be nearly 0.71eV for the first peak of Eu concentration of 10^{-1} and 10^{-2} percentage and for the concentration of 10^{-3} to 10^{-6} percentage of Eu, the activation energy comes out to be 0.67eV, with the help of this formula the activation energy for second peak was calculated and values come out to be 0.83eV.

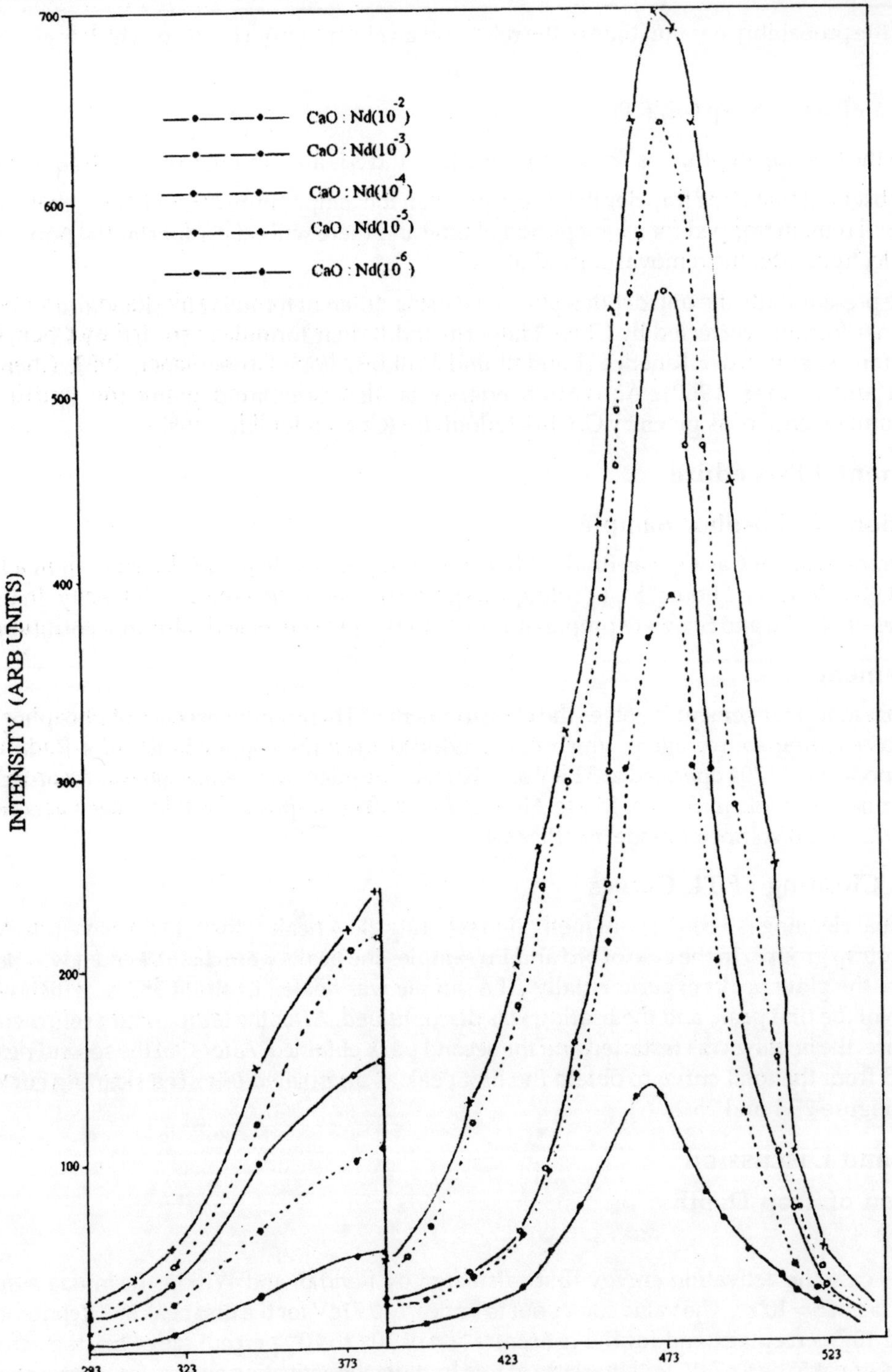

Figure 25.1: Thermal Cleaning

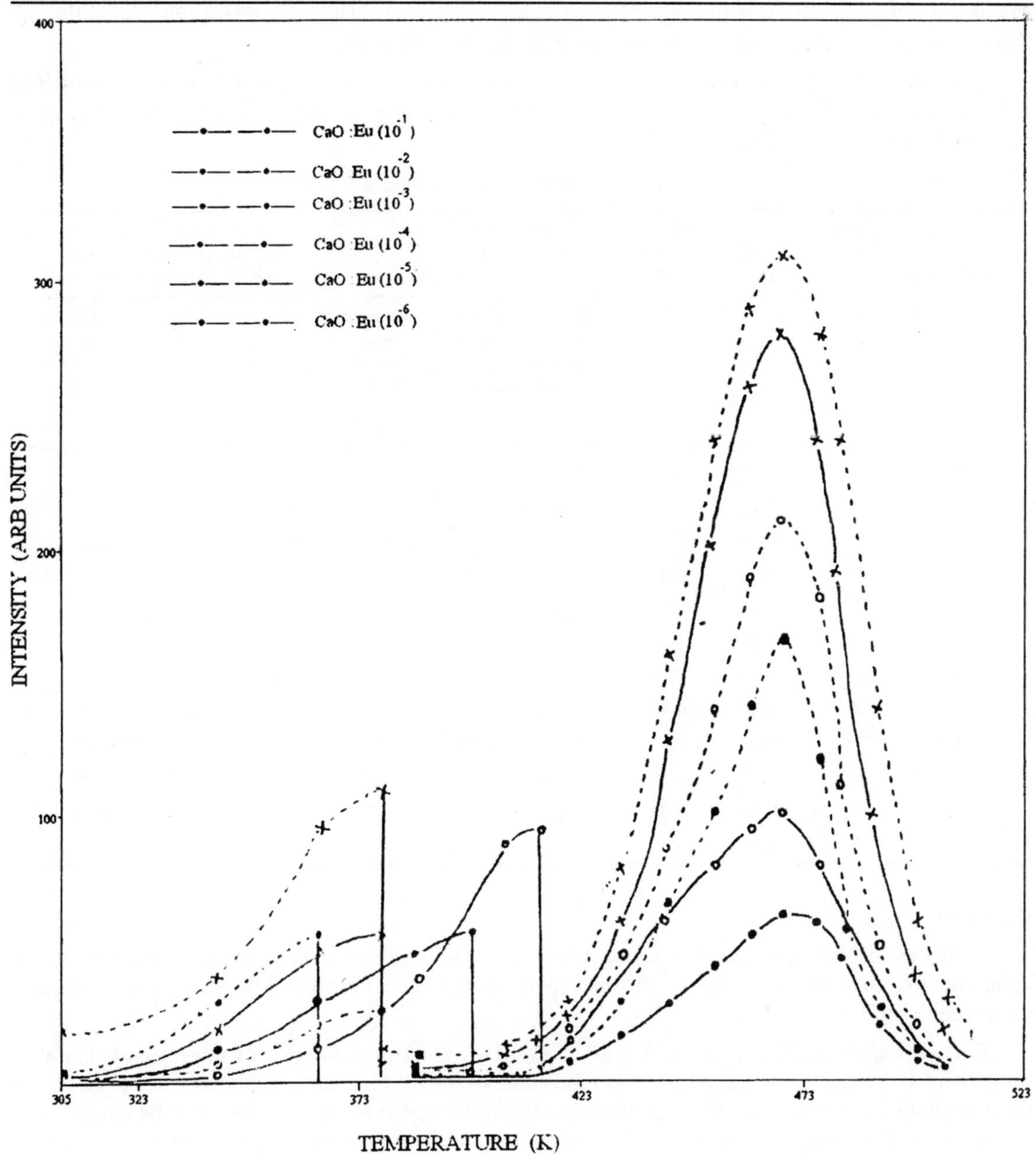

Figure 25.2: Thermal Cleaning

Grosswiener's and Halperin and Braner's formula, which are corrected by Chen, also calculated the activation energy for first order kinetics. The value of activation energy for the first peak calculated to be less than that the one by Randall and Wilkin's formula expect in two cases. However, the values from the two formulae agree with each other. The activation energy for the second peak was also calculated using these two formulae and the activation energy is nearly the same in both cases although the values for different concentration do not come out to be constant.

Activation energy was also calculated using Chen's formulae for general order kinetics. The values are given in Table 25.1. It is seen that the values calculated by the three formulae do not agree with each other in almost all cases.

Table 25.1: CaO : Eu Phosphors

Conc. of Eu (Per cent)	*Peak No.*	*Peak Temp. (k)*	*Symmetry Factor μg*	*Activation Energy Assuming First Order Kinetics in eV*			*Activation Energy Assuming R. Chen General Order Kinetics in eV*		
				Randall and Wilkins's Formula S = 10^9 sec^{-1}	*Gross-wiener Formula Corrected by Chen*	*Halperin and Braner Formula Corrected by Chen*	E_τ	E_δ	E_ω
10^{-1}	1	399	0.39	0.71	0.51	0.50	0.46	0.45	0.77
	2	471.6	0.32	0.84	0.82	0.82	0.66	0.47	0.63
10^{-2}	1	414.5	0.35	0.74	0.76	0.83	0.71	0.52	1.33
	2	468	0.39	0.83	1.2	1.2	1.2	1.0	1.1
10^{-3}	1	364.5	0.58	0.65	0.54	0.54	0.69	0.62	0.61
	2	468	0.39	0.83	1.2	1.2	1.2	1.0	1.1
10^{-4}	1	377.6	0.39	0.67	0.68	0.68	0.46	–	1.28
	2	468	0.40	0.83	1.1	1.2	1.1	1.0	1.1
10^{-5}	1	377.6	0.43	0.67	0.51	0.50	0.51	0.23	0.72
	2	468	0.35	0.83	0.87	0.87	1.2	0.56	0.69
10^{-6}	1	377.6	0.44	0.67	0.53	0.52	0.54	0.57	0.74
	2	468	0.45	0.83	0.96	0.74	1.6	1.0	1.0

CaO : Nd

In the case of Nd doped CaO phosphors the activation energy using Randall and Wilkin's formulae comes out to be 0.69eV for the first peak, the activation energy for the second peak was calculated and values comes out to be 0.86eV.

Grosswiener's and Halperin and Braner's formulae, which are corrected by Chen, also calculated the activation energy for first order kinetics. The value of activation energy for the first peak calculated to be less than the one by Randall and Wilkins' formulae except in two cases. However the values from the two formulae agree with each other. The activation energy for the second peak was also calculated by these two formulae and the activation energy is nearly the same in both cases.

The activation energy was also calculated by Chen's formulae for general order kinetics. The value are given in Table 25.2. It is seen that they agree with two formulae for E_τ and E_δ in some extent but do not agree with the third formulae for E_ω.

Table 25.2: CaO : Nd Phosphors

Conc. of Eu (Per cent)	Peak No.	Peak Temp. (k)	Symmetry Factor μg	Activation Energy Assuming First Order Kinetics in eV			Activation Energy Assuming R. Chen General Order Kinetics in eV		
				Randall and Wilkins's Formula $S = 10^9$ sec^{-1}	Gross-wiener Formula Corrected by Chen	Halperin and Braner Formula Corrected by Chen	E_τ	E_δ	E_ω
10^{-1}	1	–	–	–	–	–	–	–	–
	2	–	–	–	–	–	–	–	–
10^{-2}	1	388.2	0.46	0.69	0.65	0.65	0.69	0.71	0.86
	2	470.3	0.45	0.84	0.96	0.97	1.0	1.0	1.0
10^{-3}	1	388.2	0.42	0.69	0.65	0.65	0.64	0.66	0.94
	2	482.2	0.29	0.86	1.0	1.0	0.77	1.5	0.56
10^{-4}	1	388.2	0.31	0.69	0.59	0.59	0.60	–	0.82
	2	482.2	0.30	0.86	1.2	1.2	0.99	1.3	0.75
10^{-5}	1	388.2	0.44	0.69	0.36	0.34	0.35	0.40	0.50
	2	482.2	0.31	0.86	0.85	0.85	0.67	0.99	0.53
10^{-6}	1	388.2	0.42	0.69	0.30	0.27	0.27	0.32	0.43
	2	482.2	0.41	0.86	0.82	0.81	0.79	1.2	0.80

CaO : Sm

The glow curve for this phosphor are complex in nature and they are resolved by a method suggested by Rao as well as the one proposed by Khare and Ranade (1975). The activation energy was calculated by Chen's formulae using δ, τ and ω, which are shown in Table 25.5–25.8.

Table 25.3: Activation Energy, using the Widths of TL Peaks, given by Christodoulides, C. Formulae (Obtained by Eight Equation) of CaO : Eu Phospors

Conc. of Eu (Per cent)	Eqn. (1)	Eqn. (2)	Eqn. (3)	Eqn. (4)	Eqn. (5)	Eqn. (6)	Eqn. (7)	Eqn. (8)
10^{-1}	0.85	1.3	0.99	1.0	1.3	1.1	1.0	1.4
10^{-2}	0.84	1.1	0.95	0.86	1.0	0.91	0.95	0.84
10^{-3}	1.2	1.4	1.3	1.1	1.1	1.1	1.0	0.91
10^{-4}	1.2	1.2	1.2	1.1	1.2	1.1	1.0	1.2
10^{-5}	0.87	1.1	0.98	1.1	1.0	1.1	1.2	0.92
10^{-6}	0.96	0.85	0.92	1.0	0.90	0.91	1.0	0.93

The activation energies calculated do not show consistency. For *e.g.* in the case of 10^{-1} per cent of Sm, the activation energy for the three peaks are 0.47eV for first peak, 0.78eV for second peak and 1.30eV for the third. The trap depths show a regular increase with the shallower traps emptying at lower temperatures. The case of CaO : Sm (10^{-4} per cent) is similar. However, in the case of 10^{-2} per cent and 10^{-3} per cent of Sm the activation energies come out to be very different. In the case of 10^{-2} per cent of Sm the activation energy comes out to be 0.64eV for the first peak, 0.61eV for second peak and 1.43 for the third one.

Table 25.4: Activation Energy, using the Widths of TL Peaks, given by Christodoulides, C. Formulae (Obtained by Eight Equation) of CaO : Nd Phospors

Conc. of Nd(per cent)	*Eqn. (1)*	*Eqn. (2)*	*Eqn. (3)*	*Eqn. (4)*	*Eqn. (5)*	*Eqn. (6)*	*Eqn. (7)*	*Eqn. (8)*
10^{-1}	–	–	–	–	–	–	–	–
10^{-2}	0.97	0.86	0.92	0.99	0.63	0.82	1.0	0.41
10^{-3}	1.0	1.8	1.2	1.4	1.2	1.3	2.1	0.81
10^{-4}	1.2	2.0	1.5	1.6	1.2	1.1	1.0	0.69
10^{-5}	0.85	1.3	1.0	0.70	0.95	1.5	0.59	0.71
10^{-6}	0.82	0.86	0.83	0.69	0.84	0.74	0.59	0.84

Table 25.5: Trapping Parameters of TL Glow Peaks, Resolved by Khare *et al.* Method of CaO : Sm (10^{-1}) for X-ray excitation of room temperature (20°C)

Sl.No.	*Tm (K)*	ω *(K)*	δ *(K)*	τ *(K)*	$\mu g = \delta/\omega$	E_ω *(eV)*	E_δ *(eV)*	E_τ *(eV)*	*Randall and Wilkins' Formula $S = 10^9$ sec^{-1} (eV)*
1.	459.6	33.3	14.3	19	0.42	1.30	1.24	1.32	0.82
2.	421.5	45.3	19.1	26.2	0.42	0.78	0.78	0.77	0.75
3.	370.3	64.5	29.8	34.5	0.46	0.47	0.50	0.44	0.66

Table 25.6: Trapping Parameters of TL Glow Peaks, Resolved by Khare *et al.* Method of CaO : Sm (10^{-2}) for X-ray Excitation of Room Temperature (20°C)

Sl.No.	*Tm (K)*	ω *(K)*	δ *(K)*	τ *(K)*	$\mu g = \delta/\omega$	E_ω *(eV)*	E_δ *(eV)*	E_τ *(eV)*	*Randall and Wilkins' Formula $S = 10^9$ sec^{-1} (eV)*
1.	465.6	48.7	27.4	21.3	0.56	1.63	1.36	1.52	0.83
2.	408.4	54.7	23.8	30.9	0.43	0.61	0.63	0.60	0.73
3.	344.1	36.9	15.	21.4	0.42	0.64	0.64	0.63	0.61

This means that the first and second trap groups have nearly the same depth although they empty at different temperatures. In the case of CaO with 10^{-3} per cent Sm,the depths for the lowest temperature peak calculated using the three formulae are 1.00eV, 1.77eV, 1.94eV which represents a wide divergence. When compared with the values calculated for the peak at 448K these are twice as large. In two cases these trap depths are even larger then those for the 483K peak. This would indicate that although the trap depths are higher they empty at a lower temperature. This would be possible only if the frequency factor is very much higher for the low temperature trap group.

Table 25.7: Trapping Parameters of TL Glow Peaks, Resolved by Khare *et al.* Method of CaO : Sm (10^{-3}) for X-ray Excitation of Room Temperature (20°C)

Sl.No.	Tm (K)	ω (K)	δ (K)	τ (K)	$\mu g = \delta/\omega$	E_ω (eV)	E_δ (eV)	E_τ (eV)	Randall and Wilkins' Formula $S = 10^9$ sec^{-1} (eV)
1.	483.4	30.6	13.1	17.5	0.42	1.57	1.50	1.60	0.86
2.	448.6	44.7	19.1	25.6	0.42	0.56	0.79	0.81	0.80
3.	397.7	25.0	13.1	11.9	0.52	1.00	1.77	1.94	0.71

Table 25.8: Trapping Parameters of TL Glow Peaks, Resolved by Khare *et al.* Method of CaO : Sm (10^{-4}) for X-ray Excitation of Room Temperature (20°C)

Sl.No.	Tm (K)	ω (K)	δ (K)	τ (K)	$\mu g = \delta/\omega$	E_ω (eV)	E_δ (eV)	E_τ (eV)	Randall and Wilkins' Formula $S = 10^9$ sec^{-1} (eV)
1.	475.1.	45.2	19.0	26.2	0.42	1.00	1.00	1.00	0.84
2.	423.9	61.9	26.2	35.7	0.42	0.56	0.57	0.54	0.75
3.	353.7	40.5	14.3	26.2	0.35	0.28	0.35	0.46	0.63

In the present study only two peaks were observed with the second one being more prominent. The samples were excited at room temperature and most of the shallow traps may have got emptied by the time the glow curve recording had started. This could explain why the height of the first peak since to be less then that of the later one.

In the case of Eu and Nd′ activated CaO phosphors the two glow peaks were not completely isolated. These phosphors were subjected to experimental thermal cleaning (Kathuria and Sunta, 1982; Ratnam and Bose, 1966) in order to separate the glow curves. The TL glow curves are shown in Figures 25.3 and 25.4. The separated glow curves were then used to calculate trap parameters using Randall and Wilkins′, Grosswiener's Halperin and Braner's, Chen's (general order kinetics) and Christodoulide's formulae. The frequency factor in Randall and Wilkins′ formula was taken as 10^9 sec^{-1}.

It was found that there is good agreement between the values of trap depths calculated by Grosswiener's and Halperin and Braner's formulae as corrected by Chen. For the first peak, the values from Randall and Wilkins′ formula are consistently higher. For the second peak, the values from the first two formulae are generally higher. Chen's formula for E_τ, Ed_δ and E_ω for general order kinetics yield values, which show wide variation. There is also very little agreement between the three values. Activation energy was calculated using the widths of TL peaks, given by Christodoulides, C. formulae. The values are given in Tables 25.3 and 25.4. Activation energies calculated using Christodoulide's formulae for the second peak show fair consistency except those from equation (1), which shows

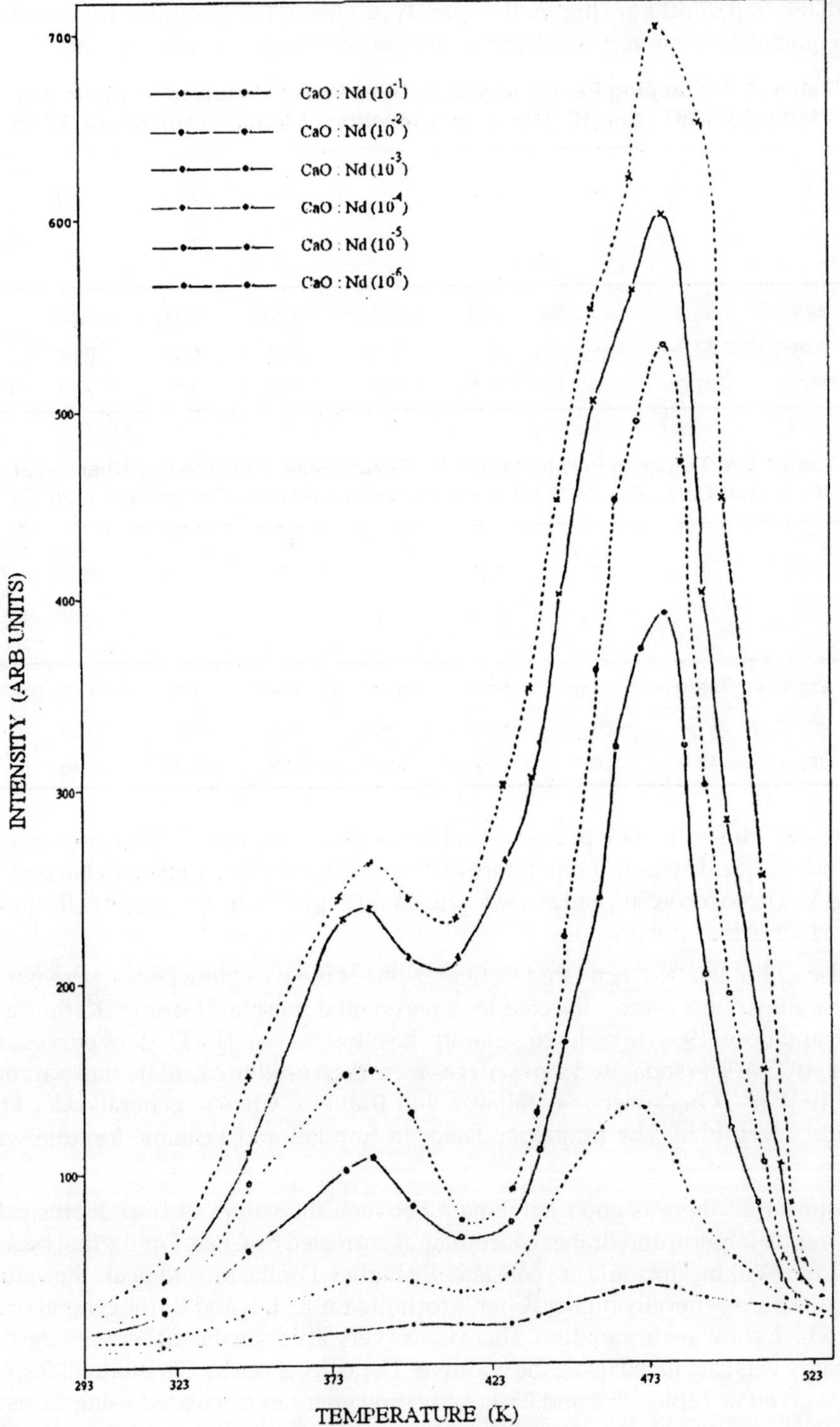

Figure 25.3: TL Glow Curve

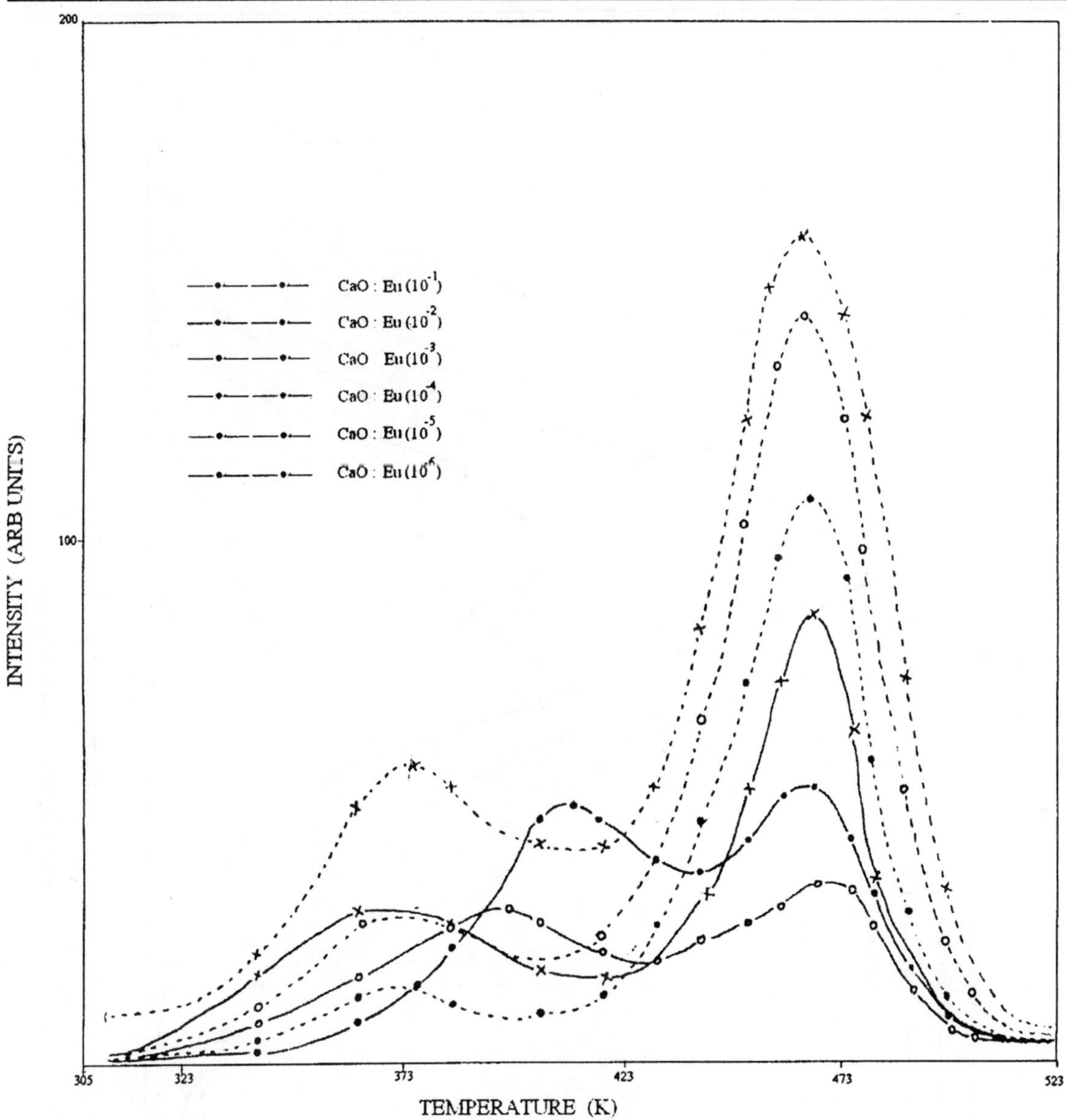

Figure 25.4: TL Glow Curve

remarkable agreement with the values obtained from Grosswiener's formula. For the first peak, the values obtained from formulae other than that of Randall and Wilkins' quite often are close to 0.50eV and sometimes even less. This would correspond to a mean lifetime, which is less than two seconds. In the phosphorescence decay studies, it was observed that the decay could be explained as a super position of two exponential processes involving traps of depth around 0.50eV. The observed decay was quite fast and at room temperature the duration of decay was less than 10 seconds. Comparison of the induction deficiency and the light sum released during the first few seconds of decay showed

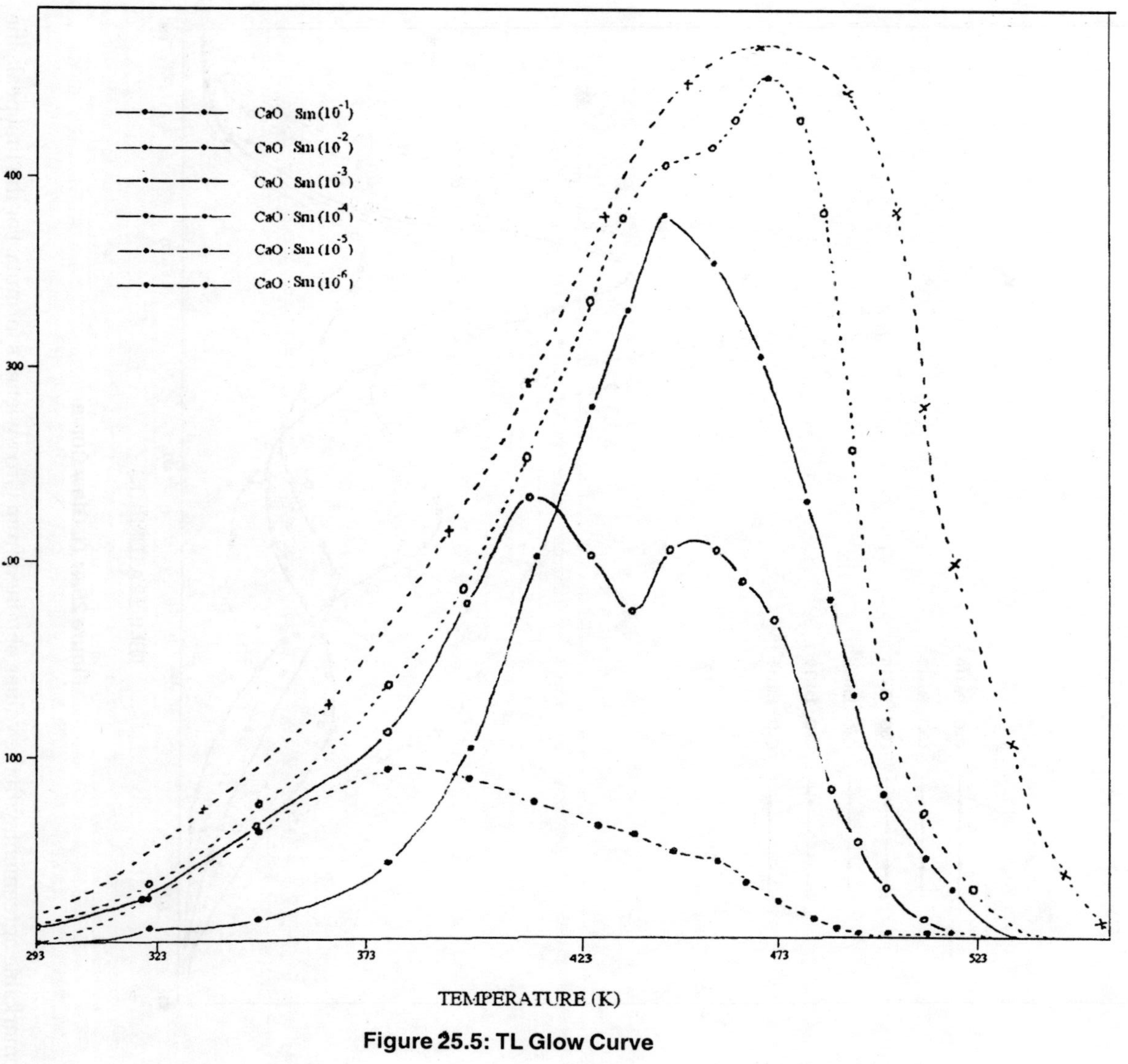

Figure 25.5: TL Glow Curve

that almost all the energy stored in the shallow traps is released during this period. Since, the glow curve experiment was begun after a time delay of one minute after cessation of excitation, the chance that traps with a mean lifetime of less than two seconds would be involved in the glow curve is negligible. On the other hand the values calculated from Randall's formula correspond to mean life times of the order of eight to ten minutes; for the traps corresponding to the first peak.

It has been observed that whenever one is dealing with glow peaks, which are not well isolated, the mutual interference leads to widely divergent and inconsistent E values (Sivaraman, 1965). It was found that there was a large variation in E values for small changes in the geometrical factors, and these cannot in general be estimated with the designed accuracy in the absence of isolated peak, it was not found possible to rely on the values calculated by these formulae. In the present study, since the two interfering peaks were isolated experimentally, there has been some measure of consistency in the values calculated for the second peak although the values for the first peak are somewhat inconsistent.

The glow curve of CaO : Sm were complex in nature. The TL glow curve is shown in Figure 25.5. An attempt was made to resolve the curves into its component peaks using the methods described by Rao and Khare. The E values were calculated by formulae, which take into account the geometrical factors of the curves. The obtained values shows not only wide divergence but also are often contradictory. Some times the traps emptying at a higher temperature have lower E values. This would need the assumption of higher S values for the low temperature peaks. This result is not also consistent. It therefore appears that resolving the thermoluminescence curves theoretical making convenient assumptions, is not without inherent dangers. Replacing the rising and falling off part of the glow curve by a straight line and subtracting the resultant curve from the total curves were found to lead to highly inconsistent results in the case of CaO : Sm.

References

Chen, R., 1969. *J. Appl. Phys.*, 40(2): 570.

Christodoulides, C., 1985. *J. Appl. Phys.*, 18: 1501.

Grosswiener, L.I., 1953. *J. Appl. Phys.*, 24: 1306.

Halperin, A. and Braner, A.A., 1960. *Phys. Rev.*, 117: 408.

Kathuria, S.P. and Sunta, C.M., 1982. *J. Phys. D. Appl. Phys.*, 15: 497.

Khare, R.P. and Ranade, J.D., 1975. *Physca Stat. Solidi* (a), 32: 221.

Randall, J.T. and Wilkins, M.H.F., 1945. *Proc. Roy. Soc. (London)* A184: 366, 390.

Rao, D.R., 1974. *Physca Stat. Solidi* (a), 22: 337.

Ratnam, V.V. and Bose, H.N., 1966. *Physca Stat. Solidi* (a),15: 309.

Sivaraman, S., 1965. *Ph.D. Thesis*, Sagar University, Sagar (M.P.).

Chapter 26

Topology and Matrix Design for Multiprocessor System in Robotics

Ranjna Bohre, Ashish Verma*, Alka Dubey, Shalini Jain, R.S. Verma and Fazal Noorbasha

Laboratory of Embedded Systems, Department of Physics and Electronics, Dr. Harisingh Gour University, Sagar – 470 003, M.P.

ABSTRACT

Topology is studied for multiprocessor system and their communication link. In present study the three-processor topology is used which are controlled by the one microprocessor. This microprocessor controls the three processor, which are further connected to the one arm and two joints of a robot. The rotation matrix is also solved for arm and two joints.

Keywords: *Topology, Multiprocessor, Microprocessor, Robot, Euler's rotation.*

Introduction

In digital electronics topology is used for multiprocessor system. The topology of a multiprocessor system is the physical arrangement of the co-processor and their communication link. The number of inter connection is given by the following relation for *n* processor system.

$$\sum_{i=1}^{n-1} i = \frac{n(n-1)}{2} \tag{1}$$

* E-mail: vermaashish31@.rediffmail.com

In this equation the number of interface required processor is (n–1). The link can be implemented with serial ports, parallel ports, a shared memory or a FIFO. In all these links topology would be the same. A robot arm is design using point-to-point serial connection.

In present study the robot is consists of one arm and two joints and are separately controlled by the three micro controllers. All these three micro controllers are controlled by a single microprocessor. This is an example of star topology as shown in Figure 26.1 (Bowen and Burr, 1980; Fathi and Knegen, 1983).

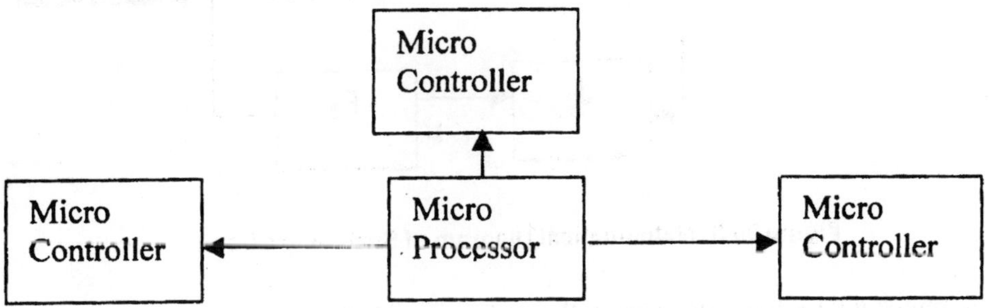

Figure 26.1: Block Diagram of Star Topology

Experimental Setup

In our study the design of the circuit used is as shown in Figure 26.2. According to Figure 26.2 a mathematical diagram is design to calculate the equation obtained from this diagram for further calculation of their rotation given in Figure 26.3.

According to Boolean postulate the function of the diagram as shown in Figure 26.3 is written as:

$$X.Y\,(J_1F_1 + J_2F_2 + J_3F_3) = B \quad (2)$$

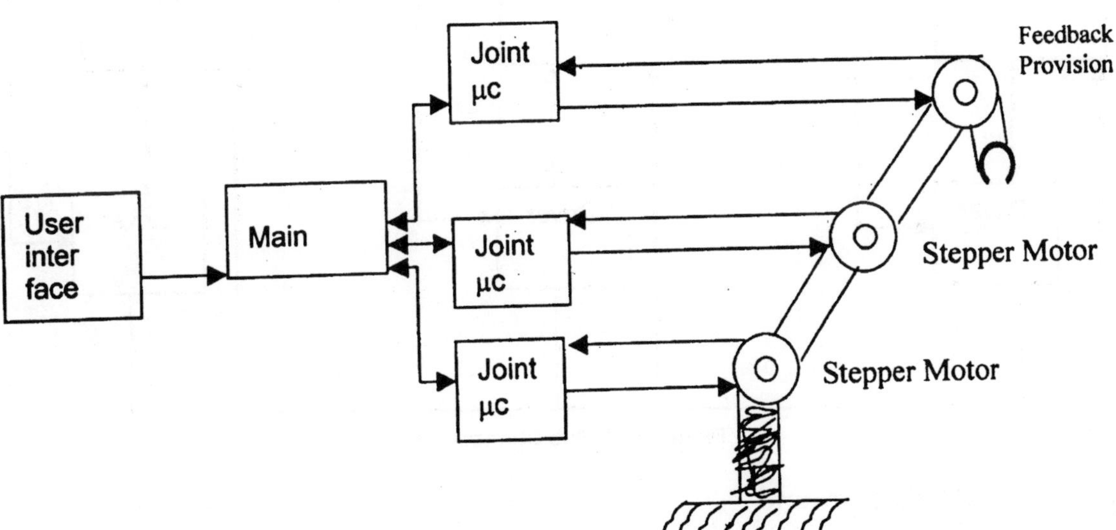

Figure 26.2: Circuit Diagram of Microprocessor System with Robotic Control

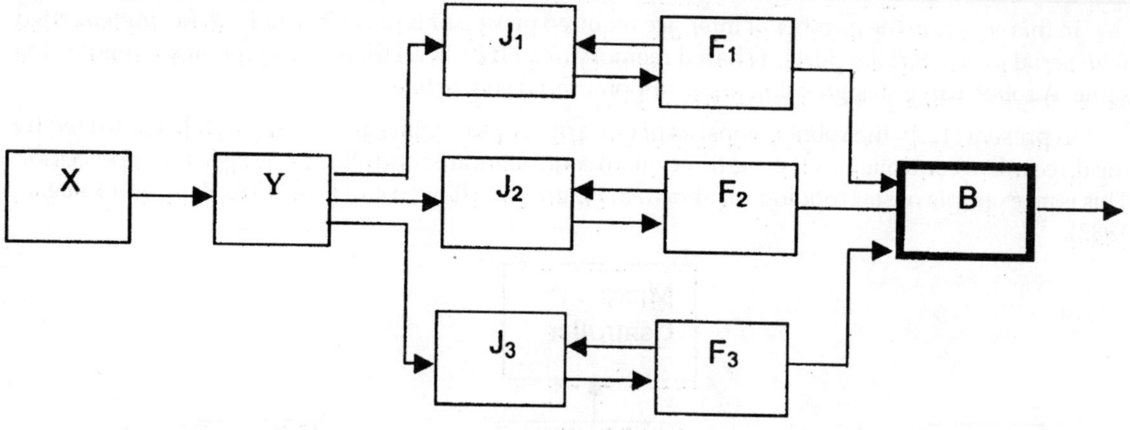

Figure 26.3: Mathematical Diagram of Switching Circuit

The bus connection for the processing of the system used is this study is as shown in Figure 26.4.

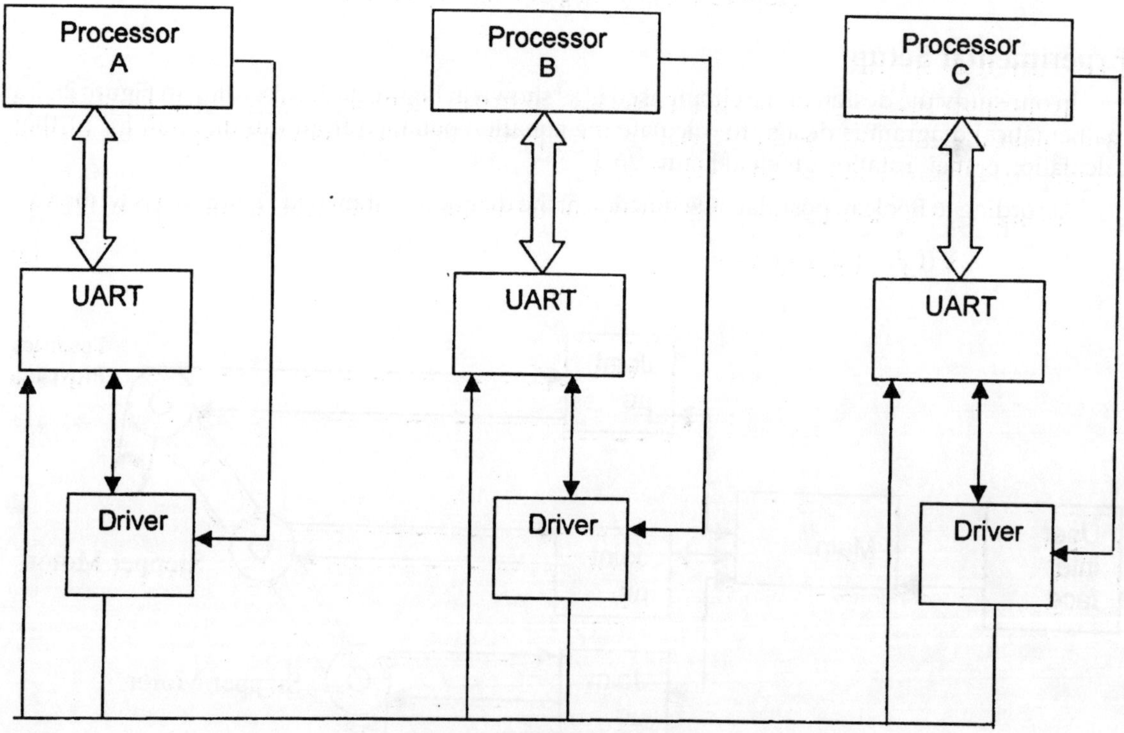

Figure 26.4: Bus Connection

Mathematical Background

A three-dimensional rotation matrix is used in the present robotic system. Let x, y, z be a Cartesian co-ordinate system in a three dimensional space and Let $\mathbf{u} = \{u_1\ u_2\}$. Be a vector in the x-y plane. A rotation of the co-ordinate system is considered about the z-axis through an angle θ in the anticlockwise and new co-ordinate system x′, y′, z′. If the same vector **u′** has the components u_1' and u_2' relative to the new system, it is evident that

$$\left.\begin{aligned} u_1' &= u_1 \cos\theta + u_2 \sin\theta \\ u_2' &= -u_1 \sin\theta + u_2 \cos\theta \end{aligned}\right\} \quad (3)$$

This could be written in the matrix rotation as

$$\begin{pmatrix} u_1' \\ u_2' \end{pmatrix} = \begin{pmatrix} \cos\theta & \sin\theta \\ -\sin\theta & \cos\theta \end{pmatrix} \begin{pmatrix} u' \\ u' \end{pmatrix} \quad (4)$$

If we denote the rotation operator by $R_z(\theta)$, we could write the transformation of the vector **u** to **u′** = $\{u_1' \; u_2'\}$ symbolically by

$$\mathbf{u} \rightarrow \mathbf{u}' = R_z(\theta)\mathbf{u} \quad (5)$$

Comparing eq's (4) and (5) the effect of the operator $R_z(\theta)$ is to multiply the column vector $\{u_1 \; u_2\}$ by the 2 × 2 matrix appearing on right hand side of eq. (4)

This matrix represents the operator $R_z(\theta)$ in the two dimensional subspace (x, y).

It is often expressed by giving an equally sign and writing

$$R_z(\theta) = \begin{pmatrix} \cos\theta & \sin\theta \\ -\sin\theta & \cos\theta \end{pmatrix} \quad (6)$$

Which is an orthogonal matrix (Pipes, 1958; Gantmacher, 1959; Artkin, 1968).

It should be realized that the representation of an operator depends on the choice of the co-ordinate system. Let us now consider a vector **u** = $\{u_1 \; u_2 \; u_3\}$ and suppose the same rotation $R_z(\theta)$ is applied on the system so that the vector changes to **u′** = $\{u_1' \; u_2' \; u_3'\}$

Since, the rotation is about the Z-axis the Z-component of the vector will remain unchanged.

The new components are given in terms of the old components by the equations

$$\left.\begin{aligned} u_1' &= u_1 \cos\theta + u_2 \sin\theta \\ u_2' &= -u_1 \sin\theta + u_2 \cos\theta \\ u_3' &= -u_3 \end{aligned}\right\} \quad (7)$$

This shows that the operator $R_z(\theta)$ is represents by the matrix.

$$R_z(\theta) = \begin{pmatrix} \cos\theta & \sin\theta & 0 \\ -\sin\theta & \cos\theta & 0 \\ 0 & 0 & 1 \end{pmatrix} \quad (8)$$

in the three dimensional space (x, y, z).

Let us now consider a rotation about any arbitrary axis passing through the origin of the co-ordinate system. It is known that any general rotation can be obtained by three successive rotations each performed about one of the co-ordinate axes.

In the Euler's method, α general rotation is the result of the following three successive rotations:

(a) a rotation through α about the z-axis followed by (b) a rotation through β about the new y axis followed by (c) a rotation through γ about the hew z axis. The angles α, β, γ are called the Euler's angles and the rotation denoted by the operator R (α, β, γ). If $\mathbf{u} = \{u_1\ u_2\ u_3\}$ is the original vector and $\mathbf{u}' = \{u_1'\ u_2'\ u_3'\}$ the transformed vector then

$$\mathbf{u}' = R(\alpha, \beta, \gamma)\, \mathbf{u}$$

The matrix representing R (α, β, γ) can be found as follows. According to the above definition the operator $R_z(\alpha)$ acts on u first, giving rise to the vector $R_z(\alpha)$.

Here, $R_z(\alpha)$ has a matrix similar to that in Eq. (8). On this vector acts as $R_Y(\beta)$ giving the vector $R_Y(\beta)\, R_Z(\alpha)\mathbf{u}$

Since $R_Y(\beta)$ is a rotation about the y-axis, its matrix representation can be obtained in a similar way as above and is found to be

$$R_Y(\beta) = \begin{pmatrix} \cos\beta & 0 & \sin\beta \\ 0 & 1 & 0 \\ -\sin\beta & 0 & \cos\beta \end{pmatrix} \tag{9}$$

Finally $R_z(\gamma)$ acts on the new vector giving the final vector

$$\mathbf{u}' = R(\alpha, \beta, \gamma)\, \mathbf{u} = R_z(\gamma)\, R_y(\beta)\, R_z(\alpha)\, \mathbf{u} \tag{10}$$

$$\begin{pmatrix} u_1' \\ u_2' \\ u_3' \end{pmatrix} = \begin{pmatrix} \cos\gamma & \sin\gamma & 0 \\ -\sin\gamma & \cos\gamma & 0 \\ 0 & 0 & 1 \end{pmatrix} \begin{pmatrix} \cos\beta & 0 & \sin\beta \\ 0 & 1 & 0 \\ -\sin\beta & 0 & \cos\beta \end{pmatrix} \begin{pmatrix} \cos\alpha & \sin\alpha & 0 \\ -\sin\alpha & \cos\alpha & 0 \\ 0 & 0 & 1 \end{pmatrix} \begin{pmatrix} u_1 \\ u_2 \\ u_3 \end{pmatrix} \tag{11}$$

Working out the matrix product we find

$$R(\alpha,\beta,\gamma) = \begin{pmatrix} \cos\gamma\cos\alpha - \sin\gamma\cos\beta\sin\alpha & \cos\gamma\sin\alpha + \sin\gamma\cos\beta\cos\alpha & \sin\gamma\sin\beta \\ -\sin\gamma\cos\alpha - \cos\gamma\cos\beta\sin\alpha & -\sin\gamma\sin\alpha + \cos\gamma\cos\beta\cos\alpha & \cos\gamma\sin\beta \\ \sin\beta\sin\alpha & -\sin\beta\cos\alpha & \cos\beta \end{pmatrix}$$

This is known as three dimensional rotation matrix. It is the most general orthogonal matrix of order 3, which is used in robotics control (Grewal, 2001; Joshi, 1995; Craig, 2005).

Conclusion

Theoretical back ground of the present study where three microcontrols which are control the robotics arm and two joints are used be chosen from star topology and the rotation for arm and joints where the steeper motor is used, their rotation are calculated by solving the matrices. In the present case the motion are used randomly for anticlockwise direction but can also be calculated for both the directions and also in different angles.

References

Artkin, G., 1968. *Mathematical Methods for Physicists.* Academic Press, New York.

Bowen, B.A and Burr, R.J.A., 1980. *The Logical Design of Multiple-Processor Systems.* Prentice-Hall, Englewood Cliffs, N.J.

Craig, John J., 2005. *Introduction to Robotics Mechanics and Control,* 2nd edn. Pearson Education.

Fathi, Eli T. and Knegen, Mosha, 1983. *Multiple Microprocessor Systems: What, Why and When Computer,* p. 23–32.

Gantmacher, F.R., 1959. *Application of the Theory of Matrices,* Wiley, New York.

Grewal, B.S., 2001. *Higher Engineering Mathematics.* Khanna Publisher, p. 47–48.

Joshi, A.W., 1995. *Matrices and Tensors in Physics,* 3rd edn. Wiley Eastern Limited.

Pipes, L.A., 1958. *Applied Mathematics for Engineers and Physicists.* McGraw Hill, New York.

Chapter 27

ESR Dating of the Neolithic Pottery Recently Exacavated in Banahalli, Karnataka State, India

G. Velraj and A. Mohamed Musthafa
Department of Physics, Periyar University, Salem – 636 011, Tamil Nadu

ABSTRACT

Electron spin resonance (ESR) spectroscopy is a sensitive tool to determine the age of bone, fossil, teeth and also pottery. The determination of age is purely based on radiation defects. A new attempt has been made to analyze the age of the pottery sample excavated from the historical place namely Banahalli in the Karnataka state, India. The powdered samples of interest were divided in to three parts with equal weighed mass. The sample was irradiated using gamma ray at room temperature at two different doses. The stimulation of unpaired electron concentrations leads to Electron spin resonance study. The ESR spectra were recorded for the three sample showed similar spectra. The Lande's g factor was calculated by using the absorption peak of the ESR spectra. The absorption peak intensity ofthe spectra were used for determining the age of the pottery sample. The Dose rate is assumed to be 1.12 mGy/a. The calculated age of the sample was 5000 years which corresponds to Neolithic period (New Stone age) as reported by the archaeologist.

Keywords: *ESR dating, Neolithic pottery, Gamma irradiation.*

Introduction

The origin of human race on the planet earth took place several thousands of years ago. The early history of human beings is called the pre-historic age. Written records are not available for the pre-historic period. However, the pre-historic people had left many things such as pieces of pottery, stone and metal tools, simple drawings, bones and skeletons. These materials provide some clue to know the history of the pre-historic period. Material remains belonging to the pre-historic period have also been unearthed in many places. Determining the age of the artifacts like pottery plays an eminent role in the science and technology. An attempt has been made to determining the age of pottery sample (BNH) collected from the archaeological place Banahalli, Karnataka by Electron Spin Resonance technique. An archaeological site Banahalli was excavated by Archaeological Survey of India (ASI), Chennai. During the excavation, number of potsherds, bricks and human skeleton and burial urns were found.

During about 3 decades of developments, electron spin resonance (ESR) dating, based on the measurement of radiation induced defects have been expanded rapidly, concerning the variety of suitable materials as well as dating range (Ikeya 1993)). There is great hope that the upper dating limit for quartz, mainly determined by the lifetime of the radiation induced defects, will reach 2 million years and more in future (Gruen, 1989)). The sensitivity limits of the present ESR spectrometers are often the reason for difficulties in dating of the sample containing only a small member of radiation induced defects.

Sample Preparation and Artificial Dosage

The sample BNH was taken from the 20th layer during excavation. A received state sample BNH has been taken and cleaned well before sample preparation. One larger pottery fragment (about 50 g) was treated as follows. The fragment was crushed carefully not destroy the quartz grains. After a treatment with hydrochloric acid (36 per cent) over night, the strong magnetic components were removed. The sample was sieved and the mass of the individual grain &actions was ensured in order to calculate the grain distribution. Samples were etched by 40 per cent hydrofluoric acid over 40 min to remove feldspars and the several micrometers of the surface of the quartz grains that contain the part influenced by external beta radiation and diffused impurities. Beforehand, the grains larger than 500 /lm were crushed carefully under water to such a degree that all grains became smaller than 500 μm (Bartoll and Ikeya, 1997). Then the sample above prepared kept in a hot air oven at 40°C for 20 mins by etching 40 per cent of HF. Finally Sample was again kept in a hot air oven for 21 hrs at 60°C. Aliquots of 100 mg were irradiated by a 60°C source. The sample was undergone additive dose of 600Gy, 1000Gy. During the irradiation process, the samples were sandwiched by plastic plates to improve the secondary electron equilibrium. The unpaired electrons are created in the samples by α, β and γ-rays. The concentration of the unpaired electrons can be determined directly by microwave absorption giving the EPR signal intensity.The Equivalent dose obtained is generally smaller than that determined by non linear fitting because the signals tend to saturate with dose, and this can arise in a slightly different way if two different types of defects are present (Ikeya 1975). In some cases, non-linear fitting to saturation curves is necessary. If the dose rate D is obtained, the age T_{age} may be expressed (in kilo annum) using the relation as suggested by Aitken (1985).

$$T_{age} = [ED(Gy)/D(Gy)]\ ka \quad (1)$$

ESR Experimental Analysis

The sample was measured with a commercial X-band ESR spectrometer (JEOL TES100) using a cylindrical cavity and a magnetic field modulation at 100 kHz with the modulation width of 0.1 mT, with a field scan of 25 mT, a microwave power of 5mW and a writing time of about 2 mT/min. The sample was rotated in the magnetic field before each measurement. A Mn2*/MgO standard, fixed with adhesive at the bottom of the sample tube, was used to calibrate the spectrometer sensitivity to compare different measurement series.

Results and Discussion

For all the ESR dating studies, the contribution of radioactivity inside the quartz grains should be considered more carefully. The ESR has shown the advantages that all measurements can be done several times with the irradiated samples and the radiation induced centres can be studied directly (Bartoll and Ikeya 1997). In the present study, the ESR spectra were recorded for the three samples as shown in Figure 27.1 and the 'g' value for the sample was 4.4758 by the formula:

$$g = hV/\mu_B B$$

where,

h: Plank's constant (6.625×10^{-34} Js),

v: Microwave frequency in Hz,

μ_B: Bohr magnetron (9.273×10^{-24} ampere-m^2)

B: Applied Magnetic field in Tesla.

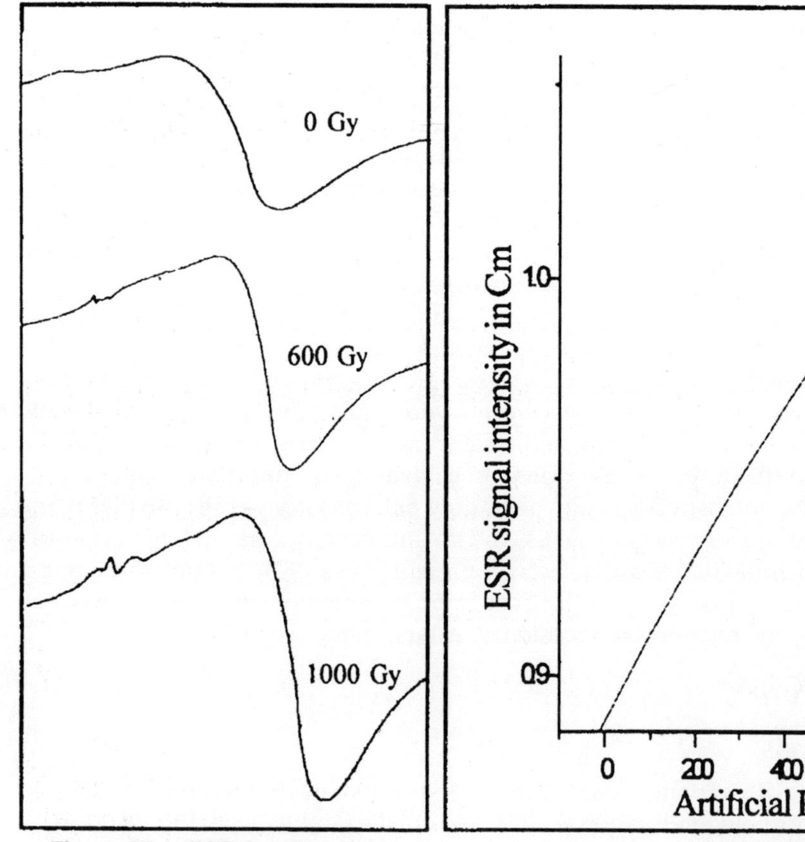

Figure 27.1: ESR Spectra of the Pottery Sample BNH

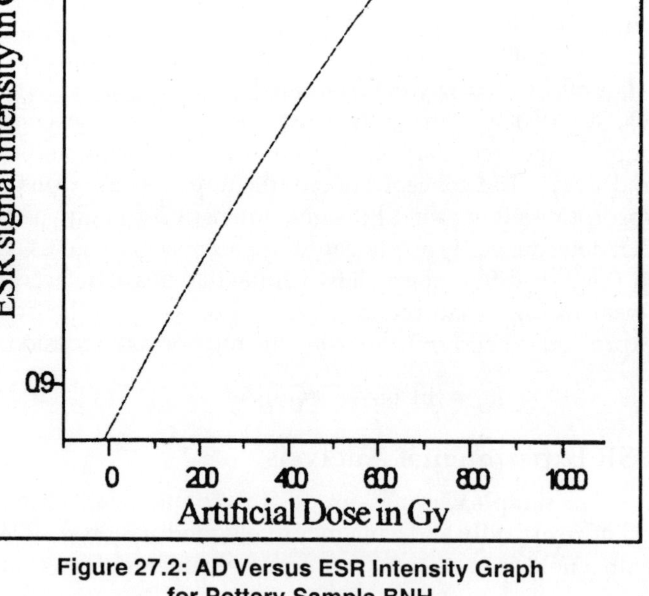

Figure 27.2: AD Versus ESR Intensity Graph for Pottery Sample BNH

The ESR intensity versus the artificial dose plot is towards the negative direction and allowing it to intersect the negative X-axis as shown in the graph. The negative value of the dose is Equivalent Dose (ED) of the sample and this value is substituted in equation (1) and the age determined by ESR studies found to be 5000 years back which corresponds to Neolithic period (New stone age) confirms the possibilities of human occupations.

Acknowledgements

The author acknowledge with thanks to Dr. B. Narasimhaiah, Site Excavator and Assistant Superintending Archaeologist (Retd.), Dr. P.S. Sriraman, Assistant Superintending Archaeologist, Archaeological Survey of India (ASI), Chennai for the timely help to collect the sample. The author thanks to Dr. D. Ponraj, Scientific Officer-F, Indira Gandhi Center for Atomic Research (IGCAR), Kalpakkam for his help to do gamma irradiation to the samples. Also thanks to Prof. (Dr.) P. Sambasiva Rao, Department of Chemistry, Pondicherry University granted permission to record ESR spectra of the irradiated samples in time.

References

Aitken, M.J., 1985. *Thermoluminescence Dating*. Academic Press, London, Oxford.

Bartoll, J. and Ikeya, M., 1997. ESR dating of Pottery: A trial. *Appl. Radiat. Isot.*, 48: 981–984.

Gruen, R., 1989. *Die ESR: Alterbestimmungsmethode*. Springer, Berlin (in German)

Ikeya, M., 1975. Dating a stalacite by electron paramagnetic resonance. *Nature*, 255: 48–50.

Ikeya, M., 1993. *New Applications of Electron Spin Resonance: Dating, Dosimetry and Microscopy*. World Scientific, Singapore.

Chapter 28

Analysis of Vibrational Spectra of 2,5-Dinitronaphthalene

P. Raju, A. Ramalingam, V. Karunakaran and C. Arunagiri
Department of Physics, Government Arts College, Ariyalur – 621 713, T.N.

ABSTRACT

The vibrational (infrared and Raman) spectra of 2,5-dinitronaphthalene have reported in the region 50– 4000cm^{-1} 2,5-dinitronaphthalene compound play an using KBr pellet. Observed fundamental bands were assigned assuming the molecules under point-group. The polarized infrared and Raman spectra of a 2,5-dinitronaphthalene have been recorded to determine the symmetries of the molecular vibration. These bands have been correlated with various modes of vibrations reported in the literature.

Keywords: *FTIR, Raman spectrum, KBr and Polyethylene pellets.*

Introduction

The vibrational assignments of infrared and Raman spectrum of 2,5-dinitronaphthalene have been extensively studied in present investigation. In the present work a detailed study of the polarized infrared and Raman spectra of 2,5-dinitronaphthalene and essential complete vibrational assignments of frequencies of the said molecule. Recent spectral investigation on 2,5-dinitronaphthalene find interesting application in commercial production and pharmacy. The studied of the vibrational assignments in light of previous work.

Material and Methods

Freshly crushed molecules of the 2,5-dinitronaphthalene were mixed with KBr in pellet. Using a hytrolic press and than subjected to the FTIR and FT Raman studies. FTIR and FT Raman of these compound were used recorded on BRUKER IFS 66Vmode1 FTIR and Raman spectro photometer. Its purity was confirmed by elemental analysis and melting point determination.

Results and Discussion

The molecular structure of the compound 2,5-dinitronaphthalene is shown in Figure 28.1. FTIR and Raman spectrum in KBr and Polyethylene pellets is shown in Figure 28.2. The observed fundamental vibrational frequencies and their proposed assignments of 2,5-dinitronaphthalene is given in Table 28.1.

Table 28.1: Assignments of the Frequency of 2.5-dinitronaphthalene

Observed Frequency (in cm^{-1})		*Intensity*	*Assignment*
FTIR	*FT Raman*		
3100	3058	W	v(C-H)
2992	–	W	v(C-H)
2988	–	W	v(C- H)
1605	1566	MS	v(C-N)
1519	1532	S	v(C- N)
1347	1352	S	v(C-C)
1231	–	MS	v(C-C)
1208	–	MS	v(C-C)
890	–	S	β(C-H)
791	803	VS	$\beta_s(NO_2)$
752	–	VS	$\beta_s(NO_2)$
705	–	S	γ_s(C-H)
571	601	MS	γ_s(C-H)
471	–	MS	$\gamma_w(NO_2)$
399	–	W	$\gamma_w(NO_2)$
374	349	MS	$\gamma_T(NO_2)$
312	–	W	γ_s(C-C-C)
221	220	VW	γ_s(C-C-C)
187	–	W	γ_s(C-N)
167	150	W	γ_s(C-N)

v: Stretching; β: In plane bending; W: Weak; S: Strong; β_s: Scissoring; γ_s: Out plane bending; MS: Medium Strong; VS: Very Strong; γ_w: Wagging; γ_T: Torsion; VW: Very weak.

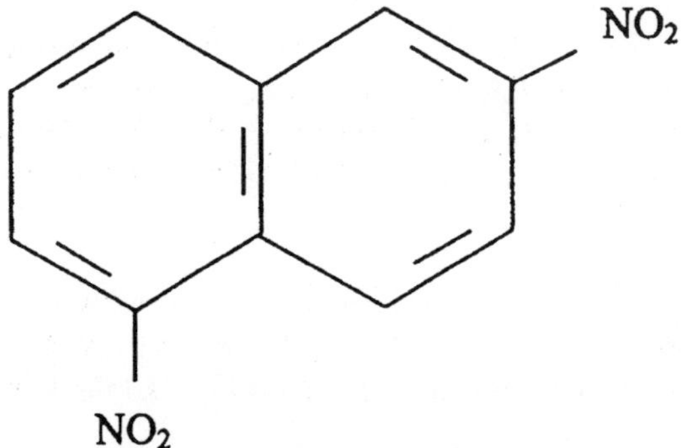

Figure 28.1: Molecular Structure of the 2,5-dinitronaphthalene

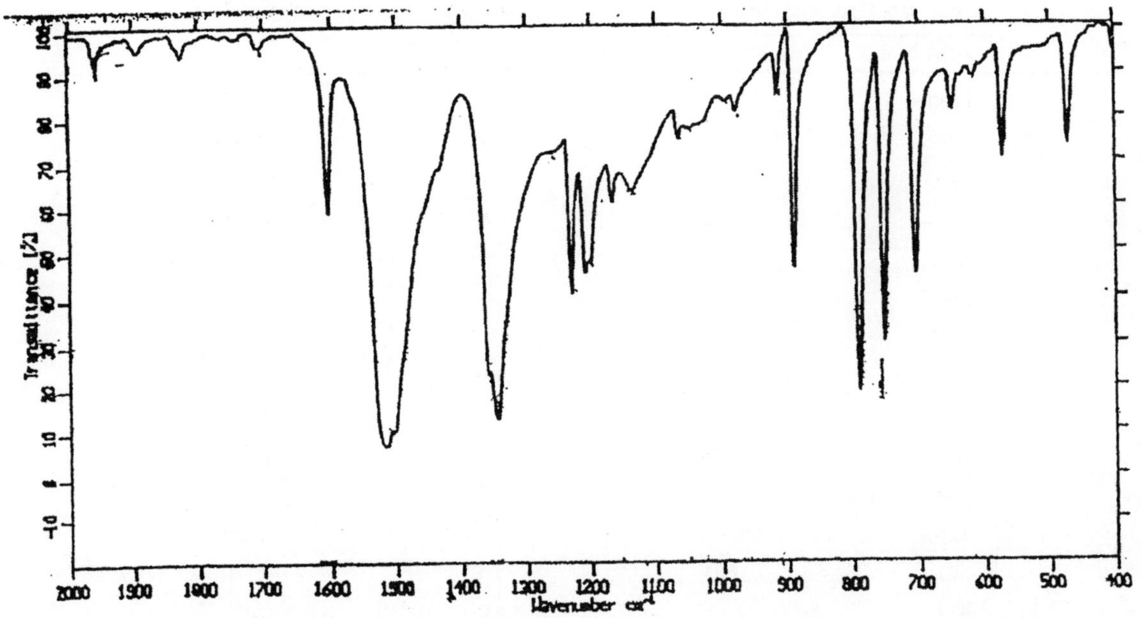

Figure 28.2a: FTIR Spectrum of 2,5-dinitronaphthalene

Group Vibration

NO_2 Group

The frequencies at 791 cm^{-1}, 752 cm^{-1} (FTIR) and 803 cm^{-1} (Raman) in 2,5-dinitronaphthalene have been assigned to scissoring. There frequencies are find in good agreement with the literature value. NO_2 group wagging has been assigned at 471 cm^{-1}, 399 cm^{-1} (FTIR) in the compound. The NO_2 group Torsion has been assigned at 374 cm^{-1} (FTIR) and 349 cm^{-1} (Raman) in the compound. This assignment is supported by the work of Mooney (Green *et al.*, 1970; Rao, 1963).

The C-N stretching has been assigned to the band observed at 1605 cm^{-1}, 1519 cm^{-1} (FTIR) and 1566 cm^{-1}, 1532 cm^{-1} (Raman) in 2,5-dinitronaphthalene compound (Yadav, 2007; Mohans, 1991).

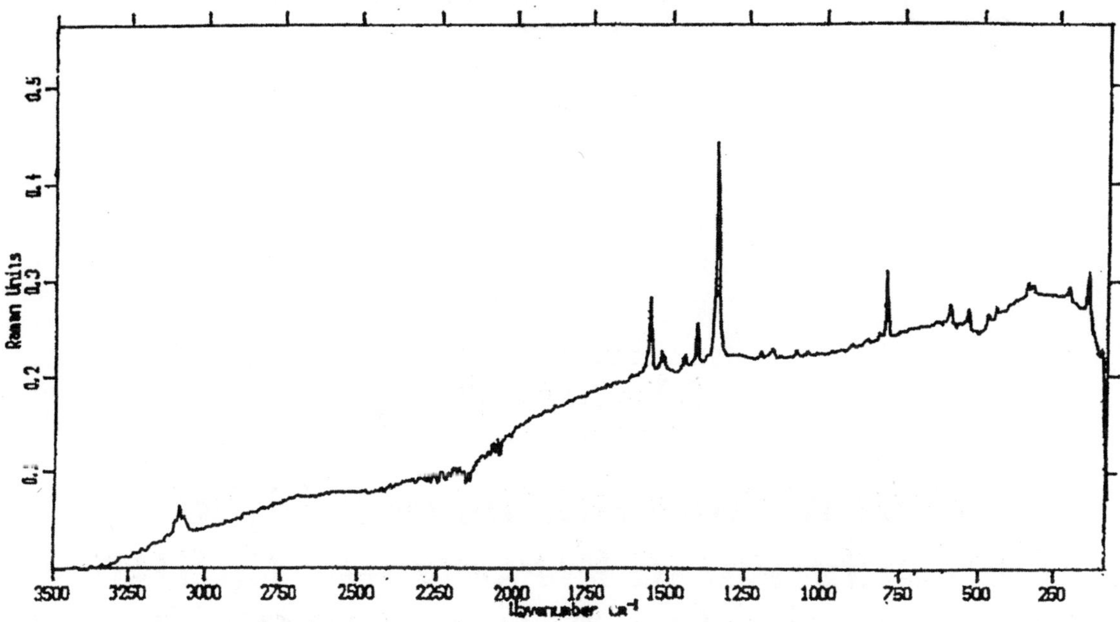

Figure 28.2b: FT Raman Spectrum of 2,5-Dinitronaphthalene

The C-H stretching vibration has been assigned the band observed at 3100 cm^{-1} to 2988 cm^{-1} (FTIR) and 3058 cm^{-1} (Raman) in 2,5-dinitronaphthalene components (Arunagiri, 2008; Yadav, 2007).

In the present investigation of the C-C-C out plane bending modes 312 cm^{-1}, 221 cm^{-1} (FTIR) with the counter part of Raman band at 220 cm^{-1} (Raman) in 2,5-dinitronaphthalene respectively.

References

Arunagiri, C. and Seshadri, S., 2008. *Acta Ciencia Indica,* 34(1): 117.

Green, J.H.S., Harrison, D.J. and Kynaston, W., 1970. *Spectochim. Acta,* 264: 1503.

Mohans, Sundaraganesan, N. and Mink, J.,1991. *Spectochim. Acta,* 47(4).

Rao, C.N.R., 1963. *Chemical Application of IR Spectroscopy*. Academic Press, New York.

Yadav, B.S. Yadav, Sanjay and Siddiqui, Seema I.A., 2007. *Acta ciencia Indica,* 33(1): 61.

Yadav, B.S., Yadav, Amit Kumar, Chand, Subhash, Yadav, M.K. and Siddiqui, L.A., 2007. *Acta Ciencia Indica,* 33(1): 119.

Chapter 29

Glutathione Status and Lipid Peroxidation in Freshwater Catfish *Clarias batrachus* Exposed to Synthetic Pyrethroid Lambda Cyhalothrin

R. Saravanan[1]*, K. Revathil[2] and P. Balakrishna Murthy[1]

[1]International Institute of Biotechnology and Toxicology, Padappai – 601 301, Kancheepuram District, Tamil Nadu

[2]Post Graduate and Research Department of Zoology, Ethiraj College for Women, Chennai – 600 008

ABSTRACT

The present study was to find out the effect of synthetic pyrethroid lambda cyhalothrin in liver, muscle, gills, kidney and testis of freshwater catfish *Clarias batrachus*. The fishes were exposed to the pesticide for a period of 45 days a sublethal concentration. Biochemical analysis of total proteins and total free amino acids were carried out on the 15th, 30th and 45th day of exposure to assess alteration in glutathione status and lipid peroxidation due to toxic stress caused to the fishes as a consequence of exposure to the compound. A decrease in the activity of glucose 6-phosphate dehydrogenase was noticed in the tissues through out the study along with increase in reduced glutathione on 30th day of exposure followed by a decrease on the 45th day. Lipid peroxides and oxidized glutathione showed an increase throughout the study.

Keywords: *Lambda cyhalothrin, Glutathione, Lipid peroxides, Glucose-6-phosphate dehydrogenase, Clarias batrachus.*

---

* E-mail: rsaravanan2@yahoo.com

Introduction

Synthetic pyrethroids were introduced in India in 1976. Discovery of new synthetic pyrethroids have sparked exciting advancement in the control of insect pests (Bradbury and Coats, 1989). Synthetic pyrethroids are most widely used group of insecticides in the world as pest control agents in agriculture and public health programmes and preferred over other pesticides because of their short residual life (Anees, 1975). However, the, use of such pesticides in excess of the recommended dose which ultimately enters the water bodies have a fatal effect on fishes and other aquatic organisms (Metelev *et al.*, 1983). The toxicity of pyrethroids pesticide to aquatic organism depends largely upon their solubility in water. All pyrethroids are not equally soluble in water. The gills of fish quickly absorb pyrethroids because of the lipophilic nature of the pesticide. Many authors have reported that pyrethroids are extremely toxic to fish and other aquatic fauna at acute sublethal levels (Agnihothrudu, 1998; Maund *et al.*, 2002; Weston *et al.*, 2004).

Synthetic pyrethroids are effective at very low application rates and remain stable in the environment. However, their rapid decomposition in soil through hydrolysis under aerobic conditions and restricted hydrolytic breakdown under aquatic environment was reported by Roberts and Standen (1977). Although these compounds have the advantage of rapid degradation they lack target specificity and have acute toxicity towards many non-target invertebrate and vertebrate species. Thus, many terrestrial and aquatic organisms may be at a risk of intoxication caused by the exposure to these compounds in the environment. Despite these facts synthetic pyrethroids are still the most abundantly used pesticides (Aspelin and Grube, 1999). The photostable pyrethroids are extremely effective as an agricultural insecticide. The metabolism of pyrethroids like delmethrin, cypermethrin and cyhalothrin in mammals involves ester cleavage due to carboxyesterase by hydrolysis of the central ester bond and oxidation whereas metabolism in fish is mostly oxidative (Demounte, 1989). The enzyme systems required to hydrolyse pyrethroids is deficient in fishes (Edward *et al.*, 1987; Bradbury and Coats, 1989). Reported half life ranges from 4 to 12 weeks. Its field half life is about 30 to 45 days (US Environment Protection Agency, 1988, 1992; Wauchope *et al.*, 1992). Lambda cyhalothrin has extremely low water solubility and is tightly bound to soil. It is expected to be prevalent in surface waters and its degradates from soil were minimal (WHO, 1990). One possible source of infiltration into surface water would be surface run off, the compound binds to soil particles and settles at the bottom and may have an effect on bottom dwelling organisms (Wauchope *et al.*, 1992; Hornsby *et al.*, 1995). Laboratory studies indicate that cyhalothrin has the potential to bioconcentrate in fish (WHO, 1990). Both cyhalothrin and lambda cyhalothrin are very toxic to fishes in clean water under laboratory conditions. Its bioconcentration has already been reported in channel catfishes (US Environmental Protection Agency, 1992).

Lipid peroxidation is a biochemical process which is mediated through free radicals. Lipid peroxidation is initiated when reactive oxidising species such as superoxide (O_2), hydrogen peroxide (H_2O_2), hydroxy radicals (OH) or singlet oxygen (O_2) interact with hydrogen atom from polyunsaturated fatty acids. Lipid peroxide formed in the primary site would be transferred through blood to other organs or tissues where the enhancement of damage occurs by the propagation of lipid peroxidation (Doramny, 1983). Severe oxidative stress may lead to cell injury due to several metabolic dysfunctions like peroxidation of membrane lipids caused by increased susceptibility of unsaturated bonds of membrane fatty acids, damage to DNA and proteins, depletion of nicotinamide nucleotide and cytoskeletal disruption (Ahmed, 2002).). Glutathione participates in reactions that destroy free radicals and certain foreign compounds. It brings about thiol transfer reaction that protect cell membrane and proteins specifically disulfide group under conditions of oxidative stress (Nicotera and Arrenius,

1986). Oxidized glutathione (GSSG) is harmful to the tissues and converted to reduced (GSH) which is required to maintain cell integrity (Lenartova *et al.*, 1997). The present study was designed to investigate the effect of lambda cyhalothrin in bottom dwelling catfishes as these organisms are more susceptible to the action of these pesticides because of their lipophilic and hydrophobic nature. The study would reveal the status of glutathione and lipid peroxidation in various tissues of these catfishes and how the fishes combat oxidative stress.

Materials and Methods

The technical grade synthetic pyrethroid, lambda cyhalothrin with 95 per cent purity used in the present study was supplied by Rallis India Ltd., Bangalore, India for evaluation of its toxic effects on nutritionally important freshwater catfish, *Clarias batrachus*.

Healthy adult male catfishes *Clarias batrachus* weighing 200–225 g and about 30–35 cms was used as the experimental model to evaluate the toxicity of lambda cyhalothrin, a synthetic pyrethroid widely used for agricultural applications in various sectors. The fishes were procured from the local fish market at Maduravoyil, Chennai, Tamil Nadu. They were brought to the laboratory and acclimatised under laboratory conditions for a period of three weeks and fed ad libitum.

The fishes were maintained in rectangular plastic tubs (64cm × 44cm × 29.5cm) filled with 20 litres of dechlorinated tap water The tubs were disinfected with potassium permanganate solution and washed thoroughly prior to introduction of fishes to prevent any fungal infection. Feeding was stopped 24 hours before the commencement of the toxicity test to keep the animals more or less in the same metabolic state. The water quality was determined periodically and physico-chemical characteristics such as temperature, pH, salinity, dissolved oxygen, total hardness and alkalinity were analysed following standard procedures (APHA, 1998).

Preliminary toxicity tests were carried out to find the median lethal tolerance limit of experimental fishes to synthetic pyrethroid, lambda cyhalothrin for 96 hours. To determine 96 hour LC_{50} static renewable bioassay method was adopted (Sprague, 1971). Five groups of 10 fishes each were set up in triplicates for the LC_{50} calculation. The LC_{50} bioassay method involved the exposure of the five groups of fishes to a range of five different concentrations of lambda cyhalothrin. Each group containing 10 fishes were maintained in 20 litres of tap water and the survival rate was noted for a period of 96 hours. The concentration at which around 50 per cent survival/mortality occurred after 96 hours was taken as the median lethal concentration.

The 96 hour LC_{50} was determined by Probit analysis method (Finney, 1971).The LC_{50} concentration for 96 hour was found to be 28.84 ppm. One-fifth of the LC_{50} concentration was taken as the sublethal concentration. The fishes were maintained for a period of 45 days at a sublethal concentration (5.768 ppm). Group I served as control while Group II was exposed to sublethal concentrations of lambda cyhalothrin for a period of 45 days. At the end of every 15, 30 and 45 days six fishes were sacrificed by cervical decapitation. The required tissues liver, muscle, kidney, gills and testis were dissected out and washed thoroughly with 0.9 N saline solution. Tissues were weighed and homogenized in Tris 0.1 M HCl buffer using Potter Elevenhjem homogenizer. The homogenate of the tissues were centrifuged at 2500 rpm for 15 minutes in a refrigerated high speed centrifuge and clear supernatant was used for biochemical analysis. Reduced glutathione was estimated by the method of Moron *et al.* (1979). Oxidised glutathione was estimated by the method of Beutler and Keller (1963). Glucose-6-phosphate dehydrogenase was assayed according to the method of Komberg and Horecker (1955). Lipid peroxide content was determined by thiobarbituric acid reaction as described by Okhawa *et al.*, (1979).

The data collected on the different parameters of the experimental study were subjected to statistical analysis (Snedecor and Cochran, 1989) by one way analysis of variance (ANOVA) followed by Duncan multiple range tests and the statistical significance was tested at 1 per cent and 5 per cent levels. The percentage change in experimental groups over controls was calculated to determine per cent elevation (+) or per cent reduction (-) during the different days of exposure.

Results

Glucose-6-Phosphate Dehydrogenase (G6PDH)

The results in changes in activity of glucose-6phosphate dehydrogenase is illustrated in (Table 29.1). This enzyme of the HMP shunt pathway showed an overall significant change ($P < 0.01$) during the different periods of the exposure to the sublethal concentration of the pesticide. The fishes exposed to sublethal concentration showed an elevation in the activity of the enzyme in the liver followed by a drastic decrease during 30th and 45th day of exposure by 11.36 per cent and 36.09 per cent when compared with control fishes. There was no significant change observed in the muscle during 15th and 30th day of exposure in the sublethal concentration. There was a significant decrease ($P < 0.05$) in the enzyme activity during 45th day of exposure in the fishes exposed to lambda cyhalothrin.The G6PDH activity in the kidney was increased by 16.14 per cent and 6.09 per cent in the groups exposed to sublethal concentration respectively during 15th day of exposure, followed by decrease of 27.29 per cent and 40.60 per cent in the group exposed to sublethal concentration.A similar trend in the G6PDH activity was observed in the gills and testis. The activity of G6PDH decreased constantly from 15th to 45th day of exposure in the fishes exposed to sublethal concentration.

Table 29.1: Effect of Lambda Cyhalothrin at Sublethal Concentration (5.768 ppm) on the Activity of Glucose-6-phosphate Dehydrogenase in Various Tissues of *Clarias batrachus*

Tissue	*Control*	*Experimental Groups*			*F-Value*	*P-Value*
		15 Days	*30 Days*	*45 Days*		
Liver	23.05±0.63^{b}	28.03±0.03^{c} (+21.61)	20.43±1.10^{b} (–11.36)	14.73±0.85^{d} (–36.09)	304.11	0.000
Muscle	4.18±0.81^{c}	3.69±0.77^{b} (–11.72)	3.19±0.08^{b} (–23.68)	1.99±0.64^{a} (–52.39)	9.218	0.000**
Kidney	16.60±0.91^{c}	19.28±0.80^{d} (+ 16.14)	12.07±0.68^{b} (–27.29)	9.86±0.41^{a} (–40.60)	208.23	0.000**
Gills	2.59±0.46^{c}	1.71±0.49^{b} (–33.98)	1.64±0.33^{b} (–36.68)	1.05±0.09^{a} (–59.46)	16.74	0.000**
Testis	11.47±1.08^{c}	10.34±0.75^{b} (–9.85)	9.43±0.52^{b} (–17.79)	7.63±1.00^{a} (–0.33)	21.13	0.000**

Values are Mean±SD (n = 6).

Specific activity of enzyme expressed as μmoles of $NADP^{+}$ reduced/min/mg ptn.

P Value < 0.01; **: Denotes significance at 1 per cent level (highly significant).

Since P value is less than 0.01 there is significant difference between the different days of exposure in higher as well as lower concentration.

Different alphabets in means between days in a row denotes significance at 5 per cent level.

Means carrying atleast one common superscript do not differ significantly (P<0.05).

Values in parentheses in experimental groups are per cent reduction (-) or per cent elevation (+) over control.

Reduced Glutathione (GSH)

The changes in the levels of antioxidant reduced glutathione in various tissues of fishes exposed to lambda cyhalothrin is depicted in (Table 29.2). In experimental group of fishes there has been a significant decrease ($P < 0.05$) in the GSH content in various tissues like liver, muscle, kidney and testis on 15th day, followed by an increase on the 30th day and thereafter a decrease on the 45th day. In the gills of these fishes a significant decline was not observed during the 30th and 45th day of exposure.

Table 29.2: Effect of Lambda Cyhalothrin at Sublethal Concentration (5.768 ppm) in the Levels of Reduced Glutathione in Various Tissues of *Clarias batrachus*

Tissue	*Control*	*Experimental Groups*			*F-Value*	*P-Value*
		15 Days	*30 Days*	*45 Days*		
Liver	65.99±0.62[d]	43.03±0.63[b] (–34.79)	57.66±0.54[c] (–12.62)	32.79±0.58[a] (–50.31)	3752.63	0.000**
Muscle	16.16±0.64[d]	14.52±0.65[b] (–10.14)	15.29±0.65[c] (–5.38)	11.90±0.52[a] (–126.3)	53.60	0.000**
Kidney	37.99±0.59[d]	18.79±0.60[b] (–50.53)	27.40±0.67[c] (+27.87)	11.91 ±0.52[a] (–68.64)	2147.92	0.000**
Gills	3.95±0.42[b]	4.20±0.66[b] (+6.32)	3.12±0.54[a] (–21.01)	2.49±0.53[a] (–36.96)	12.443	0.000**
Testis	10.86±0.41[d]	9.08t055[c] (–16.39)	5.81±0.53[b] (–46.50)	4.08±0.36[a] (–62.43)	243.08	0.000**

Values are Mean±SD (n = 6)

Values expressed as μgram of GSH/g wet tissue.

P value < 0.01; **: Denotes significance at 1 per cent level (highly significant)

P value > 0.05. Not significant (NS)

Since P value is less than 0.01 there is significant difference between the different days of exposure in higher as well as lower concentration.

Different alphabets in means between days in a row denotes significance at 5 per cent level.

Values in parentheses in experimental groups are per cent reduction (-) or per cent elevation (+) over control.

Oxidised Glutathione (GSSG)

The alterations in antioxidant status of oxidised glutathione of the different tissues of fishes exposed to lambda cyhalothrin are tabulated in (Table 29.3). The oxidised glutathione levels in liver, muscle, kidney and testis showed a similar trend of result in experimental groups. There was a significant elevation ($P < 0.05$) in the levels of oxidised glutathione on 1st, 30th and 45th day in the fishes exposed to the pesticide. The gills however differed in their response when compared to other tissues. There was an increase on the 45th day.

Lipid Peroxides (LPO)

The formation of lipid peroxides in various tissues of fishes exposed to lambda cyhalothrin are presented in (Table 29.4).The levels of lipid peroxide in liver and muscle tissue showed a highly significant ($P < 0.05$) increase in the animals exposed to sublethal concentration of the pesticide from 15th day to 45th day of exposure. In the muscle the content of the lipid peroxides increased drastically after the 15th day of exposure but were found to be reduced ($P < 0.05$) during the 45th day. Lipid peroxidation was highly significant ($P < 0.05$) in the kidney during 30th and 45th day of exposure in the animals exposed to the higher sublethal concentration. The gills did not show any significant changes during the different periods of exposure. The lipid peroxidation in testis increased to a maximum after the 30th day of exposure of the animals to the sublethal concentration of pesticide after which the change was insignificant by the end of 45th day.

Table 29.3: Levels of Oxidised Glutathione in Various Tissues of *Clarias batrachus* Exposed to Lambda Cyhalothrin at Sublethal Concentration (5.768 ppm)

Tissue	*Control*	*Experimental Groups*			*F-Value*	*P-Value*
		15 Days	*30 Days*	*45 Days*		
Liver	99.01±0.64[a]	108.67±0.53[b] (+9.75)	122.40±0.51[c] (+23.62)	143.24±0.49[d] (+44.67)	7069.08	0.000**
Muscle	27.73±0.64[a]	27.68±0.56[b] (±0.18)	32.86±0.13[c] (+ 18.49)	41.67±0.62[d] (+50.27)	1286.46	0.000**
Kidney	71.25±0.51[a]	85.58±0.49[b] (+20.11)	86.39±0.55[c] (+21.24)	96.91±0.54[d] (+36.01)	2439.23	0.000**
Gills	2.78±0.59[c]	2.89±0.61[c] (+3.95)	3.13±0.69[a] (+12.58)	4.04±0.78[b] (+45.32)	4.270	0.017*
Testis	14.59±0.33[a]	29.00±0.67[b] (+98.76)	30.99±0.46[c] (+ 112.4)	35.75±0.62[d] (+ 145.03)	1736.47	0.000**

Values are Mean±SD (n = 6).

Values expressed as µg of GSSG oxidised/g wet tissue.

P value < 0.01 **: Denotes significance at 1 per cent level (highly significant)

P value 0.011 to 0.05 *: Denotes significance at 5 per cent level

P value > 0.05 6 Not significant (NS)

Since P value is less than 0.01 there is significant difference between the different days of exposure in higher as well as lower concentration.

Different alphabets in means between days in a row denotes significance at 5 per cent level.

Values in parentheses in experimental groups are per cent reduction (-) or per cent elevation (+) over control.

Discussion

Lipid peroxidation is a problem in cells with high content of unsaturated lipids. It involves rearrangement and destruction of double bonds in unsaturated lipids by propagation of lipid free radicals formed during lipid peroxidation (Hermes Lima, 1998; Halliwell and Gutteridge, 2001). Free radicals promote lipid peroxidation which may alter membrane permeability. Glutathione is important in the regulation of redox state and protection of cells from oxidative damage (Gallagher *et al.*, 1992). Acute and subchronic exposure to pollutants seems to increase reduced glutathione contents in fish cells (Thomas and Wofford, 1984), while fishes subjected to heavy and chronic exposure to pollutants seem to be unable to maintain high constitutive glutathione and total glutathione levels. The present study shows depletion of tissue GSH in liver, muscle, kidney and gills in fishes exposed to lower and higher sublethal concentrations of lambda cyhalothrin during 15th and 45th day. In all the tissues studied the increase in GSH content on the 30th day may have reduced the deleterious effect of free radicals and increased lipid peroxides to combat the oxidative stress condition of the fishes. The thiolic group of GSH can react directly with hydrogen peroxide, superoxide anion, hydroxy radical and scavenge them (Comporti, 1987). Reports of Gallagher *et al.* (1992) support the hypothesis that exposure to environmental chemicals utilise or deplete GSH in tissues and predispose the fish to acute toxicity. Decline in GSH may result in increasing the lipid peroxidation causing stimulation of free radical formation rendering tissues to peroxidative damage and make the system inefficient. The level of oxidised glutathione is increased in all the tissues throughout the study. The higher levels of free radicals convert more reduced glutathione to its oxidised form. Reduced glutathione could not have maintained its supply towards the end of the exposure period because of its inability to regenerate due to decrease in NADPH production as a result of decline in G6PDH activity in various tissues.

Table 29.4: Lipid Peroxide Content in Various Tissues of *Clarias batrachus* Exposed to Lambda Cyhalothrin at Sublethal Concentration (5.768 ppm)

Tissue	*Control*	*Experimental Groups*			*F-Value*	*P-Value*
		15 Days	*30 Days*	*45 Days*		
Liver	9.66 ± 0.52^{a}	10.70 ± 0.50^{b} (+10.76)	17.29 ± 0.80^{c} (+78.98)	18.92 ± 0.70^{d} (+95.85)	314.91	0.000^{**}
Muscle	3.90 ± 0.61^{a}	4.58 ± 0.4^{b} (+17.43)	6.81 ± 0.45^{d} (+ 174.6)	8.37 ± 0.55^{d} (+114.6)	92.45	0.000^{**}
Kidney	3.47 ± 0.61^{a}	3.67 ± 0.62^{a} (+5.763)	5.62 ± 0.75^{b} (+6.195)	6.47 ± 0.74^{c} (+86.45)	27.70	0.000^{**}
Gills	1.78 ± 0.78^{a}	1.88 ± 0.75^{a} (+5.617)	1.72 ± 0.54^{a} (–3.370)	2.49 ± 0.66^{a} (+39.88)	1.585	0.224^{NS}
Testis	2.34 ± 0.52^{a}	3.24 ± 0.75^{b} (+38.46)	4.86 ± 0.76^{c} (+107.6)	5.15 ± 0.67^{c} (+ 120.0)	22.84	0.000^{**}

Values are Mean±SD (n = 6); Values expressed as nanomoles of TBA reactants formed/mg ptn.

P Value < 0.01- **: Denotes significance at 1 per cent level (highly significant)

P value > 0.05. Not significant (NS)

Since P value is less than 0.01 there is significant difference between the different days of exposure in higher as well as lower concentration.

Different alphabets in means between days in a row denotes significance at 5 per cent level.

Means carrying atleast one common superscript do not differ significantly (P < 0.05).

Values in parentheses in experimental groups are per cent reduction (-) or per cent elevation (+) over control.

In the present study lambda cyhalothrin could have caused cell destruction by the cyano groups reacting with sulfhydryl groups and thus leading to depletion of reduced glutathione towards the end of the exposure period in animal exposed to both higher and lower sublethal concentration. It is well established that glutathione plays a crucial protective role against cellular injury induced by oxidative stress or alkylation derived from a number of toxic compounds and their intermediates (Reed, 1990; A'Brien, 1991). A decrease in reduced glutathione and an increase in oxidised glutathione attribute to a reduction in total glutathione, hampering the ratio of reduced and oxidised glutathione. The decrease may have also been due to the decrease in the synthesis of glutathione because of the decrease in protein contents in all the tissues which was evident in the present study.

It has been indicated that the production of GSH in spermatogenic cells in dependent upon Sertoli cells and is based on the interaction between cell types and involved in sperm membrane maturation (Li *et al.*, 1989). The reducing equivalents necessary for regeneration of GSH are provided by NADPH from G6PDH activity which has been demonstrated to be the limiting step in the overall reductive pathway in the testis (Bauche *et al.*, 994).

The degeneration of spermatogenic cells due to the effect of lambda cyhalothrin in the testis could have reduced the production of GSH towards the end of the exposure period and as evidenced from the present study, testicular G6PDH activity has decreased which may have reduced the regeneration of GSH. NADPH required for the reaction has declined impeding the whole mechanism of GSH production. On the contrary, there has been an increased production of oxidised glutathione in the testicular cells. The alteration in the GSH content has reduced the protective effect of this molecule on the lipid components of testicular cell membrane indicative of an increase in lipid peroxidation process in the testis in the present study as the testis is rich in polyunsaturated fatty acid composition (Godeas *et al.*, 1994). This may cause reduction in spermatogenesis and GSH-dependent mechanisms against oxidative stress in the spermatids (Den Boer *et al.*, 1990).

ROS, free radicals and reactive metabolites are detoxified by conjugating with glutathione or reacting with tissue macromolecules to initiate the damage. Lambda cyhalothrin could be metabolised to toxic metabolites which at lower concentration can be detoxified by conjugating with glutathione. There was an increase in glutathione levels initially during the exposure period in various tissues but at the end of 45th day it could have been unable to bring about detoxification due to depletion of tissue glutathione levels. In the present study oxidative stress in fishes exposed to lambda cyhalothrin was evident by induction of lipid peroxidation, an increase in lipid peroxides in liver, kidney, muscle and testis, but gills did not show any significant change. The extent of lipid peroxide was determined by the balance between production of oxidants and the removal and scavenging of these oxidants by antioxidants. Many studies have demonstrated that lipid peroxidation and oxidative stress increases in tissues of different species of aquatic organisms as a result of being exposed to environmental stressors (Wilson and Digiulio, 1991). Lipid peroxidation may ultimately lead to disintegration of membrane phospholipid. This can eventually lead to change in membrane properties and loss of its bound enzymes (Baker *et al.*, 1997; Angeles *et al.*, 2005).

References

Agnihothrudu, V., 1998. Pyrethroid: Their future and toxicity. In: P.K. Gupta, V. Ragivarknsh (Eds.) *Advances in Toxicology and Environmennntal Health*, Proc. 4th Ann. Conf. Soc. Toxicol., Guwahati, p. 65–69.

Ahmed, R.G., 2002. Is there a balance between oxidative stress and antioxidant defense system during development? *Med. J. Isl. Acad. Sci.*, 5(2): 55–63.

Anees, M.A., 1975. Acute toxicity of insecticides in freshwater *Channa punctatus*. *Pak. J. Zool.*, 7: 135–141.

Angeles Silvia, P., Guillermo, R., Vazquez, C.M., Moreno, I., Prieto, A.I. and Camean A.M., 2005. Toxic cyanobacterial cells, containing microcystins induce oxidative stress in exposed Tilapia fish (*Oreochromis mossambieus*) under laboratory conditions. *Aquat. Toxicol.*, 72: 261–271.

APHA, 1998. *Standards Methods for the Examination of Water and Wastewater*, 20th edn. American Public Health Association, American Water Works, Association Water Environment Federation. Taras, T.J., Greenberg, A.E., Hoak, RD. and Rand, M.C. (Eds) Washington DC.

Aspelin, A.H. and Grube, A.H., 1999. Pesticide industry sales and usage 1996 and 1997. Market estimates 733-R-99-001 D.S Environmental Protection Agency, Washington, DC.

Bauche, F., Fouchard, M.H. and Jegou, B., 1994. Antioxidant systems in rat testicular cells. *FESS. Lett.*, 349: 392–396.

Baker, R.T.M., Handy, RD., Davies, S.L. and Shook, J.C., 1997. Chronic dietary exposure to copper affects growth, tissue lipid peroxidation and metal composition of grey mullet. *Mar. Environ. Res.*, 45: 357–365.

Beutler, E. and Keller, B.M., 1963. The effect of sodium nitrate on red cell glutathione. *Experientia*, 19: 96–97.

Bradbury, S.P. and Coats, J.R., 1989. Comparative toxicology of the pyrethroid insecticides. *Review Environ. Contam. Toxicol.*, 108: 134–177.

Comporti, M., 1987. Glutathione depleting agents and lipid peroxidation. *Chem. Physics. Lipids*, 45: 143–169.

Demounte, J.P., 1989. A brief review of environmental fate and metabolism and pyrethroids. *Pestic. Sci.*, 27: 375–385.

Den Boer, P.J., Mackenbach, P. and Grootegoed, J.A., 1990. Glutathione metabolism in cultured Sertoli cells and spermatogenic cells from hamster. *J. Reprod. Fertil.*, 87: 391–400.

Doramny, T.L., 1983. An approach to free radicals. *Lancet*, 2: 1010–1014.

Edwards, R., Milburn, P. and Hutson, D.H., 1987. The toxicity and metabolism of pyrethroidscis and transcypermethrin in rainbow trout, *Salmo gairdneri. Xenobiotica*, 17: 75–119.

Finney, D.J., 1971. *Probit Analysis: A Statistical Treatment of Sigmoid Curve*, 3rd edn. Cambridge University Press, London, pp. 568.

Gallagher, E.P., Canada, A.T. and Diguilio, R.T., 1992. The protective role of glutathione in chlorothalonin induced toxicity to channel catfish. *Aquat. Toxicol.*, 23: 155–168.

Godeas, C., Sandri, G. and Panfili, E., 1994. Distritution of PHGPX in rat testis mitochondria. *Biochem. Biophys. Acta.*, 1191: 147–150.

Halliwell, B., and Gutteridge, J.M.C., 2001. *Free Radicals in Biology and Medicine*, 3rd edn. Clerendon Press. Oxford.

Hermes Lima, M., Storey, J.M. and Storey, K.B., 1998. Antioxidant defenses and metabolic depression: The hypothesis of preparations for oxidative stress in land snails. *Comp. Biochem. Physiol.*, 120B: 437–448.

Hornsby, A.G., Wauchope, R.D. and Herner, AE., 1995. *Pesticide Properties in the Environment*, Springer Verlag, New York, pp. 132.

Kornberg, A. and Horecker, B.L., 1955. Glucose-6-phosphate dehydrogenase. *Methods in Enzymology*, 1: 322.

Lenartova, V.A.V., Holovska, K., Pedrajas, J.R., Martine, Z., Lara, E., Peinacto, J., Lopez-Barca, J., Rosival, I. and Kosuth, P., 1997. Antioxidant and detoxifying fish enzymes as biomakers of river pollution. *Biomarkers*, 2: 247–252.

Li, L., Seddon, A.P., Meister, A. and Risley, M.S., 1989. Spermatogenic cell-somatic cell interactions are required for maintenance of spermatogenic cell glutathione. *Biol. Reprod.*, 40: 317–331.

Maund, S.J., Hamer, M.J., Lane, E., Farrell J.H., Goggin, U.M and Gentle, W.E., 2002 Partioning, bioavailability and toxicity of the pyrethroid insecticide cypermethrin in sediments. Environ. Toxicol. Chem., 21(1): 1683–1689.

Metelev, V.V., Kanara, A.I. and Dzasokhova, N.G., 1983. *Water toxicology*. Amerinol Publishing Company Pvt. Ltd., New Delhi.

Moron, M.S., Beperre, J.W. and Mannervik, B., 1979. Levels of glutathione, glutathione reductase and glutathione-S-transferase in rat lungs and liver. *Biochem. Biophys. Acta.*, 582: 67–68.

Nicotera, P. and Orrenius, M., 1986. Role of thiols in protection against biological reactive intermediates. *Adv. Exp. Med. Biol.*, 187: 47–51.

Okhawa, H., Ohishi, N. and Yagi, K., 1979. Assay for lipid peroxides in animal tissues with thiobarbituric acid reaction. Anal. Biochem., 95: 351–345.

O'Brien 1991. Molecular mechanisms of quinone cytotoxicity. *Chem. Bioi. Interact.*, 80: 1–41.

Reed, D.J., 1990. Glutathione: Toxicological implications. *Ann. Rev. Pharmacal. Toxicol.*, 30: 603–631.

Roberts, T.R. and Standen, M.E., 1977. Degradation of pyrethroid insecticide WL 41706 (+)-Lcyano-3-phenoxy benzyl, 2,2,3,3, tetramethyl cyclopropane carboxylate in soil. *Pestic. Sci.*, 8: 600–610.

Snedecor, G.W. and Cochran, W.G., 1989. *Statistical Methods*, 8th edn. The Iowa State College Press, Iowa, pp. 135.

Sprague, J.B., 1971. Measurement of pollution and measurements, effects toxicity of fish III: Sublethal effects and safe concentration. *Water Res.*, 5: 245–266.

Thomas, P. and Wofford, H.W., 1984. Effects of metals and organic compounds on hepatic glutathione, cysteine and acid-soluble thiol levels in grey mullet (*Mugil cephalus*). *Toxicol. Appl. Pharmacol.*, 76: 172–182.

US Environmental Protection Agency, 1988. Fact sheet Number 171: Karate Washington, DC pp. 321.

US Environmental Protection Agency, 1992. *Environmental Fate and Effects: Pesticide Environmental Fate Online Summary: Lambda Cyhalothrin*, Washington, DC.

Wauchope, R.D., Buttler, T.M., Hornsby, A.G., Augustijn Beckers, P.W.M. and Burt, J.P., 1992. SCS/ARS/CES. Pesticide properties database for environmental decision making. Review of environmental decision making. *Rev. Environ. Contam. Toxicol.*, 123.

Weston, D.P., You, J.C. and Lydy, M.J., 2004. Distribution and toxicity of sediment associated pesticides in agricultural dominated water bodies of California's Central Valley. *Environ. Sci. Tech.*, 38(10): 2752–2759.

Wilson, G.W. and Digiulio, R.T., 1991. Proxidant and antioxidant mechanism in aquatic organisms. *Aquat. Toxicol.*, 19: 137–161.

World Health Organisation, 1990. Cyhalothrin–Environmental health criteria 99, Geneva, Switzerland.

Chapter 30

Diversity Indices of Selected Macrobenthos in Karanja Creek (District Raigad), Maharashtra, West Coast of India

Prabhakar R. Pawar[1]* and Balasaheb G. Kulkarni[2]

[1]**Veer Wajekar Arts, Science and Commerce College, Phunde, Uran (District Raigad) – 400 702, Maharashtra**

[2]**The Institute of Science, 15, Madam Cama Road, Mumbai – 400 032, Maharashtra**

ABSTRACT

The diversity indices can be used as a good measure for studying the effect of industrial pollution because industrial wastes and sewage almost always reduce the diversity of natural systems into which they are discharged. A measurement of diversity often provides a better index of pollution than a direct measurement of pollutants. Assessment of macrobenthos diversity with respect to diversity indices reflects the marine population, habitat disturbance and serves as an important indicator of environmental conditions. The present study is designed to study the diversity indices of selected macrobenthos at two ecologically distinct locations of the Karanja creek (District Raigad) Maharashtra, west coast of India. Results on various diversity indices like Index of Frequency (F) Or Importance Probability (Pi), Index of Dominance (c), Rarity Index (R), Shannon's Index of General Diversity (H'), Margalefs Richness Index (R') and Evenness

* Corresponding author. E-mail: prabhakar_pawar1962@yahoo.co.in

Index (e) does not vary significantly. This demonstrates that at present Karanja creek harbors varied forms of macrobenthic community showing no effect of human disturbance but in future measures must be taken for protection and conservation of macrobenthic community of the creek.

Keywords: *Diversity indices, conservation, macrobenthos, Container freight stations, Jawaharlal Nehru Port Trust (JNPT).*

Introduction

Macrobenthos constitutes a dominant functional component in marine ecosystem and reveals a wide range of adaptability to different marine and coastal habitats. Diversity indices serve as a valuable tool to quantify diversity in a community and to describe its numerical structure. In biodiversity investigations, assessment of diversity indices is essential to determine the community structure of a particular ecosystem (Dash, 2002). Many reports have been published on macrobenthos diversity in and around Mumbai (Govindan, 2002; Quadros and Athalye, 2002; Venkatachalam and Kale, 2002; Jaiswar and Kulkarni, 2005; Jaiswar *et al.*, 2006), but analysis of diversity indices of macrobenthos has not been reported adequately (Annie and Govindan, 1995; Venkatachalam and Kale, 2002).

Karanja creek (Lat. 18°50′15″ N and Long. 72°57′ 15″ E) located along the eastern shore of Mumbai harbour opposite to Coloba, encircles the village Karanja. The creek is continuous with Dharamtar and Pen-Khopoli creek joining Arabian Sea near Uran (Figure 30.1) on west coast of India. Recently whole coastal belt of Uran is under heavy process of urbanization due to number of industries settled on the coastal line ofUran. A major port called Jawaharlal Nehru Port Trust (JNPT) lies near Uran, which is 10 km away from Karanja. The JNPT is one of the busiest international ports in the country and supports a variety of maritime activities. The area of Uran shore and Karanja creek became the ground for hectic activities of many Container Freight Stations that are created along the stretch of Karanja creek. As a result of these activities, the macrobenthic fauna of the creek is affected. Hence, this area has been identified for the study of diversity indices of selected macrobenthos in relation to changing environmental conditions. Further, no earlier reports are available on macrobenthic diversity in Karanja creek, so the data presented can take as a baseline data for better management of the creek in future.

Materials and Methods

The Study Area

During present investigation, a detailed monitoring of macrobenthos diversity was done along the 17 km stretch of Karanja creek for two consecutive years (January 2004 to December 2005). Two study sites representing the wider pan (seaward end) and narrow pan were selected along the creek and at each site, intertidal area was measured and marked into Upper littoral zone (ULZ), mid littoral zone (MLZ) and Lower littoral zone (LLZ) following Bhatt (1959) and Parulekar (1973). The selected sites were visited fortnightly at spring low tide from January 2004 to December 2005 and macrobenthos were collected and processed as per the recommendations of Holme and McIntyre (1984). Identification of macrobenthos was done following the work of Subrahmanyam *et al.* (1949, 1951, 1952), Chhapgar (1957, 1958), Apte (1988), Fish and Fish (1996) and Khan and Murugesan (2005). Macrobenthos collected from all sites were analyzed for diversity indices following standard methods (Eleftheriou and Holme, 1984; Parulekar *et al.*, 1976).

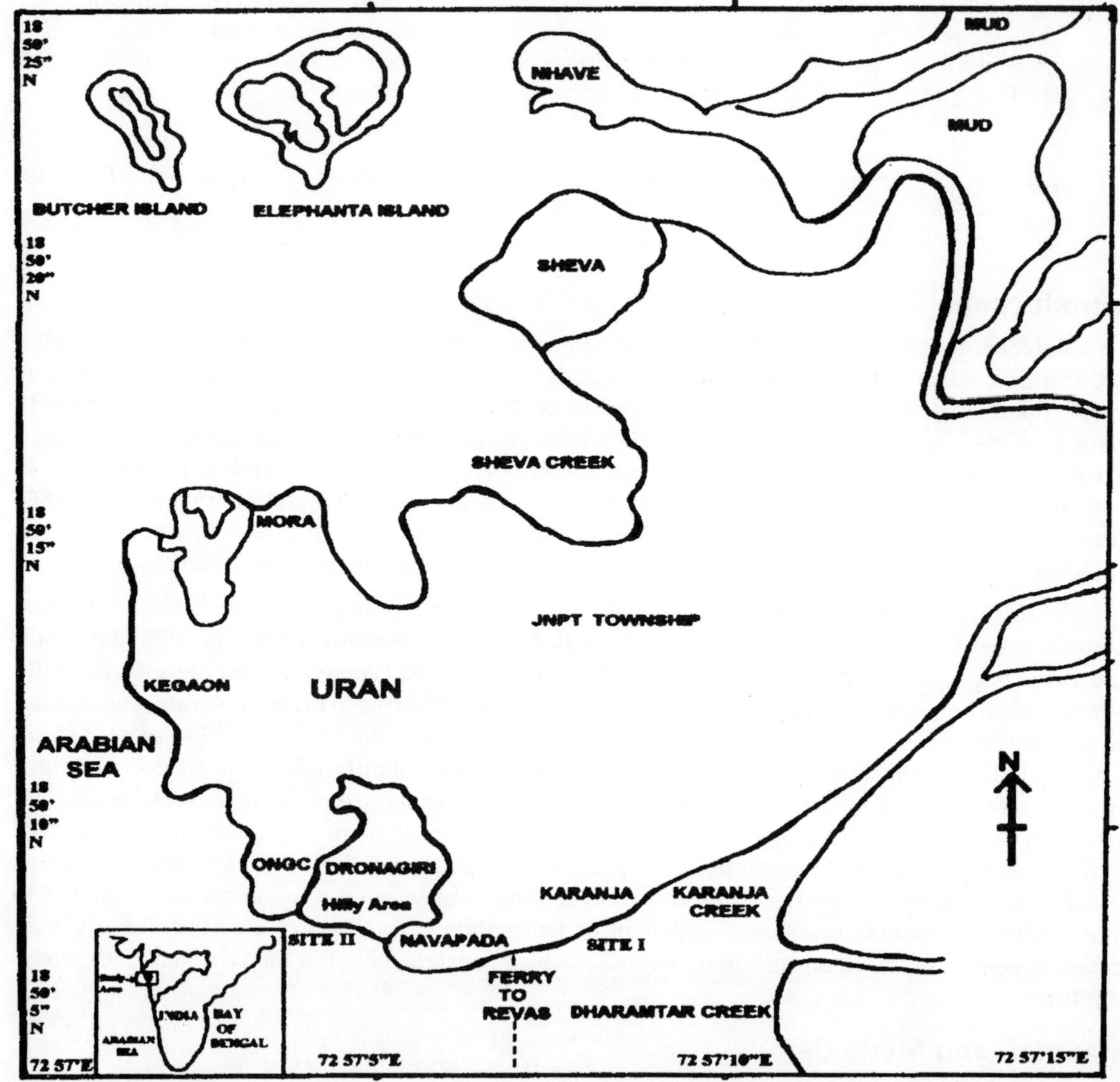

Figure 30.1: General Map of the Karanja Creek Showing the Sampling Sites I and II

Diversity Indices

Following indices were calculated for the quantification of biodiversity and comparison of species diversity.

1. **Index of Frequency (F) OR Importance Probability (Pi) [Smith and Smith, 1988]**

$$\text{Pi or F} = \frac{ni}{n}$$

where,

ni: Number of individuals of each species in the community

N: Total number of all individuals of all species in the community

2. Index of Dominance (c) [Simpson, 1949]

$$c = \Sigma (ni/Ni)^2$$

where,

ni: Number of individuals of each species in the community

N: Total number of all individuals of all species in the community

3. Rarity Index (R) [Ludwig and Reynolds, 1988]

$$R = \frac{1}{F}$$

where,

F: Index of frequency

4. Shannon's Index of General Diversity (H') [Shannon and Weaver, 1949]

$$H' = -\Sigma\, Pi.\, Log_e\, Pi$$

where,

Pi: Importance Probability,

log_e: In (log natural) OR = $log_{10} \times 2.303$

5. Margalef's Richness Index (R_1) [Margalef, 1958]

$$R_1 = \frac{S-1}{Log\, N}$$

where,

S: Number of species in the community

N: Total number of all individuals of all species in the community

6. Evenness Index (e) [Pielou, 1969]

$$e = \frac{H'}{Log\, S}$$

where,

H': Shannon's Index of General Diversity

S: Number of species in the community

Results

The faunal groups encountered at Karanja creek were Polychaetes, gastropods, crustaceans, pelecypods and echinoderms. Since species like *Bursa tuberculata, Nerita oryzarum, Trochus radiatus* of gastropods, *Arca symmetrica* of pelecypods, *Eupagurus prideauxi* of crustaceans and *Perinereis cultrifera* of polychaetes were recorded abundantly, their diversity indices were assessed during present investigation.

The index of frequency (F) or Importance Probability (Pi) of the selected macrobenthos shows seasonal variation and the values were high in premonsoon and postmonsoon than the monsoon (Figure 30.2). Highest F was recorded for *P. cultrifera* during premonsoon (March 2004, 0.741) followed by *B. tuberculata* during postmonsoon (January 2004, 0.158); whereas lowest values were recorded for *A. symmetrica* (May 2005, 0.007). The index of frequency can be placed in order of *P. cultrifera* > *B. tuberculata* > *N. oryzarum* > *E. prideauxi* > *T. radiatus* > *A. symmetrica.*

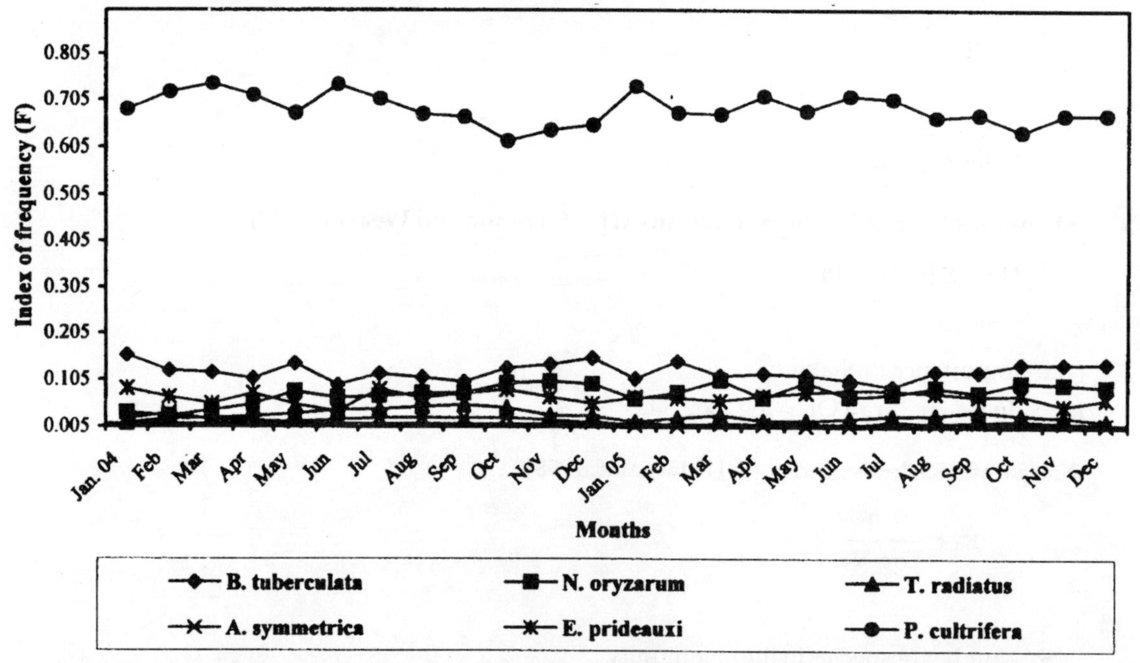

Figure 30.2: Index of Frequency (F) or Importance Probability (Pi)

The index of dominance was uniform at the three sites where the dominance of the species can be placed in order as *P. cultrifera* > *B. tuberculata* > *E. prideauxi* > *N. oryzarum* > *T. radiatus* > *A. symmetrica.* The data on index of dominance shows that the pelecepod species *A. symmetrica* dominated over by other species (Figure 30.3).

The rarity index was according to index of dominance (Figure 30.4). In this case, *A. symmetrica* showing highest rarity index, found to be rare with respect to population density as well as biomass. Rarity index of polychaetes *i.e.* as *P. cultrifera*, found to be very common at all the three sites. Lower Rarity index of polychaetes marks the higher density of it. Among other species, Bursa species found to be dominant over *Trochus* and, *Nerita* species assessed. The pattern of Rarity index was not

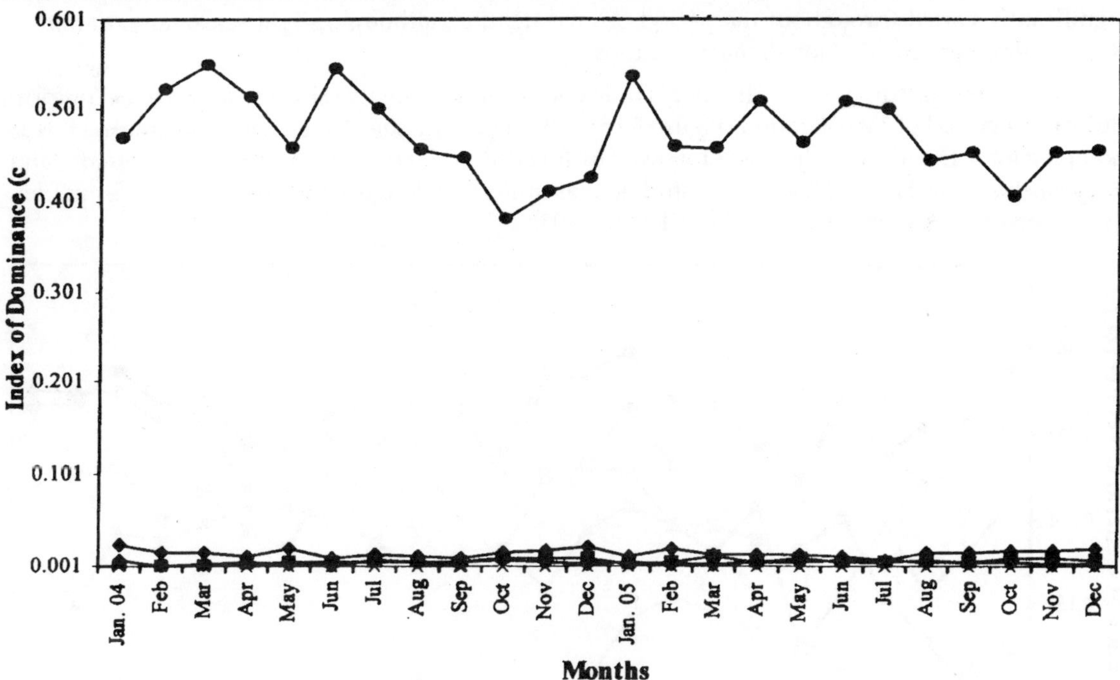

Figure 30.3: Index of Dominance (c)

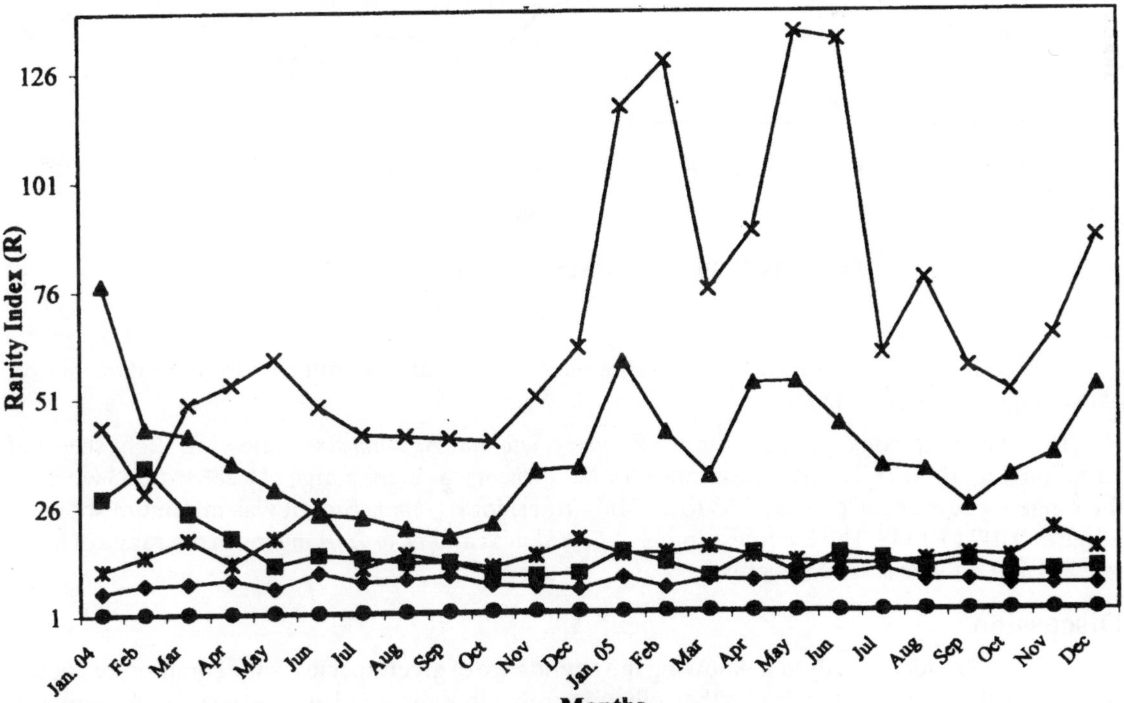

Figure 30.4: Rarity Index (R)

significantly varied throughout the period of investigation, although, slight seasonal variation of Rarity index was recorded for all macrobenthos.

Among the gastropods, the Shannon's Index of General Diversity (H'), was found to be uniform during the period of investigation (Figure 30.5). Among all the selected species, maximum H' was observed for *B. tuberculata*, which was followed by *P. cultrifera*, *E. prideauxi*, *N. oryzarum*, *T. radiatus* and *A. symmetrica*. The H' of all the species studied was uniform throughout the period of investigation. The diversity of *A. symmetrica i.e.* H' was less than 0.05.

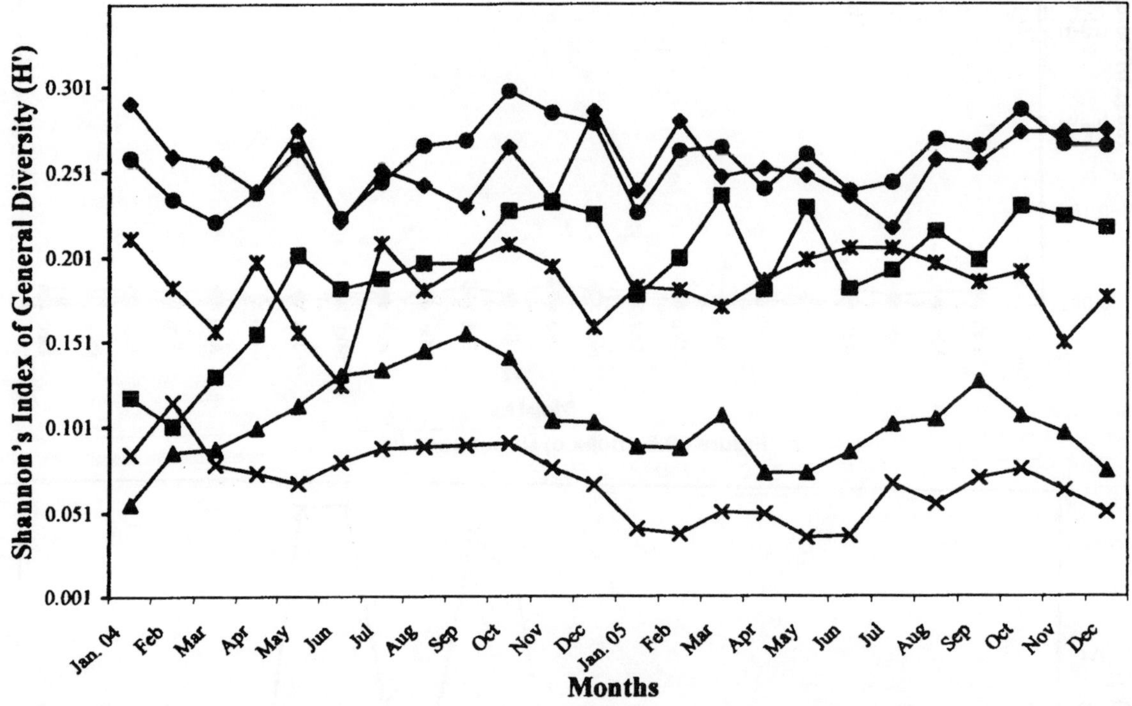

Figure 30.5: Shannon's Index of General Diversity (H')

Data on Richness index recorded is in agreement with H' and is uniform throughout the period of investigation (Figure 30.6).

The Evenness index was uniform for *B. tuberculata* and *A. symmetrica* showing slight seasonal variation (Figure 30.7). The Evenness index for *P. cultrifera* was in the range of 0.286 to 0.383 whereas the range for *B. tuberculata* was 0.253 to 0.374. Further, for *A. symmetrica*, it was minimum showing range of 0.047 to 0.148. The Evenness index of *E. prideauxi* and *N. oryzarum* was in the range of 0.160 to 0.273 and 0.130 to 0.305 respectively.

Discussion

Index of Frequency is used in knowing the abundance of given species in the community. Figure 30.2 shows abundance of *P. cultriftra* than other macrobenthos recorded at selected sites. A minimum index of frequency was noted for blood clam *A. symmetrica*. Non-significant variation of indices of

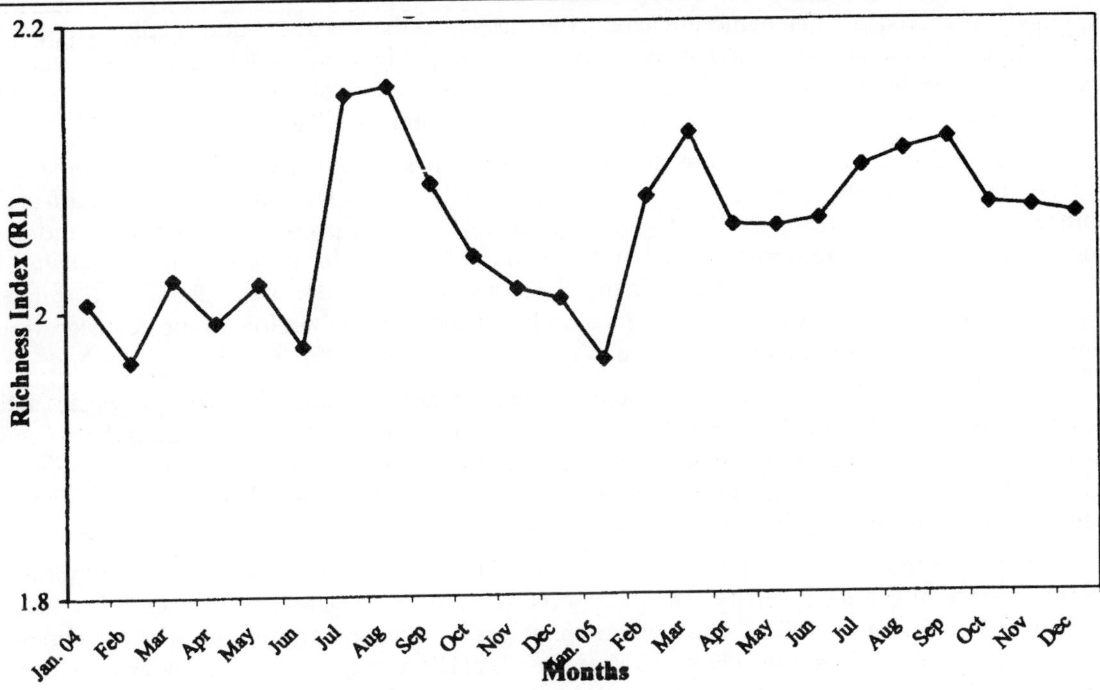

Figure 30.6: Margalef's Richness Index (RI)

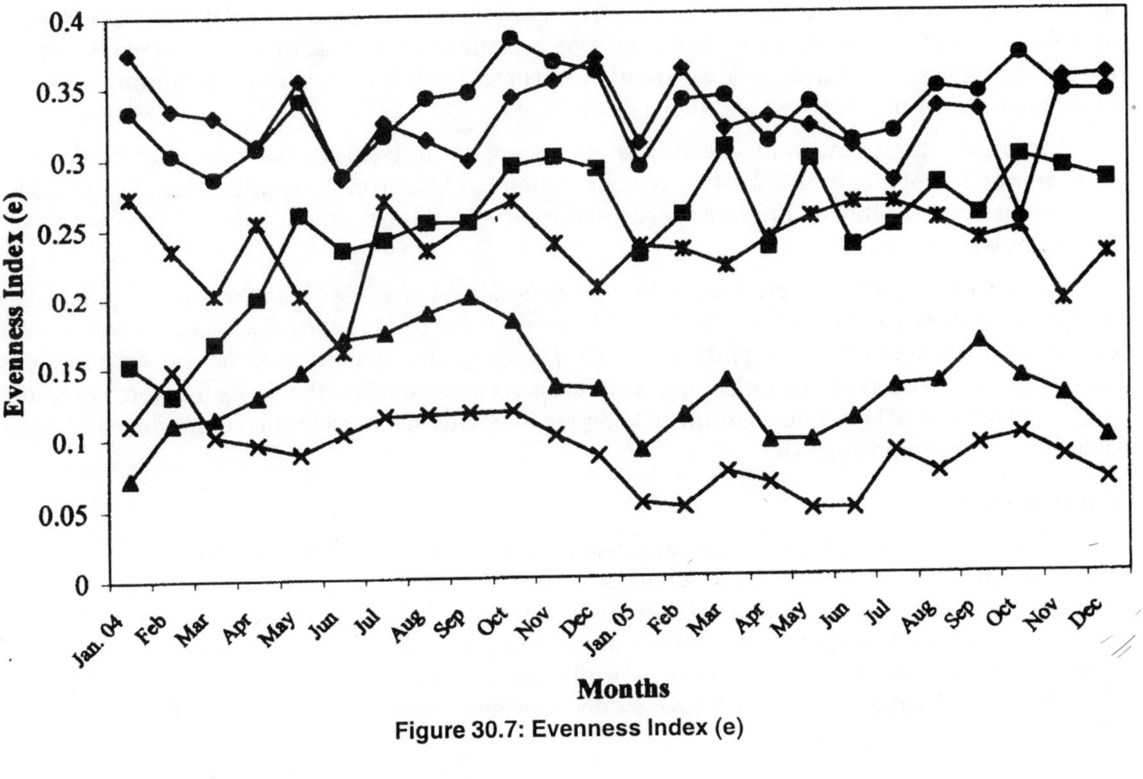

Figure 30.7: Evenness Index (e)

frequency observed in different months at both the sites shows that the distribution of different species was uniform throughout the period of investigation. Further, a higher index of frequency of *P. cultrifera* was also supported by its index of dominance. The data on index of dominance shows a positive relationship in between index of frequency and index of dominance (Figure 30.3).

Index of dominance is the degree to which dominance is concentrated in one, a few or several species in a community. The species that largely control the energy flow and/or strongly affect the other species are described as "Ecological dominants". During present investigation, index of dominance was not significantly varied and was found to lowest for *B. tuberculata, T. radiatus, N. oryzarum, A. symmetrica* and *E. prideauxi* during the monsoon. High index of dominance of *P. cultrifera* was observed during monsoon. The results of index of dominance therefore, indicate insignificant effect of monsoon conditions on *P. cultriftra.*

The Rarity index, which is the expression of how uncommon or rare, is a particular species in the community. The result of Rarity index (Figure 30.4) shows an inverse relationship in between its Rarity index and its level of dominance. *e.g.* the rarity index of *A. symmetrica* was highest whereas its dominance index and evenness index was lowest. The data on rarity index also reveals dominance of the *P. cultrifera* throughout the period of investigation.

Shannon's Index of General Diversity is used as an indication of richness of the community in terms of general species diversity. The data on H' (Figure 30.5) shows that species richness of *P. cultrifera, B. tuberculata* and *E. prideauxi* was more or less same whereas minimum species richness of *T. radiatus,* and *A. symmetrica* was observed. These results of H' are in agreement with index of frequency, index of dominance, Rarity index and was also not varied significantly.

The richness index is a measure of number of different kinds of organisms present in a particular area or number of species present per sample. The more species present in a sample, the 'richer' is the sample. A community with one or two species is considered to be less diverse than one in which several different species have a similar abundance. Therefore, the macrobenthos community present at Karanja creek is more diverse and rich (Figure 30.6).

Evenness index is a measure of relative abundance of the different species making up the richness of an area or it compares the similarity of the population size of each of the species present. Figure 30.7 shows that macrobenthos of different species at Karanja creek are evenly distributed and making up the richness of an area.

Barring few reports on west coast of India in general and on coast of Maharashtra in particular, analysis of diversity indices of coastal macrobenthos has not been reported adequately (Annie and Govindan, 1995; Venkatachalam and Kale, 2002). Therefore, the community structure of a particular intertidal area is not available for comparison. In such circumstances, the data on diversity indices presented here could be helpful in future in knowing the status of macrobenthic community and effect of industrial development on it.

Conclusion

In conclusion, it is stated that, the data on diversity indices of selected macrobenthos at Karanja creek indicates that at present the creek harbors varied forms of biodiversity. Since baseline data on macro benthic community in intertidal areas of Karanja creek is not available, effect of human disturbance to it cannot be seen. The data on diversity presented here, could be helpful in future as a baseline data in knowing the status of macro benthic community and effect of industrial development on it. The present study reveals that higher values of evenness index, the better condition of the creek and is suitable for the growth of biota present therein. Maximum values of Shannon's index of general diversity reveals that all the species in Karanja creek are equally abundant.

Acknowledgements

The corresponding author is thankful to Director, Board for College and University development, University of Mumbai, for financial assistance. The author is grateful to Principal, Dr. R. B. Bawdhankar, Head and faculty members, Department of Zoology, Veer Wajekar Arts, Science and Commerce College, Phunde (Uran) for encouragement and support.

References

Annie, M. and Govindan, 1995. Macrobenthos in the near shore coastal system of Bombay. *Proc. Nat. Acad Sci. India*, 65(B)4: 412–430.

Apte, D.A., 1988. *The Book of Indian Shells.* Bombay Natural History Society; Oxford University Press, India, pp. 115.

Bhatt, Y.M., 1959. A study of intertidal organisms of Bombay. *Ph.D. Thesis*, University of Mumbai.

Chhapgar, B.F., 1957. On the marine crabs (Decapoda–Brachyura) of Bombay state. *J. Bombay Nat. Hist. Soc.*, 54: 399–439.

Chhapgar, B.F., 1958. More additions to the crab fauna of Bombay state. *J. Bombay Nat. Hist. Soc.*, 65(3): 608–617.

Dash, M.C., 2002. Systems concept in ecology. In: *Fundamentals of Ecology*, (Ed.) M.C. Dash. Tata McGraw-Hill Publishing Company Limited, New Delhi, pp. 35–144.

Eleftheriou, A. and Holme, N.A., 1984. Macrofauna techniques, pp. 1–344, In: *Methods for the Study of Marine Benthos*, (Eds.) N.A. Holme and A.D. McIntyre. IBP Handbook No. 16, Blackwell Scientific Publications, Oxford, U.K.

Fish, J.D. and Fish, S., 1996. *A Students Guide to the Seashore*, 2nd edn. Cambridge University Press, Cambridge.

Govindan, K., 2002. Marine benthos: A future prospective. In: *Proceeding of the National Seminar on Creeks, Estuaries and Mangroves: Pollution and Conservation*, 28th–30th November, pp. 28–30.

Jaiswar, A.K. and Kulkarni, BG., 2005. Conservation of molluscan biodiversity from intertidal area of Mumbai coast. *J. Natcon.*, 17(1): 93–105.

Khan, S.A. and Murgesan, P., 2005. Polychaetes diversity in Indian estuaries. *Indian J. Mar. Sci.*, 34(1):114–119.

Ludwig, J.A. and Reynolds, J.F., 1988. *Statistical Ecology: A Primer on Methods and Computing*. John Wiley and Sons, New York, 377 pp.

Margalef, R., 1968. *Perspective in Ecological Theory*. University of Chicago Press, Chicago, pp. 111.

Parulekar, A.H., Nair, S.A., Harkanta, S.N. and Ansari, Z.A., 1976. Some quantitative studies on the benthos of Bombay. *Mahasagar*, 9: 51–56.

Parulekar, A.H., 1973. Studies on intertidal ecology of Anjidiv Island. *Proc. Indian National Science Academy*, 39(B): 611–631.

Pielou, E.C., 1966. The measurement of diversity in different types of biological collection. *J. Theor. Biol.* 13: 131–144.

Quadros, G. and Athalye, R.P., 2002. Intertidal macrobenthos from Ulhas river estuary and Thane creek. In: *Proceeding of the National Seminar on Creeks, Estuaries and Mangroves: Pollution and Conservation*, 28th–30th November, pp. 216–222.

Shannon, C.E. and Weaver, W., 1949. *The Mathematical Theory of Communication*. University of Illinois Press, Urbana, pp. 117.

Simpson, E.H., 1949. Measurement of diversity. *Nature*, 163: 681–688.

Smith, R.L. and Smith, T.M., 1988. *Elements of Ecology*. Wesley Longman Inc. C.A.

Subrahmanyam, T.V., Karandikar, K.R. and Murti, N.N., 1949. The marine Pelecypoda of Bombay. *J. Univ. of Bombay*,, 17: 50–81.

Subrahmanyam, T.V., Karandikar, K.R. and Murti, N.N., 1951. Marine gastropods of Bombay, Part I. *J. Univ. of Bombay*, 20(1): 21–34.

Subrahmanyam, T.V., Karandikar, K.R. and Murti, N.N., 1952. Marine gastropods of Bombay, Part II. *J. Univ. of Bombay*, 21(3): 26–73 New Series.

Venkatachalam, S. and Kale, P.G., 2002. Benthic macrofauna of Thane creek: Seasonal variations. In: *Proceeding of the National Seminar on Creeks, Estuaries and Mangroves: Pollution and Conservation*, 28th–30th November, pp. 186–195.

Chapter 31

Amla Juice: A Biochemical Perspective of the Indian Gooseberry

Jyoti D. Vora and Padma Lakshminarasimhan
Department of Biochemistry and Food Science and Qualiy Control, Ramnarain Ruia College

ABSTRACT

Amla or Indian Gooseberry is one of the most precious gifts of Nature to Man. Originally an exotic Tibetan fruit, it belongs to the family Euphorbiceae Its botanical name is *Emblica officinalis*. The fruit is ready spherical and known to contain significantly high amounts of Ascorbic acid (vitamin C), nearly twenty-fold of that of orange. In this research endeavor, an in-depth investigation has been attempted in the biochemical profile of this significant nutraceutical component. The studies are backed up by exhaustive sensory and organoleptic evaluation of the food products under study.

Three different samples of amla juice were used for the study and estimation of the following proximate principles–Vitamin C, Vitamin A, Carbohydrates, Proteins, Tannins and Mineral analysis which included Calcium, Phosphorus and Iron using standard biochemical methods. The sensory evaluation of the above samples was conducted with 41 semi-trained panelists and the findings were analyzed using standard biostatistical analysis.

The results obtained by the analysis of the above mentioned proximate principles were much lower than the fruit content found in original amla fruit. The ayurvedic *amla* powder sample showed comparatively better nutritive values than the other two samples. This difference in nutritive values may be attributed to various reasons, including the processing of the original amla fruit to commercially viable juice form and experimental errors. Post-Harvest Technology plays a significant role in nutritive quality.

The sensory evaluation of the three samples showed that the commercially prepared *amla* juice samples were much more preferred than the ayurvedic sample due to their favorable organoleptic properties including visual qualities.

Introduction

Amla or Indian Gooseberry is one of the most precious gifts of nature to man. It is originally an exotic Tibetan fruit which has been used extensively for various purposes.

Botany

Amla belongs to the family *Euphorbiceae*, alongwith the other gooseberries namely tomato, orange and grapes. Its botanical names include *Emblica officinalis* and *Phyllanthus emblica.*

Tree

Amalaki is small-medium sized plant with a crooked trunk and spreading branches, the greyish-green bark peeling off in flakes. The leaves are simple, sessile and closely set along branchlets, light green, resembling pinnate leaves. The flowers are greenish-yellow borne in axillary fascicles, giving way to a globose fruit with a greenish-yellow flesh and six furrows enclosing a stone with six seeds.

Constituents

Amla fruit contains a series of diterprenes referred to as the gibberllins, as well as the triterprene lupeol, flavonoidsteg. Kaempherol-3-o-β-D-glucosidel and polyphenosteg. Emblicanin A and B, punigluconin and pedunculagin. Also present are the phyllantine and zeatin alkaloids and a number of benzenoids including amlaic acid, corilagin, allagic acid, 3-6-di-o-galloyl-glucose, ethyl gallate, 1,6-di-o-galloyl-β-D-glucose, 1-di-o-D-galloyl-D-glucose, putramivain A, digallic acid phyllembic acid, emblicol and alactric acid. The fruit are also stated to contain significantly high amounts of ascorbic acid(vitamin C), nearly twenty times that of the content of oranges.

Nutritive Value

Table 31.1: Nutritive Value (per 100 g of *Amla*)

Sl.No.	*Specifications*	*Quantity*
1.	Protein	500 mg
2.	Energy	58 kilo calories
3.	Vitamin A	151 int. units
4.	Carotene	9 micro grams
5.	Vitamin C	600 mg
6.	Calcium	50 mg
7.	Iron	1.2 mg
8.	Phosphorus	20 mg
9.	Fibre	3.4 g

Basic Significance

Amla has many medicinal properties extensively researched upon, and its medicinal properties include: Anti-oxidant, Anti-inflammatory, Anti-microbial, Anti-viral, Hepatoprotective, Hypolipidemic, Anticlastogenic, Antacid, Dyspepsia, Anticancer etc.

Materials and Methods

Estimation of Carbohydrates by Anthrone Method

Materials

1. Standard Glucose (stock solution): Dissolve 100 mg of glucose in 100 ml distilled water
2. Standard Glucose (working solution): Dilute 10 ml of stock solution to 100 ml with distilled water. Store in refrigeration after adding a few drops of toluene.

Procedure

Weigh 100 mg sample in a test-tube. Hydrolyse by keeping it in a boiling water bath or heating mantle for three hours with 5 ml 2.5 N HCl and cool to room temperature. Neutralise it with solid sodium carbonate (Na_2CO_3) until effervescence ceases. Make up the volume to 100 ml. Centrifuge. Collect the supernatant and take 0.5 and 1 ml aiiquots for analysis. The analysis was done according to routine experimental procedure.

Estimation of Protein by Folin-Lowry Method

Materials

1. Standard Protein (Stock solution): Weigh accurately 50mg of BSA (Fraction V) and dissolve in distilled water and make up to 50 ml in a standard flask.
2. Standard Protein (Working solution): Dilute 10ml of the stock solution to 50 ml with distilled water in a standard flask. One ml of this solutions contains 200μg protein.

Procedure

Extraction is usually carried out with buffers used for the enzyme assay. Weigh 500mg of the sample and grind, well with the pestle and mortar in 5–10 ml of the boffer. Centrifuge and use the supernatant for protein estimation. The analysis was done according to routine experimental procedure.

Estimation of Vitamin C by Colorimetry

Materials

1. Standard Ascorbic acid (stock solution): Dissolve 100 mg ascorbic acid in 100 ml of 4 per cent oxalic acid solution.
2. Standard Ascorbic acid (working solution): Dilute 10 ml of the stock solution to 100 ml with 4 per cent oxalic acid. The concentration is 100 μg/ml.

Procedure

Grind 0.5–5 g of sample material either mechanically or by using a pestle and mortar in 250 ml 45 oxalic acid solution. Centrifuge or filter and collect the liquid. Transfer an aliquot (10 ml) to a conical flask and add Br_2 drop-wise with constant mixing. The enolic H^+ in ascorbic acids are removed by bromine. When the extract turns yellow due to excess Br_2, expel it by blowing in air. Make up to a known volume (25–50 ml) with 4 per cent oxalic acid solution into dehydro form by bromination. The analysis was done according to routine experimental procedure.

Estimation of Vitamin A by Carr-Price Method

Materials

1. Standard Vitamin A (stock solution): 10000 IU/ml
2. Standard Vitamin A (working solution): 1000 IU/ml

Procedure

Set of tubes was prepared, the reagents used were $CHCl_3$ and Carr-Price reagent.

The tubes are mixed well and blue colour is read colorimetrically at 720 nm.

Estimation of Tannins by Folin-Denis Method

Materials

1. Standard tannic acid (stock solution): Dissolve 100 mg tannic acid in 100 ml distilled water.
2. Standard tannic acid (working solution): Dilute 5 ml of the stock solution to 100 ml with distilled water. One ml contains 50 µg tannic acid.

Procedure

Weigh 0.5 g of the powdered material and transfer to a 250 ml conical flask. Add 75 ml distilled water. Heat flask gently and boil for 30 minutes. Centrifuge with 2000 rpm for 20 minutes and collect the supernatant in 100 mi standard volumetric flask and make up the volume. The analysis was done according to routine experimental procedure.

Preparation of Samples for Mineral Analysis by Ashing's Method

Procedure

Take a known weight (10 g) of an oven dried sample, incinerate and extract using HCl. Warm in water bath and make up the volume to 100 ml in a standard flask. Suitable aliquots of this extract are then used for the estimation of minerals like Ca, P, S, Fe, Cu, etc.

Estimation of Calcium (Ca) by EDTA Method

10 ml of the sample solution was used for analysis. The analysis was done according to routine experimental procedure.

Sample A: 40.08 mg per cent calcium

Sample B: 40.07 mg per cent calcium

Sample C: 40.07 mg per cent Calcium

Estimation of Phosphorus (P) by Fiske-Subbarow Method

Materials

1. Standard phosphorus (stock solution): Dissolve 0.351 g of KH_2PO_4 in 10 ml 10N H_2SO_4. Dilute to 1 L with distilled water.

2. Standard Phosphorus (working solution): Dilute 1 ml of stock solution to 100 ml with distilled water.

Procedure

5 ml of the solution was used for estimating Phosphorus colorimetrically. The analysis was done according to routine experimental procedure.

Estimation of Iron by Wong's Method

Materials

1. Standard Iron (stock solution): Dissolve 0.498 g of ferrous sulphate in distilled water. Add 1 ml of concentrated Hydrochloric acid (HCl) and dilute to 1 L with distilled water. Concentration is 0.1 mg/ml
2. Standard Iron (working solution): Dissolve 10 ml of stock solution to 100 ml with distilled water.

Procedure

3 ml of the solution was used for estimating iron colorimetrically. The analysis was done according to routine experimental procedure.

Observations

See Figures 31.1–31.6.

Discussion

For the project Amla juice: A biochemical perspective of the Indian Gooseberry, three different samples of amla juice, namely Vitamla juice, amla juice from P.P. Food and ayurvedic amla power

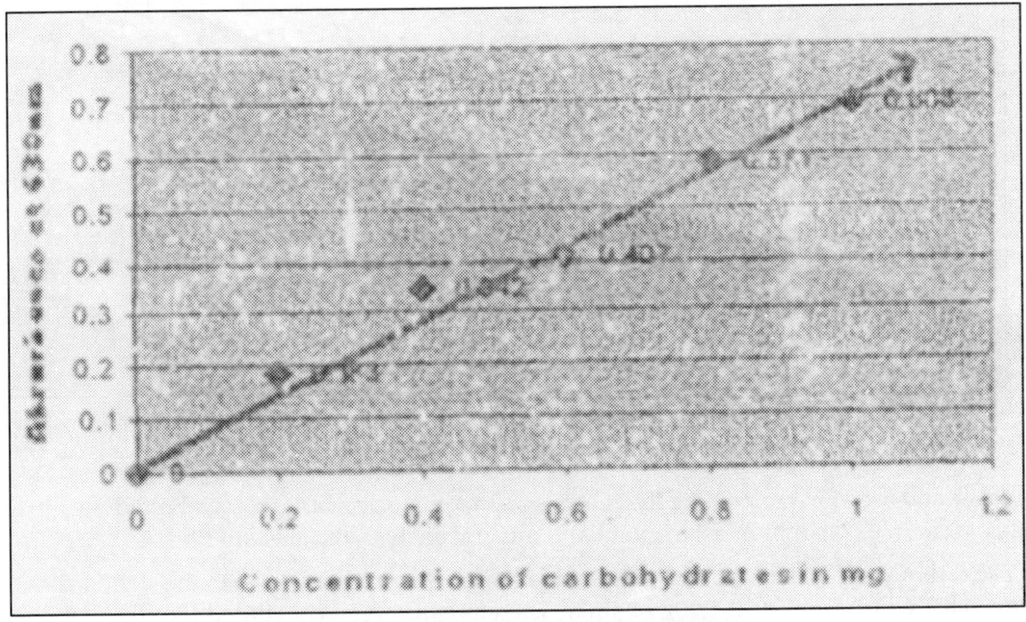

Figure 31.1: Estimation of Carbohydrates

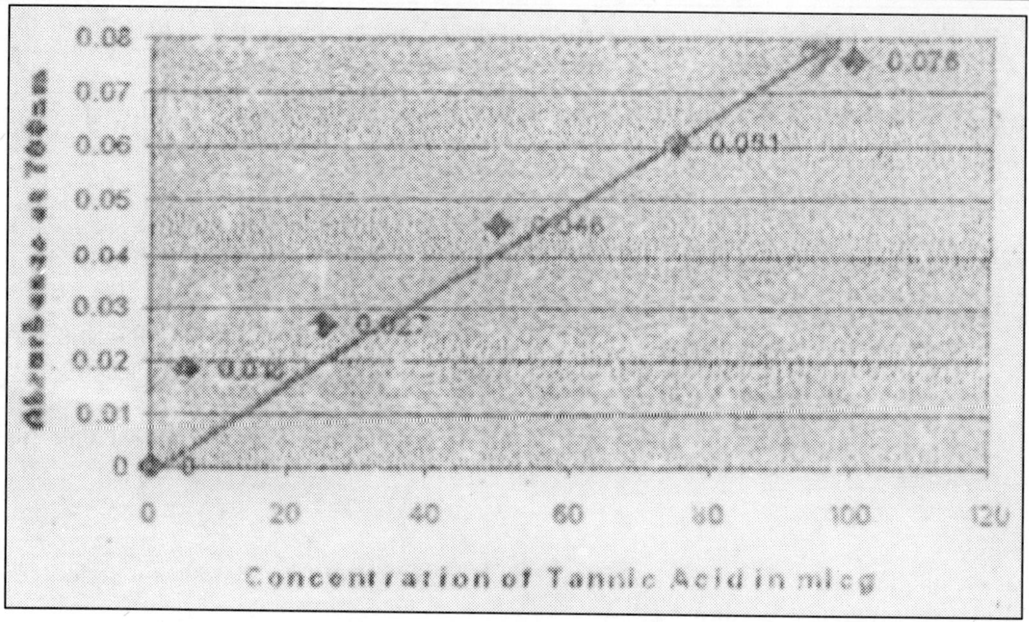

Figure 31.2: Estimation of Tannins

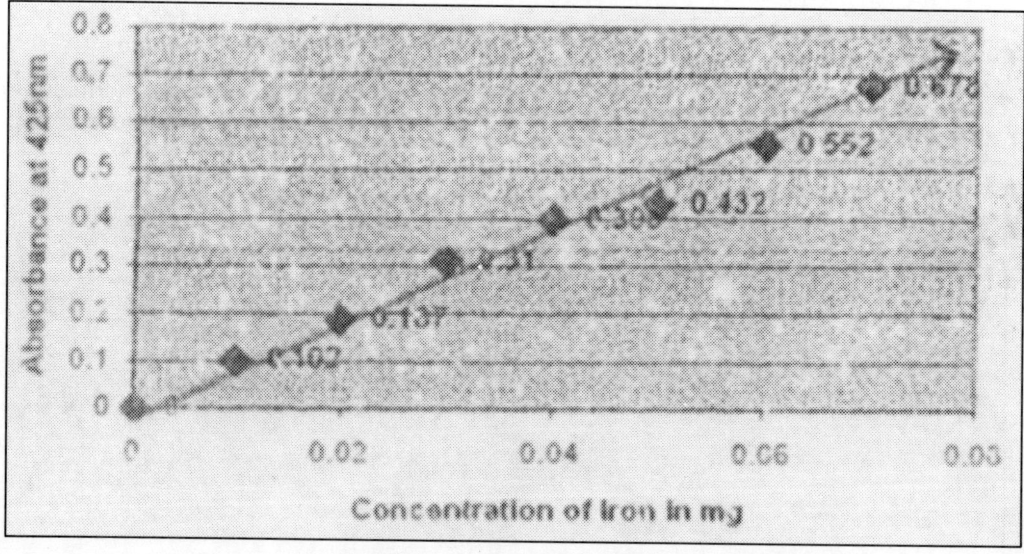

Figure 31.3: Estimation of Iron

were taken and used for, performing the analysis of proximate principles *viz.*, carbohydrates, proteins, tannins, vitamin C, vitamin A, ash content and mineral analysis (Ca, Fe and P).

The carbohydrate content in the given three samples is much lower than the standard nutritive energy value which is 58 kCal/100g of amla. Therefore, it can be concluded that the carbohydrate

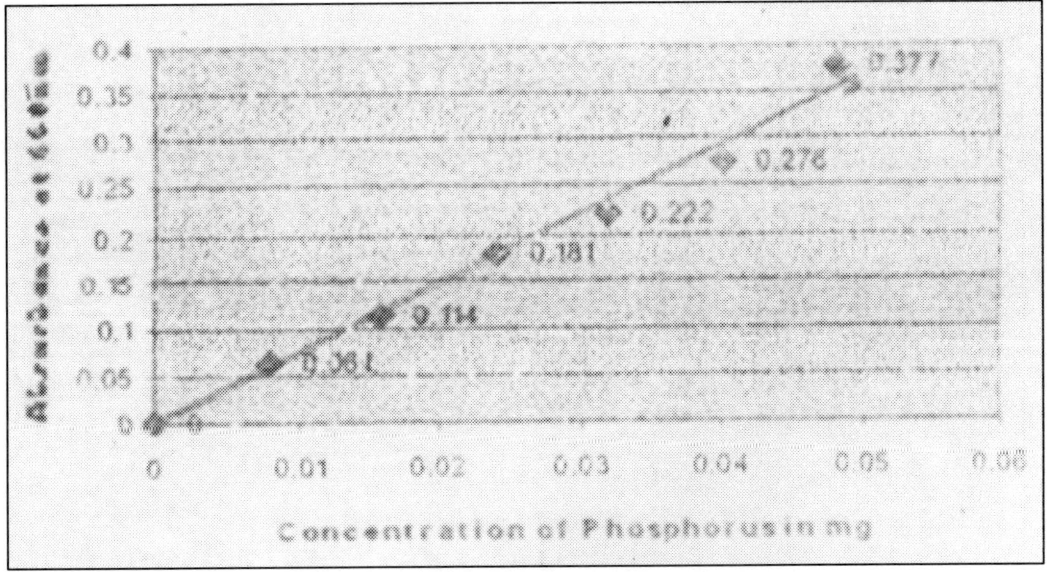

Figure 31.4: Estimation of Phosphorus

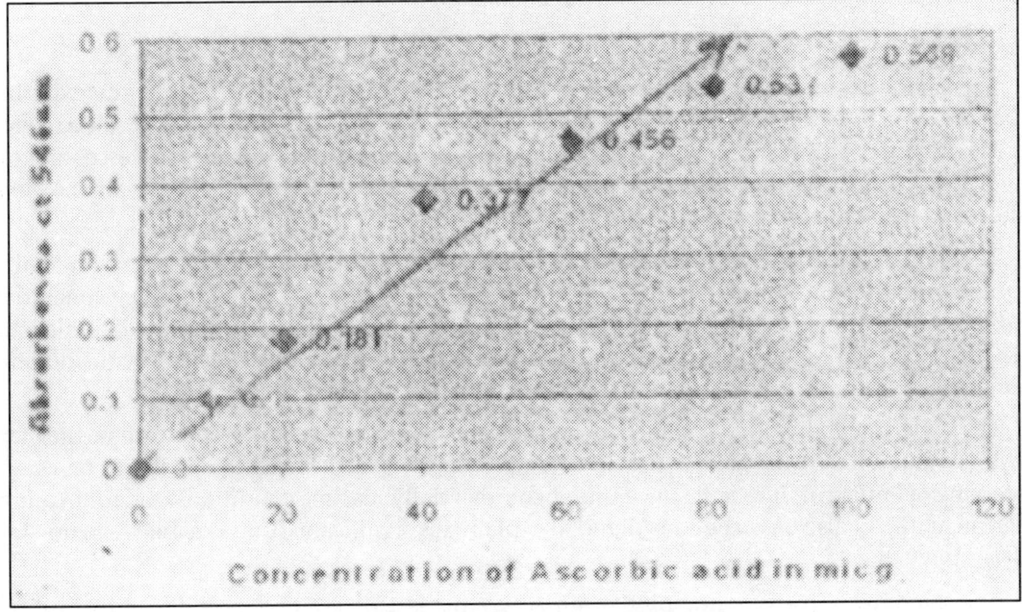

Figure 31.5: Estimation of Vitamin C

content may have been decreased during the processing of amla fruit of amla juice, Also, it should be noted that the standard energy value of amla would have been inferred using lipids also in consideration. Some amount of experimental error is also to be considered to explain the difference in the standard and sample values.

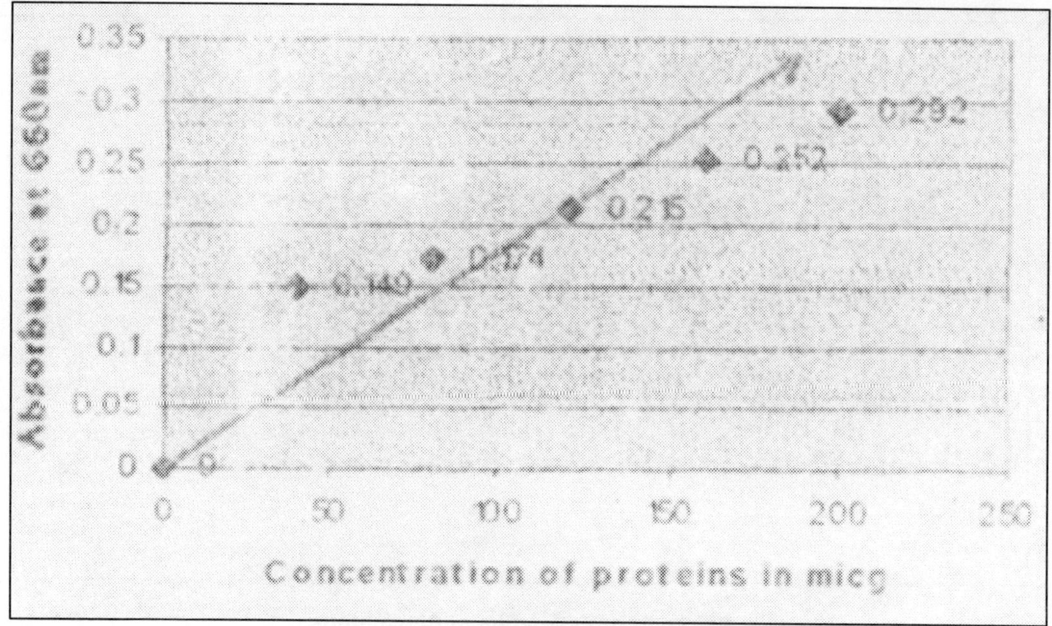

Figure 31.6: Estimation of Proteins

The standard nutritive protein value of amla is 500mg/100g of *amla*. Since, the experimental values obtained from the samples are considerably lower than the standard values, it can be concluded that the protein concentration in amla juice may have been decreased during the processing of amla fruit to amla juice. Experimental errors also have to be taken into consideration to explain the difference in the standard and sample values.

The standard nutritive tannin value of amla is 2.73 g/100g of *amla*. Since, the experimental values obtained from the samples are considerably lower than the standard values, it can be concluded that the difference in values may be due to the commercial preparation and processing of amla fruit into amla juice. There may be experimental errors also which have to be taken into consideration for inference.

The experimental values obtained from the three samples are much lower than the standard nutritive value of ascorbic acid which is 600mg/100g of *amla*. This difference may be due to the processing of amla fruit into amla juice under commercially feasible conditions. A certain amount of experimental error also has to be considered to explain this significant difference between the standard and sample values.

The standard nutritive value of vitamin A in amla is 151 IU/ml. hence, it can be said that the vitamin A content in the three samples fall in the normal range. The differences that are observed can be attributed to experimental errors.

The standard nutritive value of calcium is 50mg/100g of *amla*. The difference in value may be attributed to the dilution involved in preparation of the juice. Experimental errors also have to be considered for proper inference.

The experimental values of the samples obtained were lower than that of the standard nutritive value of iron in amla, which is 1.2mg/100g of *amla*. The difference in values may be.due to the commercial preparation and processing of *amla* fruit into *amla* juice. Experimental errors also have to be considered for inference.

The standard nutritive value of phosphorus in amla is 20mg/100g of *amla*. Therefore, the experimental values obtained were much lower than the standard values. This difference may be attributed to the commercial preparation of amla fruit into amla juice Experimental errors also have to be considered for inference.

Finally, it can be said that the three Amla juice samples which were taken for analysis were lower in nutritive value than their expected concentrations. This may be due to various factors and errors. Hence, for a confirmed inference, all the tests must be repeated in a more rigid environment under better controlled conditions, which will serve to give a better insight into the ingredients of Amla juice, the commercial preparation, processing and packaging of Amla juice and the advantages as well as the disadvantages of the same.

References

Grover, I.S. and Kaur, S., 1989. *Indian Journal of Experimental Biology*, 27(3): 207–209.

Nisha, P., Singhal, R.S. and Pandit, A.B., 2004. *International Journal of Food Science and Nutrition*, 55(5): 415–422.

Panda, S. and Kar, A., 20003. *Pharmazie*, 58(10): 753–755.

Perianayagam, J.B., Sharma, S.K., Joseph, A. and Christina, A.J., 2004. *Journal of Ethnopharmacology*, 95(1): 83–85.

Sadasivam, S. and Manickam, A., 1996. *Biochemical Methods*, Second Edition. New Age International (P) Ltd.

Sancheti, G., Jindal, A., Kumari, R. and Goyal, P.K., 2005. *Asian Journal of Cancer*, 6(2): 197–201.

Satyanarayana, U., 2002. *Biochemistry*, Second Revised Edition. Books and Allied (P) Ltd.

Sawhney, S.K. and Singh, Randhir, 2001. *Introductory Practical Biochemistry*, First Reprint. Narosa Publishing House.

Sultana, S., Ahmed, S., Sharma, S. and Jahangir, T., 2004. *Journal of Pharmacology*, 56(12): 1573–1579.

Suryanarayana, P., Kumar, P.A., Saraswat, M., Petrash, S.M. and Reddy, G.B., 2004. *Journal for Molecular Vision*, 12(10): 148–154.

Tasduq, S.A., Kaisar, P., Gupta, D.K., Kapahi, B.K., Maheshwari, H.S., Jyotsna, S. and Jomi, R.K., 2005. *Journal of Phytotherapy Research*, 19(3): 193–197.

Thorat, S.P., Rege, N.N., Naik, A.S., Thatte, U.M., Joshi, A., Panicker, K.N., Bapat, R.D. and Dahanukar, S.A., 1995. *Journal of HPB Surgery*, 9(1): 25–30.

Chapter 32

Impact of Degradation of Vegetation Cover on Quantum of Water Resources at Vizianagaram District, Andhra Pradesh, India

T. Byragi Reddy*, M.Vijaya Chandran and G.S.S.H.V. Prasad
Department of Environmental Sciences, Andhra University, Visakhapatnam, A.P.

ABSTRACT

Future water availability in Vizianagaram district of Andhra Pradesh located with in the geographical coordinates of 17°50′ and 19°15′N83°00′ and 83°45′E is likely to decrease due to the degradation of vegetation cover. District vegetation cover in its 34 mandals declining rapidly due to shifting cultivation in parts comprising of Eastern ghats and due to Aquaculture in coastal regions.

In this chapter the authors categorized 34 mandals of the district in to three zones *i.e.,* 1.High vegetation zone, 2. Moderate vegetation zone, 3. Low vegetation zone based on the statistics of forest vegetation area and sparse vegetation area. A critical analysis was made on surface and groundwater resources using the discharge flows in river catchments, Rainfall Data, Groundwater table depths, and groundwater yields to estimate the potential impact of vegetation cover degradation on the quantum of water resources in Vizianagaram. Increased volume of discharge flows, decline in the rainfall were highest during the period 1991–97 comparing to the

* E-mail: tbyragireddy@yahoo.com

statistics of 1989–90 indicating the decline in the quantum of water resources in recent times. Intensity of increase in the amount of discharge flows in all seasons were relatively very high in Low and Moderate vegetation zones comparing to the High vegetation zone. The groundwater table depths were also showed remarkable changes showed remarkable changes during the study. Highest groundwater table depths were found in Low and moderate vegetation zones comparing to the High vegetation data. Groundwater yields were showed decline in all zones demonstrating indispensable relationship between vegetation cover and water resources in the study area.

Keywords: *Degradation, Vegetation cover, Vegetation zones, Discharge flows, Quantum.*

Introduction

Vegetation cover plays vital role in preserving the environment and the natural resources. Rapid destruction of vegetation cover will affect a number of natural processes and cycles such as carbon, oxygen, nitrogen and food chains at grass root level (Vitousek, 1983). Along with this, it also affect the climatic components like temperature, Rainfall, and humidity and also the quality of land, water and air significantly (Spanns *et al.*, 1989; Burrows, 1990). In recent years Statistical abstracts of India are showing decrease in the extent of forests cover of India. 74.77 million hectares of forests cover recorded had been decreased to 67.42 million hectares. This declining trend is because of developmental activities such as establishments of industries, urban structures, mining and irrigation projects taking place at the cost of vegetation cover. Keeping the above aspects under consideration, an attempt was made in this paper to analyze the impact of Vegetation cover degradation on surface and under groundwater resources.

Study Area

Vizianagaram District situated within the geographical co-ordinates of 17°50′ and 19°15′ of the northern latitudes and 83°00′ and 83°45′ of the eastern longitudes (Andhra Pradesh, India) comprising parts of Eastern Ghats. Rivers such as Gosthani, Champavathi, Vegavathi, Swarnamukhi, Gomukhi, and Peddagadda were flowing in this district. Vizianagaram district consist of 34 mandals territories. This district was selected for the study because; the forests and sparse vegetation cover is rapidly declining due to shifting cultivation in the agency areas comprise parts of Eastern ghats and Aqua culture was also rapidly increasing in some of the coastal areas at the cost of vegetation..

Methodology

Classification of Mandals

Datasets of Mandal-wise distribution and concentration of forests areas and sparse areas (in hectares) was collected from hand book of statistics of Vizianagaram and Andhra Pradesh and computed in Table 32.1. All the Mandals of the district were categorized in to three zones based on the Mandai-wise contribution to the district total forests cover and total sparse vegetation cover. Accordingly die categorized zones are 1. High vegetation zone (>3 per cent of total contribution), 2. Moderate vegetation zone (<3 per cent– >1 per cent of total contribution) and 3. Low vegetation zone (< 1 per cent of total contribution).

Table 32.1: Mandal-wise Forests Area and Sparse Vegetation in Hectares

Sl.No.	*Mandal Territory Name*	*Forests Area in Hectares*	*Per cent*	*Sparse Vegetation in Hectares*	*Per cent*
1.	Vizianagaram	419	0.37	2274	2.33
2.	Gantyada	971	0.87	2404	2.36
3	Bhogapuram	664	0.59	1331	1.31
4	Denkada	–	–	698	0.69
5	Pusapatirega	1091	0.97	2190	2.15
6	S. Kota	272	0.24	2430	2.39
7	Jami	–	–	1474	1.45
8	Kothavalasa	1445	1.29	3290	3.23
9	Vepada	2094	1.87	1430	1.40
10	L.Kota	–	–	350	0.34
11	Nellimarla	1700	1.52	4025	3.95
12	Gurla	1385	1.24	1147	1.13
13	Cheepurupalli	–	–	1321	1.30
14	Garividi	–	–	2525	2.48
15	Merakamudidam	155	0.14	3117	3.06
16	Gaj apatinagaram	3120	2.79	1980	1.94
17	Bondapalli	1914	1.71	3495	3.43
18	Dattirajeru	1754	1.57	5532	5.43
19	Mentada	8981	8.02	6543	6.42
20	Parvatipuram	1471	1.31	2381	2.34
21	Komarada	3552	3.17	13827	13.58
22	Garugubilli	811	0.72	2244	2.20
23	Jiyyammavalasa	324	0.29	3967	3.89
24	Kurupam	23622	21.10	4049	3.98
25	Gummalakshmipuram	23288	20.81	2763	2.71
26	Badangi	–	–	1209	1.19
27	Therlam	–	–	996	0.98
28	Saluru	9187	8.20	5232	5.14
29	Pachipenta	17410	15.55	8192	8.04
30	Makkuva	2352	2.10	3443	3.38
31	Ramabhadrapuram	250	0.22	2577	2.53
32	Bobbili	2581	2.30	1024	1.00
33	Seetanagaram	984	0.88	1259	1.23
34	Balijipeta	182	0.16	1142	1.12

Surface Water Resources in the Vizianagaram

To study the impact and consequences of vegetation cover degradation in assorted vegetation zones on surface water resources were analysed using the data of volume of discharge flows over the period. Data collected for four selected rivers one for each vegetation zone. Study critically monitored the discharge flows flowing into Pddagadda River at Salur station for high vegetation zone, into Swarnamukhi River at Seetanagaram station for moderate vegetation zone, into Vegavathi River at Paradhi station for low vegetation zone. Data of discharge flows was collected on a monthly basis for all the selected rivers in three seasons *i.e.* 1. Pre-monsoon, 2. Monsoon, 3. Post-monsoon seasons for the years 1983 to 1990 and 1991 to 1997 collected by researcher along with rainfall data of High, Moderate, and Low vegetation zones collected from IMD were presented in Table 32.2. Change in discharge flows in percentage was calculated.

Undergound Water resources in the Vizianagaram

To study the impact and consequences vegetation cover degradation in various spheres in assorted vegetation zones on groundwater resources, the data collected form groundwater department, Vizianagaram district was analysed. Datasets for range of groundwater table depth (in meters) in high vegetation zone, Moderate vegetation zone, and Low vegetation zones for the years 1985–86 and 1996–97 were presented in Table 32.3. Range of groundwater yields (in GPH) for the years 1985–86 and 1996–97 were presented in Table 32.4. Using the data of Table 32.4 comparative study on groundwater yields over the period was made.

Results

Rainfall range data presented in the study are showing decline in almost all the zones during 1991–97, when compared to the figures of 1983–90. Further, the volume of Discharge flows in all the seasons in all the zones were showing remarkable increase during 1991–97 and decrease in Rainfall. Moreover, Table 32.2 demonstrates the intensity of the increase in discharge is relatively lower in high vegetation zones over the period. In respect of Moderate vegetation zone, the intensity of increase in discharges is relatively higher. Regarding Low vegetation zone, the intensity of increase in discharges is very high in almost all the seasons in respective Peddagadda, Swarnamukhi, and Vegavathi Rivers.

Data presented in Table 32.3 demonstrates the ranges of minimum, maximum and mean water Table depths. Groundwater table depths are relatively lower in the high vegetation zone, followed by moderate vegetation zone, while these are relatively higher in both the study periods. The ranges of groundwater table depths are showing increase in all most all zones in 1996–97, when compared to the figures of 1985–86. It is evident that ranges of groundwater table depth in high vegetation zone was relatively lower, when compared to the moderate and low vegetation zone. Further intensity of the increase in groundwater table depth was relatively higher in moderate and low vegetation zone, when compared to the high vegetation zone.

Relatively high value of groundwater yields found in high vegetation zone (9503–15.251 GPH), followed by moderately vegetation zone (8200–10.200 GPH), while the lowest is found in low vegetation zone (6,900–9,800 GPH)

Discussions

The findings are indicating that the volume of run-off flows have been increasing at an alarming rate and in turn, the discharges in the river flows have been increasing substantially even in pre and post monsoon seasons at lower incidence of rain fall. A significant factor responsible for this

Table 32.2: Zonal-Wise Range of Discharge Flows (m^3/sec) and Range of Rainfall (mm) Over the Period 1983 to 90 and 1991 to 97

Zone/River/ Station	*1983–1990*				*1991–997*			
	Rainfall Range (in mm per anum)	*Average Rage Range of Discharge Flows (in m^3/sec)*			*Rainfall Range (in mm per anum)*	*Average Rage Range of Discharge Flows (in m^3/sec)*		
		Pre-Monsoon Monsoon	*Monsoon*	*Post Monsoon*		*Pre-Monsoon*	*Monsoon*	*Post Monsoon*
High Vegetation Cover Zone/ Peddagaddal Salur	989.8 –1540.5 (1265)	0.60–0.79	1.39–7.14	0.62–3.75	913–1379.2 (1146.1)	0.57–368	0.51–12.25	1.44–4.17
Moderate Vegetation Cover Zone/ Vegavathi/ Paradhi	9032–1540.4 (1221.85)	0.19–1.70	0.87–17.14	0.20–8.63	884.5–1254.4 (1069.45)	0.19–6.14	0.45–23.58	0.75–9.45
Low Vegetation Cover Zone/ Suvarnamukhi/ Seetanagaram	850–1510.8 (1180.4)	0.81–4.81	7.04–41.3	2.41–15.07	851.32–1254.74 (1053.03)	2.05–14.75	5.22–61.11	6.25–23.70

phenomenon is attributable to degradation of vegetation cover and decrease in groundwater recharge. Because of increase in the coarse nature of the soil, absorption of water by the soil has not been taking place in moderate and low vegetation zones. This high discharges results in drying of surface water bodies rapidly.

Table 32.3: Ranges of Groundwater Table Depth in Meters Over the Period 1985–86 and 1996–97

Zone	*Range of the Groundwater Table Depth (in meters)*					
	1985–86			*1996–97*		
	Minimum	*Maximum*	*Mean*	*Minimum*	*Maximum*	*Mean*
High vegetation Cover zone	5.31–9.15	9.14–10.67	7.62–9.14 (838)	6.94–9.65	9.74–11.67	831–10.14 (9.23)
Moderate vegetation Cover zone	5.60–10.6	9.62–12.19	8.03–1067 (9.35)	7.01–11.63	9.80–12.93	8.65–11.98 (1032)
Low vegetation Cover zone	5.75–10.75	9.25–12.97	832–11.26 (9.80)	7.12–11.95	9.91–13.94	8.73–12.63 (10.68

Table 32.4: Ranges of Groundwater Yields (in GPH) over the Period 1985–86 and 1996–97

Zone	*Groundwater Yields, (in GPH)*			
	1985–86		*1996–97*	
	Range	*Average*	*Range*	*Average*
High vegetation Cover zone	9,554–17545	13,409	9503–15251	11,463
Moderate vegetation Cover zone	9500–11,500	1056	820–10,200	9,534
Low vegetation Cover zone	7750–11,500	9625	6900–9800	8203

Similar type of findings were made by Anon (1970), Khoshoo (1981), Dalal (1985), Astana and Astana (1996). Further these studies also conclude that major responsible factor for this phenomenon is attributable to degradation of vegetation cover.

Increased run-off flows are significantly contributing for higher degrees of siltation process in the water bodies, there by storage capacities of water bodies drastically declining. Moreover, this state of affair is also contributing to the increase in the intensity of floods. This phenomenon is predominantly found in low and moderately vegetation zones. Vahora (1987), Bajwa (1987) and Hurning *et al.* (1987) Astana and Astana (1998) also support that the storage capacities of water bodies such as lakes, tanks rivers have been declining over the period.

Quantum of water discharges in all the sampling stations are showing significant increase in the volumes over the period, even in low rain fall periods. In addition the water retaining capacity of the soil gets reduced. All these conditions lead to higher quantities of water discharge. This phenomenon is more predominant in low and moderate vegetation zone in the study area. Similar observations were made by Allen(1968), Vahora (1978), Binns (1986), Astana and Mera Astana (1996).

References

Abraham Sen, G., 1984. Effects of acidic deposition on forest soils and vegetation. *Philosophical Transactions of the Royal Society of London*, pp. 369–382.

APHA, 1992. *Standard Methods for the Examination of Water and Wastewater*. American Health Association, Washington, D.C.

Astana, D.K. and Mera Astana, 1996. *Environmental Problems and Solutions*. S. Chand and Company, New Delhi.

Basu, S. and Behra, N., 1993. The effect of tropical forest conservation and solid microbial biomass. *Bio. Ferti. Soil*, 16: 1–3.

Burrows, C.J., 1990. *Process of Vegetation Change*. Unwin Publications, Hyman, London

Centre for Science and Environment, 1985. The state of India's environment, 1984–85. *The Citizens Report*, New Delhi

Dalal, N., 1985. Are we cremating our forest. *Times of India*, June 2.

Dunne, T. and Leopold, L.B., 1978. *Water in Environmental Planning*. Freeman, San Francisco.

Flexman, E.M. *Soil Conservation Series*. Technical Service Center Advisory Eng, PO. 1.

Hornung, M., Stevens, P.A. and Reynolds, B., 1987. The effect of forestry on soils: Soil water and surface water chemistry. In: *Environmental Aspects of Plantation Forestry in Wales*, (Ed.) J. Good. LT.E.

Johnson, D.W., 1992. *Water, Air and Soil Pollution*, 64(1–2): 83.

Khoshoo, T.N., 1981. Conservation of biological diversity: The Indian experience. In: *Proc. US Strategy of Conference on Biological Diversity*, Dept. of State, Public. 9262 Washington.

Khoshoo, T.N., 1984. *Environmental Concerns and Strategies*. Indian Environmental Society, New Delhi.

Lal R., 1989. Soil degradation and conservation of tropical rain forests. In: *Changing the Global Environment*, (Eds.) D.B. Botkin, M.F. Caswell, J.R. Estes, A.A. Orio. Academic Press Inc., p. 135–153.

MOA(a), 1984. *A Report of Working Group for Soil and Water Conservation*. Ministry of Agriculture, GOI, New Delhi.

Murthy, V.V.N., 1985. *Land and Water Management Engineering*. Punjab Agricultural University, Ludhiana.

Paliwal, K.V., 1971 Quality of irrigation waters and their effect on soil properties in Rajastan: A review. *Annals of Arid Zone*, 10: 226–278.

Rao, A.A., 1986. Groundwater planning and management. *BHU–JAL*.

Vahora, B.B., 1987. Managing India's water resources, land and water management holds the key. *BHU–JAL News*, (4): 4.

Zhu, Z.L. and Wen, Q., 1992. *Nitrogen and Soil of China* (in Chinese). Jiangso Science and Technology Publishing House, Nanging, p.5.

Chapter 33

Global Warming Could Cause Marine Life Extinction

Prabhakar R. Pawar* and Hanmant V. Jadhav
Veer Wajekar Arts, Science and Commerce College, Phunde – 400 702, Tal. Uran, Dist. Raigad, Navi Mumbai, Maharashtra

ABSTRACT

Global warming is the increase of the average temperature in the atmosphere and oceans over time primarily due to human influences. Evidence of global warming includes decreased snowfall, rising sea levels arid changes to weather trends. The increased amounts of Carbon dioxide (CO_2) and other greenhouse gases (GHGs) are the primary causes of the human induced component of warming. Both primary and secondary effects of global warming may influence not only human activities but also ecosystems. Some species may be forced out of their habitats (possibly to extinction) because of changing conditions, while others may flourish. The potential effects of global warming impacts on biodiversity of oceans as the surface water of the oceans warmed, which could devastate key marine life. If the loss of marine species from over fishing and climate change continues at the current rate, all commercial fish and seafood species could collapse by 2048. The present paper highlights impact of global warming on marine biodiversity at present and if proper preventive measures were not taken, the global warming could cause marine life extinction.

***Keywords**: Acidification, Biodiversity, Global warming, Greenhouse gases, Ice caps and glaciers, Habitat loss.*

---

* E-mail: prabhakar_pawar1962@yahoo.co.in

Introduction

The phrase global warming refers to a gradual increase in the average temperatures of the Earth's atmosphere and its oceans, a change that is believed to be permanently changing the Earth's climate forever. The scientific consensus on climatic changes related to global warming is that the average temperature of the Earth has risen between 0.4 and 0.8°C over the past 100 years. The increased volumes of carbon dioxide (CO_2) and other greenhouse gases (methane, nitrous oxide, chlorofluorocarbons etc.) released by the burning of fossil fuels, land clearing, agriculture and other anthropogenic activities, are believed to be the primary sources of the global warming that has occurred over the past 50 years.

According to the 2006 Annual Report of the United States National Climatic Data Center (NCDC), in 2006 "Globally averaged land temperatures were +0.78°C (+1.40°F) and ocean temperatures +0.45°C (+0.81°F) above average, ranking 4th and 5th warmest respectively (NCDC Report, 11th January 2007). The NCDC report states that "during the past century, global surface temperature have increased at a rate near 0.06°C/decade (0.11 of/decade) but this trend has increased to a rate approximately 0.18°C/decade (0.32°F/decade) during the past 25 years. There have been two sustained periods of warming, one beginning around 1910 and ending around 1945, and the most recent beginning about 1976". The most recent NCDC quarterly report (15 November 2007) suggests continued global warming into calendar year 2007 and forecast to rise by 1.8° to 4°C by 2100. Scientist from the Intergovernmental Panel on Climate Change (IPCC) have recently predicted that the average global temperature could increase between 1.4 and 5.8°C by the year 2100 [IPCC, Summary for Policymakers (SPM), 2nd February 2007].

Amazing History of Global Warming

1. John Tyndall (1859) suggested that the most important greenhouse gas was water vapour (H_2O) but that CO_2 was also effective for climate change.
2. Svante Arrhenius (1896) predicted the decrease in amount of CO_2 in the atmosphere by half could lower the temperature in Europe by some 4–5°C (roughly 7–9°F). This confirmed the 1859 prediction that human-produced increases in CO_2 would cause the global warming.
3. French and Russian Scientists (1900) extracted a mile and half deep ice core from Antarctica and found that over the four ice-age cycles in the last 4,00,000 years, CO_2 dropped in step with temperature but only by about one third and not one half. That could easily be because other green house gases changed with CO_2. This confirms the 1896 prediction, or an amazing coincidence.

Causes of Global Warming

1. Increased volumes of Anthropogenic Greenhouse Gases.
   (*a*) Carbon dioxide (CO_2): 60 per cent
   (*b*) Methane (CH_4): 20 per cent
   (*c*) Nitrous oxide (N_2O): 6 per cent
   (*d*) Chlorofluorocarbons (CFCs): 14 per cent
2. Stratospheric ozone depletion and deforestation.
3. Diminishing biological diversity
4. Over use of land for agriculture and over irrigation.

5. Use of chemical fertilizers and insecticides.
6. Smoke, fumes and waste gases from industrial establishments.

Effects of Global Warming

1. Rising sea levels due to melting of polar ice caps and glaciers.
2. Increase in occurrence and severity of weather events *e.g.*, storms, hurricanes, floods, droughts, heat waves and tornados.
3. Land erosion, shifting in vegetation types and destructions of habitats.
4. Gradual increase in the earth's surface temperature.
5. Habitat loss, Ecological disturbance and species extinction.
6. Spreading extent of the Sahara desert.
7. Precipitation-levels and patterns, cloud cover.
8. Higher or lower agriculture yields.
9. Glacial retreat and reduced summer stream flows.
10. Increases in the ranges of disease vectors affecting human health.

Rising Global Warming Threatens the Marine Life

The ocean and its inhabitants will be irreversibly affected by the impacts of global warming and climate change. Global warming, by increasing seawater temperature, will raise sea levels, change ocean currents, increasing acidity in the oceans, habitat loss and devastate key marine life.

Ocean Currents/Ocean's Circulatory Patterns

The water in oceans is always moving–pulled by tides, blown by waves, and slowly circulating around the globe by the force of the Great Ocean Conveyor Belt (also called thermohaline circulation). The Conveyor is powered by differences in water temperature and salinity and playing an important role in the global climate. Conveyor provides an upwelling of bottom ocean nutrients, and increases the oceanic absorption of CO_2. Dilution of the ocean's salinity–from melting Arctic ice and / or increased precipitation–could switch off, slow down or divert the Conveyor. This will result in massive disruption to agriculture, climate and impact other sea currents and temperatures around the globe.

Sea Level Rise

A global sea level rise of 9–88 cm (3.5–34.6 inches) is expected over the next hundred years. Even this comparatively modest projected sea level rise will result in coastal flooding and storm damage, eroding shorelines, salt water contamination of fresh water supplies, agricultural areas, flooding of coastal wetlands and barrier islands and an increase in the salinity of estuaries. Some low lying coastal cities and villages will also be affected. Resources critical to island and coastal populations such as beaches, freshwater, fisheries, coral reefs and atolls and wildlife habitat is also at risk. The entire Antarctic ice sheet holds enough water to raise global sea level by 62 meters (203 feet). Study on the global effects of climate change has established that net sea level rise along the Arabian sea coast has been 1–2 mm per year over the past several decades. The highest net sea level rise has been recorded at Kochi (1.75 mm / year) followed by Aden in the Persian Gulf (1.37 mm), then Mumbai (1.2 mm) and finally Karachi (1.06 mm).

Increasing Acidity in the Oceans

During the last 200 years, the concentration of atmosnheric CO_2 increased from about 275 ppm to about –380 ppm. If unchecked, it could surpass 550 ppm by mid-century. The basic chemistry of the ocean is being altered by excess CO_2 absorption, which threatens marine life by increasing *'acidification'*. Acidification is caused by a reaction between CO_2 and H_2O, which forms carbonic acid (H_2CO_3). Carbonic acid increases the acidity of ocean waters by lowering the pH which inhibits the reaction organisms (*e.g.* Coccolithophores–one of the most abundant phytoplankton in the ocean, corals, foraminifera, echinoderms, crustaceans and some molluscs especially pteropods) use to secrete skeleton and shells made of Calcium carbonate ($CaCO_3$). Every marine species with calcarious skeleton and shells will find it more difficult to survive in the future. Such a decline in shell-producing marine organisms will affect ocean ecosystem. Increasing acidity will also affect many nervous, reproductive and/or physiological processes in other marine species. The ocean-surface pH has been reduced by about 0.1 during the past two centuries and it would decline a total of 0.31 by the end of this centaury, if CO_2 emissions continue on a trajectory to ultimately stabilize at 1,000 ppm.

Habitat Loss

Temperature rises are impacting on the entire marine food web ego phytoplanlaton, which feeds small crustaceans including krill, grow under sea ice. A reduction in sea ice implies a reduction in krill-and krill feeds many whale species, including the great whales. Whales and dolphins strand themselves in high temperatures. The great whales also risk losing their feeding grounds because of the melting and collapse of ice shelves. Whole species of marine animals and fish are directly at risk due to temperature rise-they simply cannot survive in warmer waters. Penguin population from Antarctica has decreased by 33 per cent because of habitat decline. An increasing occurrence of disease in marine animals is also linked to rising ocean temperatures.

Devastate Key Marine Life

Satellite data revealed for the first time that global warming could devastate key marine life and could cause marine life extinction.

1. Changes already visible on migration patterns of *Birds, Fish and Turtles.*
2. Sea level rise destroying beach nesting sites for *sea turtles e.g.* nearly a third of beaches used by turtles in the Caribbean would be lost in this centaury.
3. *Seals and Wading birds* also face destruction of their habitats.
4. Growing water scarcity in many regions, destroy the wetland areas on which *migratory waterfowl* depend.
5. Increased storminess damaging the breeding colonies of *albatross,* already facing heavy pressure from accidental capture on long-line fishing hooks.
6. Spreading extent of the Sahara desert could threaten *long-range traveler birds,* as they will be unable to "fuel up".
7. Warmer seas could lead to 6 species of *marine turtle* found in Australian waters becoming entirely female, as water temperature strongly affects the sex ratio of hatchings.
8. Like a piece of chalk dissolving in vinegar, *marine life with hard shells* is in danger of being dissolved by increasing acidity in the oceans.

9. The higher acidity threatens marine life including *corals and shellfish*, which may become extinct latter this century from the chemicals effects of CO_2, even if the planet warms less than expected.
10. In Canada's Gulf of St. Lawrence, as there was not enough ice to provide necessary habitat for thousands of baby *harp seals*, and nearly 100 per cent of all pups born this year drowned.
11. *Corals* are already under serious threat right now due to "*bleaching*" due to the seas growing warmer. (Corals generate 80 per cent of its energy by photosynthesis via tiny algae within its body known as zoozanthellae. Due to climate change, the coral's defense is to expel the algae, which results in death of the coral).
12. *Reel fish* and intertidal invertebrates (*sea anemones, crabs and snails*) in California are shifting towards the pole in response to warming.
13. *Phytoplankton* (microscopic marine plants inhabiting the surface layers of ocean, playa central roles in the global carbon, oxygen and nutrient cycles, direct food source of small fish, prawns and shoreline filter feeders and produce half the oxygen on the planet) biomass was declined due to warming of surface water of the oceans. As the oceans warmed and become more stratified, phytoplankton productivity declined by 190 million tons each year.
14. The increased acidity lowers the concentration of carbonate ions, a building block of the $CaCO_3$ that many *marine organisms* use to grow skeletons and create coral reef structure. Such organisms will grow slowly, their skeletons will become less dense, a process similar to osteoporosis in humans.
15. Decreased *calcification in marine algae and animals* impacts marine food webs and has potential to substantially alter the biodiversity and productivity of the ocean.
16. In India, biodiversity is seriously threatened by impact of global warming. 30 per cent of *amphibians*, 23 per cent of mammals and 12 per cent of birds are under threat of extinction.

Measures to Check Global Warming

1. Reforestation
2. Energy-efficient technologies.
3. Alternative fuel sources (*e.g.* methane).
4. Reduced use of CFCs for refrigeration, air-conditioning, aerosol propellants, production of foam packing and insulation and as solvents.
5. Control of population growth.
6. Shift from coal to natural gas and use of non-conventional sources of energy.
7. Reduce the rate of greenhouse gases emission by 50 per cent.

Conclusion

1. In conclusion it is to be stated that an accelerating decline in coastal species over the last 1000 years, resulting in the loss of biological filter capacity, nursery habitats and healthy fisheries. In addition to decline in food supply, damage to the oceans results in harmful algae blooms, oxygen depletion and ocean flooding.
2. Accurate and reliable methods must be designed for forecasting of the effects of global warming on biodiversity in general and of marine life in particular.

3. Efforts must be made at international level to bring together different ideas and solutions from countries around world to find ways of reducing greenhouse gases.
4. Steps should be taken which include action from both developed and developing countries in reducing greenhouse gas emissions.
5. Developing countries must leap frog from high CO_2 emission fuel source (*e.g.* coal) to a lower CO_2 emission fuel source (*e.g.* natural gas) in order to reduce greenhouse gases.

All countries (developing and developed) must try to use the newest technologies presently available.

Acknowledgements

The corresponding author is thankful to Head and faculty members, Department of Zoology, Veer Wajekar A.S.C. College, Phunde (Uran) for encouragement and support.

References

Annual Report, 2006. *Annual Report of the United States National Climate Data Center.*

Botkin *et al.*, 2007. Forecasting the effects of global warming on biodiversity. *Climate 2007: IPCC Fourth Assessment Report.*

Nodvin, Stephan, 2007. Global warming. In: *Encyclopedia of Earth.* (http://www.eoearth. org/article/ Global warming).

Rauber, Paul, 1997. Heat wave: Global warming. *Sierra*, 82: 34(8).

Sala, E. and Knowlton, N., 2006. Global marine biodiversity trends. *Annual Renew of Environment and Resources*, 31: 93–122.

The Times of India, Mumbai, 7th October 2005. Global warming could cause environmental extinction.

The Times of India, Mumbai, 26th October 2007. City sea level rising 1.2 mm per year: Study.

Chapter 34

Environmental Impact and Utilization of Fly Ash: A Study of IB-Thermal Power Plant

D.K. Sahoo[1], A. Behera[2] and Pramila Mishra[3]

[1]Ex-Officer, R&D, Tata Refractories Ltd., Belpahar, District Jharsuguda, Orissa - 768 218

Chief Executive, BF Orissa Environment Consultant, Barun Plaza, Sambalpur, Orissa - 768 004

[2]Regional Officer, State Pollution Control Board, Modipara, Sambalpur - 768 001

[3]Reader in Chemistry, P.G. Department, Sambalpur University, Jyoti Vihar, Burla, Sambalpur

ABSTRACT

The present piece of research deals various environmental problem related to generation of Fly Ash as negative impact and some of the important Fly Ash utilization of management as positive impact techniques with proper remedial measures.

Introduction of Fly Ash

The nation to face 21st Century needs electricity to fulfil its needs and accelerate its growth and one of the most abundant source of fuel for generation of electricity is Coal. Where coal is fire power plan compress 65 per cent of total energy generating capacity and 70 per cent of electricity output.

The inorganic residue which is left after the burning of pulverized coal is known as coal.

Combustion By Product (CCBP)

There are two main types of Fly Ash.

Furnace Bottom Ash (FBA)

This is the fraction that is collected at the bottom of the Incinerator and consist of Coarse and heavy particles.

Pulverized Fuel Ash (PFA) of Fly Ash

This is the lighter particles that has been carried away by fuel gases.

In powder fixes Coal Fired Electricity cants 75 to 80 per cent of total ash consist of fly ash. Fly ash readily accumulate and causes enormous problems for its disposal. The indiscriminate disposal of fly ash causes large volume of land, water and energy resources world over and also causes great environmental concern.

In India three-fourth of total electricity is generated is thermal out of which almost 90 per cent is coal base. At present 90 million tons of fly ash is produced annually in India. And most of it is dumped in ash pond consuming nearly 65000 acres of land though most of it is recoverable and can be utilized if proper techniques are employed.

IB-Thermal Power Plant

IB-thermal power plant started as a capacity of 960MW electricity plant near Banaharpali 25 KM from Jharsuguda is situated in the coal mining area of Mahanadi Coal Field, as subsidiary of Coal India Limited. Coal are mainly brought from BOCM open cast project and Lakhanpur open cast project are semi-bituminus type. The typical analysis of coal and fly ash is given in Table 34.1. Water is used from the Hirakud Dam reservoir. Before the establishment of the plant the surrounding area at a core distance of 10 km. radius was environmentally clean area. The sulphur dioxide was in average 65 microgram per cubic meter in summer 62 in winter 38 in rainy season, and SPM values are 298, 288, 178 microgram per cubic meter in summer, winter and rainy season respectively. Now it is found the sulphur dioxide values are 91, 82, 52 and SPM values are 310, 304, 211 microgram per cubic meter respectively in summer, winter and rainy season with a large amount of fly ash and bottom ash are produced in the power plant.

Table 34.1: Proximate Analysis of Coal: Average

Area	*V.M.*	*F.C.*	*Ash*
Talcher Area	30.74	49.08	12.18
BOCM of IB	36.56	28.76	34.68
Lakhanpur OCP of IB	34.11	27.79	38.10

In general fly ash is a fine Alumino-Silicate material which consist of solid and spherical particle. It mainly consist of oxides of which Al, Fe, Si are the main constituent. Apart from this many trace elements some of which may be radioactive and toxic are found. The relative percentage of these oxide determine the character and use of the fly ash.

Mineralogically it consist of small sphere of glass of complex chemical composition and crystalline constituent mainly quartz, mullite, hematite and unburnt carbon. Fly ash from different source varies in 3 basic properties like lime reactivity, specific surface density and unburnt carbon content. The fly ash dry hopper content lesser unburnt carbon content and higher lime reactivity surface area as compared to pond ash from the same source.

Chemical Analysis of Fly Ash of BOCM Coal (Sub-bituminus type)

Table 34.2(a): Analysis of Fly Ash from BOCM Coal

Moisture	0.02
LOI	0.26
SiO_2	59.20
Al_2O_3	29.87
Fe_2O_3	7.06
CaO	0.46
MgO	0.17
Na_2O	0.46
K_2O	0.60
SO_2	0.12

Table 34.2(b): Composition of Fly Ash

LOI	0.3
SiO_2	40–60
Al_2O_3	20–30
Fe_2O_3	4–10
CaO	1–5
MgO	0.2
Na_2O	0.5
K_2O	0.5
SO_2	0.2

Challenges Faced by 21st Century

Air Pollution

The world wide increase in burning of coal has increased the percentage of CO_2 in the atmosphere thus causing global warming due to the greenhouse effect (since 1995). It is the main cause of ozone depletion. Fine particle of fly ash in the form of suspended particulate matter (SPM), produced as residue after burning of coal causes several lung and other related health problems in humans and animals. Fly ash is reported to cause ailments like allergy bronchitis, silicosis, and asthma. The low particle density and definite selecting gradient.

Water Pollution

Fly ash contaminate water bodies in its contact and also indirectly. Water pollution can adversely affect human and environmental health. Fly ash contains heavy metals and such inorganic materials (nitrate, cadmium etc.) are especially known to pollute the water bodies. The combination of toxic substances and heavy deposits of materials at the bottom cause serious threat to marine environment (Pervez and Pandey, 1994). Fly ash pollutes both ground and surface water. This can be checked by contouring and lining fly ash files, diverting drainage and selecting proper soil sites, which are not thermal plant affects the aquatic life also, it increases the temperature of water bed and decreases oxygen content of water causing extinction of rare species. The effluents also result inland fish productivity of water bodies (Singh and Singh, 1994).

Pollution Caused by Radioactivity and Other Toxic Materials

Coal contains a number of trace elements in the form of impurities. These trace elements have long durability as and when it react will heat in thermal power plants. Sometimes it may be converted into other elements but most of the time it is in the present form, which will be hazardous when contaminated with water of soil. Fly ash contains radioactive materials and other toxins thus causing threat to the environment and giving rise to high risk of radiation related health problems.

Land and Soil

Huge land is required for disposal of fly ash. Fly ash which get incorporated in soil above a certain threshold limit can change the physical, chemical and biological characteristics of soil but below that level it is useful for crop yield. Fly ash contaminates land resources to a very large extent. Most of the fly ash produced in the power plants is dumped into landfills and as ponds. This converts large areas of land into useable and other ash tips. In this way large areas of agricultural and other land is wasted. Fly ash dumped in this way makes the land barren and pollutes the topsoil (Sing *et al.*, 2000).

Biodiversity

The disposal of fly ash in fly ash ponds renders the redundant and devoid of nutrients thus making conditions unfavourable for some plants to grow. Also fly ash contains many heavy and toxic materials unfit for plant and animal consumption. Fly ash also polluted ground and surface water endangering the marine life. All these factors accumulate to reduce the biodiversity of nature and thus posing a severe threat to the environment.

Effect of Human

The finer particles of fly ash of 0.1 mm, are normally harmful to the human being residing close to the power plant as these get deposited in the pulmonary tissues of respiratory tracks. The fine particles inhaled by human being normally in early morning and evening hour due to temperature variation. These small fractions are impact found to the most mutagenic and carcinogenic.

Table 34.3 shows the advantages and disadvantages of fly ash disposal.

Table 34.3: Advantages and Disadvantages of Wet and Dry Disposal of Fly Ash

Disposal Method	*Advantage*	*Disadvantage*
Wet (100 per cent Recycling)	1. Ash handling is simple 2. Less costly	1. No scope for ash utilization. 2. Land requirement is high (1.5 to 2.0 Ac/MW to last for 20 years) 3. There is a chance of groundwater pollution. 4. Often complete recycling of water budgeting.
Dry	1. Land requirement is less (0.8 to 1.00 Ac/MW) to last for 20 years 2. Water conserved saving from cess. 3. Segmented use of land helps better rehabilitation.	1. Require a good management practice failure can cause nuisance in the locality. 2. Cost of ash handling is high which may increase the operating cost of production. 3. Environmental problem during transportation.

The quality and huge amount of coal deposited in Mahanadi Coal Fields area some new thermal plants are under construction rapidly. Table 34.4 shows the future plants of Orissa and Table 34.5 shows the ash disposal status of present plants.

Disposal

Most of the fly ash is still disposed of in landfills and storage causing serious ecological problems. But much of this fly ash can be recovered and utilized as a resource (Adriano *et al.*, 1980).

Ash Utilization

1. Dense packing of minefills.
2. Roller compacted concrete technology for hydraulic structure.
3. Reclamation/Structural filling of low laying areas.
4. Building components.
5. Dyke raising.
6. High value added applications like extraction of alumina, cenospheres etc.
7. Long term studies for use of fly ash in agriculture and forestry.

Table 34.4: Future Power Plants of Orissa

Plant	Place	Capacity (MW)
NTPC (Expansion)	Kaniha, Angul	2000
IB-Thermal (OPGC) Expansion	Banharpalli, Jharsuguda	420
AET Trans Power	Banharpalli, Jharsuguda	1000
OEP Asia	Banharpalli, Jharsuguda	1320
CEP Asia	Hirma, Jharsuguda	3950
Samli Power	Lapanga, Sambalpur	500
Samli Power	Rengali, Sambalpur	500
Vedanta Alumine	Jharsaguda	5×135
Aditya Alumine	Lapange Sambalpur	5×130

Table 34.5: Ash Disposal Status in the Existing Power Plants

Power Plant	Installed Capacity (MW)	Coal Consumption* (Tonnes/Day)	Ash Generations (Tonnes/Day)	Disposal System**	Ash Pond Area	Discharge of Overflows
NTPC, Kaniha	1000	15,800	7300	Wet	800	100% recycle
TTPS, Talcher	490	5000	2200	Wet	38	Nandira
INDAL HPCL, Hirakud	72	1530	640	Dry	99	–
IB-Thermal, Banharpalli	420	4500	1090	Wet	246	100% recycle***
RSP, Rourkela	260	2500	1000	Wet	64	Brahmani

* The coal consumption and ash generation figures for 1995 is based on actual generation of power which may be considered as approximate figures.

** Present overflow is into Nandira. There is a proposal for complete recycling.

*** Although the industry has originally proposed 100 per cent recycling of ash pond overflow, it now proposes to discharge a part of its Ash Pond overflow to Hirakud reservoir along with the industrial effluent.

Use in Agriculture

Fly ash can be used as additive or amendment material in agricultural application. So far fly ash has been used in the manufacture of manure and even added to the soil directly to improve productivity.

These are various benefits of using fly ash in agriculture applications. The addition of fly ash alters the structure of clayey and sandy soil to loamy. Its addition is known to increase the water holding capacity of soil phenomenally. As fly ash is alkaline in nature its use increases the pH of the soil thus making it useful for neutralizing acidic soil. Also it adds essential plant nutrients and enhances the crop yield. The use of fly ash has greatly improved the physical health of the soil by improving the water holding capacity and porosity of the soil. It has also been holding capacity and porosity of the soil. It has also been observed that the resistance levels of crop towards diseases and pest infections have also increased due to the use of fly ash (Kumar, 1996).

Though not much evidence has been found regarding the radioactivity levels in fly ash but still precaution should be taken while using fly ash in agriculture applications.

Bricks Making

Different fly ash from the production of fly ash, sand lime bricks indicate that their compressive strength increases with the increase in moulding and autoclaving pressure. The strength of brick cures at atmospheric pressure with steam for six hours attend the value of 50 to 60 kg/cm. sq., which an autoclaving at 14 kg/cm. sq. for six hours increase to 100 to 150 kg/cm. sq. depending upon the characteristic of fly ash. Fly ash contain more unburnt carbon and lesser lime reactivity and fineness yield bricks of lesser strength under same production condition.

Recommendations

The land requirement for dry disposal is less and besides the dry fly ash can find alternate use. Alternate use of fly ash particularly cement making, should be encouraged. Government may invite private entrepreneur for establishment of a large cement mill in Jharsuguda District where a large number of thermal power plants are coming up as Ambuja Cement, and ACC cement has started production cement from fly ash in Gujarat and Madhya Pradesh.

Thermal power plants of very low plant load factor should be allowed to operate. Transmission and distribution loss can be minimized if small power plants are established near lead cements.

In areas where large number of power plants are coming up stricter standards for SPM emission and World Bank standards for sulphur dioxide and no emissions are required to be imposed.

Acknowledgements

The authors are thankful to Prof. (Retd.) B.K. Sinha, Environmental Sciences, Dr. Ashutosh Naik, Reader, Earth Science of Sambalpur University and Mr. Sibananda Bhoi, Computer Faculty, B.R. High School, Belpahar for their encouragement and time to time help for the preparation of the present study.

References

Ash Utilisation in National Thermal Power Plants.

Mathur, A.K. (2000). NTPC, Noida, Research Article, Seminar Collection.

Sahsame, Mohini and F. Ashoka (2000). *Cow Ash Utilization in Building and Agricultural*, RRL, Bhopal, Seminar Collection.

The Hetordorn Work Consultants, 1998, *Environmental Impact and Utilization of Fly Ash: A Study of IB-Thermal Power Plant*. Elesvier Science Ins., 16(6): 1–15.

Chapter 35

A Study of Groundwater Characteristics in the Vrishabhavathi Stretch, Bangalore, Karnataka

K.L. Prakash, P. Ravikumar, K. Venkatesh Raju and R.K. Somashekar

Department of Environmental Sciences, Bangalore University, Bangalore – 56, Karnataka

ABSTRACT

This study highlights the results of the physico-chemical analysis of 30 groundwater samples randomly collected from thirteen MWS, Eight hand pumps and remaining nine samples of PWS during the March 2007 to investigate groundwater characteristics in the stretch of Vrishabhavathi drainage system. During the investigations only important groundwater quality parameters which determine the status of groundwater characteristics were studied. It is observed that some of the groundwater samples were affected following contamination and quality. Some of parameters were observed to be more than that of permissible limit recommended by Bureau of Indian Standards (BIS).

Keywords: *Domestic wastewater, Groundwater characteristics, Vrishabhavathi.*

Introduction

The contamination of groundwater with toxic chemicals and excess nutrients from urban wastewater run-off, usage of wastewater in the farming activities, leaching of nutrients from farms have become a world wide problem in most of the Metropolis. The chemical characteristics of groundwater normally varies, and is influenced by factors such as physical, chemical and biological processes caused by human activities as well as naturally (Somasekhar and Ramaswamy, 1984 and Raghavendra *et al.*, 2002). The non point sources do not have a specific source of discharge as it comes from many invisible sources covering a large area. It is also caused by overland flows of surface runoff to river, application of wastewater for irrigation for instance, agricultural irrigation etc. (Bendahmane, 1992, Angelakis *et al.*, 1999., Niemczynowicz, 1999., and Anderson *et al.*, 2001). The water continuously applied over the ground, during the process of journey picks up pollutants and carries them into groundwaters. Generally, there are major constrains in identifying the non point sources of groundwater pollution and tracing their routes. The rapid increase in pollution, industrialization and heavy dependence on chemical products in the agricultural sector are leading to a serious deterioration of water quality in developing countries (Karanth, 1989). In view of these it is necessary to monitor the quality of groundwater around the wastewater drains as well as agricultural areas to facilitate an understanding of the importance of drinking water quality to public health and hygiene.

This study focuses on the influence of wastewater on groundwater characteristics along the stretch of Vrishabhavathi drainage system of Bangalore city.

Study Area

The study area is in the city outskirts downwards to south Bangalore and lies geographically between 77°25′28.4″ and 77°29′16.2″ east and lies between 12°48′54.7″ and 12°54′29.6″ north (Figure 35.1). Vrishabhavathi River is now and almost drainage system into which major portion of domestic wastewater generated from south-west part of Bangalore flows and finally joins a Byramangala tank after traversing about 30 km from the initial point.

Materials and Methods

Primary and secondary data pertaining to the study were collected both in the field level and using SOI topographic maps. The study area map and Sampling locations were mapped on 1 : 25,000 scale using SOI toposheets through an Arc map of GIS.

A total of 30 groundwater samples were collected covering entire area, out of which, thirteen samples belong to MWS, Eight to hand pumps and remaining nine samples to PWS. All the sampling stations fall between five hundred meters to few kilometers from the Vrishabhavathi open drainage system. The details of sampling locations are presented in Table 35.1 and Figure 35.2. The water samples were collected in 2L plastic containers for Physico-chemical analysis and 100ml sterilized bottles were used to collect samples for biological analysis. Sampling was carried out in March, 2007.

Laboratory Analysis

The samples were transported immediately after collection under freeze condition for laboratory analysis and were analysed for key Physico-chemical parameters following Standard Methods (APHA 1995).

Results and Discussion

The results of Physico-chemical analysis of groundwater samples are discussed parameter wise as follows.

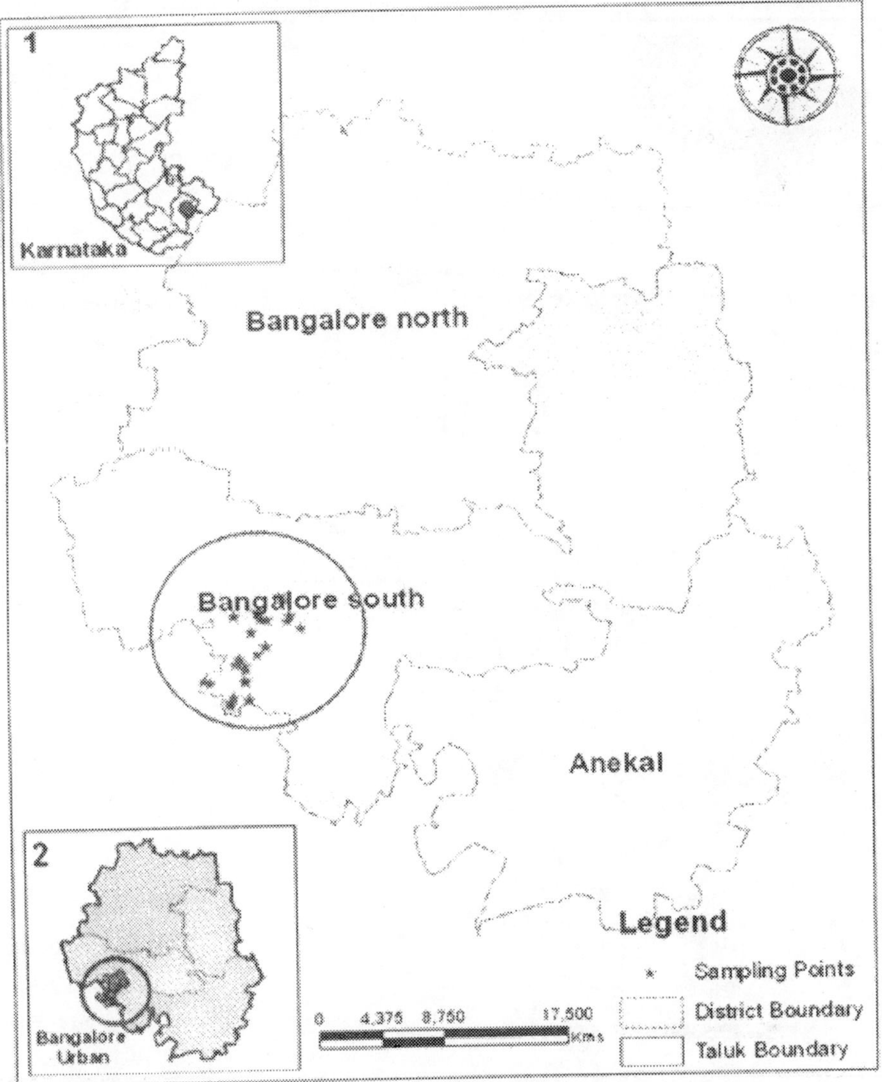

Figure 35.1: Map Showing Study Area of Vrishabhavathi Stretch

pH

A direct relationship between human health and pH of drinking water is impossible to ascertain because pH is so closely associated with other aspects of drinking water quality. The pH of water is adversely affected if it is low or higher than the prescribed limit. The pH value of the samples was observed to be in the range of 6.5-8.5, indicating that the values are within the prescribed limit by BIS and ICMR.

Colour

The determination of colour is rapid and is useful in detecting a change in the character of the water. Naturally, colour exists in water primarily as negatively charged colloidal particles. As such,

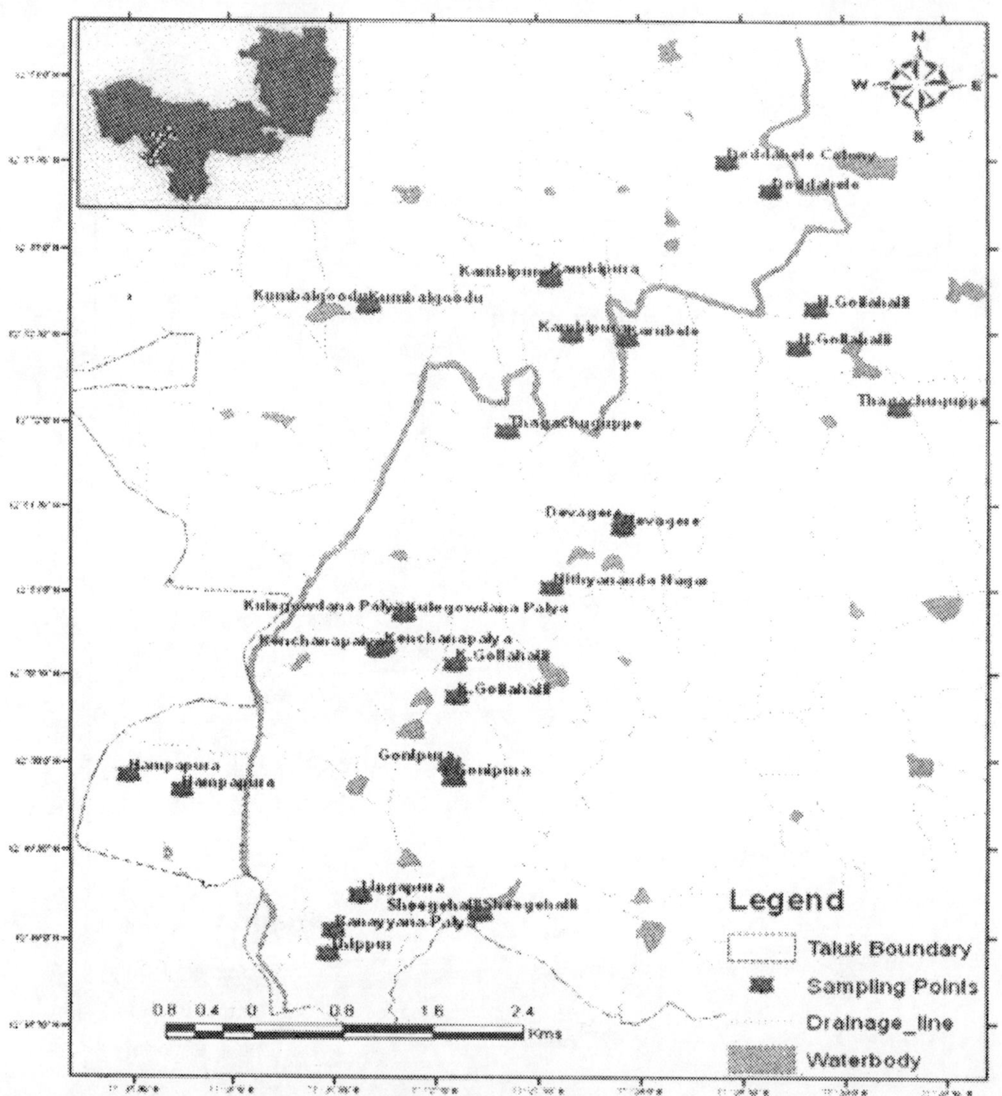

Figure 32.2: Map Showing Sampling Locations along Vrishabhavathi Stretch

its removal is usually readily accomplished by coagulation with a salt having a trivalent metallic ion, such as aluminum or iron. Colour values of water samples are observed to be within the limits (Figure 35.3).

Turbidity

Turbidity in water has a significant effect on the microbial quality. The turbidity is caused by a variety of suspended materials that range in size from colloidal to coarse dispersion, depending on the degree of turbulence. The values of turbidity are within the prescribed range (Figure 35.3).

Table 35.1: Groundwater Sampling Locations along Vrishabhavathi Stretch

Sl.No.	Location	Latitude	Longitude	Source	Elevation (MSL)
1.	Devagere	E 77°27'54.9"	N 12°51'24.7"	MWS	728
2.	Devagere	E 77°27'54.6"	N 12°51'21.9"	HP	730
3.	Nithyananda Nagar	E 77°27'33.9"	N 12°51'1.0"	PWS	751
4.	Kulegowdana Palya	E 77°26'50.9"	N 12°50'51.5"	MWS	748
5.	Kulegowdana Palya	E 77°26'51.0"	N 12°50'51.1"	HP	761
6.	Kenchanapalya	E 77°26'43.0"	N 12°50'38.6"	MWS	767
7.	Kenchanapalya	E 77°26'45.1"	N 12°50'39.8"	HP	764
8.	K. Gollahalli	E 77°27'05.6"	N 12°50'33.7"	MWS	777
9.	K. Gollahalli	E 77°27'6.0"	N 12°50'21.9"	PWS	773
10.	Gonipura	E 77°27'4.2"	N 12°49'58.8"	MWS	763
11.	Gonipura	E 77°27'5.2"	N 12°49'53.8"	HP	761
12.	Sheegehalli	E 77°27'12.7"	N 12°49'8.1"	MWS	749
13.	Sheegehalli	E 77°27'13.3"	N 12°49'8.4"	HP	749
14.	Lingapura	E 77°26'37.6"	N 12°49'14.4"	MWS	747
15.	Thippur	E 77°26'28.4"	N 12°48'54.7"	MWS	742
16.	Bandayyana Palya	E 77°26'29.9"	N 12°49'2.7"	PWS	745
17.	Hampapura	E 77°25'44.8"	N 12°49'50.4"	MWS	774
18.	Hampapura	E 77°25'28.4"	N 12°49'55.5"	HP	789
19.	Kumbalgoodu	E 77°26'41.3"	N 12°52'40.5"	MWS	792
20.	Kumbalgoodu	E 77°26'40.9"	N 12°52'39.9"	HP	784
21.	Thagachuguppe	E 77°29'16.2"	N 12°52'4.5"	MWS	793
22.	Thagachuguppe	E 77°27'21.2"	N 12°51'56.6"	PWS	785
23.	Karubele	E 77°27'56.5"	N 12°52'28.5"	MWS	793
24.	H. Gollahalli	E 77°28'52.0"	N 12°52'38.8"	PWS	801
25.	H. Gollahalli	E 77°28'46.7"	N 12°52'25.6"	PWS	800
26.	Doddabele	E 77°28'25.8"	N 12°53'29.6"	PWS	813
27.	Doddabele	E 77°28'38.9"	N 12°53'19.3"	PWS	795
28.	Kambipura	E 77°27'34.0"	N 12°52'49.0"	MWS	786
29.	Kambipura	E 77°27'33.8"	N 12°52'50.0"	HP	781
30.	Kambipura	E 77°27'40.3"	N 12°52'29.7"	PWS	774

Electrical Conductivity

It is a measure of water's capacity to convey electric current. Electrical conductivity of water is directly proportional to its dissolved mineral content, and it varies directly with the temperature of the sample. Where waters from the same source are periodically examined, the variation in the dissolved solids content is indicated by conductivity measurements. A study indicated that higher range of Electrical conductivity affects germination, growth and yield in many plants (Senthilkumar *et al.*, 2001). The Electrical conductivity value of the samples was found in the range of 936–2350 μmhos/

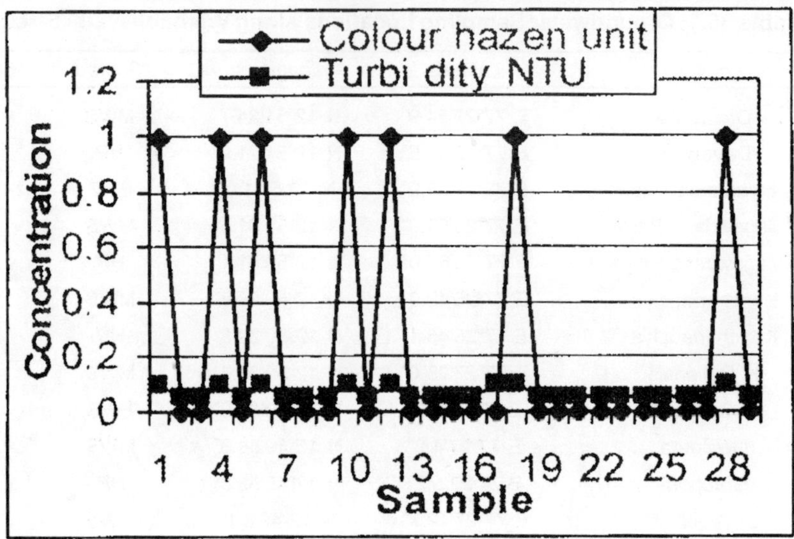

Figure 35.3: Variation in Color and Turbidity

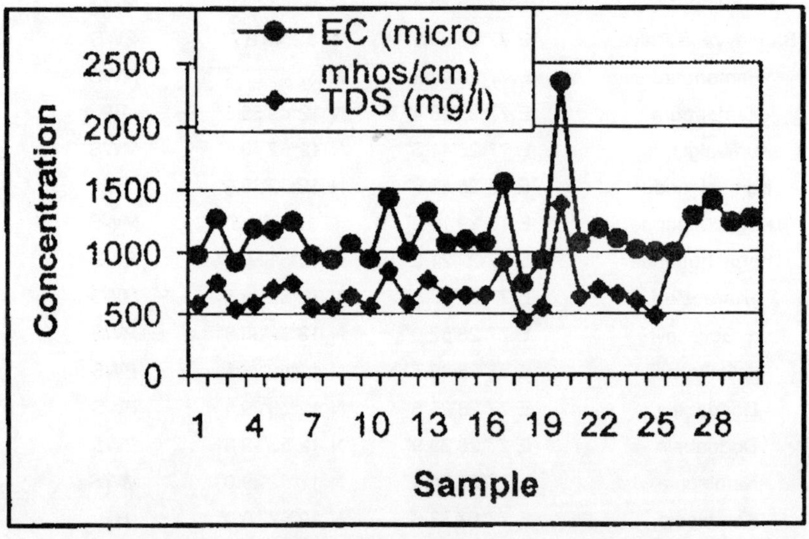

Figure 35.4: Variation in EC of Groundwater Samples

cm. The maximum EC of 2350 μmhos/cm is noticed in the Kumbalgoodu and a minimum of 936 μmhos/cm was observed in the MWS sample at Kumbalgoodu (Figure 35.4).

Total Dissolved Solids

Total dissolved solids consist mainly of bicarbonates, carbonates, sulphates, chlorides, nitrates and possibly phosphates of calcium, magnesium, sodium and potassium. The amount of dissolved solids present in water is a consideration for its suitability for drinking purpose. TDS ranged from 438 to 1380 mg/L. A maximum TDS content of 1380 mg/l was noticed in Kumbalgoodu (Hand pump) whereas a minimum TDS of 438 mg/l was noticed in the Hampapura (Figure 35.4).

A higher level of total dissolved imparts taste to the water and often has a laxative and sometimes the adverse effect upon people whose bodies are not used to higher levels.

Alkalinity

The estimation of alkalinity provides an idea of the nature of salts present. If the alkalinity is equal to hardness, calcium and magnesium salts are only present. If the alkalinity is less than equivalent hardness, neutral salts of calcium or magnesium must be present, that are not carbonates, but usually sulphates. Many materials contribute to the alkalinity of water, particularly hydroxides, carbonates and bicarbonates. About 40 per cent of the samples were found to have Alkalinity exceeding the prescribed limit. A maximum alkalinity of 667 mg/l was observed in the Kenchanapalya sample and a lowest of 412 mg/l was found in the sample of H. Gollahalli (Figure 35.5).

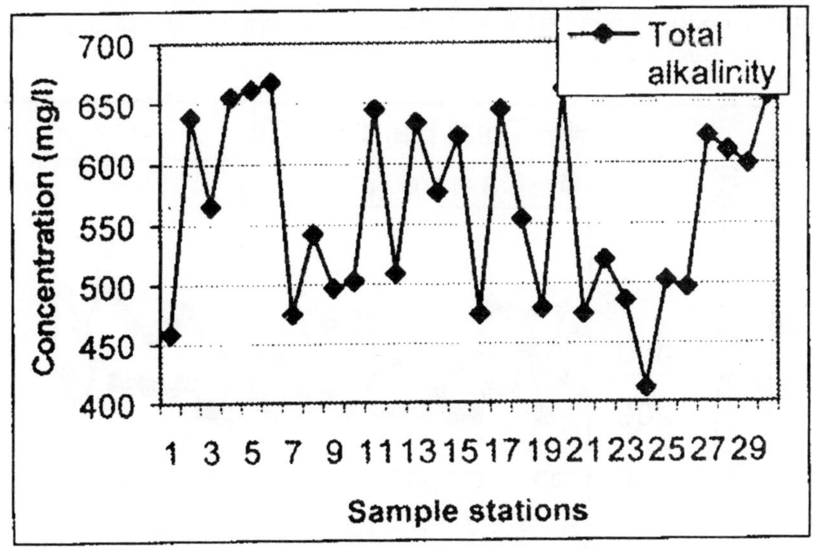

Figure 35.5: Alkalinity of Groundwater Samples

Total Hardness

Groundwater is generally harder than surface-water. The degree of hardness of drinking water has been classified in terms of its equivalent $CaCO_3$ concentration as shown in Table 35.2.

Table 35.2: Classification of Water Based on Hardness

Soft	0-75 mg/L
Medium hard	75-150 mg/L
Hard	150-300 mg/L
Very Hard	Above 300mg/L

Source: World Health Organization, 1984.

Hardness of water is due to presence of cations such as divalent calcium, magnesium, ferrous iron, and manganous ions and derived largely from contact with soil and rock formations, topsoil thick with limestone formations are present. Soft waters originate in areas where the topsoil is quite thin and limestone formations are sparse or absent. Hard waters are satisfactory for human consumption as soft waters. It was noticed that Ca^{2+} and Mg^{2+} are protective against heart ailments.

Total Hardness of the samples was found in the range between 320 and 890 mg/l. The maximum hardness was noticed in the hand pump sample of Kumbalgoodu and minimum of 320 mg/l Nithyananda Nagar.

Carbonate hardness was formerly called temporary hardness because it can be precipitated by boiling. Calcium and Magnesium is the major portion of the hardness.

The maximum calcium hardness of 750 mg/l was observed in the hand pump sample of Kumbalgoodu and minimum of 180 mg/l is observed in samples of Sheegehalli (MWS), Hampapura (HP) and Thagachuguppe (MWS) (Figure 35.6). Calcium hardness also found to be excess in only one sample of Kumbalgoodu, all other samples found to be within the stipulated limits (200 mg/l) for drinking.

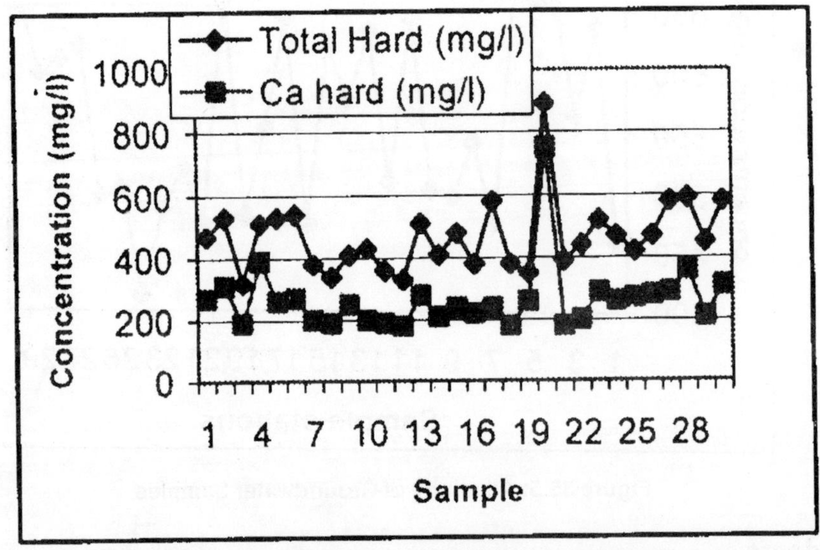

Figure 35.6: Variation of Total and Calcium Hardness

Fluoride

Fluoride is one of the public health significance in water supplies intended for human use, determination of fluoride assumes significance. In situations where fluoride is added to provide an optimum level for the control of dental caries, in some areas, industrial contamination of the atmosphere and vegetation by fluoride has been a serious problem. Control measures have to be employed to protect cattle and other herbivores animals from damage to bones and teeth. In groundwater the fluoride content naturally from the weathering of fluoride rich minerals such as primary silicates and associated minerals (Thakare *et al.*, 2005).

A fluoride content of water samples observed to be within the permissible limits (1.5 mg/l) stipulated for drinking and found in ranged from 0.2 to 1.1 mg/l. (Figure 35.7).

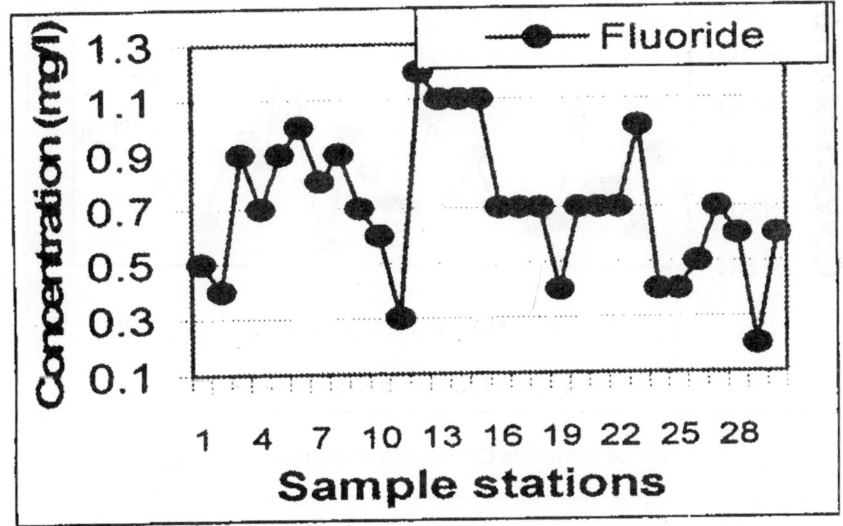

Figure 35.7: Fluoride Level in Groundwater Samples

Chloride

In natural waters, it is attributed to be dissolution of salt deposits, effluents from chemical industries, sewage discharges, irrigation drainage, and contamination from refuse leachates. These sources result in local contamination of both surface and groundwater. Chloride is in reasonable concentrations is not harmful to humans and the sources containing as much as 2000 mg/l are also used for the domestic purposes, without the development of serious adverse effects.

The chloride content found in the range of 85 and 593 mg/l. Maximum chloride content of 593 mg/l was observed in hand pump sample of Kumbalgoodu and minimum of 85 mg/l is observed in hand pump sample of Hampapura (Figure 35.8).

Sodium

It has important considerations for the irrigation waters, but for drinking and industrial purposes has very less importance. However, excessive of sodium in drinking water is harmful to persons suffering from cardiac, renal, and circulatory diseases. The ratio of sodium to total cations (per cent sodium) is important in human pathology. Sodium concentrations vary considerably, depending on regional and hydrological and geological conditions, the time of year, and salt utilization patterns. Sodium levels in groundwater vary widely but normally range between 6–130 mg/l.

The concentration observed in the ranges between 52.5 and 172.5. Maximum sodium concentration of 172.5 was noticed in hand pump water sample of Kumbalgoodu and a minimum of 52.5 mg/L in the MWS water sample of Devagere (Figure 35.9). All the samples were found to be within the permissible limit of 200 mg/l.

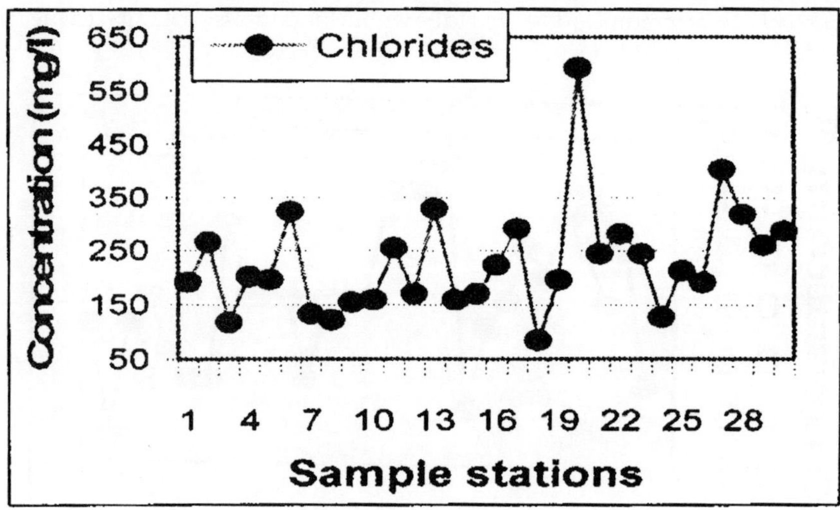

Figure 35.8: Chloride Content in Groundwater Samples

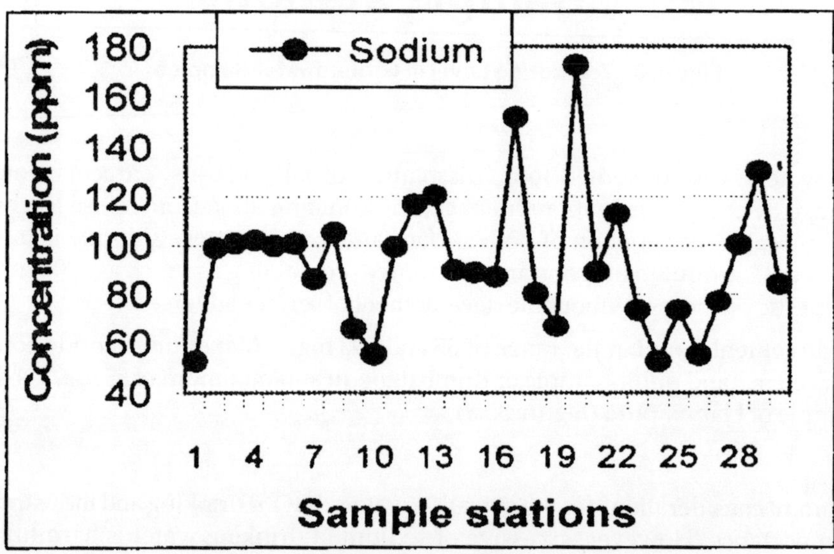

Figure 35.9: Sodium Level in Groundwater Samples

Potassium

Potassium is an essential element, but in excessive amounts it acts as a cathartic. The maximum permissible limit for potassium is 10 mg/l. The concentration ranges between 3.2 and 16.6. Maximum sodium concentration of 16.6 was noticed in MWS water sample of Kambipura and a minimum of 3.2 mg/l in the MWS water sample of Sheegehalli (Figure 35.10). Of the 30 samples, 40 per cent (around 12 samples) showed potassium concentration well within the permissible limit of 10 mg/l.

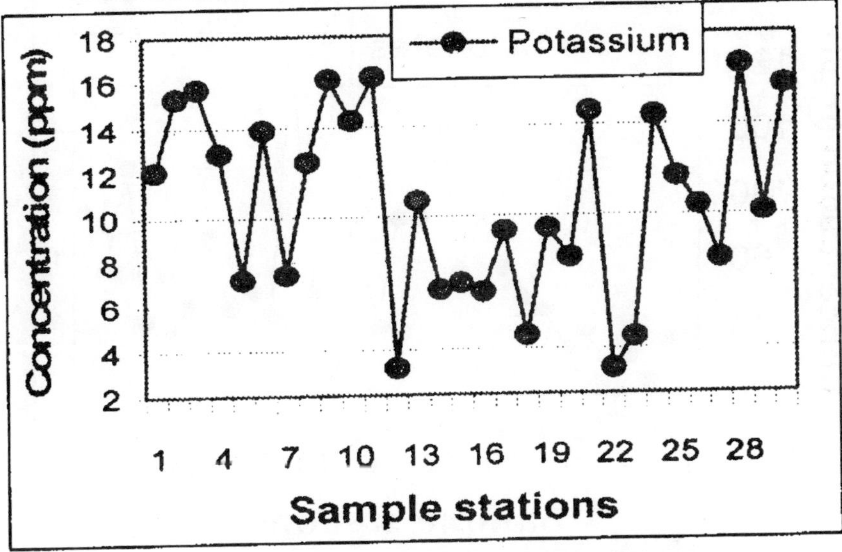

Figure 35.10: Potassium in Groundwater Samples

Sulphate

Sulphate occur naturally in water as result of leaching from gypsum and other common minerals and it contribute to the total solids content. Higher level of sulphate imparts a bitter taste to water. Sulphate is poorly absorbed from human intestine, and it slowly penetrates the cellular membranes of mammals and rapidly eliminated through the kidneys.

Highest concentration of sulphate was found to be 120.6 mg/l in the sample (hand pump) collected from Kumbalgoodu and the minimum concentration of 21.2 mg/l of sulphate was found in the sample (hand pump) collected from Hampapura. The remaining samples were found to contain sulphate concentration within the above range and were in the permissible limit of drinking water (Figure 35.11).

All surface water supplies support growth of minute aquatic organisms. The free swimming and floating organisms are called plankton and are of great interest because of their influence on water quality. Where both nitrogen and phosphates are plentiful, algal blooms occur which may produce a variety of disturbing conditions in the ecosystem: Permissible limit of phosphate for drinking water is 0.3 mg/l, the higher concentration of phosphates of 0.46 mg/l was noticed in the sample of Doddabele and minimum of 0.07 mg/L was found in Hampapura (Figure 35.12). Except for the samples from Doddabele colony (0.36 mg/l), Doddabele (0.46 mg/l) and Kambipura (0.31 mg/l), all other samples (90 per cent) were found to have phosphate within the prescribed limit.

Nitrates

Nitrates are products of oxidation of organic nitrogen by the bacteria present in soils and in water where sufficient oxygen is present. Application of fertilizers to lands and leaching from cesspools contribute nitrate to groundwaters. The results of nitrate observed to be, about 23 per cent of the samples are found to have nitrate concentration more than the permissible limits for drinking water. Maximum of 200 mg/l nitrate was found in the sample of Kumbalgoodu (Figure 35.13). The excess

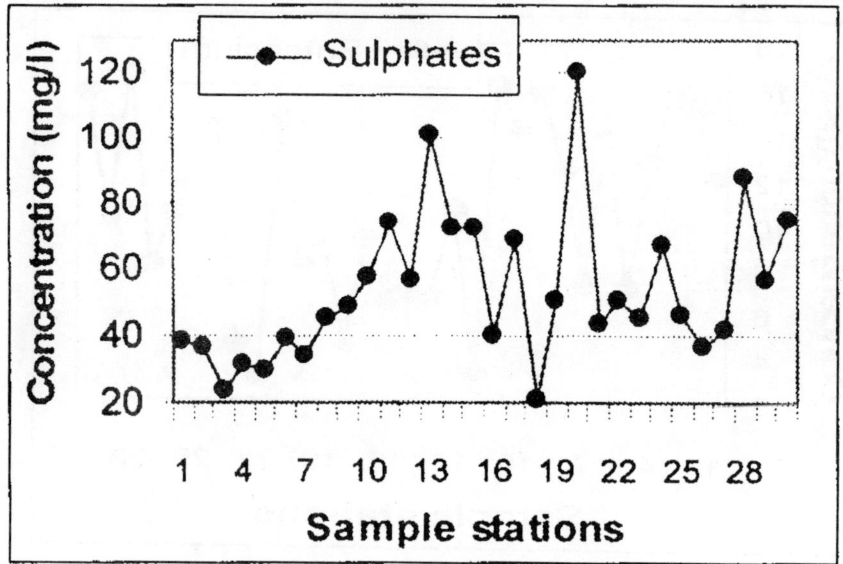

Figure 35.11: Sulphate in Groundwater Samples

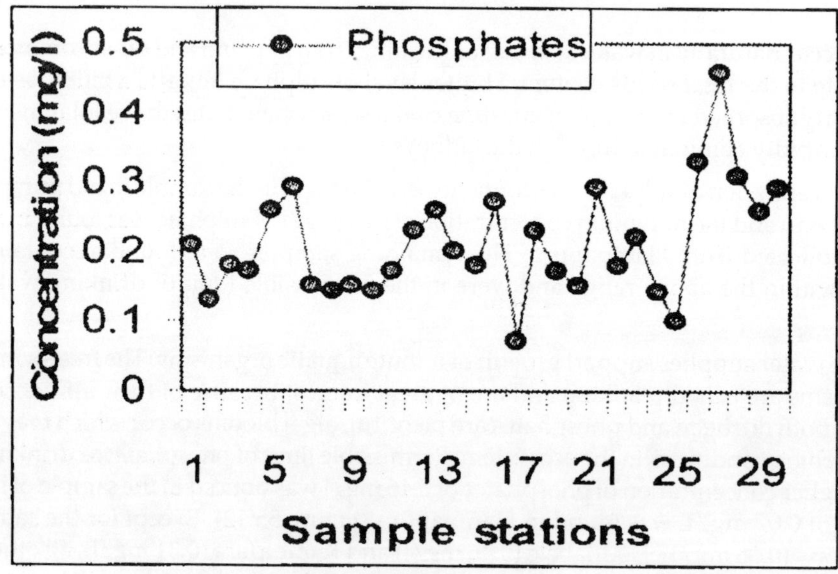

Figure 35.12: Phosphate Content in Groundwater Samples

concentration results methamoglobinemia in the infants. Nitrate forms nitrosoamine in the stomach and causes gastric cancer (Deepashika Garg *et al.*, 2007).

The samples analysis confirmed that except Nitrate and Potassium the remaining parameters pH, EC, TDS, Color, Suspended solids, Sodium, Total Hardness, Total Alkalinity, Sulphate, Phosphates, Chlorides, Fluoride and Nitrate values are within the BIS prescribed standards. Physicochemical

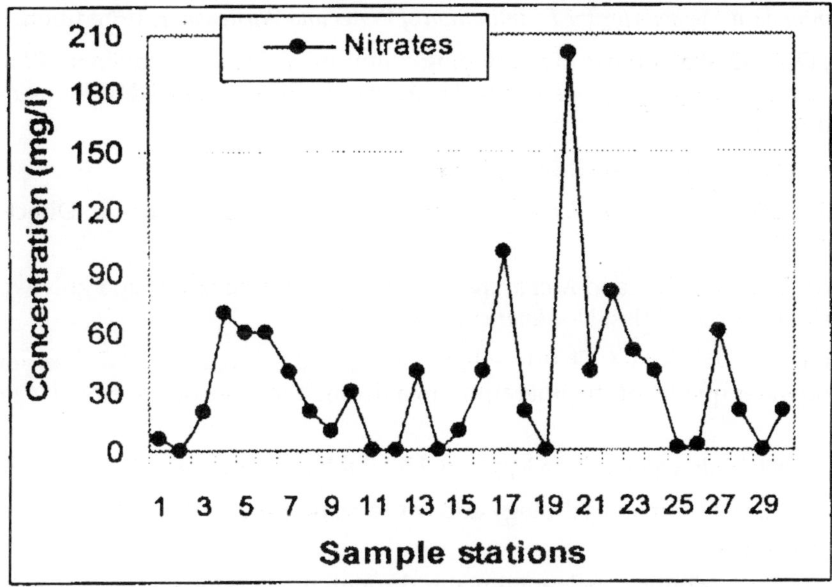

Figure 35.13: Nitrate Content in Groundwater Samples

results have shown a significant level of variation and slightly higher concentration in the samples, particularly the samples close to the wastewater drainage system. Based on the above study, the following suggestions are made.

1. Maintenance of open drainage system to reduce infiltration and recharge of sewage into the groundwater sources.
2. The need of complete treatment of domestic sewage which flowing in the vrishabhavathi valley.
3. Maintaining hygienic conditions by regulating and restricting dumping of domestic wastes around the groundwater source.
4. Mass awareness to the public to keep their water resources in good sanitary conditions.

Thus, it is suggested to control and the practices of conventional method of discharging of domestic wastewater in to the natural open drainage system necessarily stopped. As it once groundwater sources polluted, the persistency of pollutants in the groundwater sources is longer period in addition further pollution. Hence it is also recommended to monitor periodically for sources which helps to check the further pollution.

References

Anderson, J., Adin, A., Crook, J., Davis, C., Hultquist, R., Jimenez-Cisneros, B., Kennedy, W., Sheikin, B. and Vander Merwe, B., 2001. Climbing the ladder: A step by step approach to international guidelines for water recycling. *Water Sci. Technol.*, 43(10): 1–8.

Angelakis, A.N., Marecos Do Monte, M.H.F., Bontoux, L. and Asano, T., 1999. The status of wastewater reuse practice in the Mediterranean basin need for guidelines. *Water Res.*, 33(10): 2201–2217.

APHA, 1995. *Standard Methods for the Examination of Water and Wastewater*, 19th edition, USA.

Bendahmane, D., 1992. Water reuse in developing countries (including guidelines). Water Sanitation for health Project fact sheet, Sponsored by US Agency for International development operated by CDM and Associates.

BIS, 1991. *Analysis of Water and Wastewaters*, BIS India.

Censes of India, 2001. *Provisional Population Totals*. Paper of 2001 Censes–30, Director of Census Operations, Karnataka.

Christova-Boal, D., Eden, R.E. and McFarlanc, S., 1996. An investigation into grey water reuse for urban residential properties. *Desalination*, 106: 391–397.

Garg, Deepashika, Goyal, Sunita, Chaturvedi, Samiksha and Singh, R.V., 2007. Positional survey of the groundwater quality of the Bharatpur area during the monsoon season. *J. Curr. Sci.*, 10(1): 131–136.

Karanth, K.R., 1989. *Hydrogeology*. Tata McGraw Hill Publ. Co. Ltd., New Delhi, pp. 1–455.

Niemczynowicz, J., 1999. Urban hydrology and water management: Present and future challenges. *Urban Water*, 1: 1–14.

Okum, D.A., 1997. Distributing reclaimed water through dual systems. *Am. Water Works Assoc. J.*, 89(11): 52–64.

Ragavendra, K., Prakash, K.L. and Somashekar, R.K., 2002. Quantification of dry weather flow of wastewater in the Vrishbhavathi Valley, Bangalore. In: *Proc. of National Conference on Pollution Prevention and Control*, pp. 277–280.

Senthilkumar, R.D., Narayanasamy, R. and Ramkrishan, K., 2001. Pollution studies on sugar mill effluent: Physico-chemical characteristics and toxic metals. *Poll. Res.*, 20: 93–97.

Somashekar, R.K. and Ramaswamy, 1984. Biological assessment of water pollution: A case study of river Kapila. *Int. J. Env. Studies*, pp. 261–267.

Thakare, S.B., Parwate, A.V. and Rao, M., 2005. Analysis of fluoride in the groundwater of Akola district: A case study. *Indian J. Environ and Ecoplan.*, 10: 657–660.

Chapter 36

Some Marketing Aspects of Tuberose Flower in West Bengal: A Case Study

Md. H. Ali[1] and B.N. Banerjee[2]

[1]Reader, [2]Former Professor,

Department of Agricultural Economics, Faculty of Agriculture,

Bidhan Chandra Krishi Viswavidyalaya, Mohanpur, Nadia, West Bengal

ABSTRACT

Tuberose plant is perennial. Its flower is available throughout the year. This flower is commercially cultivated in some parts of West Bengal. Marketing study covers two major marketing channels. Maximum Price levels are observed in the months of February and June. Producer-sellers fetch higher prices at maximum price levels. Producer's share in consumer's rupee varies from one-fourth to three-fourths.

Introduction

Tuberose flower is widely grown in some specific traditional and non-traditional producing tracts of West Bengal on commercial basis. Wide agro-climatic adaptability, easy cultivation technique and high remunerativeness of this flower on the producer's side, and versatile uses, moderate prices and easy availability throughout the year on the consumer's front advocate the cultivation of the flower.

Mitra *et al.* (1989) have found that in case of Tuberose cut flower marketing, producer's gets about 55 per cent of consumer's rupee irrespective of channels studied. But in case of sticks, producer's share out of consumer's rupee has declined to about 48 per cent in Channel I and 52 per cent in Channel II. They also have observed that the producers are influenced by the seasonal variations of prices in both trades. The variations in prices found to be relatively high in case of marketing of cut flowers than sticks trade.

Mandal (1991) in his study has found that total marketing margin in case of Tuberose (cut flower) worked out to be Rs. 8.52 which is 47.36 per cent of consumer's rupee and trader's profit has ranged between 58 per cent to 65 per cent of marketing margin. But in case of sticks marketing margins have worked out to be Rs. 10.79 *i.e.,* 49.31 per cent of consumer's rupee and trader's profit between 79 per cent to 82 per cent of the marketing margin. He further has observed that retail prices of cut flower and consumer's prices-are lower in Kanchrapara market in comparison to Kolkata market but the converse in true for flower sticks.

Duttagupta (1995) has pointed out that how a number of flower growers and intermediaries earn their livelihood from flower. According to him Jagannathghat (Mullickghat) at Howrah is the main flower market where the marketing starts form early morning. According to Dattagupta's study, a producer of Dalhia flower can earn 15 to 16 thousand rupees per *bigha* per year as net profit by investing Rs. 7000–8000. A Balsam flower grower can fetch 200 per cent profit.

This chapter is designed to explore some economic implications of marketing aspects of Tuberose flower in West Bengal. The specific objectives of the study are as follows:

1. To analyses the variations in seasonal fluctuations in the monthly wholesale prices of Tuberose flower.
2. To work out the price spread and marketing margin of this flower and
3. To study the functional analysis of the marketing margin and its implications.

Materials and Methods

Midnapur (East) District is purposively selected because Tuberose flower is widely cultivated in this district. Panskura-II Block of Midnapur(East) where Tuberose is cultivated intensively and commercially, has been selected purposively. In this block, one Tuberose-growing village is randomly selected and two villages adjacent to it are added up to form a cluster of three villages. Following the probability proportional to number, 75 sample farmers have been selected in each cluster with the help of SRSWR Method. In order to examine the monthly fluctuations in wholesale prices, price spread and marketing margin, one primary wholesale market (Mullickghat), one secondary wholesale market (New Market) and two retail market (New Market and College Street Market)–all situated in Kolkata are selected purposively. Selection of markets is based on the volume of flowers transacted through the markets concerned. Twenty (20) producer-sellers coming to Mullickghat Whole-sale Market, ten (10) Wholesalers (5 from each market) and twenty (20) retailers (10 from each market) have been randomly selected as market intermediaries. Pre-tested survey schedule has been used to collect data with the help of Survey Method. This study is related to the agricultural year 2005–06.

The following two marketing channels–transacting major share of the total production of flower in Midnapore (East) District are being studied:

Channel-I

Producer-seller → Primary Wholesale Market (Mullickghat) → Secondary Wholesale Market (New Market) → Retail Market (New Market) → Consumers.

Channel-II

Producer-seller → Primary Wholesale Market (Mullickghat) → Retail Market (New Market) → Consumers.

Results and Discussion

Table 36.1 presents variations in monthly wholesale prices of the flower Tuberose grown in Panskura-II Block of Midnapur (East) District and sold in Mullickghat (Kolkata) wholesale market. From table it is noted that maximum price is found to prevail in the month of June and the minimum in the month of September. No trend in wholesale price is, however, observed.

Table 36.1: Seasonal Variations in Monthly Wholesale Prices of Tuberose in Mullickghat (Kolkata) Wholesale Market (2005–06)

Rs. per kg

Month	*Jul*	*Aug*	*Sep*	*Oct*	*Nov*	*Dec*	*Jan*	*Feb*	*Mar*	*Apr*	*May*	*Jun*
Price range	8.75–13.00	8.75–13.00	4.75–5.75	6.50–7.50	6.00–13.00	6.00–13.00	13.00–18.00	30.00–36.00	26.00–30.00	12.00–24.00	8.00–13.00	36.00–42.00

Table 36.2 depicts price spread and marketing margin of flower tuberose sold through Channel-I and Channel-II. From table it is observed that at producer's level major items of marketing cost are transport, packing, loading and unloading, and helping hands. Table amply reveals that at this level the producers reap markedly higher profit at maximum price level in comparison to other traders operating at different markets in both the channels. It is interestingly noted that at minimum price level the producers incur loss. Even the minimum price received at primary wholesale market is much less than cost of production. In spite of it, the producers market their produce because firstly the price received sufficiently covers the cost of marketing and secondly the surplus over cost of marketing adds something to the total revenue on one hand and on the other hand cost of production once incurred cannot be deduced or debarred and hence the alternative way out to the producers is to improve net return by selling their produce at those prices which at least cover the cost of marketing. This is more so as there are no alternative uses of flower *i.e.* processing industry.

At Mullickghat (Kolkata) wholesaler's level, major items of marketing cost are packing, loading and unloading, helping hands, and storage and maintenance. No transport cost is incurred at this level as it is met by producer-seller. Trader's profit, irrespective of levels, are relatively high in channel-II in comparison to that in Channel-I in absolute as well as percentage terms.

At secondary wholesaler's level (New Market, Kolkata), major items of cost are transport, packing, loading and unloading and helping hands. Trader's profit at maximum price level is relatively high in comparison to that at minimum one in absolute terms.

At retailer's level both at New Market and College Street Market, Kolkata, major items of cost are packing, loading and unloading, and helping hands. No transport cost is incurred in channel-I (New Market), but in channel-II (College Street Market), transport is the most important item of cost. Trader's profit at both price levels are relatively high in channel-I while comparing with those in channel-I. This is due to higher retail price and lower cost of marketing. But consumer's price in channel-II is lower than that in channel-I. This is mainly due to market location. Channel-I is situated in a prosperous area.

Table 36.2: Price Spread and Marketing Margin of the Flower Tuberose (2005–06)

Rs. per kg

Sl.No.	Items	Channel I Price Level Maximum	Channel I Price Level Minimum	Channel II Price Level Maximum	Channel II Price Level Minimum
1.	At producer's level:				
	(a) Cost of production	6.06 (12.21)	6.06 (63.33)	6.06 (12.48)	6.06 (65.16)
	(b) Cost of marketing	1.13 (2.28)	1.03 (10.76)	1.13 (2.32)	1.03 (11.07)
	(i) Packing	0.24	0.22	0.24	0.22
	(ii) Loading and unloading	0.22	0.20	0.22	0.20
	(iii) Helping hands	0.17	0.18	0.17	0.18
	(iv) Transport	0.29	0.26	0.29	0.26
	(v) Commission and others	0.12	0.10	0.12	0.10
	(vi) Storage and maintenance	0.09	0.07	0.09	0.07
	(c) Producer's profit	27.44 (55.30)	–4.46 (–)	27.44 (56.50)	–4.46 (–)
	(d) Price received by producer	34.63 (69.70)	2.63 (27.48)	34.63 (71.30)	2.63 (28.28)
2.	At primary wholesaler's level (Mullickghat, Kolkata)				
	(a) Cost of marketing	1.15 (2.32)	0.93 (9.72)	1.15 (2.37)	0.93 (10.00)
	(i) Packing	0.25	0.22	0.25	0.22
	(ii) Loading and unloading	0.20	0.18	0.20	0.18
	(iii) Helping hands	0.18	0.12	0.18	0.12
	(iv) Transport	0.00	0.00	0.00	0.00
	(v) Commission and others	0.12	0.12	0.12	0.12
	(vi) Storage and maintenance	0.24	0.15	0.15	0.15
	(vii) Spoilage	0.16	0.14	0.16	0.14
	(c) Wholesaler's profit	4.48 (9.03)	1.29 (13.48)	6.02 (12.39)	2.17 (23.33)
	(d) Price received by wholesaler	40.26 (81.14)	4.85 (50.68)	41.80 (86.06)	5.73 (61.61)
3.	At secondary wholesaler's level (New Market, Kolkata)				
	(a) Cost of marketing	1.66 (3.34)	1.30 (13.58)		
	(i) Packing	0.31	0.25		
	(ii) Loading and Unloading	0.27	0.22		
	(iii) Helping hands	0.25	0.18		

Contd...

Table 36.2–Contd...

Sl.No.	Items	Channel			
				II	
		Price Level		Price Level	
		Maximum	Minimum	Maximum	Minimum
	(iv) Transport	0.40	0.31		
	(v) Commission and others	0.16	0.12		
	(vi) Storage and maintenance	0.17	0.12		
	(vii) Spoilage	0.10	0.10		
	(c) Sec wholesaler's profit	3.07 (6.19)	0.98 (10.24)		
	(d) Price received by secondary wholesaler	44.99 (90.67)	7.13 (74.50)		
4.	At retailer's level:	*New Market Kolkata*		*College Street Market, Kolkata*	
	(b) Cost of marketing	1.61 (3.24)	1.32 (13.79)	1.89 (3.89)	1.68 (18.07)
	(i) Packing	0.36	0.29	0.39	0.34
	(ii) Loading and Unloading	0.35	0.30	0.32	0.29
	(iii) Helping hands	0.32	0.29	0.28	0.22
	(iv) Transport	0.00	0.00	0.42	0.40
	(v) Commission and others	0.14	0.12	0.14	0.12
	(vi) Storage and maintenance	0.20	0.18	0.13	0.14
	(vii) Spoilage	0.24	0.14	0.21	0.17
	(c) Retailer's profit	3.02 (6.09)	1.12 (11.70)	4.88 (10.05)	1.89 (20.32)
	(d) Price received by retailer	49.62 (100.00) 87.79	9.57 (100.00) 36.68	48.57 (100.00) 87.52	9.30 (100.00) 34.84

Figures in the parentheses indicate the percentages to the total.

Table 36.3 presents functional analysis of marketing margin of flower Tuberose. From table it is visualized that major items of marketing cost are packing, transport, loading and unloading, and helping hands. In both the channels, trader's profits are notably high at maximum price level, but negative at minimum-ones. Total profit earned by traders is worked out to be negative because of the fact that the extent of loss incurred by producer-sellers has not been compensated by other traders. In both channels, marked difference between the marketing margins at maximum and minimum prices is promptly noticed. The study clearly reveals the urgency of setting up of agroprocessing units in the sample area in order to protect the interest of flower growers.

Summary and Conclusion

The study reveals that wide and random seasonal variations in monthly wholesale prices prevail in the market throughout the year. At the maximum price level, the producers reap strikingly higher profit which is actually the dual effect of producer's net return coupled with his profit margin as a trader farmer as well. On the other hand, the producer suffers considerable loss at minimum prices. But, on an annual average, the producers earn considerably higher net returns. Producer's share in consumer's rupee varies from 27 to 71 per cent. Packing, loading and unloading helping hands and transport are noted to the major levels of marketing costs. Higher retail price is associated with longer chain of market intermediaries.

Table 36.3: Functional Analysis of Marketing Margin of Tuberose Flower (2005–06)

Rs. per kg

Sl.No.	Items	Channel			
		I		II	
		Price Level		Price Level	
		Maximum	Minimum	Maximum	Minimum
1.	Packing	1.16 (2.66)	0.98 (27.92)	0.88 (2.07)	0.78 (24.07)
2.	Loading and Unloading	0.97 (2.23)	0.90 (25.64)	0.74 (1.74)	0.67 (20.68)
3.	Helping hands	0.82 (1.88)	0.77 (21.94)	0.63 (1.48)	0.52 (16.05)
4.	Transport	1.01 (2.32)	0.57 (16.24)	0.71 (1.67)	0.66 (20.37)
5.	Commission and other charges	0.54 (1.24)	0.46 (13.10)	0.39 (0.92)	0.34 (10.49)
6.	Storage and maintenance	0.60 (1.38)	0.52 (14.81)	0.45 (1.06)	0.36 (11.11)
7.	Spoilage	0.45 (1.03)	0.38 (10.83)	0.37 (0.87)	0.31 (9.57)
8.	Trader's profit	38.01 (87.26)	-1.07 (-30.48)	38.34 (90.19)	-0.40 (-12.34)
	Total	43.56 (100.00)	3.51 (100.00)	42.51 (100.00)	3.24 (100.00)

Figures in parentheses indicate the percentages to the marketing margin.

References

Ali, Md. H. and Banerjee, B.N., 1999. Marketing costs and margins of cock's comb flower in Midnapur (east) district of West Bengal. *Agricultural Marketing*, pp. 34–37.

Ali, Md. H. and Banerjee, B.N., 2000. An economic analysis of cultivation of Chinese Rose (*Hibiscus rosa-chinensis*) in West Bengal. *The Horticultural Journal*, 13(2): 77–81.

Ali, Md. H. and Banerjee, B.N., 2000. Some economic aspects of tuberose (*Polinathes tuberosa*) cultivation in West Bengal. *The Horticultural Journal* 13(1): 93–98

Banerjee, B.N., Ali, Md. H. and Das, K.K., 1999. An economic analysis on Aster (*Aster amellus*) flower cultivation in West Bengal: A case study. *Journal of Interacademicia*, 3(1)3: 91–96.

Banerjee, B.N. and Ali, Md. H., 1999. Economic aspects of cultivation of Antirrhinum (*Antirhinum majus*) flower in West Bengal. *J. Interacademicia* 3(3 and 4)3: 369–311.

Banerjee, B.N. and Ali, Md. H., 2000. Economics of annual *Chrysanthemum* var. Cherrygold (*Chrysanthemum carinatum*) flower in West Bengal. *Environment and Ecology*, 18(1): 167–170

Dutta Gupta, S., 1995. Jeeban-O-Jeebika Order Phool Ghire (For those whose life and livings are flowers). *Ganashakti (Bengali News paper)*; 22nd January, pp. 9–10.

Mitra, A. *et al.*, 1989. Some aspects of tuberose marketing in Nadia district, West Bengal: A case study. *Agricultural Marketing*, pp. 36–40.

Mondal, P., 1991. An economic analysis on production and marketing of tuberose in Nadia district, West Bengal. *Unpublished Dissertation. M.Sc. (Ag.) in Agricultural Economics*, Bidhan Chandra Krishi Viswavidyalaya, Mohanpur.

Subrahmanyam, K.V., 1986. Economics of production and marketing of *Chrysanthemum* flowers in Karnataka. *Indian Journal of Horticulture*, 3(3 and 4): 281–286.

Chapter 37

Acetylation/Recovery of Phenols from Phenolic Effluents as Acetate using Crown Ether as Phase Transfer Catalyst

Sunil Kumar and Bigan Ram

Department of Chemistry, Women's College, Samastipur – 848 101, Bihar

ABSTRACT

Acetylation of various phenols incorporated suitably on resin matrixes were carried out with acetic anhydride in dioxane. The reaction is carried out without/and in presence of catalytic amount of crown ether as phase transfer reagent. It has been observed that the yield of product, reaction time and temperature has significantly improved in the latter case. The present study would be highly promising as a recovery process for environmental management of phenolic effluents.

Keywords: *Acetyalation reaction, Phenolic effluents, Crown ethers, Phase transfer reagent.*

Introduction

Phenolics are the major constituents of some chemical based industries. Most important are nevertheless, the synthetic phenol plants, coke oven plants and by-product from dyes, drug, leather, etc. based industries (Kirchgeasner, 2006). Efforts to work out environmentally friendly techniques for phenolics waste management are until now, focused only on to lower the risk of primary pollutants but the current methodology have attracted controversy (Weber and Stickney, 1993). Regardless of how effective the treatment process, its benefits quickly diminish if national and international regulations tend to limit the quantity of wastes to lower and lower value. Remarkable, many of these methods came to reality where pollutants are partly destroyed (Doubla *et al.*, 2007). Here is an area that is still developing, an area where one can take a leading role by amalgamating the waste management technology and synthetic organic chemistry. Noteworthy, the use of recovery process which converts the waste materials into some useful products are promising from economic point of view as well as to match the regulation regarding discharge of effluent to various estuary systems. A number of useful work has been reviewed regarding use of supported reagents in conversion of phenolic effluents in some useful product (Ram and Sarkar, 1991) but crown ether as phase transfer reagent is visualised as a maiden approach in the present study. Application of crown ether has been motivated by the fascinating features of these macromolecules, which can catalysed reactions on bound substrates (Lehn, 1979). The most elementary crown ether with this potential is the 18-crown-6 macrocycle (1, 4, 7, 10, 13, 16-hexaoxacyclooctadecane). The advantage of using present method over previous one in term of faster reaction rate and quantitative yield of product lies in the fact that there would be possible to devise a process which is more convenient and economical from environmental point of views.

Material and Methods

Preparation of Resin Phenates

Amberlite IR (Anion Exchanger resin, Cl^- form) was converted into OH^- form by soaking the resin with 1 N Sodium hydroxide solution for 2h. The resin (OH^- form) was treated with 20 per cent aqueous solution of phenol for 6h to obtain resin in phenate form (O^-pH). Similarly phenate form of the other phenols were also obtained.

Crown ether (18-Crown-6) was recrystallised from acetonitrile and dried under vacuum for 72h at room temperature.

Acetylation Procedure

Resin phenate form (containing 13 m mol) was shaken with acetic anhydride (130 m mol) in dioxane (10 ml) with TLC monitoring in a stoppered flask. In a separate experiment, to the same reaction mixture catalytic amount of crown ether was added and the completion of the reaction is monitored as previous. After completion, the resin was filtered off, and washed with methylene chloride. Excess solvent was distilled off and further distillation of the residual material afforded phenyl acetates.

Results and Discussion

Figure 37.1 illustrate the various steps involved in the incorporation of phenols on resin matrixes and subsequently acetylation of resin phenates. The incorporation step is observed to pH dependent and optimum pH = 8–9 was adjusted using strong alkaline medium.

Table 37.1 shows the number of entries of phenols in our experiment. The entries are made on the basis of most common phenols which are most often present in phenolic effluents. Acetylation under usual condition results in 45–55 per cent yield of corresponding acetates. But the yield of product is 90–95 per cent, when catalytic amount of crown ether was added to the reaction mixture. This almost double increase in the reaction product and faster reaction rates as evident from the summary Table 37.1, the use of crown ether in acetylation of phenol is a good advantage over the normal procedure. The advantage of using crown ether is most likely due to fact that phenate ions on the resin matrixes is well exposed and immoblised and thereby the phase transfer catalyst facilitates the easy assess of acetylating reagent to the resin substrates (Sarkar and Rao, 1979).

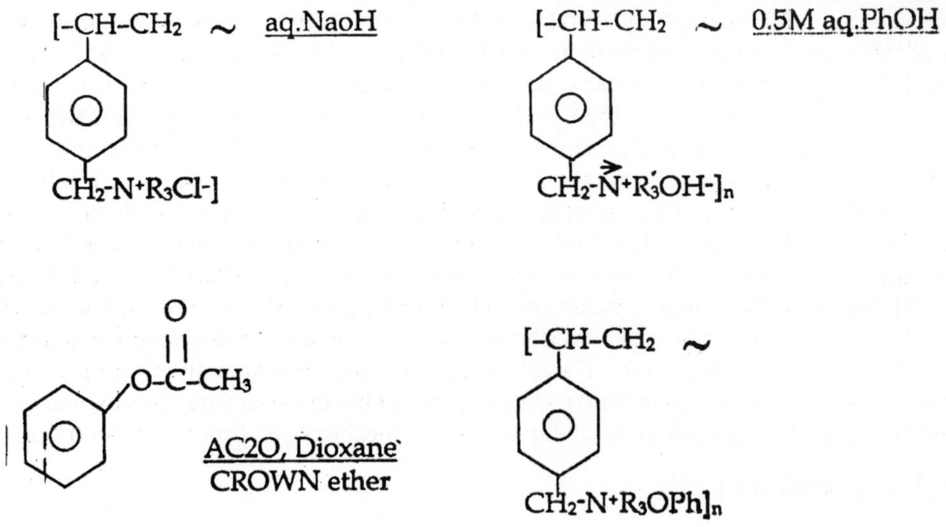

[-CH-CH2 ~ aq.NaoH

CH2-N+R3Cl-]

[-CH-CH2 ~ 0.5M aq.PhOH

CH2-N+R'3OH-]n

O

||

O-C-CH3

AC2O, Dioxane

CROWN ether

[-CH-CH2 ~

CH2-N+R3OPh]n

Figure 37.1: Mechanism of Acetylation of Polymer Supported Phenoxide

Table 37.1: Acetylation of Polymer Supported Phenoxide Anions with Acetic Anhydride Using Crown Ether as Phase Transfer Catalyst at Room Temperature

Sl.No.	*Anion*	*Reaction Time (h)**	*Yield (Per cent)**	*b.p./m.p.(Oc)*
1.	$C_6H_5O^-$	24 (0.5)	51 (94)	195.5/760 mm
2.	$CH_3C_6HO^-$ (O)	24 (1.5)	48 (88)	208/760 mm
3.	$CH_3C_6H_4O^-$ (m)	24 (1.5)	48 (92)	212/760 mm
4.	$CH_3C_6H_4O^-$ (p)	24 (1.5)	51 (92)	212/760 mm
5.	α-Naphthoxide	24 (2)	52 (91)	48–49
6.	β-Naphthoxide	24 (2)	59 (95)	71–72
7.	O-$C_6H_4O^-$ (m)	24 (2)	54 (94)	153/10 mm
8.	O-$C_6H_4O^-$ (o)	24 (2)	54 (93)	64–65
9.	Thymolate	24 (12)	Nil (72)	1180/10 mm

Values given in parantheses indicates reaction carried out with crown ether (PTC).

Conclusion

Acetylation of supported phenoxides in presence of catalytic amount of phase transfer catalyst results in quantitative yield of corresponding acetates. The simplicity and utility of the process lies in the fact that this method can easily take up to the control of phenolic effluents from various phenol based plants.

Acknowledgement

Dr. Sunil Kumar acknowledge sincere thanks to Prof. I.S. Jha, Head of Department of Chemistry, L.N.M.U, Darbhanga for providing valuable suggestions to carry out present research work.

References

Doubla, A. *et al.*, 2007. Destruction of alkaline cyanides by electric discharges. *Res. J. Chem. Environ.*, 11(1): 92–96.

Kirchgeasner, N.H., 2006. Treatment of phenolic plastic waste. *Sew. Ind. Water*, 27: 1375–1378.

Lehn, J.M., 1979. Use of crown-ether in diverse field of practical application. *Pure Appl. Chem.*, 51: 979–988.

Ram, B. and Sarkar, A., 1991. O-methylation of polymer-supported phenoxides with dimethyl-sulphate. *Organic Preparation and Procedure International (OPPI)*, 23(2): 208–210.

Sarkar, A. and Rao, B.R., 1991. Reduction of cyclic ketones with supported reagents. *Tetrachedron Letters*, 32(9): 1247–1251.

Weber, E.J. and Stickney, V.C., 1993. Conventional methods for treatment of primary pollutants. *Water Res.*, 27: 63–66.

Chapter 38

Microscopic and Porosimetry Studies to Estimate the Firing Temperature of Archaeological Potteries of Perur Site in Tamil Nadu, India

G. Velraj and K. Janaki
Department of Physics, Periyar University, Salem – 636 011, Tamil Nadu

ABSTRACT

Ceramics are amongst the artifacts that are most frequently unearthed by the archaeologist. These utilitarian goods are generally found in a fragmentary state. Such fragments of potteries collected from the archaeological site Perur in Coimbatore district was examined in the present work using Scanning Electron Microscopy (SEM with EDS) and porosimetry method. From these two measurements useful information regarding the artisans of Perur site and firing temperature achieved by the artisans at the time of manufacture has been obtained.

Keywords: SEM-EDS, Porosity, Firing temperature, Archaeological potteries.

Introduction

The examination of ancient pottery with an analytical Scanning Electron Microscope SEM is valuable for characterizing and distinguishing between the different traditions in ceramic technology in antiquity (because information is obtained on both the extent of vitrification and the firing temperature). The extent of vitrification provides a useful property for characterizing the quality of a pottery, since it influences several micro structural and physical properties which are relevant to its suitability for the various uses to which it might have been put (Sandrolino *et al.*, 1993). The determination of apparent porosity of the archaeological potteries is another valuable method to estimate the firing temperature, to identity skills employed by the artisans to manufacture the potteries (Maniatis, 1984; Palanivel and Velraj, 2003). In the present study an attempt has been made to identity the firing temperature achieved by the artisans belonging to Perur site in Coimbatore district at the time of manufacture. This site was excavated by the State department of archaeology, Chennai, Tamil Nadu.

Materials and Methods

The microphotographs of the samples were recorded using SEM JSM 561OLU JEOL make. The maximum magnification possible in the equipment is 3,00,000 times with a resolution of 3nm. The elemental analysis was done using the OXFORD INCA Energy dispersion X-ray fluorescence Spectrometer (EOS). The semiquantitative analysis for major and minor elements (Na, Mg, Al, Si, K, Ca, Ti, Mn, Fe, Cl and Cu) were carried out using the energy dispersive X-ray fluorescence Spectrometer with an accelerating voltage of 15Kv attachment to the SEM.

The open or apparent porosity of the archaeological sample is defined as the ratio of the volume of the open pores to that of the total volume of the sample (Sankaran, 1995). The water absorption method was adopted to measure the porosity values of the pottery samples taken for the present work. The open or apparent porosity is thus calculated using the relation given by (Palanivel and Velraj, 2003).

Results and Discussion

The SEM examination of the potteries in the received state (ARS) provides information on the internal morphology developed during the firing in antiquity at the time of their manufacture In SEM examination of the fracture surface of the fired clay the appearance of isolated smooth surfaced areas or filaments of the glass is the first definite stage in the development of vitrification. The structure being referred to subsequently as the initial vitrification (IV) stage. With increasing firing temperature, the extent of the glassy phase increases until the total vitrification stage (IV) is reached (Maniatis and Tite, 1978). In the total vitrification stage the continuous smooth vitrified layer is formed over the entire fracture surface. The extensive vitrification (IV) stage is in between initial and total vitrification. The presence of some slight buckling and rounding of edges of the clay plates can also be identified in an earlier stage in the development of vitrification. Such a stage is also known as no vitrification stage (NY).

The clays have the concentrations of silica, (SiO_2) alumina (Al_2O_3) and fluxes (K_2O, Fe_2O_3, CaO, MgO and TiO_2) as the composition. If the fluxes are more than 9 per cent the clays are classified as low refractory and classified as high refractory if the fluxes in the sample are less than 9 per cent. On the basis of the subsequent development in vitrification the clays can be divided into two main types as non-calcareous (NC) and Calcareous (C). The clays containing Calcium oxide (CaO) above 6 per cent are known as calcareous clays and less than 6 per cent are known as non-calcareous clays. According to (Maniatis and Tite, 1975), the development of vitrification also depends on the atmosphere in which the clay is fired. Hence, the elemental analysis was also done in the present study to identify the type of clay minerals present in the Perur site samples (Rathossi *et al.*, 2004) have stated that the firing temperature, the atmosphere in the kiln and the chemical composition of the fired material especially its Calcium and iron contents determine the extent of the vitrification of clays. On the basis of the composition of the shreds and the vitrification structure developed during firing, the tiring temperature attained by the artisans of Perur was estimated. The SEM photographs of the Perur samples collected at five different depths are shown in the Plates 38.1–38.5 and their EDS spectra are given in the Figures 38.1 to 38.5 respectively.

From the Table 38.1, it is clear that the EDS spectra of the sample shows clay composition of Na_2O, Al_2O_3, SiO_2, K_2O, CaO3, TiO_2, FeO, MnO_2 and MgO in different percentage. The flux MnO_2 is not present in the samples PRR2 and PRR4. The composition ofPRR5 alone is peculiar. It contains in addition to above fluxes, the flux ChO_3 and CuO which are not present in any of the sample taken from Perur site, with major composition of Alumina, Quartz and Iron oxides, The percentage of K_2O is more in all Perur samples. All the samples belong to low refractory and non-calcareous type except PRR1,

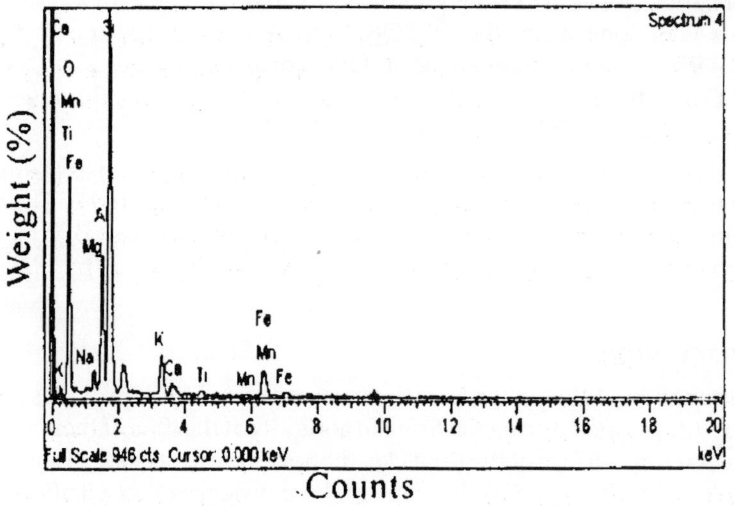

Figure 38.1: EDS Spectrum of the Sample PRR1

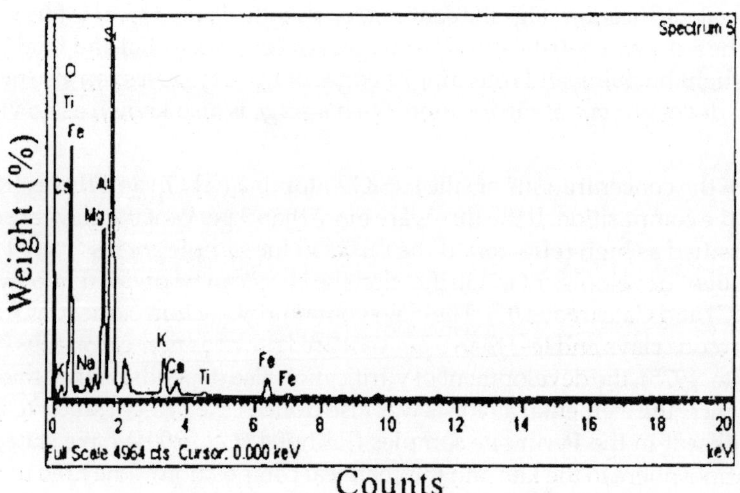

Figure 38.2: EDS Spectrum of the Sample PRR2

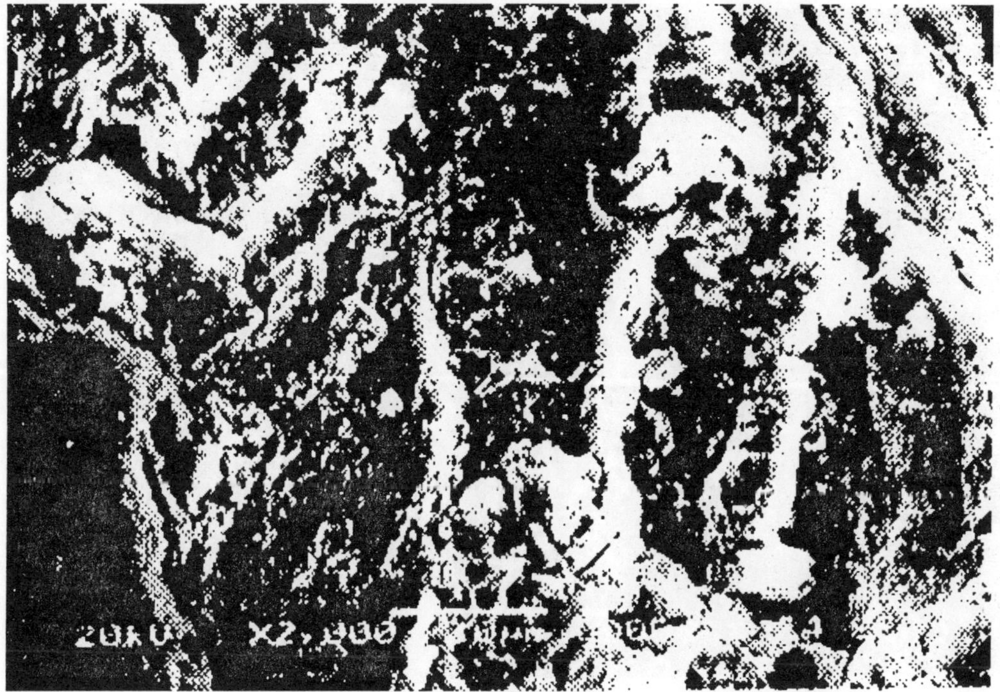

Plate 38.1: Microphotograph of the Sample PRR1

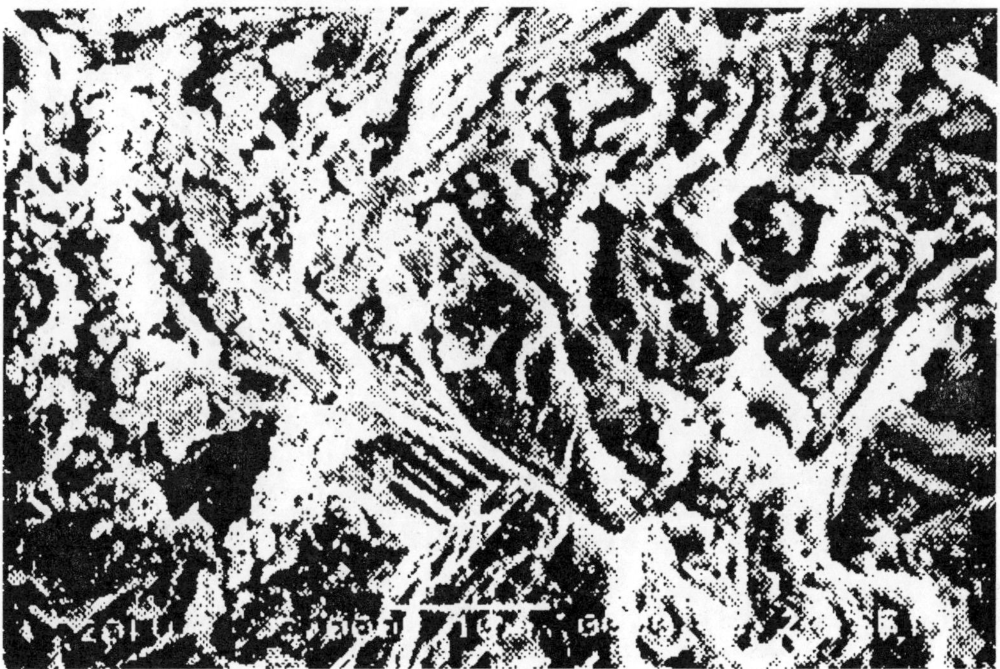

Plate 38.2: Microphotograph of the Sample PRR2

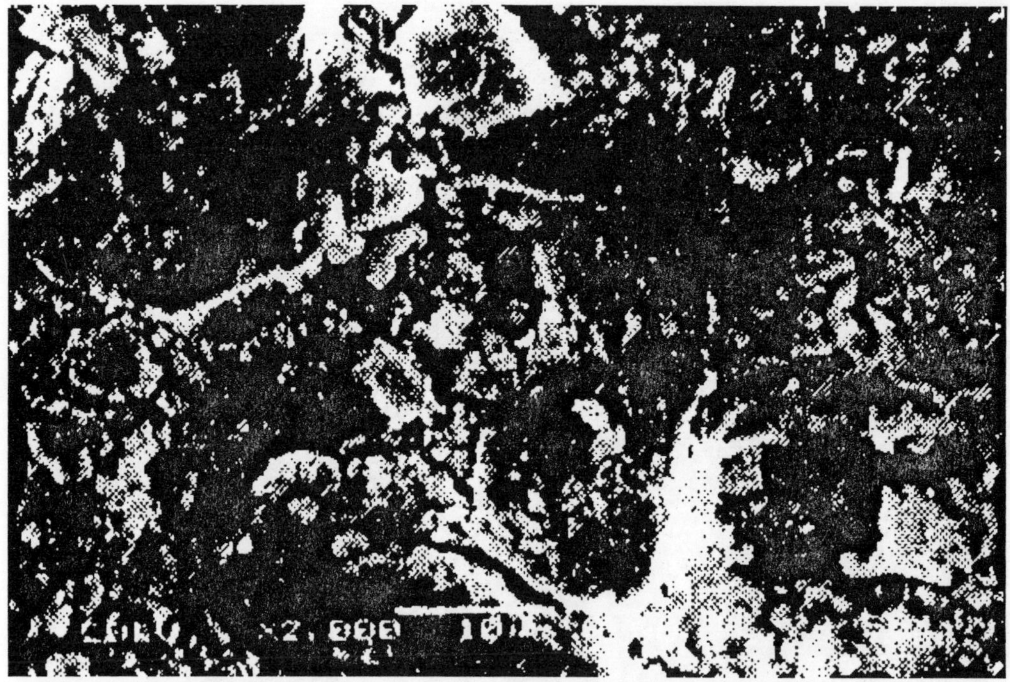

Plate 38.3: Microphotograph of the Sample PRR3

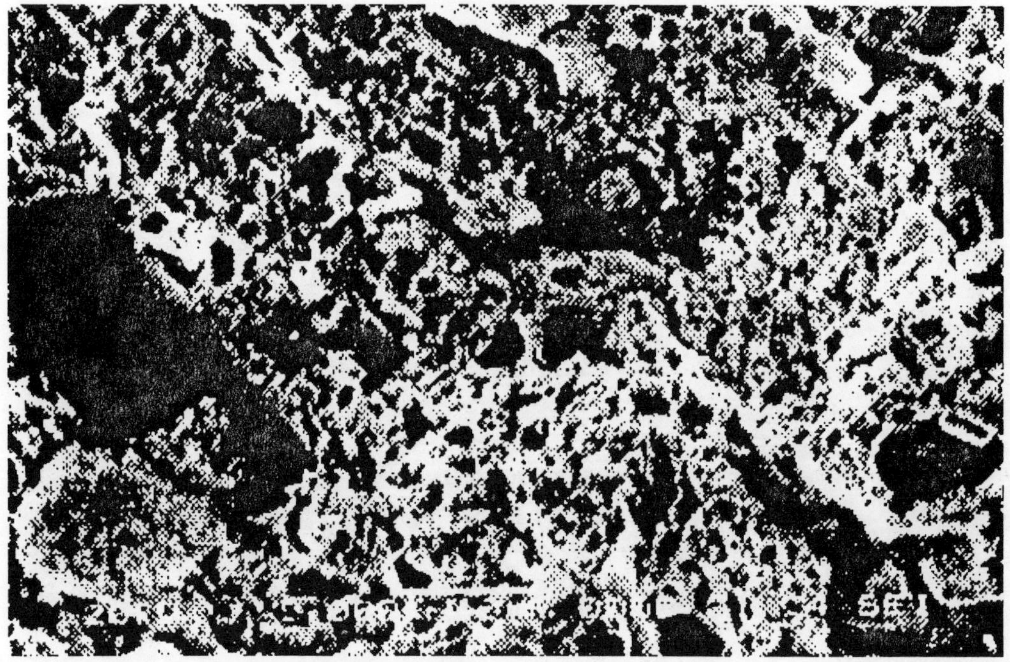

Plate 38.4: Microphotograph of the Sample PRR4

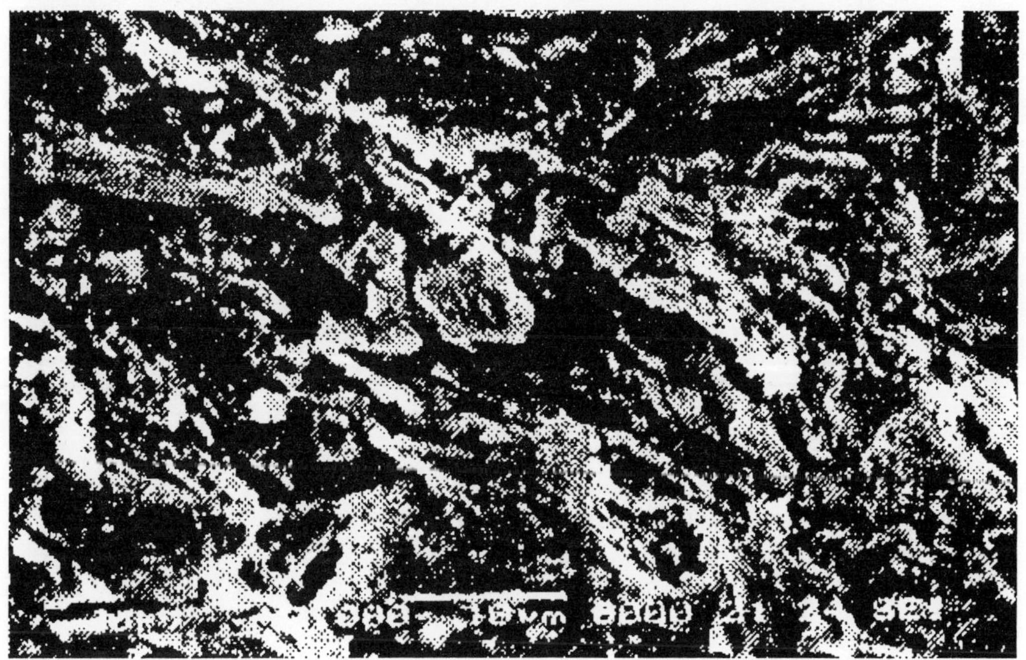

Plate 38.5: Microphotograph of the Sample PRR5

which has CaO content of 7.35 per cent, hence it is of calcareous type, The SEM microphotographs of the samples PRR1 and PRR2 shows that they were fired in no vitrification stage.

Table 38.1: Chemical Composition of Potteries using SEM with EDS Analysis in Percentage

Parameters	*PRR1*	*PRR2*	*PRR3*	*PRR4*	*PRR5*
Na_2O	0.62	2.71	1.02	2.41	0.97
MgO	2.81	2.16	3.20	1.87	1.89
Al_2O_3	14.79	18.16	15.72	16.53	15.70
SiO_2	49.08	57.06	54.46	48.60	49.15
K_2O	8.70	6.36	8.11	6.88	6.61
CaO	7.15	4.07	3.71	3.47	2.12
TiO_2	1.55	1.04	1.05	1.43	1.11
MnO_2	0.58	–	0.27	–	0.81
FeO	14.52	10.40	12.47	18.81	20.32
Cl_2O_3	–	–	–	–	0.17
CuO	–	–	–	–	1.07

The elemental analysis shows that they were fired in oxidizing atmosphere and PRR1 is calcareous clay and PRR2 is non-calcareous clay. As suggested by (Maniats and Tite, 1981), low refractory

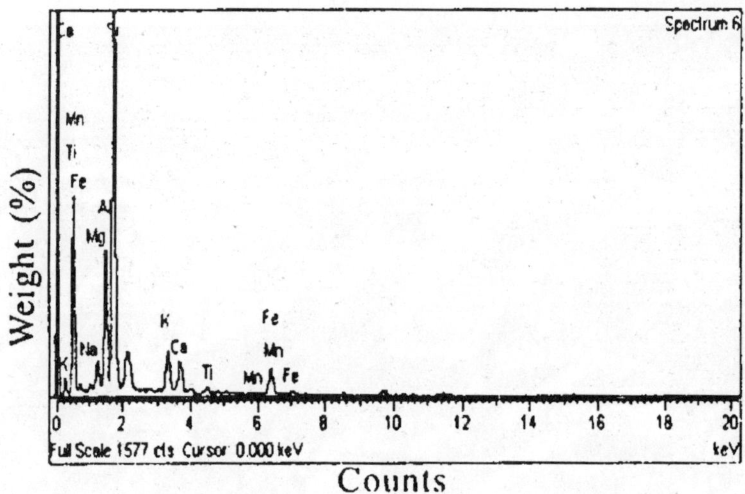

Figure 38.3: EDS Spectrum of the Sample PRR3

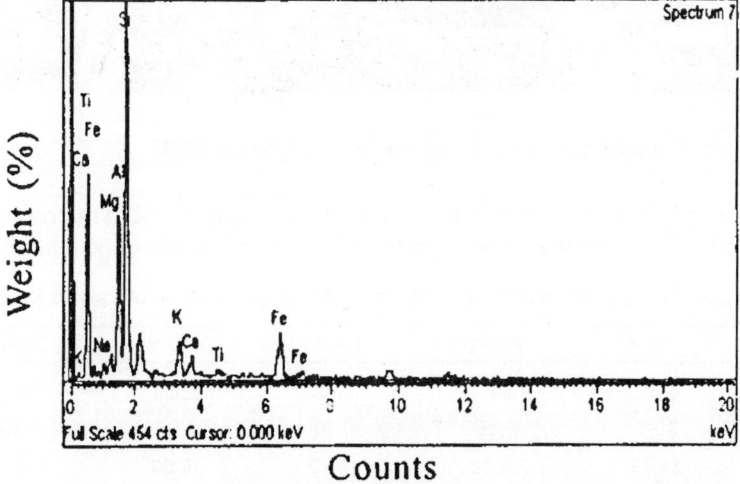

Figure 38.4: EDS Spectrum of the Sample PRR4

calcareous (or) Non-calcareous clay fired in oxidizing atmosphere may be fired below 800°C. The samples PRR3 and PRR4 are in extensive vitrification stage low refractory non-calcareous clay, hence fired in the range 850–950°C. One more sample taken from Perur site PRR5 shows different characteristics *i.e.*, it is low refractory, non-calcareous clay but is in continuous vitrification stage, fired in reducing atmosphere, hence fired in the range 850–950°C (Maniats and Tite, 1981), it may be due to different composition of clay minerals. The results are tabulated in Table 38.2. Hence the SEM pattern with EDS analysis helps us to know the mineral composition of Perur site samples and the range oftemperature to which they might have been fired.

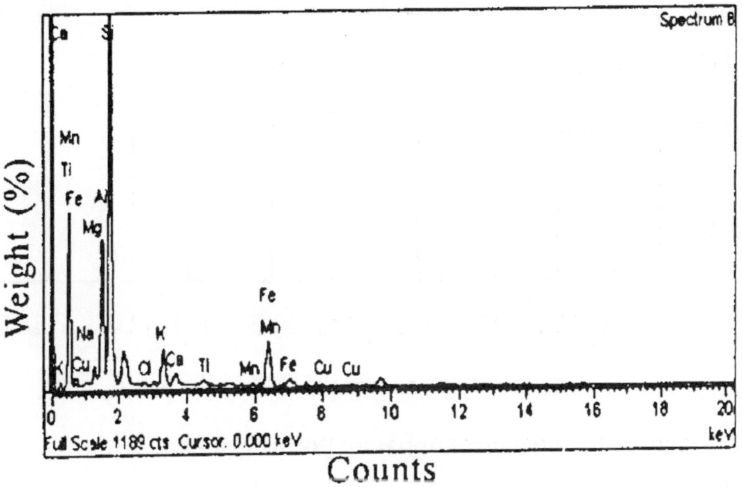

Figure 38.5: EDS Spectrum of the Sample PRR5

Table 38.2: Vitrification Stage and Firing Temperature of the Archaeological Potteries of Perur

Sample Code	*Colour*	*Atmosphere*	*Clay Type*	*Vitrification*	*Estimated Firing Temperature*
PRR1	Red slipped ware	Oxidizing	C	NV	<800°C
PRR2	Red ware	Oxidizing	NC	NV	<800°C
PRR3	Red and Black ware	Oxidizing	NC	EV	850–950°C
PRR4	Red slipped ware	Oxidizing	NC	EV	850–950°C
PRR5	Black ware	Reducing	NC	CV	850–950°C

The complementary technique to SEM *i.e.,* porosity is also utilized to verify these results. The nature of particle and percentage of porosity is correlated with the colour of the potteries to find the firing temperature achieved and the results are tabulated in Table 38.3.

Table 38.3: Estimation Firing Temperature of the Archaeological Potteries of Perur using Porosity Measurement

Sample Code	*Colour*	*Atmosphere*	*Clay Type*	*Per cent of Porosity*	*Estimated Firing Temperature*
PRR1	Red slipped ware	Fine	C	28.08	<900°C
PRR2	Red ware	Fine	NC	26.94	<900°C
PRR3	Red and Black ware	Coarse	NC	40.82	<1000°C
PRR4	Red slipped ware	Coarse	NC	32.39	<1000°C
PRR5	Black ware	Coarse	NC	35.97	<1000°C

The percentage of porosity for Perur site samples vary from 26.94–40.82. It is highest in the sample PRR3. (Maniatis, 1984)) have pointed out that the pottery samples of Non-calcareous clay of

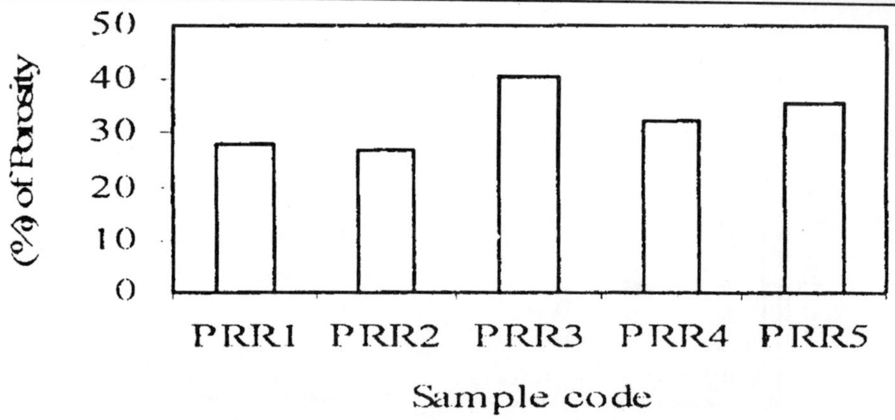

Figure 38.6: Variation of Apparent Porosity of Perur Samples

fine particles sizes having lower range of porosity values were subjected to a firing temperature <900°C. The potteries of fired clay made of coarse particle of both calcareous and non-calcareous clays subjected to a firing temperature below 1000°C have porosity values in the higher range. Hence, according to (Maniatis, 1984), the Perur samples PRR1 and PRR2 are of fine particle in nature and have percentage of porosity of 28.08 and 26.94 respectively, hence fired below 900°C but the samples PRR3–PRR5 are of coarse particle nature hence fired below 1000°C. These results coincide very well with that of SEM with EDS analysis.

Conclusion

Pottery samples at five different depths have been collected from the Perur site in Coimbatore district, Tamil Nadu. These samples were analyzed by SEM with EDS analysis and porosity measurement to view the temperature attained at the time of manufacture. From the two analyses it has been verified that PRR1 and PRR2 may be fired below 800°C and PRR3–PRR5 in the range 850–900°C.

Acknowledgements

The authors acknowledge with thanks to the Director, State Department of Archaeology, Chennai for their timely help for sample collection and also thanks to the Staff-in-charge, Annamalai University, Annamalai Nagar to record SEM pattern in time. Gratitude to The Head, Department of Chemistry, Periyar University to use lab facility for porosity measurements.

References

Christina Rathossi, Panagiota Tsolis-Katagas, Christos Katagas, 2004. Technology and composition of roman pottery in north-western Peloponnese, Greece. *Applied Clay Science*. 24: 313–326.

Maniatis, Y. and Tite, M.S., 1975. *Trans. J. of British Ceram. Soc.*, 74(7): 229.

Maniatis, Y. and Tite, M.S., 1978. *Proc. of Conferences of Thera and the Aegean World I*, London, pp. 483.

Maniatis, Y. and Tite, M.S., 1981. Technological examination of Neolithic-bronze age pottery from central and southeast Europe and from the near east. *Journal of Archaeological Science*, 8: 59–76.

Maniatis, Y., 1984. Firing conditions of white on dark ware from Easter Crete East Cretan white on Dark ware (Ed.) P.P. Betan Court, University of Pennsylvania, pp. 75.

Palanivel, R. and Velraj, G., 2003. Porosity measurements to estimate the firing temperature of ancient potteries. *Journal of Current Sciences*, 3(1): 63–66.

Sandrolino, F., Moriconi, G., Veniali, F., Zapia, G. Haynes, 1993. Principles and applications of pore structure characterization. In: *Proc. Int. Symp. RILEM/CNR*, (Eds.) J.M. Rossi and P. Doria, Milan, pp. 291–297.

Sankaran, S., 1995. Spectroscopic and magnetic studies of Neyveli sedimentary deposits and some archaeological artifacts. *Ph.D. Thesis*. Annamalai University, India.

Chapter 39

Morphological Traits of Cocoon and Non-enzymatic Antioxidant Status in the Diapausing Pupa of *Antheraea mylitta*

A. Sahoo, G.C. Patra, and J. Dandapat*
Department of Zoology, North Orissa University, Takatpur, Baripada – 757 003, Orissa

ABSTRACT

The tropical tasar silkworm *Antheraea mylitta* is an important component of non-mulberry sericulture. Daba and Jatadaba are two ecoraces of this species, respectively, reared in low and high altitudes. The present communication accounts for some of the morphological traits of cocoons and non enzymatic antioxidant status in the diapausing pupae of the above two ecoraces. Results clearly indicate that Daba cocoons are superior for most of the morphological traits. In biochemical makeup particularly the non-enzymatic antioxidants (glutathione and ascorbic acid), Daba pupa is also superior to Jatadaba. The probable reasons for such variation in the different ecoraces have been discussed in this study.

Keywords: *Tasar silkworm, Antheraea mylitta, Non-enzymatic antioxidants, Morphological traits.*

Introduction

Oxygen is indispensable for aerobic life, at the same time it is also a threat because of its partial reduction during respiration which produces several reactive oxygen species (ROS). ROS, like

* E-mail: jdandapat.nou@gmail.com

superoxide radicals ($O_2^{\cdot-}$), hydrogen peroxide (H_2O_2), hydroxyl radical ($^{\cdot}OH$), etc. are highly reactive. Improper neutralization of ROS or their excessive accumulation in the cell leads to oxidative stress, which can damage cellular macromolecules like lipids, proteins and nucleic acids and may result in cell death (Halliwell and Gutteridge, 1989). Insects, like other animals possess a suite of antioxidant system for the removal of damaging ROS (Felton, 1995). In phytophagous insects, antioxidant defence mechanisms are of paramount importance because they are exposed to ROS promotive environments as well as allelochemicals present in the host plants.

The tropical tasar silkworm, *Antheraea mylitta*, belonging to family saturnidae is a phytopolyphagous insect and has the second largest capacity for silk production among all the silk spinning insects (Akai, 2000). Daba and Jatadaba are two ecoraces of this species, respectively, reared in the low and high altitudes of the forest. There is no information about the antioxidant potential of these two ecoraces of this silkworm and therefore in this investigation we report the non-enzymatic antioxidant status in the diapausing pupa of Daba and Jatadaba, which are reared on different host plants in different altitudes.

Materials and Methods

Experimental Animals

Antheraea mylitta cocoons collected from low and high altitudes rearing field (Daba and Jatadaba, respectively) were procured from State sericulture research station, Baripada, Orissa. The cocoons were maintained in the laboratory at room temperature, till the completion of the experiment. The primary host plant, Asana (*Terminalia tomentosa*) and Sal (*Shorea robusta*) leaves were collected from sericulture rearing field of the state government, Orissa.

Morphological Traits of Cocoons

To ascertain various morphological traits biometrical analyses of the cocoons were done as per the standard procedures (Nayak and Guru 1998). The linear measurements such as length and breadth of cocoons, length of peduncle etc., were taken with slide calipers. The thickness of peduncle was measured with screw gauge. The weight of cocoons was recorded by a monopan digital balance. For volumetric measurement, water displacement method was adopted. The circumference of the cocoon was measured with the help of a cotton thread.

Chemicals

Bovine Serum Albumin (BSA), Glutathione (GSH), 5,5′-Dithiobis 2-nitrobenzoic acid (DTNB), Ascorbic acid and other reagents were procured from Sisco Research Laboratory, Mumbai and were of analytical grade.

Tissue Preparation and Biochemical Analyses

After the removal of the outer shell the Pupae were sacrificed and tissue samples (hemolymph and fat body) were collected. Fat body was washed in ice-cold physiological saline (0.67 per cent), wiped dry and weighed on a digital monopan balance. For the estimation of protein, ascorbic acid and glutathione, 10 per cent homogenate of the tissue was prepared in 5 per cent metaphosphoric acid with the help of hand homogenizer (borosilicate glass).

For the estimation of foliar protein content fifty grams of each leaf sample was homogenized in 250ml ice-cold extraction buffer (0.1 M Tris-HCl, pH 8.0,1 mM EDTA, 5mM $Na_2S_2O_3$ and 20mM β-mercaptoethanol). Soluble protein concentration was determined in the supernatant after centrifuging the homogenate at 10,000 rpm for 10 minutes at 4°C.

Estimation of Protein Content

Protein content of the tissue samples were measured according to the method of Lowry et al. (1951) using Bovine Serum Albumin as standard.

Assay of Ascorbic Acid (ASA) and Glutathione (GSH)

Tissue homogenates (indicated earlier) was centrifuged at 10,000 x g for 20 min at room temperature. The deproteinized supernatant was used for assay of ascorbic acid by monitoring the reduction of phosphomolybdate by the method of Jagota and Qani (1982) and for the estimation of GSH using DTNB by the method of Ellman (1959).

Results

Morphological Traits

Morphological traits of Cocoons collected from high (Jatadaba) and low altitudes (Daba) were presented in Table 39.1. Results indicate that peduncle length, peduncle ring diameter and Cocoon circumference breadth of Jatadaba were higher than Daba. However, Daba Cocoon had higher value in its cocoon length, breadth, volume, weight and circumference length.

Table 39.1: Comparison of Different Morphological Traits of Cocoons from Low and High Altitude (Jata daba and Daba). Data are mean±SD(n=8)

Ecorate	*Cocoon Weight in gm*	*Cocoon Circum-ference Length in cm*	*Cocoon Circum-ference Breadth in cm*	*Cocoon Length in cm*	*Cocoon Breadth in cm*	*Peduncle Length in cm*	*Peduncle Ring Diameter in cm*	*Cocoon Volume in ml*
Daba (Low altitude)	11.63 ±2.05	13.66 ±0.66	10.1 ±0.59	5.28 ±0.24	3.21 ±0.23	4.2 ±0.36	0.65 ±0.13	30.12 ±4.31
Jata daba (High altitude)	11.24 ±0.74	13.27 ±0.25	10.21 ±0.25	4.95 ±0.13	3.13 ±0.09	8.33 ±2.26	0.8 ±0.27	26.5 ±1.32

Protein and Non-enzymatic Antioxidants

Hemolymph and fat body protein content in diapausing pupae of *Antheraea mylitta* from low (Daba) and high altitude ecorace (Jatadaba) was presented in Figure 39.1. It was observed that the protein content of hemolymph is almost similar in both the ecoraces however the fat body from Jata daba contain lower (8.6 per cent) amount of protein than Oaba. Figure 39.2 represents foliar protein content of different host plants. Asana (*Terminalia tomentosa*) leaves contain higher amount of protein 194.2±9.23 mg/g followed by Sal (*Shorea robusta*) which contain 175.72±19.8 mg/g of protein. Figure 39.3 represents AsA content in the hemolymph and fat body of Daba and Jatadaba pupae. It was revealed that AsA content was lower in hemolymph (32 per cent) and fat body (10 per cent) of Jatadaba than that Daba ecorace. GSH content of diapausing pupae collected from different altitudes was presented in Figure 39.4. In Jatadaba, hemolymph GSH content was marginally high, however, in fat body the GSH content is substantially more in case of Daba ecorace.

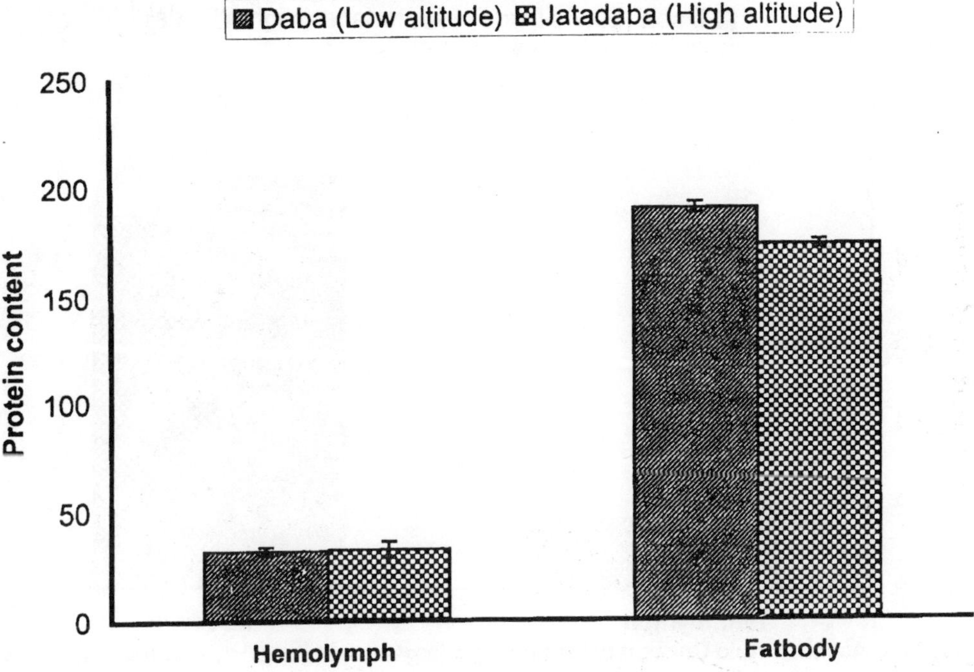

Figure 39.1: Protein Content of Hemolymph (mg/ml) and Fat Body (mg/g tissue wet wt.) in Diapausing Pupa of *Antheraea mylitta* from Low and High Altitudes (Daba and Jata daba). Data are mean±SEM (n = 4).

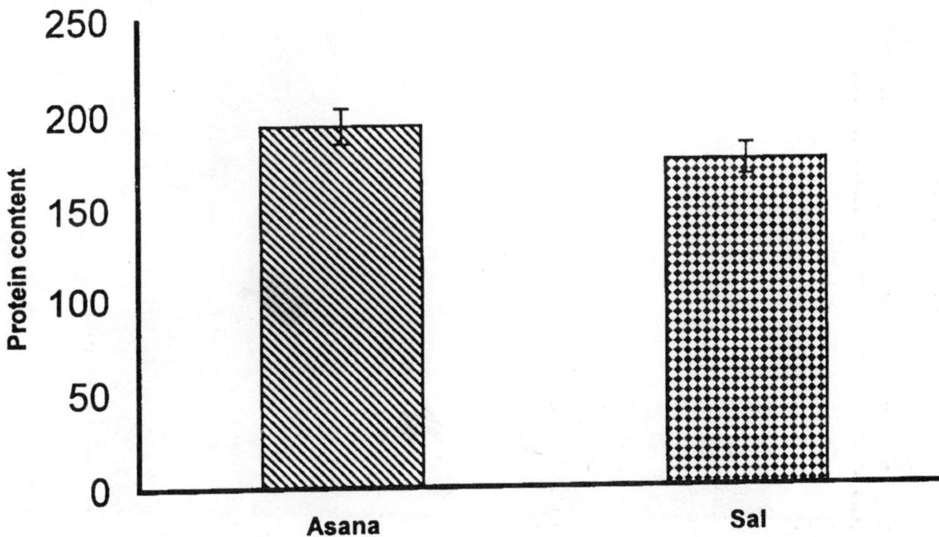

Figure 39.2: Protein Content (mg/g tissue wet wt.) of Different Host Plant Leaves (Asana and Sal). Data are mean±SEM (n = 4).

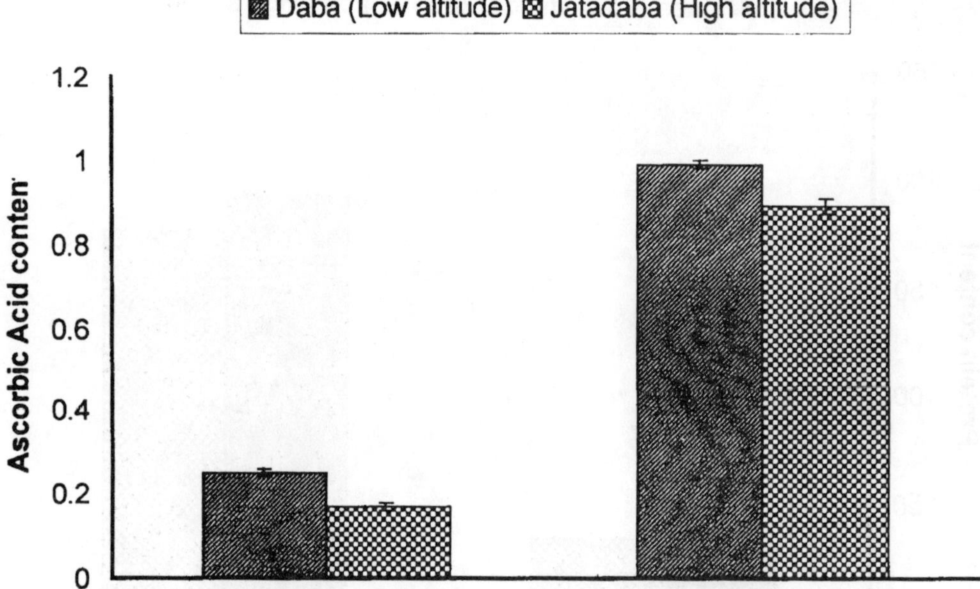

Figure 39.3: Ascorbic Acid Content of Hemolymph (mg/ml) and Fat Body (mg/g tissue wet wt.) in Diapausing Pupa of *Antheraea mylitta* from Low and High Altitudes (Daba and Jata daba). Data are mean±SEM (n = 4).

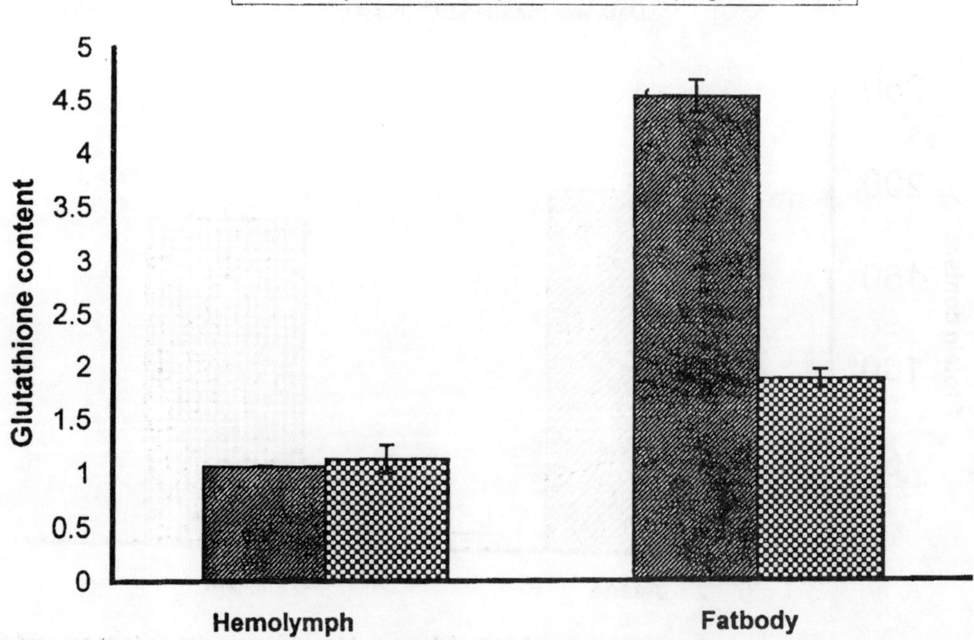

Figure 39.4: Glutathione Content of Hemolymph (μ mol/ml) and Fat Body (μ mol/g tissue wet wt.) in Diapausing Pupa of *Antheraea mylitta* from Low and High Altitudes (Daba and Jata daba). Data are mean±SEM (n = 4).

Discussion

Results of the present investigation suggest that the cocoons collected from low altitude (Daba) are superior for most of the traits, *viz*, cocoon length, breadth, volume, weight and circumference length in comparison to the high altitude ecorace (Jatadaba). In spite of numerous studies on phytopolyphagous insects in relation to feeding and antioxidant defense status (Mataruga *et al.*, 1997; Siansky, 1990) virtually no work has been done on Tasar silkworm, *Antheraea mylitta*. The results of the present investigation suggest that though the hemolymph protein content are almost similar in both the ecoraces, Daba and Jatadaba, the protein content in the fat body of Daba is relatively higher in comparison to Jatadaba. This may be attributed to the fact that the former ecorace (Daba), which generally inhabit in lower altitude feed on abundantly on the leaves of *Termanalia tomentosa*, which contain higher protein in comparison to *Shorea robusta*. The ascorbic acid content was also found to be higher in both the tissues (hemolymph and fat body) of Daba in comparison to Jatadaba. However, the other non-enzymatic antioxidant, GSH, though exhibit almost similar value for hemolymph, in case of fat body it is substantially higher in case of Daba. The variation in the antioxidant status may be due to the variable antioxidant reserve of the respective host plants. Besides, the presence/abundance of prooxidants in the host plant may lead to differential induction of antioxidant defence in the silkworm. The present findings are in good agreement with the earlier works in other phytophagus insects like gypsy moth, *Lymartria dispar* (Mataruga *et al.*, 1997). Nevertheless, as antioxidant components of the host plant has not been analyzed in this study the influence of host plant on the antioxidant status of the silk worm is only speculative and needs further study to ascertain the fact.

Acknowledgements

The authors wish to thank Dr. S.K. Dutta, Professor and Head, Department of Zoology, North Orissa University, Baripada for providing necessary laboratory facilities.

References

Akai, H., 2000. Cocoon filament characters and post cocoon technology. *Int. J. Wildsilkmoths Silk*, 5: 255–259.

Ellman, G.L., 1959. Tissue sulfhydril groups. *Arch. Biochem. Biophys.*, 82: 72–77.

Felton, G.W., 1995. Antioxidant defences of vertebrates and invertebrates. In: *Oxidative Stress and Antioxidant Defences in Biology*, (Ed.) S. Ahmad. Chapman and Hall, New York, p. 356–434.

Halliwell, B. and Gutteridge, J.M.C., 1989. *Free Radicals in Biology and Medicines*, 2nd edition. Clarendorn Press, Oxford, UK.

Jagota, S.K. and Dani, H.M., 1982. A new colorimetic technique for the estimation of Vitamin C using folin phenol reagent. *Clinical Biochem.*, 127: 178–182.

Lowry, O.H., Resebrough, N.J., Farr, A.L. and Randall, R.J., 1951. Protein measurement with the Folin phenol reagent. *J. Biol. Chem.*, 193: 265–275.

Mataruga, V.P., Blagojevic, D., Spasic, M.B., Ivanovic, J. and Hladni, M.J., 1997. Effect of the host plant on the antioxidative defence in the midgut of *Lymantria dispar* L. caterpillars of different population origin. *J. Insect Physiol.*, 43(1): 101–106.

Nayak (Jr.), B.K., Das, S.K. and Guru, B.C., 1998. Studies on cocoon biometry of bivoltine tasar silk moth, *Antheraea mylitta* D. (Lepidoptera: Saturniidae) reared at Khunt gaon, Sundergarh district Orissa. *Bull. Ind. Acad. Seric.*, 2(2): 50–53.

Slansky, F., 1990. Insect nutritional ecology as a basis for studying host plant resistance. *Florida Entomol.*, 73: 359–378.

Chapter 40

Length Weight and Breadth Weight Relationship in Gastropod Slug, *Vaginulus marginalis*

M. Kavitha and K. Thirumathal
P.G. Department of Zoology, A.P.A. College for Women, Palani – 624 614, Tamil Nadu

ABSTRACT

Studies have been conducted between animal length to body weight and breadth to body weight in Gastropod slug, *Vaginulus marginalis*. The studies revealed the significant correlation ($P > 0.05$). Different size groups of *Vaginulus marginalis* will be measured for total body Length, total body breadth and total body weight. Mean, Standard Deviation, Standard error, Correlation (r) and regression were calculated.

Keywords*: Vaginulus marginalis, Body length, Weight and breadth weight relationship.*

Introduction

In Indian gastropod species morphometry of the shell were observed in *Bellamya bengalensis* (Raut, 1980 and Khan and Chaudhuri, 1984). Similar studies were carried out in *Achatina fulica* (Rees, 1950 and Ghose, 1963). *Ariophanta madera spatna* (Masurekar and Bagalkote, 1976), *Opeas gracile* (Subha Rao *et al.*, 1981) and *Indoplanorbis exustus* (Nagomi, 2004). No such work is available with regard to the slug *Vaginulus marginalis*. Hence, the present investigation aims to correlate the length and breadth of the slug *Vaginulus marginalis*, for five different size groups namely I, II, III, IV and V.

Materials and Methods

Vaginulus marginalis were collected during the rainy season from June, 2004 to March, 2005 from its natural habitat with different sizes and length and they were released and maintained in the terrarium. Total length (cm) of each slug was taken from the tip of snout to the extreme tip of the caudal region using a meter scale. Simultaneously the body weight of each animal is also recorded with the help of electronic balance. Then they were grouped into five size groups *viz.*, group I (0.030–1.20/gm) group II (1.20–2.40/gm) group III (2.40–3.60/gm) and group V (4.80–6.0/gm).

The correlation between animal length and body weight, animal breadth and body weight and regression equation and test of significance of the five size groups were analyzed from the observed data.

Results

Table 40.1 shows the relationship between the body weight of the animal (gm) and length and breadth of the animal (cm) of the five selected groups of *V. marginalis.*

Table 40.1: Mean Weight of the Length and the Breadth of Slug *Vaginulus marginalls*

Size Group	*Weight of the Animal (gm) Mean±S.E.*	*Length of the Animal (cm) Mean±tS.E.*	*Breadth of the Animal (cm) Mean±S.E.*
I	0.618±0.084	1.35±0.101	0.95±0.054
II	1.776±0.081	2.56±0.072	1.12±0.038
III	2.95±0.075	3.53±0.070	1.37±0.037
IV	3.99±0.077	4.39±0.102	1.55±0.048
V	4.97±0.066	5.78±0.147	2.02±0.07

±S.EN = 20.

Group I (0.03–1.20/gm)

The body weight ranges from 0.03/gm to 1.20/gm and the body length includes from 0.4 to 1.3 cm. In significant correlation is obtained ($P < 0.05$) between the body weight or the length of the body. The regression equation was Y = 0.66+1.47X (r=97) (Figure 40.1a).

The breadth of the animal increased significantly ($P > 0.05$) from 0.4 to 1.3 cm as the body weight of the slug *V. marginalis* increases. The regression equation was Y = 0.99+0.88X (r=98) (Figure 40.1b).

Group II (1.20–2.40/gm)

The body weight group II ranges from 1.20/gm to 2.40/gm and the body length varies from 0.8 to 1.4 cm. In significant correlation is obtained ($P < 0.05$) between the body weight of the animal and the length of the animal. The regression equation was Y = 0.58+0.59X (r=93) (Figure 40.2a).

The breadth of the animal increased significantly ($P > 0.05$) from 0.8 to 1.4 cm as the body weight of the slug *V. marginalis* increases. The regression equation was Y = 0.30+0.46X (r = 96) (Figure 40.2b).

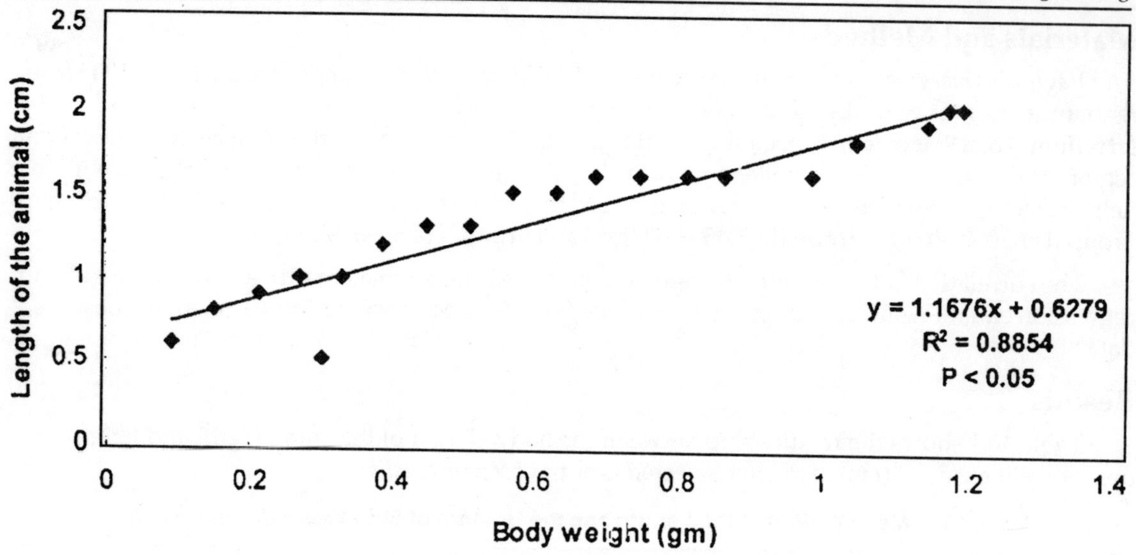

Figure 40.1a: Correlation Between Body Weight and Animal Length of *Vaginulus marginalis* in Group I

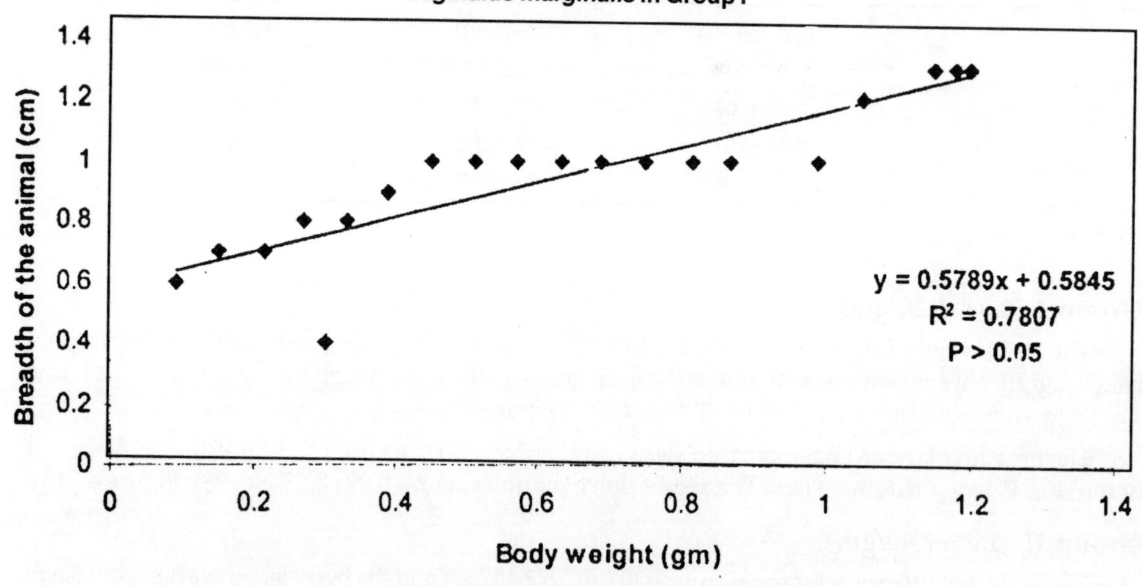

Figure 40.1b: Correlation Between Body Weight and Animal Breadth of *Vaginulus marginalis* in Group I

Group III (2.40–3.60/gm)

Group III slugs the body weight ranges from 2.40 to 3.60 gm and the body length varies from 3.0 to 4.0 cm. Significant correlation is obtained ($P > 0.05$) between the body weight of the animal and the length or the animal. The regression equation obtained between the body weight and the body length was Y = 0.85+0.91 X (r=98) (Figure 40.3a).

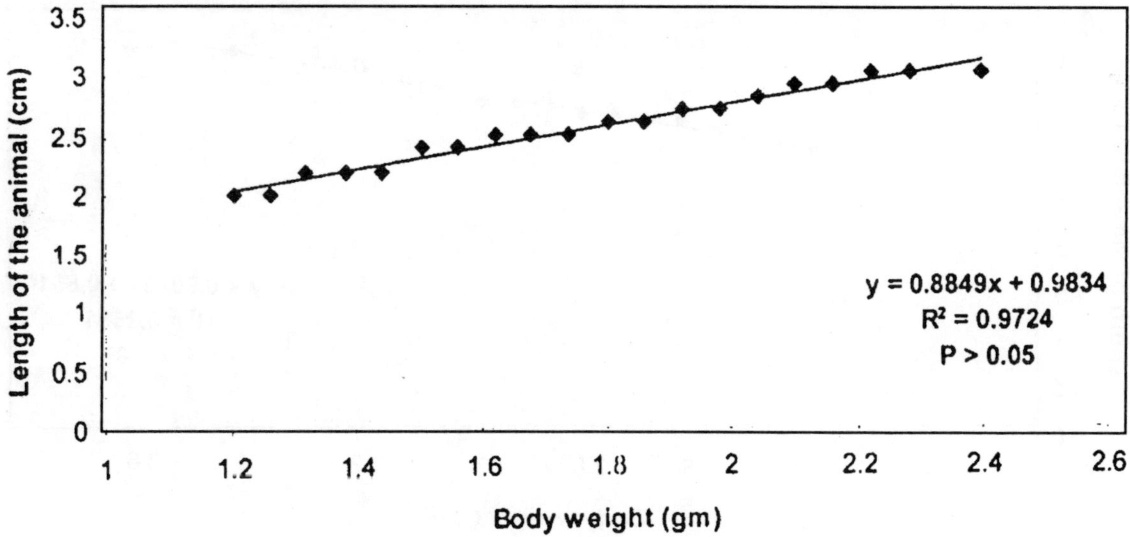

Figure 40.2a: Correlation Between Body Weight and Animal Length of *Vaginulus marginalis* in Group II

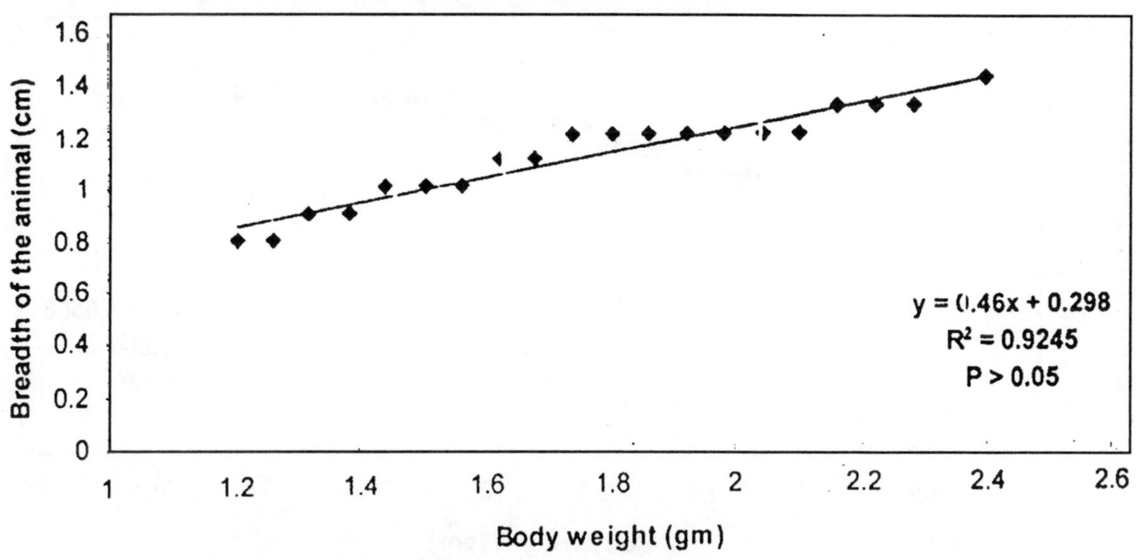

Figure 40.2b: Correlation Between Body Weight and Animal Breadth of *Vaginulus marginalis* in Group II

The breadth of the animal increased significantly ($P > 0.05$) from 1.0 to 1.5 cm as the body weight of the slug *V. marginalis* increases. The regression equation was $Y = 0.050+0.45X$ (r=89) (Figure 40.3b).

Group IV (3.60–4.80/gm)

The body weight ranges from 3.6 to 4.80 gm and the body length includes 3.5 to 5.0 cm. Significant

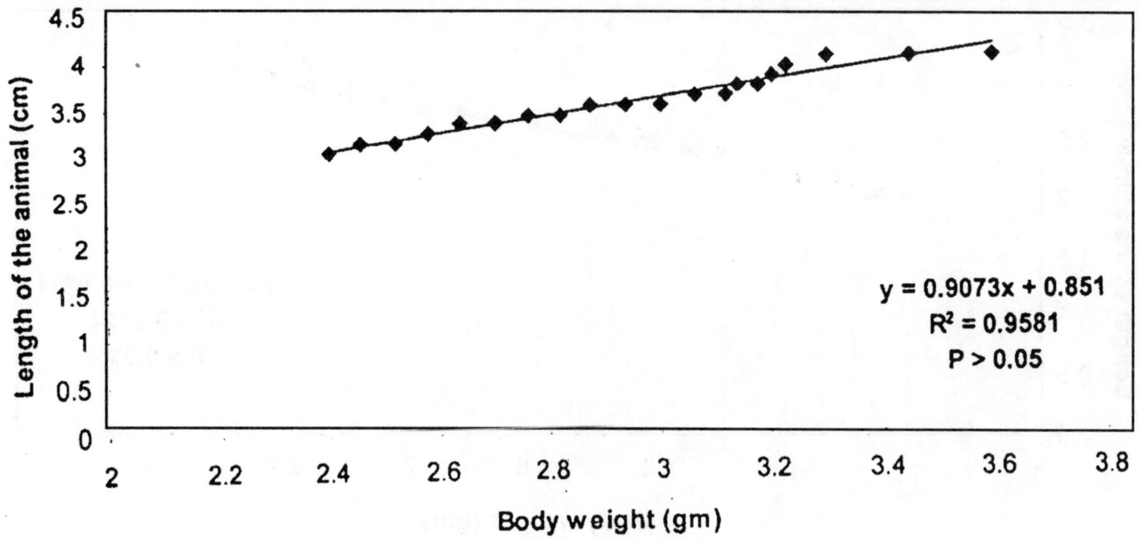

Figure 40.3a: Correlation Between Body Weight and Animal Length of *Vaginulus marginalis* in Group III

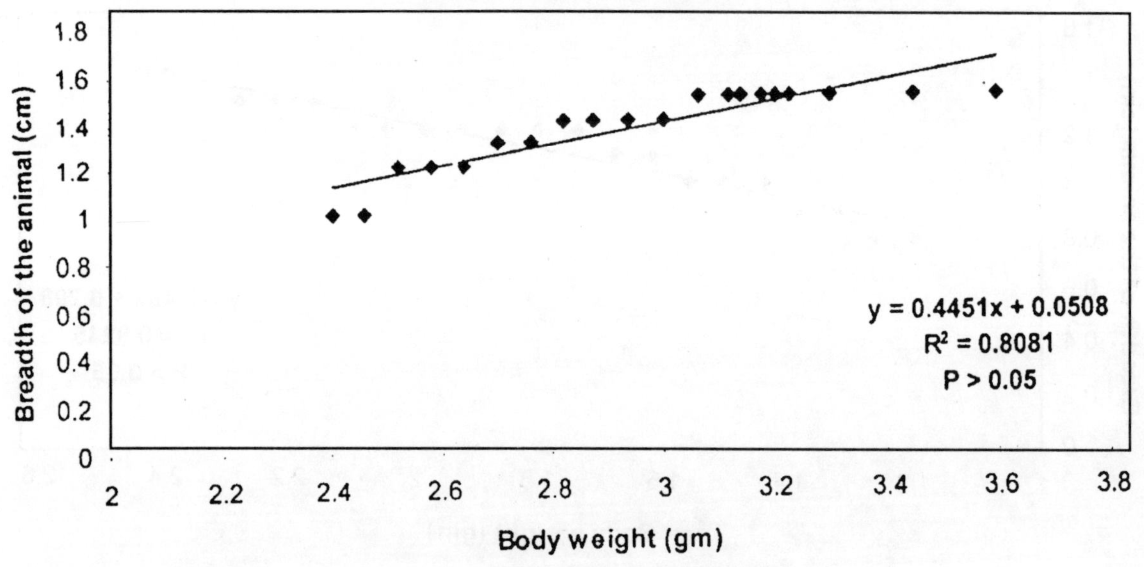

Figure 40.3b: Correlation Between Body Weight and Animal Breadth of *Vaginulus marginalis* in Group III

correla n ($P > 0.05$) is obtained between the body weight of the animal and the length of the animal. The regression equation was $Y = 0.32+1.18X$ ($r=0.89$) (Figure 40.4a).

The animal breadth also increases significantly ($P > 0.05$) from 1.2 to 2.0 cm as the body weight of the animal increases. The regression equation was $Y=0.81+0.59X$ ($r=0.95$) (Figure 40.4b).

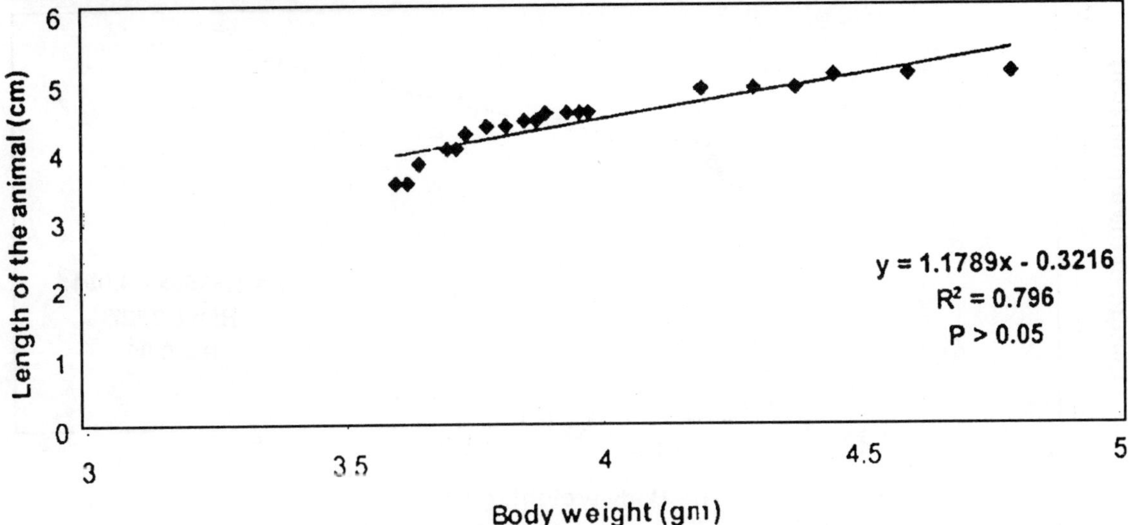

Figure 40.4a: Correlation Between Body Weight and Animal Length of *Vaginulus marginalis* in Group IV

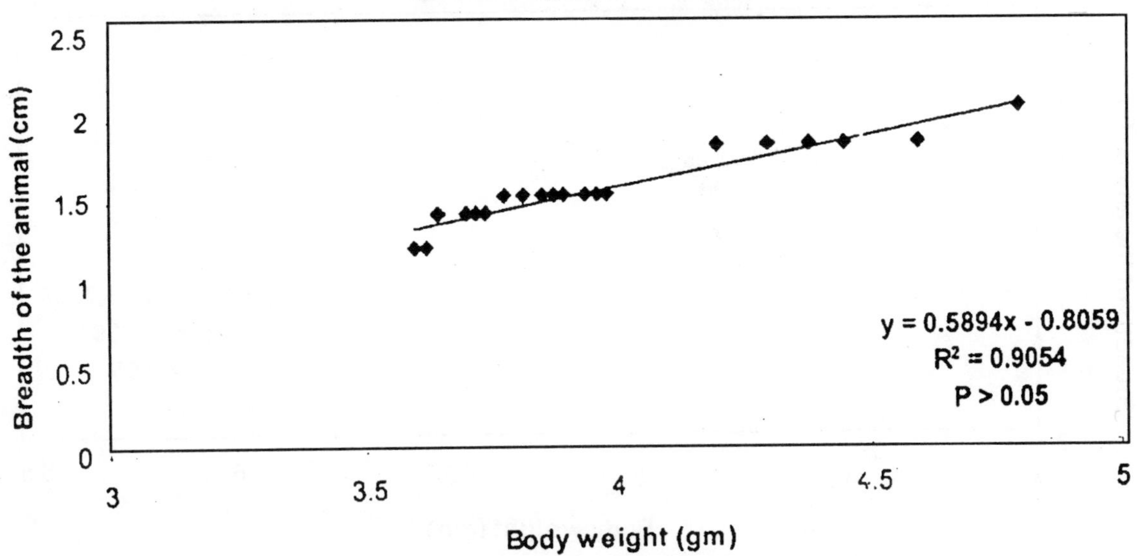

Figure 40.4b: Correlation Between Body Weight and Animal Breadth of *Vaginulus marginalis* in Group IV

Group V (4.80–6.0/gm)

The body weight of group V ranges from 4.80 to 6.0 gm and the body length varies from 1.5 to 2.5 cm. Significant correlation ($P > 0.05$) was obtained between the body weight of the animal and the length of the animal. The regression equation was Y=0.32+1.18X (r=0.89) (Figure 40.5a).

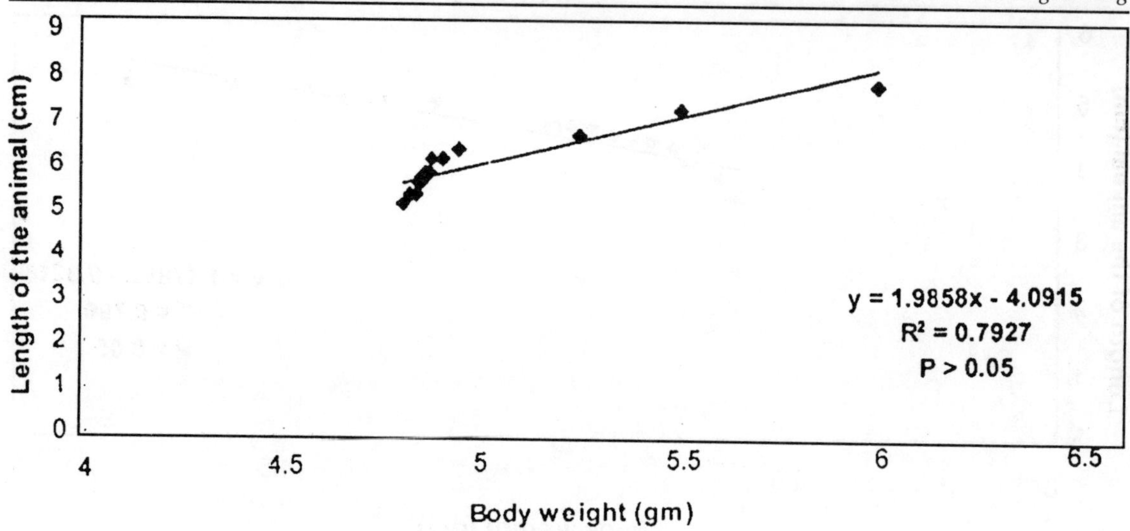

Figure 40.5a: Correlation Between Body Weight and Animal Length of *Vaginulus marginalis* in Group V

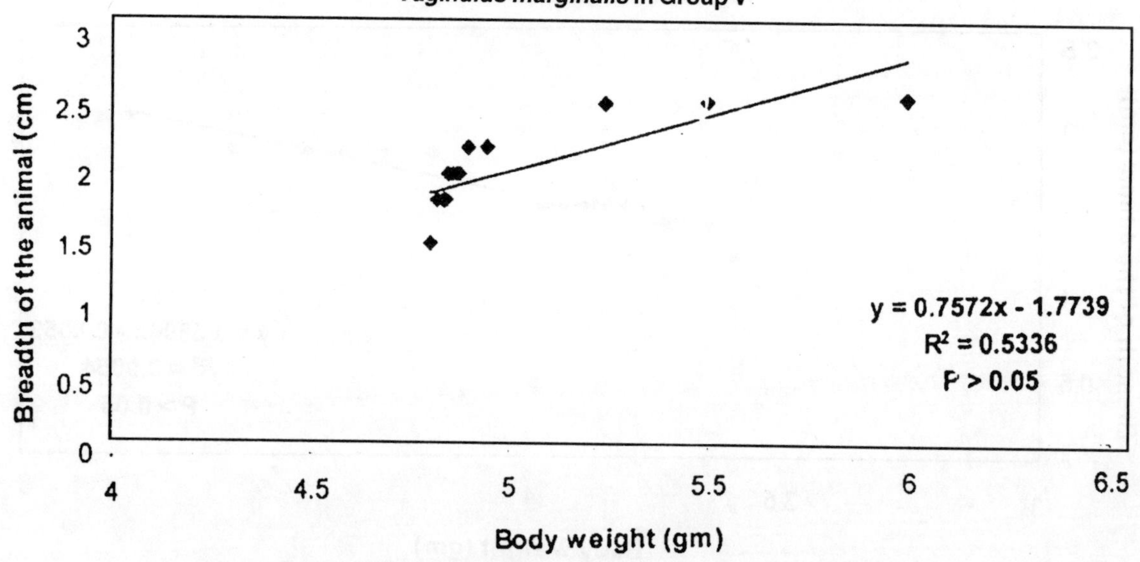

Figure 40.5b: Correlation Between Body Weight and Animal Breadth of *Vaginulus marginalis* in Group V

The animal breadth also increases significantly ($P > 0.05$) from 1.5 to 2.5 cm as the body weight of the animal increases. The regression equation was Y=1.77+0.76X (r=0.73) (Figure 40.5b).

The current investigation was also revealed that all the five size groups (Group I to Group V) chosen for this study show an increasing trend in the length and breadth of the animal as the body weight of the animal increases (Table 40.1).

Discussion

In the present study it has been observed that the length and breadth of the animal increases with increasing body weight in *V. marginalis*, Khan and Chadhuri (1984) also observed that the length and the breadth of the shell increases with increasing body weight in *Bellamya variala.*

Similar results were also reported by Mahilini (2000) in *T. vittata, P. globosa* and *I. exustus* and also in *Indoplanorbis exustus* (Nagomi, 2004). Terrestrial snails are diverse and easily sampled and serve as the indicators of leaf litter biodiversity (Boycott, 1934). Molluscs play a significant role as links in food chain as detritus feeders, improving bottom sediments and soil condition. They are the 'reducers' in natural ecosystem (Barker, 1989; Martin, 1991 and Reddy, 1995). Hence, the above study related to taxonomy is useful for the studies of biodiversity and is also pave way for the identification of distributional ranges. In near future it will be useful for the planning of slug conservation strategies.

References

Barker, G.M., 1989. Slug problem in New Zealand Pastoral agriculture. In: *Slugs and Snails in World Agriculture,* (Ed.) I. Henderson. British Crop Protection Council, Thornton Health, pp. 59–68.

Boycott, A.K., 1934. The habitats of lemd mollusca in Britain. *J. Ecol.,* 20[11(1)]: 1–38.

Ghose, K.C., 1963. The early stages of development in *Achntina fulica bowdich* (Mollusca : Grastropoda). *J. Bombay. Nat. Hist. Soc.,* 60: 228–232.

Khan, R.A. and Chaudhuri, S., 1984. The population and production ecology of a freshwater snail *Bellamya bengalenis* (Lamarcle) (Gastropoda: Viviparidae) in an artificial lake of Calcutta, India. *Bull. Zool. Surv., India,* 5(2 and 3): 59–76.

Mahilint, H.M., 2000. Studies on the hematology of chosen molluscs. *Ph.D. Thesis,* Manonmaniam Sundaranar University, India.

Martin, A., 1991. *Molluscs as Agricultural Pests: Outlooks on World Agriculture.*

Masurekar, V.B. and Bagalkote, M.S., 1976. Some observations on the ecology of the land snail, *Ariophanta maderaspatana* (Gray). *J. Bombay. Nat. Hist. Soc.,* 73: 36–41.

Nagomi, J., 2004. Studies on shell morphology, haemogram, biochemistry and microbiology of *Indoplanorbis exustus* (Dehayes), *M.Phil., Dissertation,* Manonmaniam Sundaranar University.

Raut, S.K., 1980. Preliminary observations on the breeding of the banded pond snail, *Viviparous bengalensis* (Lamarcle) (Gastropoda: Viviparidae) in West Bengal. *Bull. Zool. Surv. India,* 3(1 and 2): 47–50.

Rees, W.J., 1950. *Achatina's Odyssey: Ravages of a Globe Trotting Giant Snail, Loris.* 5(4): 159–161.

Subba Rao, N.V., Rout, S.K. and Mitra, S.C., 1981. Observations on growth rate of the common garden snail, *Opeas gracile* (Stylommatophora : Subulinidae). *Bull. Zool. Surv., India,* 4(2): 205–207.

Vikram Reddy, M., 1995. *Soil Organisms and Litter Decomposition in the Tropics.* Oxford and Co. Pvt. Ltd., pp. 272.

Chapter 41

Antimycotic Characteristics of a Medicinal Herb: An *In vitro* Study

B.K. Dutta, A. Naglot, S. Karmakar, I. Rahman and A.B. Srivastava
Defence Research Laboratory, Post Bag No. 2, Tezpur – 784 001, Assam

ABSTRACT

In the present study *Terminelia chebuta* dry fruit aqueous extract was tested in the laboratory conditions at six concentrations ranging from 0.25 to 10 per cent against five dermatophytes *viz.*, *Trichophyton tonsurans, T. rubrum, T. mentagrophytes, Microsporum fulvum* and *M. gypseum* and one opportunistic yeast *Candida albicans.* Except *C. albicans,* a resistant species, remaining five dermatophytes revealed minimum inhibitory concentration (MIC) at 0.25 per cent but in case of *C. albicans* it was at 5 per cent concentration.

Keywords: Terminelia chebuta, Dermatophytes, Opportunistic fungi, Aqueous extract, Antimycotic property.

Introduction

Dermatomycosis is a superficial skin infection in man and animals caused by a group of pathogenic fungi known as dermatophytes. They attack the outer horny layer of skin as well as hair and nails. According to the natural distribution, they are classified as anthropophilic, zoophilic and geophilic. Dermatomycosis is not fatal but highly contagious in favorable conditions which lead to discomfort with intense itching/scaling and ultimately decreases the work performance in both man and animals.

In Danish Veterinary School, a pig and a horse as well as students have got infected after arrival of two ringworm infected calves. *Trichophyton verrucosurn* was identified in the skin scrapings from the animals and affected studens (Jorgen and Soland, 2003). A total 5544 samples were collected from

3751 directly suspected dermatomycosis patients in Crete, Greece. Skin, hair and nail samples were examined by direct microscopy and culture, in which dermatophytes were isolated from 520 patients (13.9 per cent) and *Trichophyton rubrum* was the most frequent accounting 48 per cent (Maraki *et al.*, 2007). A 52-yearold man from Pakistan first reported with progressive bilateral breast enlargement for breast pseudomycetoma due to *Trichophyton rubrum* (Ali *et al.*, 2006).

The genus *Terminalia* (Myrobalan) has 13 species, commonly found in Indian sub-continent and SARC countries. They have wide range of medicinal applications in various ailments such as heart disease, tonic, asthma, piles, laxative, astringent, skin diseases, etc. An in vitro study revealed the antifungal property of *Terminafia chebula* dry fruit aqueous extract against *Trichophyton mentagrophytes, Microsporum gypseum* and *M. fulvum*. Highest zone of inhibition was recorded (24 mm) against *T. mentagrophytes* which was greater than the two antibiotics, Clotrimazole and miconazole (Dutta *et al.*, 2006). Antifungal and antibacterial property was determined with 10 crude extracts (water and methanol) from four different plant species used in Maxican folk-medicine for the treatment of infectious diseases. *Laeselfa mexicana* methanol extract showed best antifungal property against *T. mentagrophytes* and *T. rubrum* (Garcia *et al.*, 2006). A comparative study was conducted to find out the antidermatophytic property among *Terminalia arjuna, T. bellirica* and *T. chebula* leaf and bark aqueous extracts. Bark extracts of three *Terminalia* species were found more effective than the miconazole at 300 µg/ml concentration. (Dutta *et al.*, 2007).

Due to nontoxic and eco-friendly characteristics, the scientists throughout the world are emphasizing on herbal preparation, than any chemical formulations. A 2-year-old girl developed tinea incognito after applying methylprednisolone acepopnate for three months and a zoophilic dermatophyte *Trichophyton mentagrophytes* as causative organism was confirmed by histopathologic and mycological examination (Sanchez-Castellanos *et al.*, 2007). The aim of the present study was to evaluate the *in vitro* antimycotic activity of *Terminalia chebula* dry fruit aqueous extract against five dermatophytes and one opportunistic fungus.

Materials and Methods

The test fungi were collected from School of Tropical Medicine, Kolkata for conducting the study. The pathogens were:

Trichophyton tonsurans: An anthropophilic species in the nature and causing chronic superficial infection of hairs.

T. rubrum: Anthropophilic species which is commonly isolated from the glabrous skin infection in man.

T. mentagrophytes: A zoophilic species causing inflammatory ringworm infection in foot, hand and glabrous skin particularly in the intertriginous areas and also causes tinea pedis and tinea cruris.

Microsporum fulvum: Geophilic in nature and occasionally found causing tinea pedis.

M. gypseum: Universally distributed as a geophilic species. It causes infection of the scalp in man, occasionally causes inflammatory infections.

Candida albicans: It is a diploid fungus and opportunistic causal agent of oral and genital infection in humans. Infections are often in intertriginous areas in the human body.

Aqueous extract was prepared by grounding the clean *T. chebula* dry fruit boiling with double distilled water at a ratio 1 : 4 (w/v) for ten minutes. Cooled extract was filtered by using Whatman filter paper No. 1 and the filtrate was used in the study. Filtrate was incorporated in Sabouraud's Dextrose

Agar (SDA) at five concentrations 10, 5, 2, 1, 0.5 and 0.25 per cent and slants were prepared by autoclaving at 151bs pressure for 15 minutes. Test pathogens were inoculated in slants and incubated for three weeks into BOD incubator at 28±2°C. Control was maintained by inoculating the test cultures in plain SDA media in the same manner. Data was recorded by the extent of the mycelial growth on the exposed surface of the slants at four point scale *viz.* 1 denotes nil, 2 denotes trace (like pinhead), 3 denotes normal (less than 50 per cent) and 4 denotes maximum (more than 50 per cent) mycelial growth.

Results and Discussion

In the present investigation, antimycotic activity of the aqueous extract against five pathogenic fungi is recorded Table 41.1. *T. tonsurans, T. rubrum, T. mentagrohytes, M. fulvum* and *M. gypseum* revealed that the minimum inhibitory concentration (MIC) was lowest at 0.25 per cent concentration. But *C. albicans*, an opportunistic yeast, prevalent in NE region showed MIC at 5 per cent concentration and normal growth at 2 per cent concentration.

Due to nontoxic and eco-friendly characteristics of herbal formulation *T. chebula* dry fruit aqueous extract may play a significant role in controlling growth of pathogenic fungi.

Table 41.1: Minimum Inhibitory Concentration of Extract against Pathogenic Fungi

Test Dermatophytes	*Concentration (Per cent)*						
	10	*5*	*2*	*1*	*0.5*	*0.25*	*Control*
T. tonsurans	1	1	1	1	1	1	4
T. rubrum	1	1	1	1	1	1	4
T. mentagrophytes	1	1	1	1	1	1	4
M. fulvum	1	1	1	1	1	1	4
M. gyseum	1	1	1	1	1	1	4
Candida albicans	1	1	3	3	4	4	4

*: Mean of three replications.

1: Nil growth; 2: Trace growth; 3: Normal growth and 4: Maximum growth.

Acknowledgements

The authors would like to convey their sincere thanks to the Director, Defence Research Laboratory, Tezpur (Assam) for his guidance throughout the study. Thanks are also due to School of Tropical Medicine, Kolkata for providing the test pathogen for the study.

References

Ali, S.S., Zafar, A., Muzaffar, S., Aslam, F. and Khan, J.A., 2006. Bilateral breast enlargement due to *Trichophyton rubrum. Breast*, 15: 263–265.

Dutta, B.K., Rahman, I. and Srivastava, R.B., 2007. Antimycotic property of *Terminalia* species. *Biomedicine.* 27: 149–181.

Dutta, B.K., Rahman, I., Karmakar, S. and Choudhury, K., 2006. Bioassay of *Terminalia chebula* dry fruit extract in comparison with antibiotics against dermatophytes. *Geobios*, 33: 33–36.

Garcia, V.M.N., Rojas, G., Zepeda, L.G. and Aviles, M., 2006. Antifungal and antibacterial activity of four selected Maxican medicinal plants. *Pharmaceutical Biology*, 44: 297–300.

Jorgen, R.J. and Soland, T.M., 2003. Ringworm at the Danish Veterinary School. *Dansk Veterinoertidsskrift*, 86: 14–17.

Maraki, S., Nioti, E., Mantadakis, E. and Tselentis, Y., 2007. A 7-year survey of dermatophytoses in Crete, Greece. *Mycoses*, 50: 481–484.

Sanchez-Castellanos, M.E., Mayorga-Rodriguez, J.A., Sandoval-Tress, C. and Hernandez-Torres, M., 2007. Tinea incognito due to *Trichophyton mentagrophytes*. *Mycoses*, 50: 85–87.

Chapter 42

Etiological Status of Leukaemic Patients at Ahmedabad

H.A. Roy[1] and R.J. Verma[2]

[1]Department of Zoology, The M.S. University of Baroda, Vadodara, Gujarat

[2]Department of Zoology, Gujarat University, Ahmedabad, Gujarat

ABSTRACT

Leukaemia (blood cancer) is the one of the cancer which ultimately cause death of patients. The aim of present study is to know etiological status of leukaemia at Ahmedabad during December 2003 to October 2004 at M.P. Shah Cancer Hospital and Research Centre, Ahmedabad. 203 patients who underwent for treatment of leukaemia were studied. After the approval of questionnaire and work, patients were interviewed and blood was collected for different blood parameters like WBC, RBC, Hb, platelets count. All patients under study (203) were categorized in nine (9) age groups (1–10, 11–20, 21–30, 31–40, 41–50, 51–60, 61–70, 71–80 and 81–90 in years). The per cent incidence of various factors was calculated. According to per cent incidence, leukaemia is age dependant and gender specific. Addiction plays major role in development of leukaemia. The per cent incidence of different blood parameters shows that it was not in normal range (above or below).

Keywords: *Leukaemia, Etiology, Blood parameter.*

Introduction

In developed countries cancer is the second leading cause of death, next to cardiovascular diseases. In developing countries cancer is ranked third as a cause of death. Death because of various cancer is increasing as alarm. Leukaemia is one of them. Leukaemia is a malignant process, which ultimately cause the death of the person.

Leukaemia is classified mainly as acute leukaemia and chronic leukaemia. Acute leukaemia is again classified into two; acute myeloid leukaemia (AML) and acute lymphoid leukaemia (ALL), whereas chronic leukaemia is classified into chronic myeloid leukaemia (CML) and chronic lymphoid leukaemia (CLL).

Several natural and artificial agents increase the frequency of leukaemia in human being. Chemicals like arsenic, benzene, busulfan cyclophosphate etc. cause leukaemia. Radiation and pollution also play major role in development of leukaemia. Genetic changes like translocation, inversion and deletion also cause leukaemia in body. Viruses like Human T cell lymphotopic virus 1 (HTLV 1), Epstein-Barr virus (EBV), Spleen focus forming virus (SFFV) etc. develop leukaemia in the human body.

Various kind of effect of leukaemia are observed on the body of patients. Leukaemic cells reproduce from bone marrow so greatly that they invade the surrounding bone, causing pain and bone fracture. Common effects in leukaemia are the development of infections, severe anaemia and bleeding tendency caused by thrombocytopenia (lack of platelets). These effects result mainly from the displacement of the normal bone marrow and lymphoid cells by the nonfunctional leukaemic cell.

No study on the prevalence of risk factors and hematological status in hospital patients of Ahmedabad has been conducted so far. Therefore, the present investigation was undertaken to study the leukaemia and its prevalence in the population of Ahmedabad. The specific aims and objective of this study was to determine the risk factors involved in leukaemia based on a survey conducted at the M.P. Shah Cancer Hospital and Research Centre, Ahmedabad.

Materials and Methods

The inclusion criteria for this study was to evaluate the prevalence of risk factors in leukaemic patients of M. P. Shah Cancer Hospital and Research Centre, Civil Hospital Campus, Asarwa, Ahmedabad during December, 2003 to October, 2004. After the approval of questionnaires and method of work by director of hospital, 203 leukaemic patients were studied.

The questionnaire was prepared to interview the leukaemic patients about their personal information regarding their age, gender, family history, type of leukaemia, habits like drinking of alcohol and tobacco consumption in the form of chewing and smoking, snuff etc. The questionnaire include age and gender to know at which age leukaemia is more specific and in which sex. Family history to know whether it is genetic or not. The questionnaire had question about type of leukaemia to know which type of leukaemia is more common in Ahmedabad, whereas to find out the role of addiction in the development of the leukaemia, addiction related question was included. Blood parameters like WBC counts, RBC counts, Hb content and platelets count also included in study.

Results and Discussion

The present investigation was an attempt to study the prevalence of risk factors in leukaemia. The data was collected randomly from 203 patients during December, 2003 to October, 2004. All patients under study (203) were categorized in nine (9) age groups (1–10, 11–20, 21–30, 31–40, 41–50, 51–60, 61–70, 71–80 and 81–90 in years). This was so done to analyze the incidence of risk factors in different age groups.

The present study clearly indicates that age is one of the risk factor for occurrence of leukaemia. The incidence of leukaemia was commonly found between 1–10 and 61–70 years of age. After 71 years of age it gradually declined. At the age of 1–10 years, per cent incidence of leukaemia was 31.53 per cent, which was highest among all the age groups. The per cent incidence of leukaemia was higher in males than females in all age groups except 51–60 years where incidence was similar in both the sexes (Table 42.1). The ratio between males and females was 2.1 : 1. The incidence was more in males because they were more exposed to environment for job purpose like in factory etc. and they are more addicted than that of females.

Table 42.1: Per cent Incidence of Leukaemia at Different Age and Sex

Age (Years)	*Per cent incidence*	
	Male	*Female*
1–10	27.5	4.03
11–20	11.33	1.97
21–30	7.39	5.42
31–40	0.35	6.89
41–50	7.88	5.42
51–60	3.45	3.45
61–70	2.96	0.98
71–80	0.49	0.0
81–90	0.49	0.0

Overall the per cent incidence of chronic leukaemia (50.73 per cent) and acute leukaemia (49.26 per cent) were almost similar. Acute leukaemia was more prevalent in age group of 1–10 years; however, chronic leukaemia was highest in 31–40 years age group (Figure 42.1).

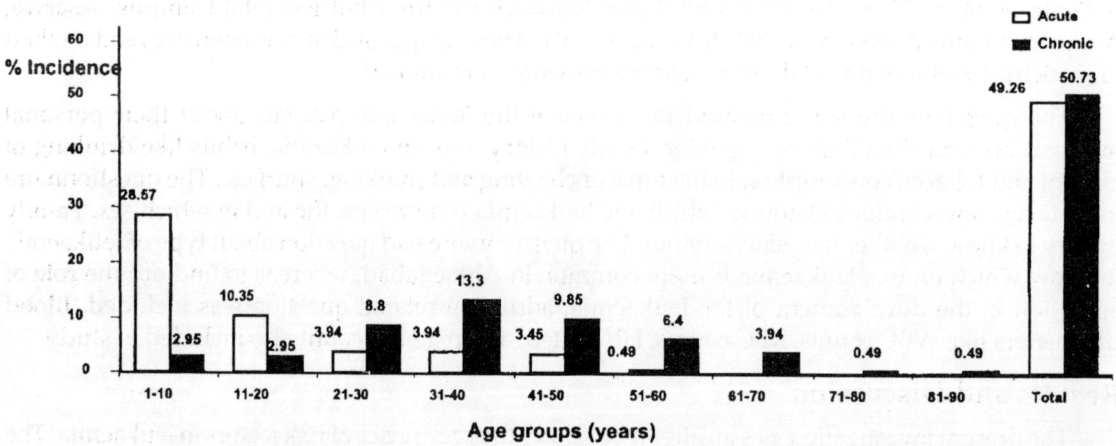

Figure 42.1: Showing Per cent Incidence of Different Types of Leukaemia in Different Age Group

Acute lymphoid leukaemia (ALL) was more found in children than that of adults. The per cent incidence of AML, CML and CLL was higher in adults than in children. The acute myeloid leukaemia was found more in age group of 11–20 years. The present investigation indicates that the chronic myeloid leukaemia (CML) was found more in age group of 31–40 years. The present study shows that the occurrence of chronic lymphoid leukaemia (CLL) was more in age group of 61–70 years (Table 42.2).

Table 42.2: Per cent Incidence of Type of Leukaemia in Adults and Children

Type of Leukaemia	*Per cent incidence*	
	Adults	*Children*
ALL	20	80
AML	71	29
CML	92	08
CLL	83	17

The highest incidence of leukaemia was for the alcohol addiction (1 to 3 times per day for 5 to 30 years, 46.6 per cent) and the second highest incidence of leukaemia was with addiction of tobacco (38.4 per cent), particularly for chewing pan-masala and gutkha (3 to 15 times per day for 5 to 45 years). The addiction of smoking (bidi, cigarette, 32.5 per cent) is the single most important environmental factor implicated in etiology of cancer of the oral cavity, lung and gradually develops into the blood cancer (leukaemia). The addiction of smoking was for 10 to 30 years at least 5 to 15 times per day. In case of snuff, the per cent incidence was 6.8 per cent, which was the lowest out of these addictions. The patients used to snuff 3 to 10 times per day for 15 to 45 years.

WBC count gets increase in leukaemic patients. The WBC count was found high before the chemotherapy; after treatment it comes into the normal range. But when the effect of chemotherapy is completed, WBC count starts increasing. WBC count is increased because bone marrow loses its normal activity. RBC count decreases in leukaemic patients. Heamoglobin content decrease in leukaemic patients. Decrease in heamoglobin content is due to the decrease in RBC count. Bleeding and anemia are also responsible for lower content of hemoglobin. The platelets count decreases in leukaemic patients. Thrombocytopenia is mainly due to the bleeding in gastrointestinal tract and other sites (Table 42.3).

Table 42.3: Per cent Incidence of WBC, RBC, Hb and Platelets Count

	Per cent Incidence		
	Normal	*Decrease*	*Increase*
WBC count	23.64	8.8	67.48
RBC count	33.49	51.17	14.78
Platelets count	7.8	3.9	88.63
Hb Content	44.33	48.28	7.4

The current understanding on the risk factors of leukaemia emphasizes that age and gender are unavoidable risk factors. The heredity is also irremediable risk factor but the modifiable risk factors such as different types of addiction can be prevented by stripping alcohol consumption, smoking and chewing of tobacco. Regular medical check up specially blood tests keeps leukaemia away. In this way the progression of leukaemia can come to a half.

References

Bennett, J.M., Catovsky, D., Daniel, M.T., Flandrin, G., Galton, D.A.G., Gralnick, H.R. and Sultan, C., 1985. Criteria for the diagnosis of acute leukaemia for megakaryocytic lineage (M7). *Ann. Intern. Med.*, 123: 460–462.

Bennett, J.M., Catovsky, D., Daniel, M.T., Flandrin, G., Galton. D.A.G., Gralnick, H.R. and Sultan, C., 1976. Proposals for the classification of acute leukaemia. *Br. J. Haematology*, 47: 451–458.

Catovsky, D., 2000. Chronic Lymphoid Leukaemia. In: *Postgraduate Haematology*, 4th edn, (Eds.) A.V. Hoffbrand, S.M. Lewis and E.G.D. Tuddenha. Butterworth Heinemann, Oxford, pp. 405–433.

Cripe, L.D. and Tallman, M.S., 2003. Acute leukaemia. In: *Clinical Haematology and Oncology, Presentation, Diagnosis and Treatment*. (Eds.) B. Furie, P.A. Cassileth, M.B. Atkins and R.J. Mayer. Churchil Livingstone, Philadelphia, pp. 521–548.

Dameshek, W., 1967. Chronic lymphoid leukaemia: An accumulative disease of immunologically incompetent lymphocytes. *Blood*, 29: 556–584.

French Cooperative Group of Chronic Lymphocytic Leukaemia, 1989. Long-term results of the CHOP regimen in stage C chronic lymphocytic leukaemia. *Br. J. Haematol.*, 73: 334–340.

Goldman, J.M., 2000. Chronic myeloid leukaemia. In: *Postgraduate Haematology*, 4th edn, (Eds.) A.V. Hoffbrand, S.M. Lewis and E.G.D. Tuddenha. Butterworth Heinemann, Oxford, pp. 434–444.

Indian Journal of Cancer. Haematology Centre, Mumbai, 2004.

Mohan, H., 2000. *Textbook of Pathology*, 4th edition. Jaypee Brothers, New Delhi, pp. 177–215.

Wickremasinghe, Hoftbrand, A.V. and R.G., 2000. The molecular basis of leukaemia and Lymphoma. In: *Postgraduate Haematology*, 4th edn, (Eds.) A.V. Hoffbrand, S.M. Lewis and E.G.D. Tuddenha. Butterworth Heinemann, Oxford, pp. 354–372.

Chapter 43

Comparative Morphometric Studies on the Honeybees *Apis florea* and *Apis cerana indica* in Mannampandal Village, Nagai District, Tamil Nadu

S. Thiripurasundari[1], K. Krishnappa[1], M. Varadarajan[1] and V. Mathivanan[2]

[1]Department of Zoology, A.V.C. College, Mannampandal – 609 305, Mayiladuthurai

[2]Department of Zoology, Annamalai University, Annamalainagar – 608 002

ABSTRACT

Morphometric studies on selected characters of the honeybees *Apis florea* and *Apis cerana indica* were studied in Mannampandal village of Tamil Nadu. A total of 125 bees including 75 *Apis florea* and 50m *Apis cerana indica* were collected and measurements were made. Morphological features measured are included (1) Body length; (2) Tongue length; (3) Forewing length; (4) Forewing width; (5) Number of wing hooks; (6) Width of hind metatarsus and (9) Length of hind leg etc. The weight of the honeybees was estimated to mg level and the result indicated the average weight of *Apis cerana indica* as 54.9 mg. All morphometric measurements indicated a higher value for the Indian hive bee *Apis cerana india* and the adaptive significance of the characters were discussed.

Keywords: *Morphometric studies, Apis florea, Apis cerana indica.*

Introduction

Honeybees belong to the Family Apidae in which other types of social bees are also included. The sub-family Apidae or honeybees, comprises single genus, Apis. The genus Apis comprises for species *viz., Apis florae* (the little bee), *Apis dorsata* (the giant bee), *Apis cerona* (the Eastern honeybee) and *Apis mellifera* (the Western honeybee). Because of the ability of honeybees to regulate high temperature, they are able to survive in temperate as well as tropical countries resulting in many geographic subspecies in each type.

It is thought that honeybees originally evolved from hunting wasps during cretaceous period, 146 million years ago. At the time of evolution of honeybees, flowering plants also evolved. Also it is suggested that *Apis flòrea* and *A. dorsata* evolved earlier than the remaining two species of the genus *Apis*. Nowadays, several subspecies of honeybees have been recognized and only in recent years, a comprehensive classification has been attempted based on differences in physical characters and their distribution between subspecies (Ruttner, 1988; Singh, 1990; Kastbergcr *et al.*, 2000).

The dwarf bee or the little bee *A. florea* has been distinguished from *A.c. indica* and reniforms be morphometric studies (Rinderes *et al.*, 1995). It is important to identify and insect of a locality upto subspecies level for a better understanding of its biology. Morphometric studies involving characters of wings and legs of honeybees provided comparison of allied species as suggested by Verma (1998). Hence, the present study was undertaken to compare the morphometric features of the co-existing honeybees species *viz., Apis florea* and *A.c. indica* in Mannampandal village of Mayiladuthurai Taluk in Tamil Nadu.

Materials and Methods

Biology of the Insects

Apis cerana indica

It is also known as Indian honeybee and is the only native domesticated honeybee species in our country. It's golden yellowish abdomen characterizes this species. It prefers to live in one place for generations and builds the comb in dark places like old buildings, broken pots, trees etc. They build a number of comps in one place. The top of the comb is meant for honey storage while the central portion is devoted for brood and pollen stores. The average honey production per year is 4–5 kg. Each colony of honeybee has a single queen who is the only fertile female laying eggs at the rate of 400 per day. She controls the whole colony activities. The life span of the queen is 2 years. There are several thousand workers in a colony who are also females but incapable of reproduction due to atrophied ovaries or they are made sterile by the pheromonal control of the queen. The number of males in a colony are less and they are known as drones. Only the act of males being to mate with the queen at the time of swarming. Both the queen and the workers are capable of stinging while the males lack the sting apparatus. The workers perform the whole activities of the colony like cleaning the hive, taking care of the queen and brood, regulating the temperature of the colony, protection against intruders and at the final age of the workers they go out for collecting the pollen and nectar from the flowers upon which the whole colony depends for their food and survival. The pollen mixed with honey is supplied to the brood and adults of the colony as "bee bread".

Apis florea

These bees could be easily distinguished by their small size and striped abdomen with black and white stripes. They are nomadic and the number of comb is one usually found in bushes or small branches of herbs and shrubs in open. The size of the comb is 15–45 cm long and 15–30 cm wide. Their feeding range is comparatively lesser than the other honeybee species. The amount of honey production is also less than a kg per year. This species is not employed in apiculture practices because of the nomadic behavior and low production of honey. The social structure of the colony of *A. florea* is the same as that of *A.c. indica* whereas the comb pattern is well defined than *A.c. indica* (Thirupurasundari *et al.*, 2007).

Collection of the Insects

The test species were collected from the field while foraging for nectar and pollen. Once a bee was located, it was captured using a sweep net and carefully transferred in to a glass container, without any damage to their body parts. After transferring 10–15 bees, the container was kept in a freezer for 5 minutes to immobilize the bees. Each bee was observed using a hand lens for any damage on the parts that were measured. Each bee was weighed in an electronic balance with 1 mg sensitivity. Seventy-five insects of *Apis florea* and fifty *A.c. indica* were measured in this way.

Methods of Measurements

The average body weight of the test honeybees are given in Table 43.1 and Figure 43.1. A total of 7 different characters were measured including 3 wing characters and 2 leg characters in addition to total length of body and tongue length. The body parts of the honeybees that were measured are given in Table 43.2 and were shown in Figures 43.2–43.8. Measurements were done using a divider and scale or by occulometer fitted in a 5x eyepiece of a compound microscope depending upon the size of the parts. The microscope was calibrated with a stage and ocular micrometer previous to measurements.

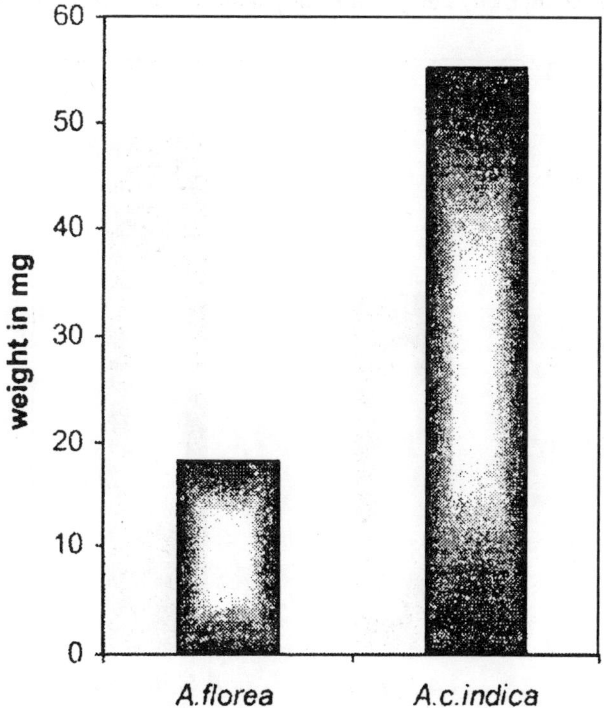

Figure 43.1: Mean Body Weight of Honeybees *A. florea* and *A.c. indica*

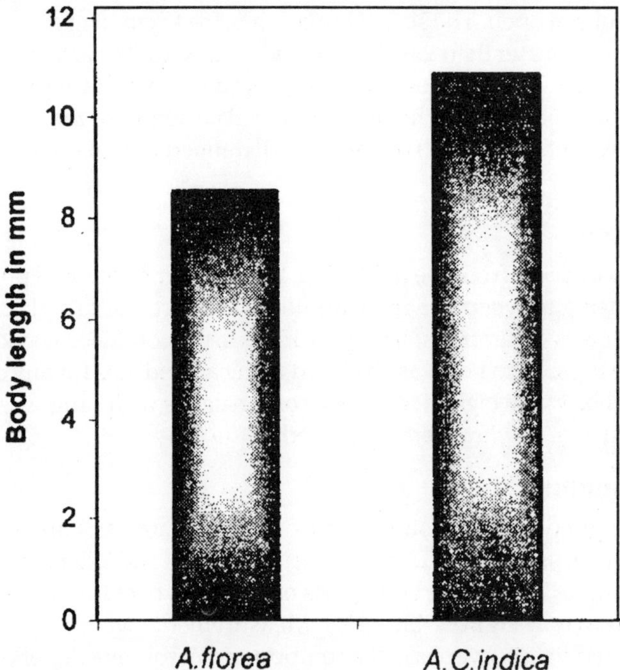

Figure 43.2: Mean Body Length of Honeybees *A. florea* and *A.c. indica*

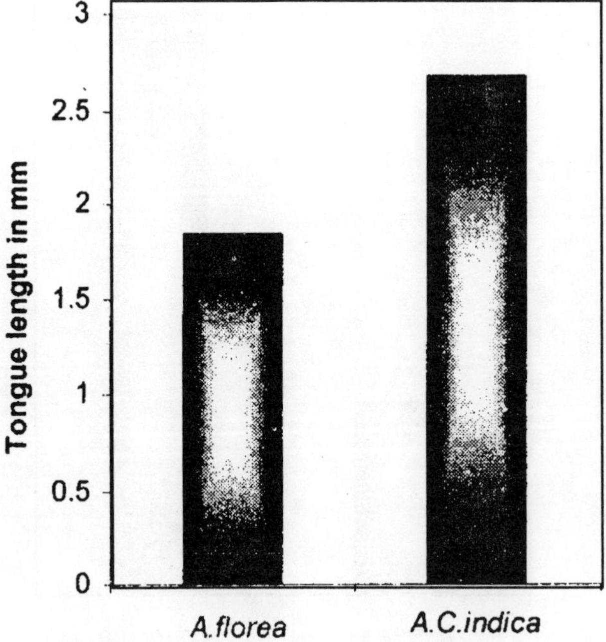

Figure 43.3: Mean Tongue Length of Honeybees *A. florea* and *A.c. indica*

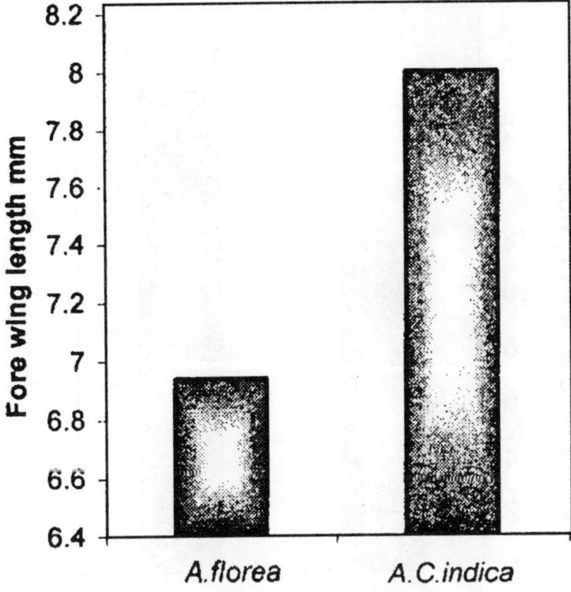

Figure 43.4: Mean Fore Length of Honeybees *A. florea* and *A.c. indica*

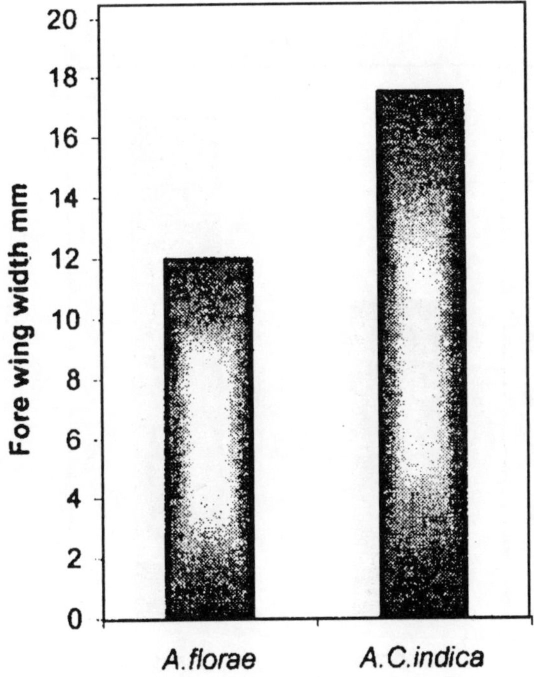

Figure 43.5: Mean Fore Wing Width mm

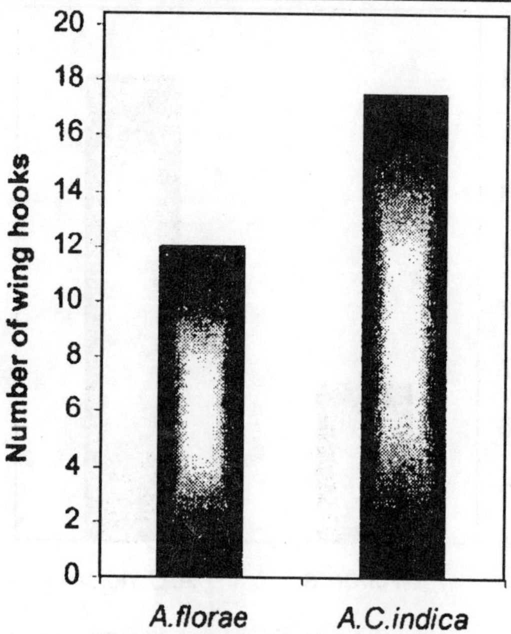

Figure 43.6: Mean Number of Wing Hooks

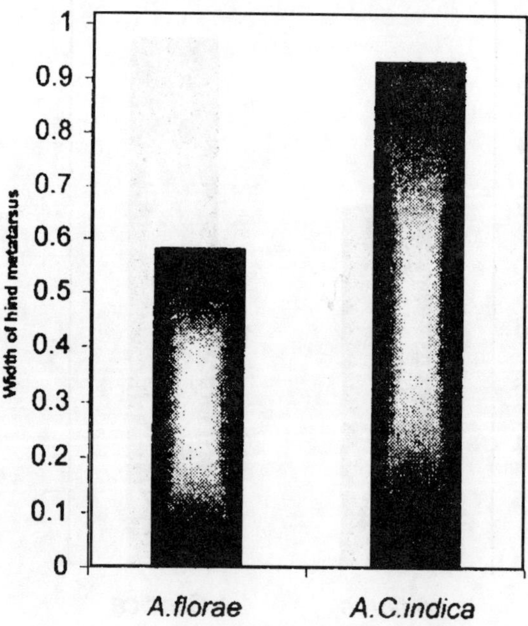

Figure 43.7: Mean Width of Hind Metatarsus

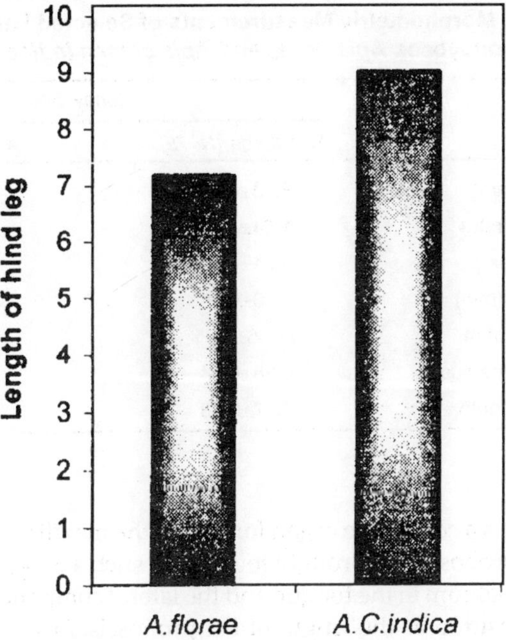

Figure 43.8: Mean Length of Hind Leg

Table 43.1: Average Body Weight of the Honeybees *A. florea* and *Apis cerana indica*

Species		*Body Weight in mg*
A. florea	Mean	18.08
(N=75)	S.D	2.34
	Minimum	12
	Maximum	24
A.c. indica	Mean	54.96
(N=50)	S.D	9.8
	Minimum	38
	Maximum	75

Description of the Parts

Body Length

The total length of the body is the length between the head and tip of abdomen. The head is triangular with the apex below, from which are appended the mouthparts. The head and mouthparts are hypognathous type and a measurement was taken by using a divider forks pointed between clypeus of head and the tip of abdomen.

Table 43.2: Morphometric Measurements of Selected Parts of the Honeybees *Apis florea* and *Apis cerana indica*

Sl.No.	Parts	Honey Bee	
		Apis florea (N=75)	*A.c. indica (N=50)*
1.	Body length (mm)	8.53±0.55	10.84±0.41
2.	Tongue length (mm)	1.84±0.06	2.67±0.41
3.	Fore length (mm)	6.94±0.22	8.00±0.00
4.	Fore wing width (mm)	2.20±0.06	2.79±0.11
5.	Number width (mm)	11.96±1.05	17.44±1.17
6.	Width of hind metatarsus	0.58±0.04	0.93±0.27
7.	Length of hind length	7.17±0.38	9.00±0.00

Tongue Length

The proboscis of the bee is a composite organ formed of the maxilla and the labium. At the act of feeding, the components of proboscis are brought together in such a way so as to form a tube through which nectar is taken in. The labium in the tongue and the lateral components are maxilla. The length of the tongue is usually measured as the length of the proboscis. From the base of the labium *i.e.,* prementum to the tip of glossa the labellum.

Forewing Length

The wings of honeybees are simple in the shape and the pattern of venation. The forewing is larger than the hind wing but in light the two wings of each side couple together. Each wing arise from the sides of the thorax between the nodal and plural plates of their segments. The length of the forewings is the actual measurement between the basal scletrites to the tip of wing.

Forewing Width

The width of the forewing is the transverse measurement of the wing after region of posterior margin of fold in the forewing. The forewing is broader only at this portion.

Number of Wing Hooks

During flight, both wings of a side act together by the wing coupling apparatus that consists of a row of minute hooks on the anterior margin of the hind wing and a narrow fold in the posterior margin of the forewing. The hooks of the wing curve upward and backward from the wings margin with a twist in the direction of the wing tip. When the wings are extended preparatory to the flight of the forewing is drawn over the upper surface of the hind wing beneath it, and the hooks of the latter engage in the fold of the front wing. The number of hooks is counted.

Width of Meta Tarsus

The legs of honeybee in workers are modified for several purposed besides that of walking. The bee is an artisan of many trades and is provided with tools for all its needs in which the leg is one of the important tools. Each leg is made up of 6 segments *viz.,* coax, trochanter, femur, tibia, tarsus and pre tarsus. The tarsus of each leg is divided into 5 segments or tarsomeres, but the basal tarsomere is much larger than the other and in the hind leg is broad and flat. This is the metatarsus or planta or basitarsus. The posterior surface of the basitarsus is broad and armed with 9 transverse rows of long stiff spines, directed distally that resembles a flat brush. It serves for the collection of pollen and propolis to be stored in the pollen basket of the tibia. The width of the basitarsus is shown.

Length of Hind Leg

It is the actual length between the coax and the tip of the pre tarsus.

Results and Discussion

The comparative body weight of the honeybee species studied is given in Table 43.1 and Figure 43.1. The mean body weight of *A. florea* is 18.08 mg and is comparatively three times smaller than the body weight of *A.c. indica* with a mean body weight of 54.96 mg; there was a wide variation in the weight of insects measured in each species since the minimum and maximum body weights were 12 mg and 24 mg for *A. florea*, and 38 and 75 mg for *A. florea* and *A.c indica*, respectively. Also it has been proved that the honeybees could carry a nectar load almost equal to it's body weight (Wells and Wells, 1985). Since the bees were collected in the foraging arena there is a possibility that the weight recorded for individual bees could also be due to the nectar load in their honey stomach. Higher body weight as well as the wide variations observed in the body weight of insects was also due to varied state of feeding among the bees.

The body length of both species was in accordance to their body weight with the mean body length of *A. florea* being smaller with 8.53 mm than *A.c. indica* with a mean body length of 10.84 mm. The body length of *A. florea* observed in the present study was similar to the observations of (Rinderer *et al.*, 1995) in Thailand and they reported that this species was slightly different from *A. anderiformes*. The body length of *A.c. indica* was within the size range reported for the plain variety of *A.c. indica* (Mishra 1998) and was lesser than the body length of hill variety (Table 43.3).

Table 43.3: Morphometric Measurements of Selected Parts of the Honeybees *A. florea* and *A.c. indica*

Species	*Name of the Parts*						
	Body Length (mm)	*Tongue Length (mm)*	*Fore Wing Length (mm)*	*Fore Width (mm)*	*Number of Wing Hooks*	*Width of Hind Metatarsus (mm)*	*Length of hind leg (mm)*
A. florea (N=75)	8.53±0.55	1.84±0.06	6.94±0.22	2.20±0.06	11.96±1.05	0.58±0.04	7.17±0.38
A.c. indica (N=50)	10.84±0.41	2.67±0.41	8.00±0.00	2.79±0.11	17.44±1.17	0.93±0.27	9.00±0.00

The tongue length of both species indicated their adaptations to feed nectar on specific plants suitable to their proboscis length. The tongue of *A. florea* was between 1.4 mm and 2.08 mm with a mean of 1.84 mm that is suitable for feeding on flowers with smaller corolla tube as reported by Kalpana and Ramanujam (1991). It has been identified that the little bee preferred flowers of several herbs like *Justicia simplex* and *Oldenlandia umbellate* (Selvakumar and Srinivasan, 2000). The tongue length of *A.c. indica* was longer than *A.c. indica* with a mean of 2.67 mm and accordingly their flower preference also varied. The nectar sources of *A.c. indica* has been identified as many shrubs and trees as for *e.g.* Datura with flowers of long corolla tube.

The wing characteristics also showed variations between the species. The forewing length and width were smaller in *A. florea* than in *A.c. indica*. The mean forewing length of *A. florea* and *A.c. indica* were 6.95 mm and 8 mm respectively. The little bee being smaller in size has a flying range of few meters for foraging whereas, the maximum foraging range of *A.c. indica* was few kilometers. Accordingly, the wing size was also adapted in both species.

One of the important features measured for subspecies identification is the number of wing hooks (Mattu and Verma, 1984; Singh, *et al.*, 1990). The number of wing hooks in *Apis florea* has been 12 in Thailand and the present result was also in accordance with the findings with the findings of Rinderer *et al.* (1995). The number of wing hooks in *A.c. indica* was 17, which resembled the description given for the plain type as presented by Verma (1998). The wing hooks were more in *A.c. indica* than in *A. florea*.

The hind leg characters such as width of metatarsus and the length of the leg playa vital role in pollen carrying capacity of the honeybees. The width of metatarsus has been studied in the identification of species. The width of metatarsus decides the quantity of pollen to be brushed from the body parts before packing into the pollen press. The width of metatarsus of *Apis florea* being smaller (0.58 mm) could collect a pollen pellet of 6 mg whereas, the metatarsus of *A.c. indica* could make a pollen pellet of 12mg. The length of hind leg in both species varied according to the differences in other morphometric features were studied. The mean length of the hind leg of *Apis florea* 7.17 mm whereas that of *A.c. indica* was 9 mm, indicating the ability of the latter species to crawl over the petals of larger flowers while feeding on the nectar.

References

Kalpana, T.P. and Ramanujam, C.G.K., 1991. A melittopalynological investigation of Honeys from *Apis florea* and *Apis dorsata* hives. *Biovigyanam*, 17(1): 12–23.

Kastberger, G., Sharma, D.K., Hultinger, E. and Krammer, G., 2000. Morphometric studies of *A. dorsata* in Assam India. In: *Proc. of the 8th Int. Conf. on Tropical Bees Mgt. and Diversity*, Thailand, pp. 129.

Mattu, V.K. and Verma, L.R., 1984. Comparative morphometric studies in the Indian honeybee of the northwest Himalayas. 3. Hind leg, tergites and sternites. *J. Apic. Res.*, 23(3): 117–121.

Mishra, R.C., 1998. *Perspectives in Indian Apiculture*. Agrobotanica, New Delhi.

Rinderer, T.E., Oldroyd, B.P., Wongsri, S., Sylvester, H.A., Guzman, L., Stelzer, J.A. and Riggio, R.M., 1995. A morphological comparison of the dwarf honeybees of southeastern Thailand and Palawan Philippines. *Apidplogie*, 26: 387–394.

Ruttner, F., 1998. *Biogeography and Taxonomy of Honeybees*. Springer-Verlag, Berlyn.

Singh, M.P., Verma, L.R. and Daly, H.V., 1990. Morphometric analysis of the Indian honeybee in the North-east Himalayan region. *J. Apic. Res.*, 29(1): 3–4.

Thiripurasundari, S., 2003. Bee flora and biochemistry of honey and pollen of the rock bee, *Apis dorsata*, at Mannampandal, Tamil Nadu. *M.Phil Thesis*, Bharathidasan University, India.

Thiripurasundari, S., Mathivanan, V. and Varadarajan, M., 2007. Studies on the insect foragers of banana with special reference to the little bee *Apis florea*. *J. Exp. Zool., India*, (1): 251–255.

Verma, L.R., 1998. *Honeybees in Mountain Agriculture.* Agro Botanica, New Delhi.

Wells, H., Hill, P.S. and Wells, P.H., 1992. Neactarivore foraging ecology: rewards differing in sugar types. *Ecological Entomology*, 17: 280–288.

Wells, P.H. and Wells, H., 1985. Ethological isolation of plants 2. Odour selection of by honeybees. *Journal of Apiculture Research*, 24(2): 86–92.

Chapter 44

Heavy Metal Concentrations in Groundwater from Kadur and its Surrounding Areas, Karnataka

E.T. Puttaiah[1] and E. Nalina[2]

[1]Department of P.G. Studies and Research in Environmental Sciences, Kuvempu University, B.R. Project – 577 115

[2]Department of Chemistry, Adichunchanagiri Institute of Technology, Chikmagalur

ABSTRACT

Heavy metals have a rather unique position among the chemicals that man encounters daily. Most of the heavy metals are toxic to many freshwater organisms and man. The objective of present work was to study the presence and concentration of heavy metals (Lead, Copper, Zinc and Chromium) in groundwater of Kadur and its surrounding areas.

It was found that the concentration of heavy metals such as Lead (Pb), Zinc (Zn), Copper (Cu) and Chromium (Cr) in the groundwater of Kadur and its surrounding areas below the maximum permissible limit of human consumption.

Keywords: Heavy metals, Pollution, Water quality.

Introduction

Water is the one of most important compounds required for very existence of life. Now-a-days due to increased human population and man made conditions, the water quality is deteriorating every where. It is therefore necessary that the quality of drinking water should be checked at regular time intervals because the use of contaminated drinking water cause a variety of water borne diseases such as diarrhoea, cholera, jaundice, typhoid, dysentery and viral infection. Without clean water neither human nor the environment, which sustain them, can survive. It is with this strong background the present study is focused on Kadur and its surrounding areas, since no such study has been carried in the past.

Increasing demand for water in domestic, agricultural and industrial sectors necessitates exploitation of groundwater. Groundwater is used directly for drinking as well as for other household purposes and hence the quality of the groundwater is extremely important. To safeguard the long-term sustainability of groundwater resources, the quality of the water needs to be continually monitored. As there is dearth of literature on water quality of the area, the present study was undertaken to assess the exact Heavy metal concentration of the groundwater of Kadur taluk in Karnataka with special reference to suitability of water for drinking purposes.

Study Area

Kadur is one of the developing urban regions and is one of the taluks of Chikmagalur district. This town is about 40 kms away from Chikmagalur. The geographical location is 12°54′42″ to 13°53′53″N latitude and 75°04′46″ to 76°21′52″E longitude. Even though it belongs to Chikmagalur district (Malnad area), it has 80 per cent of dry land area and hence surface water usage is very less compare to groundwater. Therefore, groundwater is the main source of drinking water in Kadur and its surrounding areas. There are complaints regarding the contamination of groundwater in these areas. Hence, there is a great need to study the causes for this contamination. It is with this background, the present study has been carried out.

Materials and Methods

Water samples were collected season-wise for a period of 12 months from February 2005 to January 2006 from Kadur and its surrounding areas. Samples were collected in plastic containers, which were thoroughly cleaned with diluted nitric acid and rinsed several times with distilled water. Analysis was carried out to determine the concentration of various heavy metals such as Copper (Cu), Chromium (Cr), Lead (Pb) and Zinc (Zn) by using Atomic Absorption Spectrophotometer (AAS, Perkins Elmer Analyst 105). Heavy metals like Copper (Cu), Chromium (Cr), Lead (Pb) and Zinc (Zn) were estimated by using standard methods as described by APHA (1998).

Results and Discussion

The samples were analysed and contents of heavy metals were detected in order to know the concentration of these metals in water. The results are given in Table 44.1.

In the present investigation, it was found that copper (Cu) content of water was in the range of 0.01 to 0.2 ppm. The copper content in the freshwater was estimated and studied by several workers Mule and Patil (2000), Bhosle and Patil (2001) and Mazher *et al.* (2005). They reported the increased levels of copper content when compared to permissible limit. However, in the present study, the concentration level is below the permissible limit present for drinking water.

Copper is essential for the formation of enzyme in human beings. Intake of excessively large doses of Cu leads to severe mucosal irritation and corrosion, wide spread capillary damage, hepatic and renal damage and central nervous system irritation followed by depression. Copper toxicity includes blue green diarrhoea stool and saliva Athor and Vohora (1995).

Table 44.1

Sample →	*Sample-1*				*Sample-2*				*Sample-3*			
Heavy metal →	*Cu*	*Cr*	*Pb*	*Zn*	*Cu*	*Cr*	*Pb*	*Zn*	*Cu*	*Cr*	*Pb*	*Zn*
Summer 2005	0.02	0.016	Bdl	0.08	0.09	0.08	Bdl	0.01	0.15	Bdl	Bdl	0.02
Monsoon 2005	0.06	0.014	Bdl	0.03	0.12	0.012	Bdl	0.06	Bdl	0.016	0.001	0.01
Winter 2005	0.01	0.016	0.001	0.01	0.08	0.018	Bdl	0.02	0.2	0.016	Bdl	0.01
Summer 2006	0.08	Bdl	Bdl	0.081	0.16	Bdl	Bdl	0.015	0.16	Bdl	Bdl	0.021
Monsoon 2006	0.06	0.014	Bdl	0.02	Bdl	0.012	Bdl	0.06	Bdl	0.016	0.002	0.01
Winter 2006	0.014	0.01	Bdl	Bdl	0.015	0.02	Bdl	Bdl	0.016	01	Bdl	0.001

The chromium concentration in the water samples studied ranged between 0.1 and 0.2 ppm. The data recorded during the study revealed that the chromium concentration is below the permissible limit. Higher concentration of chromium in drinking water has adverse effect and may cause liver necrosis, nephrites and gastrointestinal irritation.

In atmosphere Lead is relatively more abundant than other heavy metals. The Lead concentration in the water samples studied ranged between 0.001 and 0.002 ppm. The data recorded during the study revealed that the Lead concentration is below the permissible limit. At higher levels of Lead in the blood causes anemia and disruption of haemoglobin synthesis, damage to nervous system and kidney, brain damage, decrease in bone strength. In mild doses it causes gastrointestinal problems, restlessness, and loss of appetite.

The Zinc concentration in the water samples studied ranged between 0.01 and 0.08 ppm. The data recorded during the study revealed that the Zinc concentration is below the permissible limit. It is an essential and beneficial element for human bodies. However concentration above 5 ppm cause bitter taste.

The maximum permissible limit of copper and chromium in drinking water is 0.05 mg/l (Awashthi, 2000). In present study, copper, lead, zinc and chromium levels in the water samples were observed to be well within the desirable limit.

Metals like Zn, Cu, Fe, Mn, Mo, Ni, Cr and Co are called micro-nutrient and are toxic when taken in excess of requirement Blaylock and Huang (2000). Some metals like Cd, Hg, As, and Pb are biologically nonessential and toxic in nature. The aquatic ecosystem has been reported to be the ultimate recipient of heavy metals pollution Idodo-Umesh (2002). The present study indicates that the Godavari river system is contaminated with heavy metals (O'Brien *et al.*, 2003).

References

APHA (American Public Health Association), 1998. *Standard Methods for the Examination of Water and Wastewater*, 18th Edition. APHA, AWWA, WPCF, Washington D.C., pp. 1268.

Athar, M. and Vohora, S.B., 1995. Heavy metals and environment. In: *Man and Environment Series*, (Ed.) P.K. Ray. Wiley Eastern Ltd., New Delhi, pp. 1–195.

Bhosle, A.B. and Patil, P.M., 2001. Seasonal variation in the copper content of the river Godavari water at Nanded, (M.S.). *J. Ecotoxicol. Environ Monit.*, 11(1): 61–64.

Idodo-Umesh, G., 2002. Pollutant assessments of olomoro water bodies using physical, chemical and biological indices. *Ph.D. Thesis*, University of Benin City, Nigeria, pp. 485.

Mazher, S. and Dawood, S.S., 2005. Heavy metal contamination of the double lake (ERETTAI ERI), Chennai, Tamil Nadu, India. *J. Aqua. Biol.*, 20(1): 53–57.

Mule, M.B. and Patil, R.K., 2000. Study of metal content in water resources from Radhanagari forest. *Indian J. Environ. Prot.*, 20(1): 55–59.

Obasohan, J., Oronsaye and Obano, E.E., 2006. Heavy metal concentration in *Malapterurus electricus* and *Chrysichthys nigrodigitatus* from Ogba river in Benin city, Nigeria. *African J. of Biotechnology*, 5(10): 974–982.

O'Brien, T., Cerjak, S. and Patiemo, S.R., 2003. Complexities of *Chromium carconogenesis*, role of cellular response, repair and recovery mechanisms. *Mutat. Res.*, 533: 2–26.

Chapter 45

Antimicrobial Activity of Isolated Isoflavone from the Seeds of *Butea monosperma*

Rajiv Agrawal[1], Udita Garg[1] and Harsh Kumar Garg[2]

[1]Research Laboratory, Department of Chemistry, D.B.S. (P.G.) College, Kanpur – 06, U.P.

[2]Research Laboratory, Department of Chemistry, S.B.S. Science Mahavidyalaya, Orai, U.P.

ABSTRACT

The defatted seeds of *Butea monosperma* yielded a light yellow coloured compound, molecular formula $C_{16}H_{12}O_6$, m.p. 230–31 °C, Rf value 0.92, m/e peak at 300 (m+) BAW 4 : 1 : 5 v/v gave yellow colour with few drop of concentrated H_2SO_4, aquous NaOH and Mg and HCl; pink colour with sodium amalgam and HCl. Finally the structure of isoflavone, was established as 5,7,3′-trihydroxy 4′-methoxy isoflavone by various degradative reactions and confirmed by IR, UV and mass spectrum studies. Then antimicrobial activities have been studied.

Keywords: *Butea monosperma, Defatted seed extracts, Isoflavone and antimicrobial activities.*

Introduction

Butea monosperma is commonly known as Dhak or, Palas. It belongs to family Leguminacae. It is important indigenous plant, cultivated through out India and abundantly found in all tropical countries. The plant possesses significant medicinal value as a powerful Anthelmintic for roundworms useful for children and delicate women suffering from diarrhoea and dysentry etc. Earlier workers reported that significant number of compounds have been isolated from the different parts of plant.

Materials and Methods

The seeds of plant *Butea monosperma* were collected locally and authenticated by department of Botany DBS (P.G.) College Kanpur (U.P.), air dried and crushed seeds (2.5 kg.) were defatted with petroleum ether (60–80°C) in a Soxhlet extractor for 30 hours in different lots. The defatted seeds material refluxed with ethanol 95 per cent for 18 hours and residue of ethanolic extract, after chloroform extraction, was refluxed with hot ethylacetate and filtered. The filterate was concentrated and extracted with diethyl ether and filtered. After concentration the ethereal extract was placed over silica Get-G Column for chromatography. It was eluted with benzene: ethyl acetate (2 : 1 v/v) to give a reddish brown compound. It analysed for molecular formula $C_{16}H_{12}O_6$, m.p. 230–31°C m/e peak at 300 (m+). The compound a treatment with neutral $KMnO_4$ solution, gave ortho-hydroxy anisic acid (iso-vanillic acid m.p. 254°C) which is one of the oxidation product. On treatment with KOH it yielded degradation product phloroglucinol (m.p. 115°). The identity was confirmed by mixed melting print and co-chromatography (co-PC and co-TLC) with authentic sample.

The isoflavone (RA-III) was identified as 5,7,3',-trihydroxy 4'-methoxy isoflavone by specific colour reactions, chemical degradation and spectral studies.

UV Max

(EtoH) 258, 325 (Alcb) 341 (EtOH+$Alcl_3$) 342 (16) (EtOH+NaOAc) 335 (10).

IR Bands (K Br)

2850,1760,1660,1530,1230 and 1120 cm^{-1}.

Mass Spectrum

m/e 300 (M^+), 272, 177, 149, 148 and 123.

Thus from the above data, it is finally determined the structure of Isoflavone as follows- 5,7,3'-trihydroxy 4'-methoxy isoflavone.

The various solvents fractions of seeds extract of *Butea monosperma* containing compound (RA-III) was tested for antimicrobial activity by filter paper disc method. The result was 'expressed in terms of inhibition zone. The antibacterial activities against plant and human pathogenic bacteria were tested by "oxide nutrient agar" and by "Sikorowski *et al.* method" whereas "Sabourad Dextrose Agar media" was used to check antifungal activities.

Results and Discussion

The Isoflavone, isolated from ether extract of the seeds of *Butea monosperma* was identified as 5,7,3'-tri hydroxy 4'-methoxy isoflavone represented in Figure 45.1 (RA-III).

Table 45.1 and 45.2 represented that various solvent seed extract were found to possess remarkable antibacterial and antifungal activities. In Table 45.1 chloroform extract represents better result in comparison to control against *Bacillus pumilis*. Ethyl alcohol seeds extract showed better result in compression with control against staphylococcus typhi. From Table 45.2 it is clear that chloroform extract showed better inhibition zones than control against *Chrysoparium-tropicum* and *Keratinophyton terreum*. Ethyl acelate extract showed better result in comparison with control against *Keratinophyton terreum*.

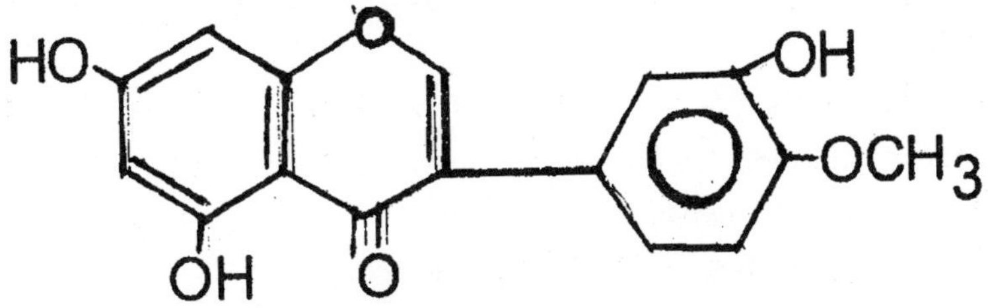

5,7,3'-trihydroxy 4'-methoxy isoflavone

Figure 45.1: RA III

Table 45.1: Antibacterial Activity of Successive Seed Extract of *B. monosperma*

Sl.No.	Bacterial Species	Diameter of Inhibition Zone (mm*) in					
		Pt.Ether	Benzene	Chloroform	Fthyl Alcohol	Ethyl Acetate	Control**
1.	*Bacillus pumilis*	24.2	32.0	37.8	15.2	18.2	37.2
2.	*Escherichia coli*	18.3	22.1	17.6	18.1	15.0	28.2
3.	*Staphylococus typhi*	19.2	15.3	16.0	22.0	15.1	22.8

Table 45.2: Antifungal Activity of Successive Seed Extract of *B. monosperma*

Sl.No.	Fungal Species	Diameter of Inhibition Zone (mm*) in					
		Pt.Ether	Benzene	Chloroform	Ethyl Alcohol	Ethyl Acetate	Control**
1.	*Aspergillus niger*	9.0	–	8.8	9.6	12.6	16.2
2.	*Chrysoparium tropicum*	8.2	10.8	18.3	11.4	12.3	18.7
3.	*Keratinophyton terreum*	11.9	9.6	27.8	18.9	23.2	24.2
4.	*Malbrachea pulchella*	8.1	–	10.8	10.0	12.7	14.6

*: Including the diameter of filter paper disc (6mm).

**: Acromycin and streptomycin (500 ppm) against gram positive and gram negative bacteria.

***: Griseofulvin 1000 (ppm).

Acknowledgements

Authors are thankful to the Director, CDRI Lucknow (U.P.) for spectral analysis and Head of Chemistry Department, D.B.S. (P.G.) College, Kanpur (U.P.) for providing necessary laboratory facilities.

References

Bible, R.H., 1965. *Interpretation of NMR Spectra*. Plenum Press, New York.

Chatterjee, A. and Pokrashi, S.C., 1991. *Treatise on Indian Medicinal Plants*, Vols. 2 and 3. Publication and Information Directorate, C.S.I.R. New Delhi, India.

Elanchezhiyan, M., Rajarajan, S., Rajendra, P., Subramanian, S. and Thyagarajan, S.P., 19930 *J. Med. Microbial.*, 38: 262.

Geissman, T.A., 1962. *The Chemistry of Flavonoid Compounds*. Pergamon Press, Oxford.

Harborne, J.B. and Mabry, T.J., 1982. *The Flavonoids Advances in Research.* Chapman and Hall, London.

Kirtikar, K.K. and Basu, B.D., 1935. *Indian Medicinal Plants*, 2nd edn., Vol. 2, Dehradun.

Muruzella, J.C. and Henary, P.A., 1958. *J. Amer. Pharm. Assoc.*, 47: 471.

Nakanish, K., 1956. *IR Absorption Spectrum*. Butterworth, London.

Pant, R., Nair, C.G., Singh, K.B. and Koshti, G.S., 1976. *Current Science*, 43(13): 773–776.

Sokorowski, P.P., Kent, A.D., Lindig, O.H., Wijgal, G. and Robinsen, J.J., 1980. *Entmol.*, 73: 106–110.

Talapatra, S.K., Mallick, A.K. and Talaptra, B., 1980. *Phytochemistry*,. 19: 1199.

Vincent, J.G. and Vincent, H.W., 1944. *Proc. Soc. Exp. Bio. Med.*, p. 55–162.

Chapter 46

Postharvest Fungal Rot of Chilli (*Capsicum annuum* L.) Fruits

V. Bali Reddy, G. Bindu Madhavi and D. Venkata Reddy
Regional Agricultural Research Station, Lam, Guntur – 522 034, A.P.

ABSTRACT

Survey was conducted in five districts of Andhra Pradesh *viz.*, Guntur, Warangal, Khammam, Prakasam and Krishna to observe the postharvest fungal fruit rot causing fungal pathogens in chilli. It was observed that *Drechslera tetramera, Bipolaris tetramera* and *Alternaria alternata* were causing major fruit rot diseases during pre-monsoon period. *Alternaria alternata, Colletotrichum capsici* were prevalent in monsoon period. Postharvest fungal fruit rot damage during post-monsoon was mainly due to *Fusarium semitectum, Aspergillus flavus, Aspergillus niger, Rhizopus* spp. and *Mucor*.

Introduction

Chilli fruits (*Capsicum annuum* L.) is one of the spice crop famous for its pungent aromatic flavour. Among the spices dry chillies contributes a major share in India growth for both home and export market. The mature green fruits are used as vegetable or in fish and meat preparations. The green fruits are very often damaged by species of *Alternaria, Fusarium, Aspergillus, Colletotrichum, Rhizopus* and *Mucor*. An attempt was made to study the post harvest fungal rots caused by various pathogens in different districts of Andhra Pradesh.

Materials and Methods

Chilli fruits were collected from vegetable market from seven districts of Andhra Pradesh during pre- monsoon, monsoon and post monsoon periods. The infected fruits were collected and pathogens were isolated and identified (Ellis, 1971,1976). Fungi associated were cultured in Potato Dextrose Agar medium. Fresh chilli fruit samples were surface sterilized with mercuric chloride (1 per cent) and plated on PD agar and incubated at 32±1°C for 5 days and confirmed pathogenecity by following Koch's postulates.

Results and Discussion

Fruit rot due to *Alternaria alternata* (Fr.) Keissl. was recorded in fruits collected during the monsoon period. The temperature was as low as 26°C during the period. This disease was characterized by the production of small dark brown spots on fruits. These sports later coalesced and covered large necrotic areas on fruit. The fruits finally rotted completely. The fungi produced abundant conidia *in vivo* and *in vitro*. This disease has been reported earlier (Wall and Biles, 1993; Prabhavathy and Reddy, 1995; Vijayalakshmi and Rao, 2003) from different parts of India. Fruit rot caused by *Fusarium semitectum* Berk and Rav was recorded during post-monsoon season. The fungus produced white hairy hyphae on the stalk, calyx and the stalk end in the early stages of infection. Later the whole surface of the fruit was covered by the fungal mycelium. The fungus produced microconidia and macroconidia on the fruit surface, the microconidia were aseptate and macroconidia were 3–5 septate and measured 24.5–34.5 × 3.5–4.0 µm and were curved with a wedge shaped but not pedicillate basal cell curved with a wedge shaped but not pedicillate basal cell and pointed apex. Micro-conidia measured 7.0–11.0 × 3.0–3.5 µm species of *fusarium* have been found to infect chilli fruits (Wall and Biles, 1993; Prabhavathy and Reddy, 1995; Vijayalakshmi and Rao, 2003; Bilgrami *et al.*, 1979, 1981, 1991; Hosagoudar *et al.*, 1996; Mukerji and Bhasin, 1986). *Aspergillus niger*, caused rot of chilli fruits during the monsoon season. Water soaked brown areas developed near the blossom end or sometimes near the stem end. As the disease advanced further, conidiophores bearing black, catenulate and spiny conidia covered the whole fruit. This disease was found to spread rapidly and affected fresh fruits in storage. Infection of dried fruits by *A. niger* was observed. Anthracnose was observed in chilli fruits during monsoon season, caused by *colletotrichum capsici* (Syd.) Bud. and Bisby. This disease was characterized by dark brown or dirty black, oval to circular depressed spots over the fruits. The mycelium penetrated the inner tissues and even the seeds and produced a peculiar pungent odour. The pathogen produced setae and conidia *in vivo* and *in vitro*. *C. capsici* infection was recorded earlier from Andhra Pradesh (Vijayalakshmi and Rao, 2003) and Hosagoudar *et al.* (1996) from Kerala Rot due to *Drechslera tetramera* (McKinney) Subram. and Jain [=*Bipolaris tetramera* (McKinney) shoemaker] was observed during the pre-monsoon season. The infected fruits developed brown areas on the fruit wall which later turned into deep brown or black and covered by sparse white mycelia. The conidiophores were brown and produced grey to brown 3-septate conidia. Colonies effuse, grey to brown, hairy and velvety. Mycelium submerged 4–9 µm wide. Conidiophores were brown, macronematous, straight or flexuous, geniculate, unbranched and smooth. Conidia solitary, four celled, strainght, brown, smooth, cylindrical with rounded ends. They measured 22–38 µm long and 9–13 µm thick. *D. rostrate* was reported earlier from other parts of India (Vijayalakshmi and Rao, 2003). These fungi are pathogenic and produced toxins in chill (Dadar, 1998; Dubey *et al.*, 1999). Seunghum *et al.* (1997) reported that *A. alternata* causing fruit rot of chilli produced toxin metabolites such as altemariol and alternariol monomethyl ether. Mathur (1995) observed the production of phytotoxin metabolites by *C. capsici* in chill fruits.

Table 46.1: Postharvest Fruit Rot Causing Fungi Observed in Chilli Fruits Collected from Markets of Different Districts

District	Variety Grown in Major Area	Fungi Observed*										
		Drechslera tetramera	*Bipolaris tetramera*	*Alternaria alternata*	*Colletotrichum capsici*	*Fusarium semitectum*	*Aspergillus flavus*	*A. niger*	*A. ohraceus*	*Curvularia lunata*	*Rhizopus spp.*	*Mucor*
Guntur	LCA–334	+++	++	+++	+++	+++	+++	+++	++	+	+	+
Khammam	LCA-335	+++	++	+++	+++	+	++	++	+	–	–	–
Warangal	G–4	++	++	+++	+++	+	+++	+++	+	+	–	–
Prakasam	LCA–206	+	++	+++	+++	+	+	+	–	–	–	–
Krishna	CA–960	++	++	+++	+++	++	+	+	++	++	++	+

*: Cumulative of pre-monsoon, monsoon and post-monsoon periods; +++: Severe; ++: Moderate; +: Low.

References

Bilgrami, K.S., Jamaluddin and Rizwi, M.A., 1979. *Fungi of India*. Today and Tomorrow's Printers and Publishers, New Delhi, India.

Bilgrami, K.S., Jamaluddin and Rizwi, M.A., 1981. *Fungi of India*. Today and Tomorrow's Printers and Publishers, New Delhi, India.

Bilgrami, K.S., Jamaluddin and Rizwi, M.A., 1991. *Fungi of India*. Today and Tomorrow's Printers and Publishers, New Delhi, India.

Datar, V.V., 1998. *Indian J. Mycol. and Pl. Pathol.*, 26: 239–242.

Dubey, L.N., Das, K.K. and Jha, R.R. 1999. *Indian J. Mycol. and Pl. Pathol.*, 29: 276.

Ellis, M.B., 1976. *Dermataceous Hyphomycetes*. CMI, Kew, England.

Hosagoudar, V.B., Abraham, T.K. and Pushpangadan, P., 1996. *Fungi of Kerala*. TBGRI, Palode, Kerala, pp. 121.

Mathur, K. 1995. *Indian J. Mycol. and Pl. Pathol.*, 29: 276.

Mukerji, K.G. and Bhasin, J., 1986. *Plant Diseases of India*. Tata McGraw Hill Co., Ltd., New Delhi, pp. 468.

Prabhavathy, K.G. and Reddy, S.M., 1995. *Indian Phytopath.*, 4: 492.

Seunghun, Y., Sohyoung, O., Hyangburn, L., Byungrgum, K., Minill, C. and Subong, P., 1997. *Koean J. Pl. Pathol.*, 13: 323–330.

Vijayalakshmi, M. and Rao, V. Umamaheswara, 2003. *Indian J. Mycol. and Pl. Pathol.*, 33(3): 451–452.

Wall, M.M. and Biles, C.L., 1993. *Phytopath.*, 83: 324–328.

Chapter 47

Integrated Management of *Helicoverpa* Pod Borer on Pigeonpea at Nalgonda and Mahaboobnagar Districts of Andhra Pradesh

A.D.G. Grace, U. Venkateswarlu, G. Penchala Raju, M.V. Reddy and A. Subba Rami Reddy

ABSTRACT

IPM on farm demonstrations for the management of *Helicoverpa armigera* were conducted at Nalgonda and Mahaboobnagar districts of Andhra Pradesh for eight seasons (1999–2000 to 2006–07). Per cent pod damage was less in IPM fields when compared to farmers practice fields. The per cent pod damage was 47.2, 23.73, 17.46, 6.89, 8.27, 3.94, 5.15 and 5.18 in IPM plots respectively in 1999–2000 to 2006–07 as against 67, 37.65, 27.76, 15.21, 26.84, 7.56, 13.8 and 12.48 respectively in farmers practice fields. The Benefit-Cost Ratio was high in IPM fields compared to farmers practice fields.

***Keywords**: Helicoverpa armigera, Per cent pod damage, Benefit cost ratio, IPM fields and Farmers practice fields.*

Introduction

Pigeonpea [*Cajanus cajan* (L) Millispaugh] is one of the major grain legume (Pulse) crops of the tropics and sub-tropics. It is now widely grown in the Indian sub-continent which accounts for about

90 per cent of the worlds pigeonpea cultivated area (3.46 m ha) and production (1.95 m t). In India it is grown in almost all the states, but UP is the leading producer (0.48 m t, 24.6 per cent of the total production) followed by Maharashtra and Gujarat (Anonymous, 1999). Insect pests represent the main constraint to yield and its stability (Singh *et al.*, 1996). Gram pod borer *Helicoverpa armigera* is considered one of the major pod borer pests of pigeonpea. On an average, single larva per plant reduced the yield to the extent of 138.5 kg/ha (Narendra Reddy *et al.*,2001). In the past few decades, over dependency and indiscriminate use of pesticides have resulted in the development of pesticide resistance in pests, secondary pest outbreaks, elimination of beneficial insects, environmental pollution (pesticide residues) and health hazards. Therefore several components of pest management approaches are being used singly or in combinations for managing this pest. An IPM approach therefore is the best mix of various pest control options such as cultural practices, biological control agents and use of chemical sprays. Hence, AP-NL Project entitled "Integrated Management of *Helicoverpa* pod borer in pigeonpea in selected villages of Nalgonda and Mahaboobnagar districts of Andhra Pradesh" has been taken up to standardize and popularize the IPM package among the farmers.

Materials and Methods

As a part of this project, on farm demonstrations were conducted for eight seasons (1999–2000 to 2006–07) in farmers fields in Nalgonda and Mahaboobnagar districts of Andhra Pradesh in 800 farmers fields covering an area of 320 hectares. IPM package demonstrated consisted of resistant/tolerant varieties (ICP 8863, LRG 41), growing locally important inter-crops (groundnut, greengram, blackgram, sorghum and sunflower), seed treatment with *Trichoderma viride*, erecting bird perches (50/ha), monitoring through pheromone traps (10/ha), use of botanical and biological pesticides (NSKE, NPV, *Bt*), collection and destruction of larvae by shaking method and finally need based application of chemical pesticides (thiodan, chlorpyriphos, spinosad). Observations on predator population, pod damage and yield were recorded in IPM and non-IPM (farmer's method) plots and benefit-cost ratios were calculated.

Results and Discussion

The per cent pod damage done by Helicoverpa recorded in IPM fields from 1999–2000 to 2006–07 was 47.2, 23.73, 17.46, 6.89, 8.27, 3.94, 5.15 and 5.18 respectively as against 67, 37.65, 27.76, 15.21, 26.84, 7.56, 13.8 and 12.48 respectively in the farmers practice fields. The per cent pod damage for each season was calculated by collecting the data on number of pods damaged by the gram pod borer, total number of pods per each plant from both IPM fields and farmers practice fields and the average was calculated. Benefit-Cost Ratio was calculated separately for each season by collecting the cost of cultivation aspects from each of the farmer in the village, gross returns and net returns were also calculated to obtain BC Ratio. The average Benefit-Cost Ratio over eight seasons at Kodangal (Mahaboobnagar district) was 0.77 in IPM and 0.068 in the Farmers practice, at Gaddipalli (Nalgonda district) for six seasons it was 0.92 in IPM against 0.78 in Farmers practice and at KK Thanda (Nalgonda district) for two seasons (2003–04 and 2004–05) it was 0.735 in IPM and 0.25 in farmers practice (Table 47.1).

It is very clear from the data that *Helicoverpa* damage was less in IPM fields compared to the Farmers practice and also the BC ratios were high in the IPM fields. In addition to the above, following of IPM strategy is a eco-friendly practice and farmers can realize more economic returns with minimum risks thus ensuring food security.

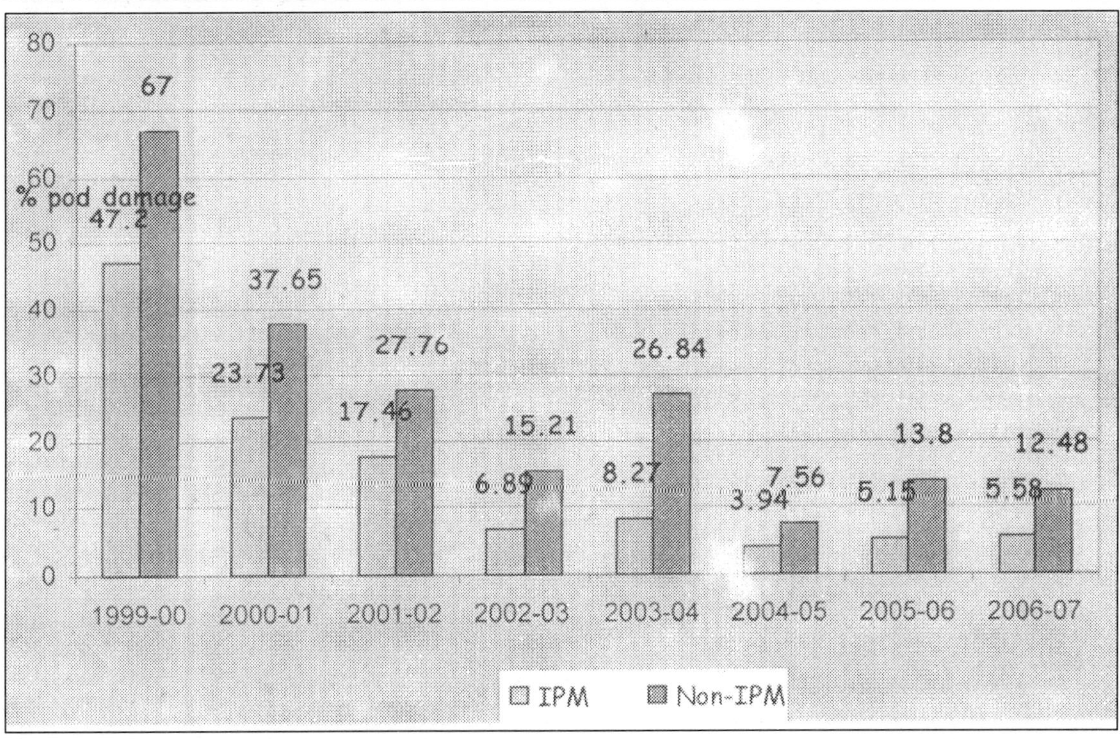

Figure 47.1: Helicoverpa Pod Damage in IPM Demonstrations at Mahaboobnagar and Nalgonda Districts of Andhra Pradesh during 1999–2000 to 2006–07

Table 47.1: Cost-Benefit Ratios for IPM for Helicoverpa Pod Borer on Pigeonpea Vs Non-IPM from 1999–2000 to 2006–07 at Nalgonda and Mahaboobnagar Districts of Andhra Pradesh

	Kodangal (Mahaboobnagar)		*Gaddipalli (Nalgonda)*		*K.K. Thanda (Nalgonda)*	
	IPM	*FM*	*IPM*	*FM*	*IPM*	*FM*
1999–00	1.08	–0.18	0.85	–	–	–
2000–01	0.04	–0.07	0.37	–	–	–
2001–02	0.9	–0.25	–0.01	–	–	–
2002–03	1.09	0.36	1.09	–	–	–
2003–04	0.91	0.47	–	–	0.93	0.43
2004–05	0.9	0.13	–	–	0.54	0.07
2005–06	0.61	0.24	2.51	1.53	–	–
2006–07	0.696	–0.156	0.751	0.032		–
Average	0.77825	0.068	0.926	0.781	0.735	0.25

References

Annual Report of AP-NL Project "*IPM in Pigeonpea*"– 2002–2003.

Annual Report of AP-NL Project "*IPM in Pigeonpea*"–2000–01.

Annual Report of AP-NL Project "*IPM in Pigeonpea*"–2003–2004.

Annual Report of AP-NL Project "*IPM in Pigeonpea*"–2004–05.

Annual Report of AP-NL Project "*IPM in Pigeonpea*"–2005–06.

Annual Report of AP-NL Project "*IPM in Pigeonpea*"–2006–07.

Anonymous, 1999. *Agricultural Statistics at a Glance*. Directorate of Economics and Statistics, Department of Agriculture and Co-operation, Ministry of Agriculture, Government of India.

Narendra Reddy, C., Yeshbir, Singh and Singh, Vijai S., 2001. Economic injury level of gram pod borer (*Helicoverpa armigera*) on pigeonpea. *Indian J. Entomol.*, 63(4): 381–387.

Progress Report of AP-NL Project "*IPM in Pigeonpea*"–1999–2002.

Singh, L., Chauhan, Y.S., Johansen, C. and Singh, S.P. (Eds.), 1996. Prospects for growing extra-short duration pigeonpea in rotation with winter wheat crops. In: *Proceedings of IARI/ICRISAT Workshop and Monitoring Tour* 16–18, October, 1995, New Delhi, India, Indian Agricultural Research Institute, Patancheru, India, International Crops Research Institute for Semi-Arid Topics, pp. 132.

Chapter 48

Effect of Soil Solarization on Damping Off Disease Management in Vegetable Nursery

V. Bali Reddy, G. Bindu Madhavi, D. Venkata Reddy, V. Chenga Reddy, K. Gurava Reddy and M. Chandra Sekhar Reddy
Regional Agricultural Research Station, Lam, Guntur – 522 034, A.P.

ABSTRACT

Vegetable nursery management through soil solarization was investigated in relation to polythene thickness moisture status of nursery beds, temperature, microbial population, seedling stand, temperature difference between atmosphere and nursery beds was recorded 6–10°C under different moisture regime. Fungal population of pathogenic and non-pathogenic was changed after solarization with and without polythene. The extent of damping off reduction was 25.8 to 57.6 per cent over control treatment.

Introduction

Most of the vegetables crops are first raised in nurseries and later transplanted in the main field. Successful nursery rising is very difficult in kharif season due to damping off and weed population. Several soil borne pathogens cause heavy losses not only in nursery but also in main field after transplanting. Repeated sowing of seeds in the same plot results in a high inoculum build up in the soil through seeds and other susceptible planting materials. Tropical and subtropical part of the India is very hot during April to June. In these months solar radiation can be easily trapped by white polythene to increase heat of the soil for partial sterilization. This approach for disease and weed management has been studied thoroughly in Israel (Horowitz, 1980; Katan *et al.*, 1976). In India, soil solarization has been studied for controlling chick pea and pigeonpea wilt (Arya and Mathew, 1993). The results from the study as effect of solarization on vegetable crops are reported here.

Materials and Methods

The experiment was conducted during 2005–07 at Krishi Vigyan Kendra, Darsi, Prakasam District, A.P. Raised nursery beds were prepared in with 3 meter length, 80 cm width and 8" height. Dry, moist, wet and flood soil moisture levels were created by adding water in nursery beds during polythene mulching. White polythene of two different thickness *i.e.* 80 μm and 130 μm were used for experimentation. Moisture percentage of nursery bed was calculated by weight (w/w) basis. Polythene was mulched between the last week of April to 2nd week of June. Polythene was tightly sealed from all sides through soil. The fungal population in soil was recorded colony forming units/g (Martin 1950) prior to polythene mulching and just after removal to determine microbial status of pre and post solarized soil. Total number of mature seedling stand was recorded in per square meter area basis after twenty-five days of sowing of seeds. Germination percentage of tomato chilli and brinjal was counted after 10 days of sowing. Each laboratory experiment was replicated six times while field experiments three times.

Results and Discussion

Polythene Thickness

Effect of polythene thickness revealed that 130 μm thick polythene is better than 80 μm for reducing the total fungal population and increasing the seedling stand of tomato and chilli in per unit area (Table 48.1). The total weed population was slightly increased from 42 to 64 under thick polythene treatment. The total fungal population, total seedling stand of tomato and chilli was lower under thin polythene. In the present finding, decrease of fungal population was recorded from thin to thick polythene mulching. Sharma and Sharma (2002) reported increase in total microbial population from 25 μm to 50 μm thickness. The difference in soil type and agro-climatic conditions might have interfered with microbial diversity. In this region 130 μm thick polythene is found better than 80 μm for reducing total fungal population. 300 gauge polythene sheets mulching resulted isolation on only six fungi from rhizosphere of pigeon pea (Arya and Mathew, 1993).

Table 48.1: Effect of Polythene Thickness on Vegetable Nursery Management in Soil Solarization

Treatments	*Solarized Beds*		*Unsolarized Beds (Control)*
	Thin Polythene (0.08 mm)	*Thick Polythene (0. 13 mm)*	
Total fungal population (cfu/g soil)	2.66×10^4	2.2×10^4	3.2×10^4
Total chilli seedling stand (area/m^2)	225.33	278.33	195.3
Total tomato seedling stand (area/m^2)	42268.3	292.3	193.0
Total weed population after seven days (area/m^2)	42	64	580.0

*: Average of three replications; **: Average of six replications.

Moisture Status of Nursery Beds

Four different moisture regime of nursery bed was calculated as 0.42, 2.9, 11.1 and 19.5 per cent respectively in dry, moist, wet and flood beds. The moisture content of control (unmulched) bed was 0.81 per cent w/w basis. The minimum temperature was recorded in morning while maximum temperature at 2:30 p.m. The study revealed that maximum temperature 52°C was recorded in moist nursery bed holding 2.9 per cent water content in w/w basis. The next higher temperature 51°C was observed in dry beds having 0.42 per cent water only. The wet and flooded solarized beds showed slightly lower temperature of 49°C and 48°C respectively. The temperature was 46°C in unsolarized control beds while 42°C at atmosphere. The minimum temperature varied from 35 to 37°C in the treatments. It indicated that minimum thermal regime in solarized beds in relation to moisture status was almost stabilized during whole period of solarization maximum temperature was variable. Tu *et al.* (1991) reported that highest temperature 57°C of the mulched plots and lowest 27°C in comparison to unmulched plots (39°C and 29°C). Dry and moist beds recorded 5°C and 6°C increase in soil temperature over control beds. Katan *et al.* (1976) reported that maximum 49.5°C temperature in mulched soil with 0.03 mm thick polythene. However, in present study the thickness of polythene was 0.13mm. Temperature increase was reported 10.3°C and 10.9°C respectively in unirrigated and irrigated soil by Sharma and Sharma (2002) under hilly region of Himachal Pradesh. The moist and dry nursery beds were equally effective for increase in temperature during solarization.

Effect of Nursery Diseases

Soil solarization of nursery beds revealed that it was very effective in reducing the damping off in tomato, chilli and brinjal. The seed germination was comparatively higher in solarized beds in comparison to unsolarized beds. Solarization has greater impact on these vegetable crops. The germination percentage of these crops varies from 74 to 82 in polythene mulched solarized beds (Table 48.2).

Table 48.2: Effect of Soil Solarization on Damping Off and Bacterial Leaf Spot Incidence in Nursery Beds

Crop	*Solarized Seedling Stand (Per cent)*	*Unsolarized (Control) Seedling Stand (Per cent)*
Tomato	78	62
Chilli	74	50
Brinjal	82	52

Effect of Soil Fungi

It is evident from the Table 48.3 that total soil fungi population was reduced in polythene mulched solarized beds. The change in total fungi was also recorded in pre-solarized beds to post solarized beds without polythene mulch. The fungi was also categorized in two groups *viz.*, pathogenic and non-pathogenic. The pathogenic fungal population was 11.8×10^3 colony forming units/g in pre-solarized bed and 10.8×10^3 cfu/g in solarized beds without polythene. It was drastically reduced with pathogen-solarized beds under three moisture regimes except flooded beds. It is clear that most of the pathogenic fungi reduced in maximum amount in moist beds and hence moist conditions of soil are effective for solarization. When non-pathogenic fungi were compared, the dry beds showed least population followed by moist beds. Interesting result observed that the total non-pathogenic fungi were more in post-solarized non-mulched beds. Because thermophilic nature of the fungi resting spores of fungi are activated resulting in increasing fungal population. Similar results were obtained with *Fusarium* sp. also where chalymodospore have activated after solarization and gave more count after planting. However, after polythene mulching, these fungi were killed and reduced. It was due to increase of latent temperature. However, total non-pathogenic fungi also decreased in polythene mulched solarized beds.

Table 48.3: Effect of Moisture Status of Nursery Beds on Fungal Population (10^3 cfu/g soil) in Relation to Solarization

		Post-Solarization				
	Before Solarization	Without Polythene	With Polythene			
			Dry	Moist	Wet	Flooded
Non-pathogenic fungi						
A. terreus	8.0	10.0	8.2	6.0	5.3	3.5
A. quadrilinatus	10.0	11.5	2.0	1.9	1.5	4.5
A. niger	8.0	4.2	1.2	3.2	3.0	4.3
A. flavus	3.5	2.6	1.0	0.8	1.2	2.3
Pencillium sp.	1.0	1.5	3.0	8.7	12.5	2.5
Unidentified	4.0	3.5	2.5	1.5	3.3	2.5
Sub total non-pathogenic	34.5	33.3	17.9	22.1	26.8	19.6
Pathogenic fungi						
S. rolfsii	3.0	2.0	0.8	0.3	1.5	1.9
Fusarium spp.	2.5	3.0	1.2	0.9	1.0	2.0
Rhizoctonia solani	2.0	2.0	2.0	0.4	0.5	0.9
Pythium spp.	3.0	2.8	2.5	2.2	2.0	2.1
Curvularia lunata	0.5	0.3	0.0	0.0	0.0	0.0
Macrophomina phaseolima	0.8	0.7	0.4	0.0	0.0	0.0
Sub total pathogenic	11.8	10.8	5.3	3.9	5.4	8.3
Total fungi	46.3	44.1	23.2	26.0	32.2	27.9

References

Arya, Arun and Mathew, D.S., 1993. Soil solarization of *Fusarium udam*. *Indian Phytopathology*, 46(2): 151–154.

Horowitz, M., 1980. Weed research in Israel. *Weed Sci.*, 28: 457–460.

Katan, J.A. Greenberger, Alon, H. and Grinstein, A., 1976. Solar heating by polythene mulching for the control of diseases cause by soil borne pathogens. *Phytopathology*, 66: 683–688.

Martin, J.P., 1950. Use of acid, rose Bengal and streptomycin in the plate method for estimating soil fungi. *Soil Science*, 69: 215–233.

Pandey, K.K., 2001. Important diseases of vegetable crops and their integrated management. In: *National Training Course on Vegetable Production Technology*, Directorate of Extension, Ministry of Agriculture, New Delhi, pp. 177–186.

Sharma, M. and Sharma, S.K., 2002. Effect of soil solarization on soil microflora with special reference to *Dematophora necatrix* in apple nurseries. *Indian Phytopathology*, 55(2): 158–162.

Tu, C.C., Cheny, Y.H. and Hwang, S.C., 1991. Effect of solar energy and green manures on the control of southern blight of tomato. *Plant Protection Bulletin (Taipei)*, 33(1): 95–102.

Chapter 49

Studies on Genetic Variability Correlation and Path Analysis for Yield and Fibre Characters in Cotton (*Gossypium hirsutum* L.)

S. Rajamani, Ch. Mallikarjuna Rao and R. Krishna Naik
Regional Agricultural Research Station, Lam, Guntur – 34, Andhra Pradesh

ABSTRACT

Genetic variability, correlation and path coefficient analysis were studied in 63 cotton genotypes revealed that fibre length, boll weight, uniformity and seed cotton yield had high heritability and high genetic advance indicating presence of additive gene action. Characters plant height, number of monopodia, number of sympodia, number of bolls per plant and boll weight had positive correlation with yield. Studies on path analysis revealed that plant height, number of monopodia, number of sympodia, number of bolls per plant and boll weight had direct positive effect on yield.

Keywords: *Cotton, Correlation, Path analysis, Variability.*

Introduction

Cotton (*Gossypium hirsutum* L.) is an important fibre crop of India which would be cultivated nearly 9.5 m.ha with a productivity of more than 30 million bales. Improvement in yield and fibre properties are the important objectives in any improvement programme cotton crop. Selection of parents with desirable yield contributing characters coupled with fiber properties is one the most important aspects while selecting the parents in hybridization programme. Knowledge on yield and yield complements and their association towards yield improvement and studies on direct and indirect effects on yield is also essential to develop successful heterotic combinations for yield improvement. Correlation and path analysis studies provide useful information regarding selection of parents based the character association and their direct and indirect effects on yield. The present investigation was carried out to study corrections and path analysis of yield and fibre properties.

Materials and Methods

The present investigation was carried on 63 cotton genotypes at Regional Agricultural Research Station, Lam, Guntur during *Kharif* 2005, in randomized complete block design with 3 replications. Each genotype was sown in two rows of 6 m length with spacing of 120 × 60 cm in black cotton soil.Data was recorded on ten randomly selected plants per genotype per replication on characters *viz.*, plant height, number of monopodia, number of sympodia, number of bolls per plant, boll weight, fibre length, fibre strength, micronaire, uniformity and yield. The genotype means were considered for estimation of genotypic and phenotypic coefficient of variations (GCV and PCV), heritability (broad scene) and genetic advance as per Burton (1952) and Allard (1960). The correlations were estimated as per Falconer (1964) and path analysis as per Dewey and Lu (1959) were estimated.

Results and Discussion

Analysis of variance indicated significant difference for all the character studied.The treatment difference are highly significant for plant height, number of sympodia, number of bolls per plant, boll weight, fibre length, fibre strength, uniformity ratio and yield. These characters were found to be significant due to presence of sufficient variation in the genotypes studied.

The results revealed that the correspondence between the phenotypic and genotypic variance for ten character is higher indicating the environmental influence on the characters studied.This was confirmed by much difference between PCV and GCV values (Table 49.1). The values of GCV to PCV were ranged from 0.38 to 0.40 in boll weight and 115034.20 to 126547.60 in yield. The characters with high magnitude of phenotypic and genotypic variances had high GCV and PCV. The PCV values for characters *viz.*, yield number of bolls per plant, plant height, micronaire and number of sympodia were much higher than GCV value indicating that these characters were interacted with environment to greater extent. Little difference between PCV and GCV indicated less inference of environment in expression of there traits. High value of PCV and GCV were found in characters like yield, number of monopodia and number of bolls per plant.High heritability with high genetic advance in characters fibre length, boll weight, uniformity and yield indicating prevalence of additive gene action where as the character like., micronaire, number of boll per plant, number of monopodia and plant height had low heritability with low genetic advance indicating prevalence of non additive gene action results in low genetic gain under selection.

Genotypic and phonotypic correlations were estimated among the yield contributing character with yield. The results revealed that plant height number of monopodia, number of sympodia, number of bolls per plant and boll weight had positive correlations with yield. But fibre length, fibre strength, micronaire and unformity had negatively corrected with yield at genotypic and phonotypic levels. None of the characters show significant positive or negative correction with yield as well as with other characters at genetic level. But in phenotypic correction plant height shown significant positive correction with number of sympodia, number of bolls per plant and with uniformity ratio but significant negative correction with number of monopodia, fibre length and fibre strength (Table 49.2).

Table 49.1: Components of Variance for Yield and Fibre Characters in Cotton

Character	*Mean*	*GCV*	*PCV*	*Heritability*	*GA (as % of mean)*
Plant height (cm)	149.49	12.02	14.41	0.697	20.68
No. of monopodia	2.26	21.83	28.72	0.57	34.18
No. of sympodia	19.27	12.96	16.51	0.61	20.95
No. of bolls per plant	47.36	16.00	21.77	0.54	24.22
Boll weight (g)	4.54	13.61	14.08	0.93	27.11
2.5 per cent span length (mm)	27.81	8.43	8.63	0.95	16.98
Fiber strength (g/tex)	22.46	7.23	7.77	0.86	13.86
Micronaire (10^{-6} g/inch)	3.89	5.02	14.71	0.11	3.53
Uniformity ratio	48.15	7.00	7.29	0.92	13.84
Yield (kg/ha)	1387.32	24.44	25.64	0.90	48.01

Table 49.2: Genotypic and Phenotypic Correlations between Seed Cotton Yield and Other Characters in Cotton

Character		*Plant Height*	*No. of Mono-podia*	*No. of Sym-podia*	*No. of Bolls per Plant*	*Boll Weight*	*2.5 per cent Span Length*	*Fiber Strength*	*Micro-naire*	*Uniformity Ratio*
Plant height	G	1.0000	–0.2521	0.4321	0.4780	0.1682	–0.3252	–0.3183	0.4797	0.2597
	P	1.0000	–0.1798*	0.3519**	0.3624**	0.1339	–0.2703**	–0.2608**	0.0390	0.2050*
No. of monopodia	G	–0.2521	1.0000	–0.0590	0.3395	–0.0027	–0.0492	0.0658	–0.2983	–0.0758
	P	–0.1798*	1.0000	–0.0865	0.1663	–0.0137	–0.0213	0.0412	0.0345	–0.0367
No. of sympodia	G	0.4321	–0.0590	1.0000	0.5636	0.1174	–0.2323	–0.1782	0.0884	0.1746
	P	0.3519**	–0.0865	1.0000	0.3577**	0.1063	–0.1760*	–0.1 053	–0.0780	0.1 075
No. of bolls per plant	G	0.4780	0.3395	0.5636	1.0000	0.1742	–0.2082	–0.1525	0.0093	0.2443
	P	0.3624**	0.1663	0.3577**	1.0000	0.0962	–0.1210	–0.0647	0.0282	0.2032*
Boll weight	G	0.1682	–0.0027	0.1174	0.1742	1.0000	0.00456	0.1437	–0.3740	0.0122
	P	0.1339	–0.0137	0.1063	0.0962	1.0000	0.0344	0.1325	–0.1286	0.0192
2.5 per cent span length	G	–0.3252	–0.0492	–0.2323	–0.2082	0.0456	1.0000	0.9284	–0.7463	–0.6196
	P	–0.2703**	–0.0213	–0.1760	–0 121 0	0.0344	1.0000	0.8183**	–0.2507**	–0.5731 **
Fiber strength	G	–0.3183	0.0658	–0.1782	–0.1525	0.1437	0.9284	1.0000	–0.7463	–0.6196
	P	–0.2608**	0.0412	–0.1053	–0.064 7	0.1325	0.8183**	1.0000	–0.2139*	–0.5467**
Micronaire	G	0.4797	–0.2983	0.0884	0.0093	–0.3740	–0.7093	–0.7463	1.0000	0.4169
	P	0.0390	0.0345	–0.0780	0.0282	–0.1286	–0.2507**	–0.2139*	1.0000	0.2420**
Uniformity ratio	G	0.2597	–0.0758	0.1746	0.2443	0.0122	–0.5973	–0.6196	0.4169	1.0000
	P	0.2050*	–0.0367	0.1075*	0.2032*	0.0192	– –0.5731 **	–0.5467**	0.2420**	1.0000
Yield	G	0.3129	0.1852	0.3501	0.5520	0.1700	–0.0517	–0.0112	–0.4684	–0.0262
	P	0.2478	0.1585	0.2636	0.3 649	0.1705	0.0508	–0.0041	–0.0945	–0.0179

The character number of sympodia had significant positive phenotypic correction with plant height, number of bolls per plant and positive correction with monopodia, fibre length, fibre strength and micronaire value.

Number of bolls per plant had significant positive phenotype correction with plant height, number of sympodia and uniformity but only positive correlation with number of monopodia, boll weight and micronaire.This character had negative correlation with fibre length and strength.

Boll weight had positive phenotypic correlation with plant height, number of sympodia, number of bolls per plant, fibre length, fibre strength and uniformity. It had negative correlation with number of monopodia, micronaire and uniformity.

Micronaire had significant positive phenotype correlation with plant height, number of monopodia, number of bolls per plant and uniformity but other character *viz.*, number of sympodia, boll weight, fibre length and strength had negative correction.

The character uniformity had significant positive phenotype correlation with plant height, number of bolls per plant, micronaire and boll weight but it had negative correction with number of monopodia, fibre length and fibre strength, These findings are III accordance with Shanti and Raveendran (1999) and Sambamurthy *et al.* (1995).

Table 49.3: Direct (*Italic*) and Indirect Effects of Different Characters on Yield in American Cotton

Character	*Plant Height*	*No. of Mono-podia*	*No. of Sym-podia*	*No. of Bolls per Plant*	*Boll Weight*	*2.5 per cent Span Length*	*Fiber Strength*	*Micro-naire*	*Uniformity Ratio*
Plant height	*0.1380*	–0.0248	0.0486	0.0500	0.0185	–0.0373	–0.0360	0.0054	0.0283
No. of monopodia	–0.0270	*0.1503*	–0.0130	0.0250	–0.0021	–0.0032	0.0062	0.0052	–0.0055
No. of sympodia	0.0436	–0.0107	*0.1238*	0.0443	0.0132	–0.0218	–0.0130	–0.0097	0.0133
No. of bolls per plant	0.0925	0.0425	0.0913	*0.2553*	0.0246	–0.0309	–0.0165	0.0072	0.0519
Boll weight	0.0149	–0.0015	0.0118	0.0107	*0.1113*	0.0038	0.0147	–0.0143	0.0021
2.5 per cent span length	0.0122	0.0010	0.0079	0.0055	–0.0016	*–0.0451*	–0.0369	0.0113	0.0258
Fiber strength	0.0006	–0.0001	0.0002	0.0001	–0.0003	–0.0018	*–0.0022*	0.0005	0.0012
Micronaire	–0.0028	–0.0025	0.0056	–0.0020	0.0092	0.0179	0.0153	*–0.0716*	*–0.0173*
Uniformity ratio	–0.0241	0.0043	–0.0239	–0.0239	–0.0023	0.0675	0.0644	–0.0285	–0.1177
Yield	0.2478	0.1585	0.3649	0.3649	0.1705	–0.508	–0.0041	–0.0945	–0.0179

Residual effect = 0.8866.

The phenotype and genotype positive correlations of yield with plant height, number of monopodia, number of sympodia, number of bolls per plant and boll weight indicating that improvement in these characters will ultimately result in improvement of seed cotton yield.The phenotype and genotype positive correlation of yield with plant height, number of monopodia, number of sympodia, number of bolls per plant and boll weight indicating that improvement in there characters will ultimately results in improvement of seed cotton yield and there finding as in agreement with Yadava *et al.* (2000).

The path coefficient analysis (Table 49.3) indicated that plant height, number of monopodia, number of sympodia, number of bolls per plant and boll weight exerted direct positive effect on yield.The results suggested that number of bolls per plant, number of sympodial branches and plant height could be found very effective in improvement of yield per plant.where as fibre length, fibre strength, micronaire and uniformity found to be direct negative effect on yield. Number of sympodia, number of bolls per plants, boll weight and fibre properties *viz.*, fibre strength, micronaire and uniformity had indirect positive effect through plant height on yield. Boll weight, micronaire and uniformity ratio also had indirect positive effect on yield through number of bolls per plant. Thus based on the above results much importance would be given to number of bolls per plant, boll weight, number of sympodial branches per plant to improve the yield component in cotton.

References

Allard, R.W., 1960. *Principles of Plant Breeding*. John Wiley and Sons Inc., pp. 75–98.

Burton, G.W., 1952. Quantitative inheritance in grasses. In: *Proceedings of the Sixth International Congress*, pp. 277–283.

Dewey, D.R. and Lu, K.H., 1952. A correlation and path coefficient analysis of component of crested wheat grass production. *Agro. J.*, 51: 515–518.

Flaconer, D.S., 1960. *Introduction of Quantitative Genetics.* Oliver and Boyd, Edinberg, pp. 365.

Sambamurthy, J.S.V., Mohan Reddy, D. and Reddy, K.H.G., 1995. Genetic Divergence for lint characters in upland cotton. *Ann. Agric. Res.*, 16(3): 357–359.

Shanti, P. and Raveendran, T.S., 1999. Genetic Variability in F_2 population in cotton (*Gossypium hirsutum* L.), PVK. *Res. J.*, 23(1): 9–13.

Yadava, O.P., Lather, B.P.S. and Dhayia, B.N., 2000. Studies on variability, correlation and path analysis in *desi* cotton (*Gossypium arboreum* L). *J. Cotton. Res. Dev.*, 14(2): 143–146.

Chapter 50

Fish Health Under Fenvalerate Stress

Poonam Kumari, Amit Kumar Jha and M.M. Jha
Research Scholar, P.G., L.N.M.D., Darbhanga
Department of Zoology, M.L.S.M. College, Darbhanga

ABSTRACT

This study attempts to evaluate the health status of the fresh water fish *Anabas testudineus* on the basis of changes in blood biochemistry under the toxic stress of synthetic pyrethroid. Fenvalerate (0.015 ppm) for 30 days significant changes were observed in the blood biochemical properties in the form of hyperglycemia, hypoproteinemia and hypercholesterolemia. The present study concludes that indiscriminate use of fenvalerate even at lower doses, may prove detrimental effect to fish health and population as such some alternative non toxic chemicals needs to formulated in order to have better fish productivity.

Keywords: *Blood glucose, Serum protein, Serum cholesterol, Anabas testudineous, hyperglycemia, hypoproteinemia and Hypercholesterolemia.*

Introduction

The manufacturing of pesticides was started in Indian in 1952 and at present 2700 units are engaged in the production and formulation of over 137 different varieties of pesticides. Based on the assessment of Planning Commission of India, the consumption of pesticides was 1.09 million metric tones in 1986–87 and it increased to 1.19 million tones in 1990 and 200 thousand tones in 2000 AD (Mehrotra).

Very scantly information is available as regards to toxic effects of pyrethroid to fish (Ghosh and Chatterjee, 1988; Reddy and Basha Mohidden, 1983; Prasad *et al.*, 1991; Dutta *et al.*, 1994; Srivastava *et al.*, 1995, Singh: 2002; Pandey, 2003; Mukhopadhyay *et al.*, 2005; Rita and Milton, 2006).

Newly synthetic pyrethroid (fenvalerate) having very long mammalian toxicity and high bio-efficacy got its discovery and are being manufactured in recent year, with a recommendation to the farmers to use this compound in pest control. In fact that physio-biochemical parameters of fresh water fishes in relation to fenvalerate toxicity are still very meager led this worker to undertake the present study.

Materials and Methods

Anabas testudineus procured live from the local fish market were washed with 0.1 per cent $KMnO_4$ solution to remove dermal infection if any. Healthy fish of average length (9–12cm) and weight (21–25 g) were acclimated for 15 days to laboratory conditions. The fish were fed with chopped goat liver every day adlibitum.

Static acute bioassays were performed to determine LC_{50} values of fenvalerate for 24, 48, 72 and 96 hours following the methods of APHA, AWWA and WPCF (1985). The LC_{50} values for these periods were 0.130 mg/l, 0.098 mg/l, 0.088 mg/l and 0.078 mg/l respectively. The sublethal concentration was determined following the formula of Hart *et al.* (1945). Fish were exposed to a sublethal concentration (0.016 mg/l) of fenvalerate for 30 days. Side by side same number of fish as that of experimental one was maintained as the control group. At the end of exposure period the fish were anaesthetized with 1 : 4000 MS222 (tricane, methane, sulfonate, sandoz) and were then processed for quantitative estimation of blood glucose (Sinha, 1990), serum protein (Varley, 1980) and serum cholesterol (Kabanls, 1986). Blood samples were collected in heparinized vials with the help of 1 ml disposable syringe equipped with 2 gauge hypodermic needle by puncturing ventral aorta.

Results

The fish of this exposure period reflects hyperglycaemic, hypoproteinemic and hypercholesterolemic response. The blood glucose level elevates to 90.64±2.98 mg/100 ml of blood against the control value of 70.83±2.49 mg/100 ml of blood. This amounts to an increase by 27 per cent and is significant at $P < 0.0\ 1$ (Table 50.1). The protein depletion shows hypoproteinemia which depleted to 4.07±0.77 g/100 ml of blood against the control value of 5.89±0.65 g/100 ml bf blood. This amount to decrease by 30 per cent and is significant at $P < 0.001$ (Table 50.1). The serum cholesterol elevation shows hypercholesterolemic response which elevates to 171.28±2.53 mg/100 ml of blood against the control value of 145.85±2.38 mg/100 ml of blood. This amount an increase by 17 per cent and is significant at $P < 0.05$ (Table 50.1).

Table 50.1: Changes in Blood Serum Biochemical Parameters of *Anabas testudineus* Exposed for 30 Days to 0.16 ppm Fenvalerate Values are mean±SE of 5 observations

Parameters	*Control*	*Fenvalerate Exposed*	*Level of Significance*
Blood glucose (mg/100ml)	70.83±2.49	30.64±2.98 (+27.96)	p< 0.01
Serum protein (mg/100ml)	5.89±0.65	4.07±0.77 (–30.89)	P < 0.001
Serum cholesterol (mg/100 ml)	145.85±2.38	171.28±2.53 (+17.43)	P < 0.05

Values in parentheses indicate per cent increase (+) or decrease (–) over control.

Discussion

Blood Glucose

Various studies have reported similar increase in the level of blood glucose in various fishes subjected to pyrethroidal and other pesticideal exposures. Singh and Srivastava (1982) observed hyperglycemia in *H. fossilis* treated with 10.4 ppm for malathion. Similarly, endosulfan induced increase in blood glucose was reported by Kalarani *et al.* (1984). The present study is in conformity with the finding of Ghosh (1990) who recorded significant elevation of blood sugar of *C. carpio* and *O. mossambicus* exposed to sublethal concentration of fenvalerate. Bhattacharya *et al.* (1987), based on their studies with single and mixture of pollutant exposure to fish, reported that hepatic glycogen is probably the source of hyperglycemia. They found that maximal glycogen depletion corresponded to a dramatic increase in blood glucose levels, suggesting that some of the hepatic glycogen via the intermediate glucose phosphate was converted to glucose and that this glucose enters the circulation (Hinson *et al.*, 1983).

In the light of available literature and the present study on blood glucose clearly reveals that there exists a high catabolic potency during pesticide exposure which have serious consequences for the general body metabolism and energetic economy of the fish.

Serum Protein

Various studies have reported similar decrease serum protein by Verma *et al.* (1979), Sastry and Siddiqui (1983), Rao *et al.* (1984), Jha (1992). The present loss in serum protein may also be explained in this light. The liver cells might have reduced or stopped the synthesis of serum protein due to the direct toxic effects of the pesticides and the serum protein would have been utilized under pesticides induced stress leading to their depletion.

Serum Cholesterol

Various studies have reported similar increase serum cholesterol by Love (1970), Jha (1995). The blood cholesterol content is usually high in fishes, often much higher than mammals. It is utilized in the conversion of bile salts and in the synthesis of steroid hormones including androgens. Present findings are conformity with Sastry and Sharma (1980) explained increased cholesterol in *Channa punctatus* exposed to mercuric chloride. Gill *et al.* (1991) have opined enhanced de novo cholesterogenesis in liver to be responsible for rise in blood cholesterol.

References

APHA, 1985. *Standard Methods for the Examination of Water and Wastewater*, 16th edn. American Public Health Association, Washington D.C.

Bhattacharya, S., Ray, A.K. and Bhattacharya, S., 1987. Blood glucose and hepatic glycogen inter relationship in *Channa punctatus* (Bloch): A parameter of non-lethal toxicity bioassay with industrial pollutants. *Ind. J. Exp. Biol.*, 25L 539–541.

Ghosh, T.K. and Chatterjee, S.K., 1988. Toxic influence of Fenvalerate in the biological parameters of the fish *Anabas testudineus*. *Environ. Ecol.*, 6(1): 107–110.

Ghosh, T.K., 1990. Toxicological evaluation of synthetic pyrethroids on the freshwater fishes. In: *Environmental Pollution and Health Hazards,* (Eds.) R. Prakash and S.M. Choubey. Society of Biosciences. Muzaffarpur, India, pp. 15–25.

Gill, T.S., Pande, J. and Tewari, H., 1991. Effects of endosulfan on the blood and organ chemistry of freshwater fish, *Barbus conchonius* (Hamilton). *Ecotoxicol. Environ. Saf.*, 21: 80–91

Hart, W.B., Dondoroff, P. and Greenbank, J., 1945. *The Evaluation of Toxicity of Industrial Wastes, Chemicals and Other Substances to Freshwater Fishes*. Atlantic Refining Company, Phil. Part (1): 317–326.

Hinson, J.A., Mays, J.B. and Cameron, A.M., 1983. *Biochemical Pharmacol.*, 32: 1979.

Jha, B.S. and Jha, M.M., 1995. Biochemical effects of nickel chloride on the liver and gonads of the freshwater climbing perch, *Anabas testudineus* (Bloch). *Proc. Nat. Acad. Sci. (India)*, 65B(1): 39–46.

Jha, M.M., 1992. Studies on the effects of nickel chloride to some organs of the climbing perch, *Anabas testudineous* (Bloch). *Ph.D. Thesis*, L.N. Mithila University, Darbhanga.

Kabara, J.J., 1986. Determination and microscopic localization of cholesterol. In: *Method of Biochemical Analysis,* (Ed.) D. Glick. Inter Science Publishers Inc., New York, p. 263–318.

Kalarani, V., Reddy, D.C. and Naidu, B.P., 1984. Impact of endosulfan toxicity on certain aspects of carbohydrate metabolism of the freshwater snail, *Pila globosa* (Swainson). *Geobios*, 11: 197–200.

Kaviraj, A. and Konar, S.K., 1982. Acute toxicity of mercury, chromium and cadmium to fish, plankton and worm. *Geobios*, 9(3): 97–100.

Love, R.M., 1970. *The Chemical Biology of Fishes*. Academic Press, New York.

Mehrotra, K.N., 1990. Pest control strategies for 2000 A.D. *Froc. Ind. Natn. Sci. Acad. B.*, 52(1): 10–16.

Mukhopadhyay, K., Bera, R. and Roy, R., 2005. Modern agriculture and environmental pollution. *Every Man's Science*, 40(3).

Pandey, Ratna, 2003.0 Pesticides and sterility. *Everyman's Science*, 38(2): 84–86.

Prasad, T.A.V., Srinivas, T.G.N. Rafi and Reddy, D.C., 1991. Effect *in vivo* of atrazine on haematology and oxygen consumption in fish. *Telapia mossambica. Biochem. Int.*, 23(1): 157–161.

Rao, K.J., Baig, M.A. and Ramamurthi, K., 1984. Effect of a systemic pesticide phosphamidon on some aspects of freshwater fish, *Tilapia mossambica. Indian J. Environ. Hlth.*, 25(1): 60–64.

Reddy, P.M. and Bashamohideen, Md., 1989. Toxicity of synthetic pyrethroid insecticides fenvalerate and cypermethrin to the fish *Cyprinos carpio. Environ. Ecol.*, 7(4): 1016–1018.

Rita, J.J. Arockia and Milton, M.C. John, 2006. Effect of carbabate pesticide (Methomyl) on the biochemical components of the fish water *Oreochronis mossambicus* (Petera). *Indian J. Environ. and Ecoplex*, 12(1): 1–8.

Sastry, K.V. and Sharma, K., 1980. Mercuric induced haematological and biochemical anomalies is *Channa punctatus* (Bloch). *Toxicol. Letters*, 5: 245.

Sastry, K.V. and Siddiqui, A.A., 1983. Metabolic changes in the snake-head fish, *Channa punctatus* chronically exposed to endosulfan. *Water, Air and Soil Pollution*, 19: 133–141.

Singh, P.R., 2002. Studies on changes in blood of an air-breathing fish exposed to O.C. and O.P. *Ph.D. Thesis*, L.N.M.U., Darbhanga.

Sinha *et al.*, 1990. *Manual of Practical Biochemistry*. Scientific Book Company, Patna.

Srivastava, A.K., Singh, N.N. and Srivastava, A.K., 1995. Biochemical change in freshwater Indian catfish: Following exposure to sublethal concentration of propoxur. *J. Freshwater Biol.*, 7(4): 257–260.

Varley, H., Gowenlock, A.H. and Bell, M., 1980. *Practical Clinical Biochemistry, Vol. 1: General Topics and Commoner Tests*. William Heinemann Medical Books Ltd., London.

Verma, S.R., Bansal, S.K., Gupta, A.K. and Dalela, R.C., 1979. Pesticide induced haematological alterations in a freshwater fish, *Saccobrailachus fossilis*. *Bull. Environ. Contam. Toxicol.*, 22: 467–474.

Chapter 51

Effect of Continuous Cultivation and Nutrient Management for Seven Years on Soil Fertility Determinants and Performance of Maize in a Terraced Agroecosystem

B.S. Chauhan and Sewak Ram

Department of Soil Conservation,
School of Agricultural Sciences and Rural Development,
Nagaland University, Medziphema Campus, Medziphema – 797 106, Nagaland

ABSTRACT

Eleven treatment combinations involving farm yard manure (FYM), N, P and K (NPK) fertilizers, lime, surface soil, Alder (*Alnus nepalensis*) litter incorporated and burned were evaluated for their effect on soil fertility determinants and maize yield in a terraced agroecosystem under continuous cultivation and nutrient management. At the end of seven years, available N content increased significantly in all the treatments. Available N in FYM + NPK and FYM + NPK + Lime (577.0 kg/ha) was highest and surpassed the original levels (564.5 kg N/ ha) that existed beforeterracing the soil. Available P not only increased significantly but also surpassed the original level (4.1 kg P/ha) in all the treatments except FYM + 1/2 NPK and control. Maximum available P (14.6 kg/ha) accumulated in NPK treatment. Available K content increased in all the treatments and was highest in FYM + NPK + Lime (253.9 kg/ha). Available K levels also surpassed the original levels (166.5 kg K/ha) in all the treatments except control. Organic C content increased in all the treatments over control and reached highest (1.56 per cent) level in FYM + NPK + Lime. However, organic C content could not reach original level in any treatment. With the present rate of accumulation in organic C, the soils of the terraced agroecosystem under different nutrient management practices would take 22.0 to 44.0 years with an average of 33.6 years to reach original organic C level (2.2 per cent) that existed at the time of terracing. Addition of NPK fertilizers significantly decreased soil pH where as application of lime caused a significant increase in soil pH. FYM + NPK + Lime produced maximum grain yield followed by NPK + Lime and FYM + NPK. Addition of NPK fertilizers alone produced significantly lower yield than when applied in combinations with lime and/or FYM.

Keywords: *Available N, P and K, Soil organic C, Soil pH, Crop performance, Terraced agroecosystem.*

Introduction

Bench terracing, though expensive, is very effective means of soil conservation on slopes higher than 15 per cent. It permits sustainable economic yield to be maintained on long term basis under more intensive agricultural use than otherwise be possible. Bench terracing is used to manipulate surface topography of hill slopes to convert them to suit agriculture. Terracing encourages permanent fanning, besides conserving fertilizers, moisture and soil. Bench terracing usually exposes infertile and biologically inert subsoil of less desirable properties for crop growth as compared to top soil. Therefore, the production potential of terraced land is rather low at initial stage. Development of soil fertility of terraced land for sustained production however, is a time taking process. To increase immediate as well as long term productivity of the terraced land, sound fertility management practices are obviously needed. These practices should ensure a steady build up of soil fertility and other physico—chemical properties desirable for plant growth.

The deficiency of most plant nutrients in subsoil can be corrected through the addition of fertilizers, manures and other amendments either alone or in combinations. These materials however, increase the cost of production. Organic residues like straw, compost and farm yard manure (FYM) have been successfully used to improve soil tilth and increase the proportion of water stable crumbs in soil. Most of the phosphate and probably all the potassium FYM contains is as effective as fertilizer phosphate, but only one-quarter of the nitrogen is as effective as fertilizer nitrogen (Russell, 1973). The incorporation of both Alder litter and its ash to improve soil fertility has been as effective as FYM during the initial year of cultivation after terracing (Chauhan and Game, 2002). Management induced changes in major soil fertility determinants like available N, P, K and organic C content and soil pH would have long term agronomic and soil fertility conservation implications. The present investigation was carried out to study the effect of continuous cropping and nutrient management for seven years on soil fertility and maize yield on terraced land in Nagaland.

Materials and Methods

A field experiment was established in 1991 on newly developed bench terraces at the experimental farm of the School of Agricultural Sciences and Rural Development, Medziphema, Nagaland to assess long term effects of various management practices on soil fertility determinants and performance of maize on terraced land. The original surface soil contained 564.5 kg/ha available N, 4.1 kg/ha available P, 166.5 kg/ha available K, 2.2 (per cent) organic C before terracing. The pH of the original soil was 4.8. After terracing, available N, P, K, organic C content and pH of the resultant surface soil decreased. It contained 393.6 kg available N/ha, 2.9 kg availble P N/ha, 138.6 kg available K/ha and 1.7 per cent organic C and had a pH value of4.4. The sources of N, P, K fertilizers were urea, single super phosphate and muriate of potash and were applied each year @ 120 kg N, 60 kg P_2O_5 and 50 kg K_2O/ha, respectively. The FYM was applied @10 t/ha/yr. Lime as calcium carbonate was added every alternate year beginning 1991 @ 4000 kg/ha. Surface soil equivalent to 2.5 cm surface soil depth was applied/plot after the construction of the terraces. Alder tree (*Alnus nepalensis*) leaves collected from the forest floor were applied @ 5.0 t/ha/yr after drying. The entire dose of P and K and one-third dose of N was applied at the time of final land preparation. Remaining two-third N was applied in two equal splits at 30 and 60 days after sowing (DAS). Alder litter, FYM and lime were applied one month before final land preparation and mixed well. The Alder litter was burnt on soil surface and the ash mixed well in the soil. Since then, all the plots continued to receive the same treatment every year till 7th year (1997). During this period, maize (*Zea mays* L.), (Vijay composite variety) was cultivated each year. After the harvest of 7th year crop, soil samples from each plot were collected and analysed for available N, P, K and organic C content and soil pH to evaluate the long term effect of continuous cultivation and nutrient management on soil fertility determinants and maize performance. The available N was determined by alkaline permanganate method of Subbiah and Asija (1956). The available P content was extracted by Bray's method No. 1 (Bray and Kurtz, 1945) and was estimated colorimetrically as described by Jackson (1973). The available K content in the soil was extracted by neutral normal ammonium acetate and was determined flame photometrically (Jackson, 1973). The organic C content of the soil was determined by Walkley and Black's method (Jackson, 1973).

Results and Discussion

Available N, P, K and Organic C Content and Soil pH

The data on the cumulative effect of different treatments on available N, P and K and organic C content and soil pH in soil after seven years of continuous cultivation of maize and nutrient management in a terraced agroecosystem, is presented in Table 51.1. The data showed that the available N content in soil increased significantly in all the treatments as compared to control which showed minimum available N. The maximum available N was recorded in FYM + NPK + Lime and FYN + NPK treatments (577.0 kg/ha). The available N content in these treatments surpassed the original level (564.5 kg N/ha) that existed before terracing. In other treatments, the available N levels were lower than original level after seven years of nutrient management and cultivation. The available N content of NPK + Lime, surface soil + FYM and NPK treatments were at par. Treatment FYM + 1/2 NPK showed a significant increase in available N level when compared with surface soil. Alder litter incorporated and FYM treatments showed significantly higher levels of available N when compared with Alder litter burned. Available N content in surface soil + FYM was significantly higher than surface soil alone. Only the addition of FYM + NPK and FYM + NPK + Lime continuously for seven years after terracing could bring about an increase in the N fertility status of the soil from medium to high. Other treatments were not effective in doing so. An analysis of the data (Table 51.1) showed that after seven years of continuous cultivation and nutrient management, available N in different treatments increased in variable amounts ranging from 83.5 to 284.3 kg N/ha over control with an average of 195.6 kg N/ha. The, rate of build up of available N in various nutrient management practices was estimated to be 11.9 to 40.6 kg N/ha/yr with an average of 27.9 kg N/ha/yr. Treatments FYM + NPK and FYM + NPK + Lime resulted in highest build up of available N. As the N mainly accumulates in organic form (Foth, 1990), the increase in available N on addition of fertilizer N together with FYM with or without lime and also on incorporation of Alder litter is expected. Significant increase in available N on addition of NPK + lime with or without FYM is in agreement with the fmdings reported earlier (Mongia *et al.*, 1997; Borthakur and Mazmudar, 1968). Powlson *et al.* (1987) also reported that incorporation of straw for 18 years caused an increase in the quantity of mineralizable N in soil by 40–59 per cent and total N by 10 per cent. The Alder litter burned showed minimum increase in available N. Fire induced changes in the nature of soil humus rendering it more mineralisable may be partly responsible for the increase in available N in this treatment.

Table 51.1: Changes in the Levels of Soil Fertility Determinants in Different Treatments After Seven Years

Treatments	*Available N (kg/ha)*	*Available P (kg/ha)*	*Available K (kg/ha)*	*Organic C (Per cent)*	*Soil pH*
Control	292.7	2.8	149.3	0.8	4.6
FYM	404.5	5.8	238.9	1.24	4.7
NPK	514.3	14.6	197.7	1.21	4.3
FYM + NPK	577.0	12.0	186.7	1.43	4.7
NPK + Lime	536.2	8.2	175.5	1.31	7.1
FYM + NPK + Lime	577.0	11.0	253.9	1.56	7.3
FYM + 1/2NPK	491.7	3.2	190.4	1.23	4.7
Alder litter incorporated	437.1	6.2	235.2	1.51	4.7
Alder litter burned	376.2	4.9	209.1	1.28	4.9
Surface soil	439.1	5.5	197.9	1.30	4.8
Surface soil + FYM	530.0	5.8	241.2	1.35	4.7
SEm±	8.1	0.5	8.0	0.03	0.06
CD (P = 0.05)	26.4	1.5	26.1	0.11	0.20

The data (Table 51.1) showed that after seven years, available P content in soil in all the treatments except NPK + 112 FYM increased significantly when compared with control which showed minimum level. The available P content in control did not differ significantly with FYM + 1/2 NPK. Application of NPK fertilizers with or without lime or FYM or both caused a significant increase in available P content. The highest available P was recorded in NPK (14.6 kg/ha) treatment followed by NPK + FYM and FYM + NPK + Lime. The available P content in NPK treatment was significantly higher than all the other treatments. The data clearly established that increase in available P was relatively higher in treatments where NPK fertilizers either alone or with FYM or with lime or both were applied. These findings are in accordance with those observed by earlier workers (Olson *et al.*, 1970; Khiani and More, 1984). The available P content in Alder litter incorporated, surface soil + FYM, FYM, surface soil and Alder litter burned did not differ significantly. The available P contents surpassed the original levels (4.1 kg/ha) in all the treatments except FYM + I/2 NPK and control. Addition of NPK, FYM + NPK and FYM + NPK + Lime continuously for seven years increased the phosphorus fertility status of terraced land from low to medium. Other treatments were not effective in doing so and caused only a slight increase in available P. The difference in the levels of available P in the Alder litter incorporated and burned treatments were not significant. However, these treatments showed significantly higher available P levels than control. During the last seven years, available P content increased in variable amounts ranging from 0.4 to 11.8 kg P/ha in different treatments over control with an average of 4.9 kg P/ha. The rate of build up of available P was estimated to be 0.06 to 1.7 kg P/ha/yr with an average of 0.7 kg P/ha. Application of fertilizer P resulted in higher available P build up as compared to small build up observed on addition of FYM and forest litter. Maximum increase in available P in NPK treatment established that addition of fertilizer P alone had resulted significantly higher build up of available P in soil. Phosphorus added without energy source has been found to accumulate mainly in inorganic (Pi) form (Chauhan *et al.*, 1981).

The data on the available K content in soil (Table 51.1) after seven years of continuous cultivation and nutrient management showed that all the treatments caused a significant increase in available K. The addition of FYM + NPK + Lime resulted maximum available K content (253.9 kg/ha) in soil followed by surface soil + FYM, FYM and Alder litter incorporation. However, the differences among these treatments were not significant. Addition of NPK fertilizers alone or with or without FYM or lime showed lower levels of available K as compared to FYM, Alder litter incorporated and burned and surface soil + FYM. This perhaps may be due to higher crop yield in treatments with NPK fertilizer (Table 51.2) as compared to non-fertilizer treatments. The higher yield obviously, is suggestive of larger amount of K used in crop production as well as its removal from soil in crop harvest. The available K levels in surface soil, NPK, FYM + 1/2 NPK, FYM + NPK and NPK + Lime were recorded at par. Alder litter incorporated showed higher available K content over Alder litter burned but the difference among these treatments were not significant. These observations are similar to those reported earlier (Chauhan and Game, 2002). Available K levels surpassed the original levels (166.5 kg/ha) in all the treatments except control. Available K content in Alder litter burned treatment was significantly higher than control. Similar findings were observed by Donnaar and Schaber (1985). They observed that burning led to small but significant increase in extractable K levels in the soil immediately after fire. Continuous addition of various treatments for seven years resulted in the build up of available K levels in all the treatments over control. However, none of the treatment could increase the fertility status of soil which remained medium after seven years. During the last seven years, the build up of available K varied from 26.2 to 104.6 kg K/ha in different treatments with an average of 63.4 kg K/ha. The build up of available K was recorded highest in FYM + NPK + Lime treatment followed by surface soil + FYM, FYM and Alder litter incorporated and burned. The rate of build up of available K in different treatments was estimated to be 3.7 to 14.9 kg K/ha/yr with an average of 9.0 kg K/ha/yr. These results together with the fact that K does not complex with organic compounds in soil suggested that the combine effect of uptake, loss, chemical transformation of added fertilizer K and the release of K present in organic residues added on its decomposition, would mainly govern the magnitude of K build up in soil under different treatments and its contribution towards the available K pool.

All the treatments had caused a significant increase in the organic C content (Table 51.1) of the soil as compared to control. The highest organic C content was recorded with the application of FYM + NPK + Lime (1.56 per cent) followed by Alder litter incorporation. However, the difference among these treatments was not significant. The difference in the organic C content in FYM + NPK and surface soil + FYM treatments was also not significant. The differences in surface soil, Alder litter burned, FYM, FYM + 1/2 NPK and NPK treatments were also not significant. The Alder litter burned treatment showed higher organic C levels than control. The fertility status in respect of organic C in all the treatments did not change after terracing and remained high. All the treatments showed accumulation of organic C over control. However, after seven years of cultivation and management, organic C levels could not reach the original levels that existed before (2.2 per cent) or after (1.7 per cent) terracing in any treatment. During the last seven years, organic C accumulated in different treatments in variable amounts ranging from 0.35 to 0.70 per cent over control with an average of 0.48 per cent. The rate of build up of organic C in these treatments worked out to be 0.05 to 0.10 per cent, organic C/ha/yr with an average of 0.069 per cent organic C/ha/yr. The data clearly established that even the organic C levels that existed after terracing could not be sustained under different nutrient management practices. This may be partly due to the fact that the organic matter crop residue returns and additions in these treatments were not sufficient to sustain the level of organic C that existed after terracing. The nature of organic C present in the resultant surface soil after terracing may also be partly responsible for its accelerated decomposition during initial years of cultivation. Continuous cultivation would contribute towards the depletion in the soil organic C levels. It has been established that the loss of soil organic C is most rapid in first 10 to 20 years of cultivation and more gradual afterwards (Voroney *et al.*, 1981). The organic C accumulation in all the treatments over control however, suggested that N and P would also be accumulating with it. The quantity of these nutrients accumulating with each molecule of organic C would vary with the treatment as the levels of labile pool affect the quality of organic matter with respect to N (Campbell and Souster, 1982) and P (Hedley *et al.*, 1982; Chauhan *et al.*, 1981). Thus the contribution of each treatment towards the available N and P pool would also vary. Further, the accumulation of available nutrients in soil is a combined effect of the processes like immobilization, mineralization, redistribution, uptake and loss. The variations in the magnitude of these processes within the treatments obviously would result in accumulation of variable amounts of available nutrients in the soil as observed in this investigation. The facts that available N increased in excess of original level present before terracing in FYM + NPK and FYM + NPK + Lime treatments and that organic C did not increase to original level, together suggested that organic matter of high quality with narrower C:N ratio would be accumulating in these treatments at this stage. Similarly, organic matter of lower C: P ratio would also be accumulating in all the treatments other than FYM + 1/2 NPK and control. The implication of this is that those treatments where organic matter of high quality with lower C:N and C:P ratio would be accumulating are relatively better than others for long term fertility conservation of terraced land in respect of N and P.

Table 51.2: Effect of Various Treatments on Grain Yield and Relative Efficiency of Crop Production

Treatments	*Grain Yield (kg/ha)*	*Relative Efficiency (Per cent)*
Control	0.0*	0.0*
FYM	1754.3	38.2
NPK	4598.0	100.0
FYM + NPK	5442.9	118.4
NPK + Lime	6848.2	148.9
FYM + NPK + Lime	7842.6	170.6
FYM + 1/2NPK	5079.3	110.5
Alder litter incorporated	1589.7	34.6
Alder litter burned	956.3	20.8
Surface soil	715.7	15.6
Surface soil + FYM	1982.3	43.1
S Em±	270.5	
CD (P = 0.05)	884.8	

*Plants could not survive beyond 60 DAS.

The data (Table 51.11) showed that addition of NPK fertilizers caused a significant decreased in soil pH (4.3) as compared to Control (4.6). The application of NPK + Lime, FYM + NPK + Lime and forest litter burned caused the significant increase in soil pH (7.1, 7.3, and 4.9, respectively). Among the treatments, NPK treatment showed the lowest pH and FYM + NPK + Lime showed the highest pH. The data led to conclude that the addition of NPK fertilizers caused a significant decreased in soil pH where as addition of lime resulted in a significant increase soil pH. Similar observations have been reported earlier (Chauhan and Game, 2002). The decrease in soil pH on addition of NPK fertilizers has been well established. The release of H^+ during biological oxidation of ammonia and ammonium ions of nitrogenous fertilizer contributes to soil acidity (Foth, 1990). The data further established that the addition of lime caused an increased in the pH of the soil. This may be due to the fact that lime would hydrolyse to produce OH^- and thereby increase the base saturation of soil resulting in an increase in soil pH. The increase in pH on burning of Alder residue is similar to the finding of Doormar and Schaber (1985). They reported that burning of residue on soil surface caused a significant increase in soil pH. After seven years, application of lime caused an increase in the pH of the soil by 2.7 units in FYM + NPK + Lime and 2.5 units in NPK + Lime with an average of 0.4 units/year.

Performance of Maize

The data pertaining to the effect of different treatments on the grain yield of maize (Table 51.2) clearly showed that the grain yield increased in all the treatments except Alder litter burned when compared with surface soil treatment which produced lowest yield (715.7 kg/ha). In control however, no yield was recorded due to non survival of plants beyond 60 DAS. Maximum grain yield (7842.6 kg/ha) was recorded in FYM + NPK + Lime which was significantly higher as compared to other treatments. The grain yield in FYM + NPK and FYM + 1/2 NPK did not differ significantly. These treatments however, produced significantly lower yield as compared to NPK + Lime. The yield in NPK and FYM + 1/2NPK was significantly higher than that observed in FYM. The yield in surface soil + FYM, FYM and forest litter incorporated treatments did not differ significantly. Also, the grain yield in surface soil and forest litter burned treatments did not differ significantly. The data established that the addition of NPK fertilizers in combination with FYM or lime or both produced higher yield than when applied alone. Among the combinations, NPK + FYM + Lime produced maximum grain yield followed by NPK + Lime. Similar results have also been reported earlier (Chauhan and Game, 2002). Further, the application of NPK alone produced lower yield than that obtained with NPK + Lime, FYM + NPK and FYM + 1/2NPK. These findings are similar to those observed by Sharma (1983) and Bhatia and Shukla (1978).

The computation of relative efficiency (Table 51.2) of crop production in different treatments based on the yield response over NPK (taking NPK as 100) established that the grain yield in FYM + NPK + Lime, NPK + Lime, FYM + NPK and FYM + 1/2NPK was 170.6, 148.9, 118.4 and 110.5 per cent higher over NPK treatment, respectively. From the relative efficiency and yield data (Table 51.2), the treatment could be grouped into three categories in respect of their productivity. First category includes FYM + NPK + Lime which produced maximum yield which was significantly higher than NPK. The second category consists of NPK + Lime, FYM + NPK and FYM + 1/2NPK which produced appreciably higher yield than NPK. The third category includes FYM, Forest litter incorporated, Forest litter burned, surface soil + FYM, surface soil and control which either produced nothing or significantly less than NPK. The findings led to conclude that FYM + NPK + Lime is the best nutrient management practice from the point of soil fertility development and maize yield in terraced agroecosystem followed by NPK + Lime, FYM + NPK. In situations, where NPK fertilizers are neither adequately available nor affordable, FYM + 1/2NPK could be a better alternative as compared to FYM, surface soil + FYM and Alder litter incorporated and burned. Addition of FYM alone, incorporation of Alder litter and surface soil + FYM could not increase soil available N, P and K contents to the levels necessary for proper crop growth and produced lower yields with relative efficiency ranging from 34.6 to 43.1 per cent.

The magnitude of development of soil fertility determinants varied among the treatments. The variations in available N, P and K, organic C content and soil pH in each treatment is perhaps because of the combined effect of the source and amount of the nutrients and amendment added, nutrients loss from soil including crop removal, nature of compounds synthesized during chemical and biological transformations of added nutrients and nature of reaction products and their contribution towards available nutrient pool and soil conditions favourable for plant growth. The combined effect of these factors would affect the amount of nutrients available and their utilization by crop for its growth and yield. The nutrients added to the soil and not utilized by the crop due to poor soil conditions would accumulate in soil till proper plant growth is favoured. A good crop on the other hand would deplete the available nutrient pool to produce higher yields and also return simultaneously larger quantity of organic residue to the soil to help maintain or increase its organic C content. Management practices in which crop performance remains relatively poor and the increase in available N, P and K contents is induced by consequences of poor soil conditions can not be considered agronomically sound and hence should not be recommended for fertility development and crop production in a terraced agroecosystem. With the present rate of accumulation of soil organic C content in various treatments over control, it is estimated that the soils of the terraced agroecosystem under different nutrient management practices will take 22.0 to 44.0 years with an average of 33.6 years, to reach original organic C level that existed at the time of terracing. At that stage, these soils would perhaps be at a dynamic equilibrium content of organic C where additions balance losses; a steady state. The C : N : P ratio at this point may be expected to stabilize the vicinity of 100 to 120 : 10 : 1. Till such time the impact of terracing induced changes in soil properties perhaps would not diminish completely and continue to influence crop performance. These findings corroborate with the observations made by Smith, et al. 1967. These authors reported that average grain yield of sorghum was 30 per cent less 30 years after desurfacing (cited from Burnett, *et al.*, 1985). Comparatively, the effect of terracing on soil properties is expected to last for longer duration as compared to desurfacing.

Liming may be practiced for maize production on the terraced agroecosystem as incorporation of 16000 kg/ha of lime in seven years have increased the pH of the soil from 4.4 to 7.1 (NPK + Lime) and 7.3 (FYM + NPK + Lime) and produced higher yield among treatments. This amount of lime which has brought about a change in soil pH value favourable to crop growth could also be applied by adopting other suitable liming schedules keeping in view the availability and affordability of lime for the cultivation of maize during the initial years on terraced land to bring about similar change in soil pH. Thereafter, liming should be practiced only to sustain the pH values.

References

Bray, R.H. and Kurtz, L.T., 1945. Determination of total, organic and available forms of phosphorus in soils. *Soil Sci.*, 9: 39–45.

Borthakur, H.P. and Mazmudar, N.N., 1968. Effect of lime on nitrogen availability in paddy soil. *J. Ind. Soc. Soil Sci.*, 16: 143–147.

Bhatia, K.S. and Shukla, K.K., 1978. Effect of continuous use of fertilizer and manure and fertility status of eroded alluvial soil. *Ind. J. Soil Conserv.*, 6: 94–97.

Burnett, E., Stewart, B.A. and Black, A.L., 1985. Regional effects of soil erosion on crop productivity: Great plains. In: *Soil Erosion and Productivity*, (Eds.) R.F. Follett and B.A. Stewart. ASA, Inc., CSSA,

Inc., SSSA, Inc., Publishers, Madison, Wisconsin, USA, pp. 294–295.

Campbell, C.A. and Souster, W., 1982. Loss of organic matter and potentially mineralizable nitrogen from Saskatchewan soils due to cropping. *Can. J. Soil Sci.*, 62: 651–656.

Chauhan, B.S. and Game, Tadu, 2002. Effect of various management practices on soil fertility and performance of maize (*Zea mays* L.) on newly constructed bench terraces. *Indian Agri.*, 46: 55–63.

Chauhan, B.S., Stewart, J.W.B. and Paul, E.A., 1981. Effect of labile inorganic phosphate status and organic carbon additions on the microbial uptake of Phosphorus in soil. *Can. J. Soil Sci.*, 61: 373–385.

Dormaar, J.F. and Schaber, B.D., 1985. Chemical properties of soil as affected by a single burn of alfalfa stubble to control insect. *Can. J. Soil Sci.*, 65: 357–361.

Foth, H.D., 1990. *Fundamentals of Soil Science*, 8th edn. John Wiley and Sons, Singapore.

Hedley, M.J., Stewart, J.W.B. and Chauhan, B.S., 1982. Changes in inorganic and organic soil phosphorus fractions induced by cultivation practices. *Soil Sci. Soc. Am. J.*, 46: 970–976.

Jackson, M.L., 1973. *Soil Chemical Analysis*. Prentice Hall of India Pvt. Ltd., New Delhi.

Khiani, K.N. and More, D.A., 1984. Long-term effect of tillage operation and FYM application on soil properties and crop yield in a vertisol. *J. Ind. Soc. Soil Sci.*, 32: 392–393.

Mongia, A.D., Singh, N.T., Mandal, L.N. and Guha, A., 1997. Effect of soil water regimes and liming on chemical changes and nutrient transformations in acid and acid sulphate soils. *J. Ind. Soc. Soil Sci.*, 45: 344–348.

Olson, R.N., Hensler, R.F. and Attoe, O.J., 1970. Effect of manure application, aeration and soil pH on soil nitrogen transformation and on certain soil test values. *Soil Sci. Soc.* Am. Proc., 34: 222–225.

Powlson, D.S., Brookes, P.C. and Christensen, B.T., 1987. Measurement of soil microbial biomass provides an early indication of changes in total soil organic matter due to straw incorporation. *Soil Biology and Biochemistry*, 19: 159–164.

Russell, E.W., 1973. *Soil Conditions and Plant Growth*, 10th edn. Longman, London.

Sharma, J.P., 1983. Economy in fertilizers use through organic manures in growing maize. *Ind. J. Agron.*, 25: 154–155.

Smith, R.M., Henderson, R.C., Cook, E.D., Adams, J.E. and Thompson, D.O., 1967. Renewals of desurfaced Austin clay. *Soil Sci.*, 103: 126–130.

Subbiah, B.V. and Asija, G.L., 1956. A rapid procedure for the determination of available N in soils. *Curr. Sci.*, 25: 259–260.

Voroney, R.P., Van Veen, J.A. and Paul, E.A., 1981. Organic C dynamics in grassland soils. 2. Model validation and simulation of the long term effects of cultivation and rainfall erosion. *Can. J. Soil Sci.*, 61: 211–224.

Chapter 52

Changes in Blood Glucose and Tissue Glycogen Level of *Clarias batrachus* Due to Lead Toxicity

Shamboo Mandal and A.N. Jha
Department of Zoology, C.M. Science College, L.N. Mithila University, Darbhanga – 846 004

ABSTRACT

During the present study, the blood glucose level for sub chronic period showed significant increase by 11.14 per cent. Where as chronic exposure showed glucose depletion by 35.91 per cent. The hepatic and gonadal gycogen showed depletion in sub chronic as well as chronic period of exposure but the significant decline of hepatic and testis glycogen was noticed under chronic exposure as compared to ovary glycogen. Hyperglycaemia is supposed to be due to glycogenolysis where as hypoglycemia can be related to the utilization of reserve glycogen and impaired glycogenesis. The tissue glycogen reduction in present study is reflection of tissue acidosis and break down of glycogen in stress.

Introduction

Work in heavy metal toxicology has been done (Forstner *et al.*, 1979; Goyer,1986; Roy and Dubey, 2001). The alterations in biochemical profiles of fish on lead exposure have also been worked out by Jha and Pandey (1989), Jha (1991) and Jha (2003). In recent years lead pollution attracted attention in view of its large scale industrial use and toxic potential. But, a critical review of previous works reveal that most of the works are confined to bioassay and T.L.M. determination (Valee an Ulmer, 1972; Sexena and Parashari, 1983). In the light of above facts the fish, *Clarias batrachus* (Linn.) commonly called "Magur" was selected for biochemical study under sub lethal concentration of lead acetate (237.80 mg/l) for the sub chronic (15 days) period and chronic (30 days) period.

Materials and Methods

A static bio-assay test was done to determine LC_{50} of lead acetate to *C. batrachus* following the methods of APHA (1985) and sublethal concentration was calculated by adopting the formula of Hart *et al*. (1945). For each experiment the fish (average length 22–24 cm and wt. 38–43 gm were exposed to a sublethal concentration of lead acetate (237.80 mg/l) for a period of 15 and 30 days. Side by side a control was also run in equal volume of water (pH–7.6, Temp.–34.6°C, hardness in mg/l as $CaCO_3$–194.22).

The exposure medium was renewed every 24 hours. At the end of exposure period that is on days 15th and 30th days the fish were anaesthetized with 1 : 4000 MS 222 (tricane methane sulphonate, sandoz) and then processed for quantitative estimation. Blood glucose was estimated by *O*-toluidine reagent (Dubowski, 1962) at 630 nm in spectrophotometer (spectronic 20). The tissue glycogen was estimated by the method of Carrol *et al.* (1956) at 620 run in spectrophotometer 1974). In present study statical analysis and t 'test' (Campbell, 1974) were done.

Results and Discussion

Blood Glucose

Both 15 and 30 days exposed fish exhibits interesting results in the blood glucose profiles. These are being mentioned below under separate subheadings.

15 Days Treatment Period

The fish of this exposure period reflects hyper glycaemic response. The blood glucose level elevates to 68.09±0.08 mg/100ml of blood against the control value of 861.26±0.28 mg/100ml of blood. This amounts to an increase by 11.14 per cent and is significant at P<0.05 (Table 52.1)

Table 52.1: Profiles of Blood Glucose in Normal and Lead Acetate Treated Fish, *Clarias batrachus* (Values are ±SE of five fish in each group)

Parameter	*Days*	*Control*	*Treated*	*Student 't' Test P Value*	*Per cent Increase (+) or Decrease (–)*
Blood glucose (mg/100 ml)	15	61.26±0.28	68.09±0.08	<0.05	+ 11.14 per cent
	30	63.63±0.14	40.78±0.13	<0.001	–35.91 per cent

30 Days Treatment Period

Contrary to the observations of 15 days exposure period, the fish of this group become hypoglycaemic denoting exhaustion of glucose to meet high energy demand under the toxic stress of lead. The extent of glucose depletion in highly significant at p<0.001 and has been calculated to be nearly 35.91 per cent. The blood glucose level in control fish is 63.63±0.14 mg/100ml and it lowers down to 40.78±0.13mg/100ml of blood.

Glycogen

The alteration in the glycogen content of liver, testis and ovary has been shown in Table 52.2. In all these tissues there has been a significant depletion in the glycogen content in both the exposure periods (15 and 30 days). The depletion has been recorded to be more pronounced in 30 days exposed fish. The liver reveals maximum depletion in both exposure period amounting to 17.62 per cent and 37.28 per cent respectively (Table 52.2).

Table 52.2: Profiles of Glycogen Content (mg/g wet tissue ±SE of five fish) in Liver, Testis and Ovary of *Clarias batrachus* following Sublethal Treatment of Lead Acetate

Parameter	*Days*	*Control*	*Treated*	*Student 't' Test P Value*	*Per cent Increase (+) or Decrease (–)*
Liver glycogen	15	35.07±0.05	28.89±0.22	<0.001	–17.62 per cent
	30	36.90±0.12	23.14±0.29	<0.001	–37.28 per cent
Testis glycogen	15	18.18±0.17	13.13±0.10	<0.001	–27.77 per cent
	30	17.16±0.18	12.98 0.14	<0.001	–24.35 per cent
Ovary glycogen	15	16.10±0.21	12.98±0.10	<0.001	–19.37 per cent
	30	17.92±0.15	12.96±0.06	<0.001	–27.67 per cent

15 Days Treatment Period

The glycogen content of liver, testis and ovary of the control fish group has been estimated as 35.07±0.05, 18.18±0.017 and 16.10±0.21 mg/g wet tissue respectively. Where as in the experimental group it depletes to 28.89±0.22, 13.13±0.10 and 12.98±0.10 mg/g wet tissue in liver, testis and ovary respectively. All these values significant depletion at $P < 0.001$ (Table 52.2).

30 Days Treatment Period

Glycogen profiles of the treated fish group have been estimated as 23.14±0.29, 12.98±0.14 and 12.96±0.06 mg/g wet tissue in liver, testis and ovary as against their control values of 36.90±0.12, 17.16±0.18 and 17.92±0.15 mg/g wet tissue respectively. The glycogen depletion has found to be highly significant ($p < 0.001$) in liver and testis as compared to that of ovary (Table 52.2).

The present study provides interesting results in blood glucose and tissue glycogen levels Table 52.1 and 52.2 reveal that there has been a significant increase in the blood glucose level accompanied by a highly significant decrease in hepatic and gonadal glycogen content in the lead treated fish sacrificed on day 15. However, 30 days exposed fish unlike their 15 days counterparts showed hypoglycaemia and highly significant glycogen depletion amounting to 37.28 per cent, 24.35 per cent and 27.67 per cent, in liver, testis and ovary respectively. The highest decrease in glycogen has been recorded in the hepatic tissue for both the chronic periods. Liver is the seat of metabolism and site of glycogenesis of storage where its maximal loss seems to be convincing under adverse situations. The significant elevation in blood glucose level in 15 days treated fish points towards a hyper metabolic state of the fish in response to lead intoxication. Thus glucose up take of the fish is increased to meat the excessive demand of energy under metalic stress. At the same time hypoglycaemic condition seen in 30 days treated fish lead us to think of the two options, one being the increased ability of fish to tolerate lead stress and the another being exhaustion of the reserve energy.

The present results on the tissue glycogen and blood glucose resemble with those of Gill and Pant (1981). Larsson and Haux (1982) observed cadmium induced significant increase in blood glucose level in rainbow trout and flounder. They correlated this with reduced insulin secretion and stimulated gluconeogenesis. This results can further be supported by the contention that metals accumulate in fish pancreas and cause necrotic lesion in the insulin producing B-cell in the islet of Langerhans (Havu, 1969).

Grag *et al.* (1989) observed hyperglycaemia in manganese treated *Heteropneustes fossilis*. Decreased glycogen content in liver and Kidney of the manganese exposed *Ophiocephalus puctatus* has also been recorded (Garg and Garg, 1990). The European eel, *Anguilla anguilla*, showed hyperglycaemia under penta chlorophenol exposure as low as 0.1 ppm and interestingly elevated blood glucose level persisted even after 55 days of recovered in clean water (Holmberg, 1972). A few explanations have been given by number of workers to accounts for glycogen depletion and blood glucose elevation. Under the observation of toxic impacts of various pollutants in general and metals in particular which would make this presentation more clear and therefore, needs to be discussed. Sastry and Gupta (1978b, 1979c) in mercury and lead treated ophiocephalus and lead cadmium treated *Heteropneustes fossilis* noted a decreased activity of glucose-6 phoshatase in their digestive system. Similar results were obtained by Chandra and Imam (1973) in gastrointestinal mucosa of guinea pig treated with manganese. Glycogen decreased as noted in present study may also be associated with inhibition of glucose-6-phosphatase activity due to lead toxicity. Particularly in case or liver it may be associated with the lost capacity of hepatocytes to mobilize glycogen normally as suggested by Orr *et al.* (1948).

Tissue hypoxia and alteration in buffering system resulting in tissue acidosis and breakdown of glycogen as the possible reasons breakdown of glycogen as. The possible reasons for glycogen (Jafri and Shaffi, 1975).

It is supposed that lead acetate might be enhancing the activity for key gluconeogenic enzymes, which may also be contributing to the hyperglycaemia in the present case. Another possibility may be that there might have occurred a reduction in calcium permeability at the level of B-cell membrane, which would inhibit insulin release since calcium is a prerequisite for insulin secretion.

Elevated blood glucose level recorded in the present observation suggests that lead influences the glucose homeostasis. In addition, it also seems to affect ion balance as suggested by Larsson *et al.* (1967). Stressful condition stimlulates the rapid secretion of both glucocorticoids (Wedemeyer, 1969; Grant and Mehrle (1973) and catecholamines (Nakano and Tomblinson, 1967) from the adrenal tissue of the fish. Both the hormones induce hyperglycaemia (Robertson *et al.*, 1963; Oguri and Nace, 1966 and Larsson, 1973)

Mazeaud *et al.* (1977) suggested that all types of stress in fish results in increase the circulating catecholamines and plasmatic corticosteroid referred to as the primary effects. These endocrine changes induce secondary biochemical responses such as pronounced increase in blood sugar level.

Similar to the present observation Bhattacharya *et al.* (1987) recorded a depletion in blood glucose level on day 30 of the mercury exposure in *Channa punctatus*, they have opined that hepatic glycogen is the source of hyperglycaemia as maximal glycogen depletion corresponded to a dramatic increase in blood glucose level. It is presumed that xenobiotic insult is responsible for glucose elevation and glycogen depletion.

The depletion of blood glucose as recorded on day 30 of exposure in the present study supports the contention that hexose is given to other tissue and glycogenolysis can follow biochemical pathway differing from those of glucose synthesis (Bhattacharya *et al.*, 1987).

The gluconeogenic activity might be mediated by cortisol (Bentley, 1980; Lidman *et al.*, 1979) and altered cortisol may sufficiently enhance gluconeogenesis. Besides, an increased influence of other hormone such as glucagon, in the situation of insulin deficiency cannot be excluded as the reason for obviously higher glycogenolytic activity in lead acetate exposed *Clarias batrachus*. Thus diverse interpretation can be given for glycogen depletion and blood glucose elevation. No single reason can, explain the changes noted during the present study. At last we can say that there exists a high catabolic potency during lead acetate exposure. This metal might have caused severe anaerobic stress resulting in breakdown of tissue glycogen.

Biochemical observation denoted a hyperglycaemic response in 15 days exposed fish whereas the 30 days treated fish became hypoglycaemic. Hyperglycaemia has been attributed to the glycogenolysis and gluconeogenesis whereas hypoglycaemia has been associated with utilization of reserve glycogen and inhibition of glycogen synthesis. The glycogen profile of liver, testis and ovary showed highly significant decrease in both the exposure periods and has been associated with tissue acidosis and breakdown of glycogen.

References

APHA, 1985. *Standard Method for the Examination of Water and Wastewater*, 16th edn. American Public Health Association and Water Pollution Control Federation, American Public Health Association, Washington.

Bentley, P.J., 1980. *Comparative Vertebrate Endocrinology*. Cambridge University Press, Cambridge, London, New York and Melbourne, pp. 145.

Bhattacharya, T., Roy A.K. and Bhattacharya, S., 1987. Blood glucose and hepatic glycogen interrelationship in *Channa punctatus* Bloch.: A parameter of non lethal toxicity Bioassay with industrial pollutants. *Ind. J. Exp, Biol.*, 25: 539–541.

Campbell, R.C., 1974. *Statistics for Biologists*. Cambridge University Press, London, pp. 385.

Carrol, N.V., Longley, R.W. and Roe, J.H., 1956. The determination of glycogen in liver muscle by use of anthrone reagent. *J. Biol. Chem.*, 220: 583–593.

Chandra, S.V. and Imam, Z., 1973. Manganese induced histochemical alteration in gastrointestinal mucosa of guinea pig. *Acta Pharmacol. Toxicol.*, 33: 449–458.

Dubowski, K.M., 1962. An O-toluidine method for body fluid glucose determination. *Clin. Chem.*, 8: 215–235.

Forsner, U. and Wittmann, G.T.W., 1979. *Metal Pollution in the Aquatic Environment*. Springer-Verlag, Berlin, Heidelberg, New York, pp. 436.

Garg, S.K. and Garg, V.K., 1990. Metabolic changes in the snake head fish, *Ophiocephalus punctatus* (Bloch.) to experimental manganese poisoning. In: *Environmental and Pollution and Health Hazard*, Environmental Series, 3: 145–149.

Garg, V.K. and Tyagi, S.K., 1989. Manganese induced haematogical and biochemical anomalies in Indian cat fish *Heteropneustes fossilis*. *J. Environ. Biol.*, pp. 349.

Gill, T.S. and Pant, J.C., 1981. Toxicity of nickel to fish *Puntius conchonius* (Ham.) and its effect on blood glucose and liver glycogen. *Comp. Physiol. Ecol.*, 6: 99–102.

Goyer, R.A., 1986. Toxic effects of metals. In: *Casarett and Doull's Toxicology: The Basic Science of Poison*, 3rd edn, (Eds.) J. Doull, G.D. Klassen and M.O. Amdur. MacMillan, New York, pp. 599.

Grant, B.F. and Mehrle, P.M., 1973. Endrintoxicosis in rainbow trout (*Salmo gairdneri*). *J. Fish Res. Board. Can.*, 30: 31–40.

Hart, W.B., Doudoroff, P. and Greenbank, J., 1945. *The Evaluation of Toxicity of Industrial Wastes, Chemicals other Substances of Freshwater Fishes*. Atlantic Refining Co. Phil. Part (1): 317–326.

Havu, N., 1969. Sulphydryl inhibitors and pancreatic islet tissue. *Acta. Endocrinol. Suppl.*, 139: 1–231.

Holmberg, B., 1972. Metabolic effects of technical pentachlorophenol (PCP) on the eel, *Anguilla anguilla. Comp. Biochem. Physiol.*, 43: 171–183.

Jafri, A.K. and Shaffi, S.A., 1975. Effects of asphyxiation on the serum and kidney alkaline phosphatase activity in *Clarias batrachus* (Linn.). *Broteria*, 44: 59–62.

Jha, A.N., 2003. Alteration induced by lead toxicity in blood glucose and tissue glycogen level of *Heteropneustes fossilis* (Bloch.).

Jha, B.S., 1991. Alteration in the protein and lipid contents of intestine, liver and gonads in the lead exposed fresh water murrel, *Channa punctatus* (Bloch.). *J. Ecobiol.*, 3: 29–34.

Jha, B.S. and Pandey, S., 1989. Alteration in the total carbohydrate levels of intestine, liver and gonads induced be lead nitrate in the fish, *Channa punctatus*. In: *Environmental Risk Assessment*, (Eds.) Y.N. Sahai, P.B. Deshmukh, T.A. Mathai and K.S. Pillai. The Academy of Environmental Biology, India, pp. 207–211.

Jha, M.M., 1992. Studies on the effects of nickel chloride to some organs of the climbing perch, *Anabas testudineus* (Bloch.). *Ph.D. Thesis*, L.N. Mithila University, Darbhanga.

Larsson, A., 1973. Metabolic effects of epinephrine and norepinephrine in the eel, *Anguilla anguilla* L. *Gen. Comp. Endocrinol.*, 20: 155–167.

Larsson, A., Bengtsson, B.E. and Haux, C., 1981. Disturbed ion balance in flounder, *Platichthys flesus* L. exposed to sublethal levels of cadmium. *Aquatic. Toxicol.*, 1: 19–35.

Lidman, U., Dave, Johansson-Sjobech, M.L., Larsson, A. and Lawander, K., 1979. Metabolic effects of cortisol in the European eel, *Anguilla anguilla* L. *Comp. Biochem. Physiol.*, 63: 339–344.

Mahajan, C.L., Juneja, C.J. and Malaisselegal, F., 1971. Effects of sulfonylurease on calcium uptake and insulin secretion of islet of Langerhans. *Arch. Int. Pharmacodyn. Therap.*, 192: 205–207.

Mazeaud, M.M., Mazeaud, F. and Donaldson, E.M., 1977. Primary and secondary effects of street in fish, some new data with a general review. *Trans. Am. Fish. Soc.*, 106: 201–212.

Nakano, T. and Tomblinson, N., 1967. Catecholamines and carbohydrate concentrations in rainbow trout (*Salmo gairdner*) in relation to physical disturbance. *J. Fish. Res. Board. Can.*, 24: 1701–1715.

Oguri, M. and Nace, P.F., 1966. Blood sugar and adrenal histology of the gold fish after treatment with mammalian adrenocorticotrophic hormone. *Chesapeake Sci.*, 7: 198–202.

Orr, J.W., Price, D.E. and Srickland, L.W., 1948. Glycogen content in rat liver after poisoning with large doses of dimethylaminobenzene. *J. Pathol. Bacteriol.*, 60: 573–581.

Robertson, O.H., Wexler, B.C. and Rinfret, A.P., 1963. The effects of hydrocortosone on immature rainbow trout (*Salmo gairdneri*). *Gen. Comp. Endocrinol.*, 3: 422–436.

Roy, D.N. and Dubey, N.K., 2001. Haematological response of trivalent chromium poisoning in Indian teleost catfish *Heteropneustes fossilis* (Bloch.). *Indian J. Environ. and Ecoplan.*, 5: 607– 616.

Sastry, K.V. and Gupta P.K., 1979b. Mercuric chloride intoxication in digestive system of *Channa punctatus*. *J. Toxicol. Environ. Health*, 4: 777–783.

Sastry, K.V. and Gupta, P.K., 1979c. Enzymological alteration in the digestive system, induced by lead nitrate. *Toxicol. Letter*, 3: 145–150.

Saxena, O.P. and Parashari, A., 1983. Comparative study of the toxicity of six heavy metals of *Channa punctatus*. *J. Environ. Biol.*, 4: 91–94.

Vallee, B.L. and Ulmer, D.D., 1972. Biochemical effects of mercury, cadmium and lead. *Ann. Rev. Biochem.*, 41: 91–128.

Wedemeyer, G., 1969. Stress-induced ascorbate depletion and cortisol production in to salmonoid fish. *Comp. Bio. Chem. Physiol.*, 29: 1247–1251.

Chapter 53

Effect of Time of Application of Phosphorus and Potassium on Yield Attributes and Yield of Sugarcane Plant Crop

T. Chitkala Devi, A. Sambasiva Reddy and M. Bharathalakshmi
**Acharya N.G. Ranga Agricultural University,
Regional Agricultural Research Station,
Anakapalle – 531 001, Visakhapatnam District, Andhra Pradesh**

ABSTRACT

The effect of time of application of phosphorus and potassium on yield attributes and yield of sugarcane plant crop was studied during 2001–02 and 2002–03 on sandy loam soil of Regional Agricultural Research Station, Anakapalle. The results of the study indicated that yield attributes *viz.*, number of millable canes, length of millable cane, girth of the cane and single cane weight were significantly higher with three split doses of phosphorus (basal, 45 and 90 days after planting) and potassium (basal, 60 and 120 days after planting) as compared to entire dose of phosphorus and potassium and found on par with two split doses of phosphorus (basal and 45 days after planting) and potassium (basal and 60 days after planting). Significantly higher cane and sugar yields were recorded with three split doses of phosphorus (basal, 45 and 90 DAP) and potassium (basal, 60 and 120 DAP) as compared to application of entire dose of phosphorus and potassium as basal during both the years of study but was comparable to two split doses of phosphorus (basal and 45 DAP) and potassium (basal and 60 DAP).

Keywords: *Time of application, Sugarcane, Phosphorus, Potassium, Number of millable canes, Cane yield.*

Introduction

Sugarcane is one of the major agro-industrial crops of the country having very great economic impact particularly on rural India. In India sugarcane is grown in an area of 4.25 million hectares with 282 million tonnes of cane production and 66.7 t ha^{-1} productivity. In Andhra Pradesh it occupies an area of 2.32 lakh hectares with 154 lakh tonnes of production and productivity of 66.2 t ha^{-1} (2002–03). In increasing sugarcane productivity mineral nutrition 'plays a key role among various management practices. Next to nitrogen, phosphorus and potassium are the two major nutrients required for increasing productivity in sugarcane. Like nitrogen, phosphorus is also subjected to leaching in sandy soils up to 90 per cent of the applied phosphorus from single super phosphate may be lost through drainage (Gillman, 1973) which can be reduced through split application of phosphatic fertilizers. Similarly, in sandy soils potassium availability is low due to poor potassium supplying power and low retention of fertilizer potassium. Hence, application of potassium in split doses may reduce the losses and help in better utilization of applied fertilizer. Moreover, recently released varieties are responding well to phosphorus and potassium.

Materials and Methods

Field experiments were conducted during 2001–02 and 2002–03 at Regional Agricultural Research Station, Anakapalle to study the effect of time of application of phosphorus and potassium on sugarcane plant crop. The soil of the experimental plots was sandy loam in texture, neutral in reaction (pH: 6.9), low in available nitrogen (180.5 kg ha^{-1}), phosphorus (21.6 kg P_2O_5 ha^{-1}) and potassium (136.8 kg K_2O ha^{-1}). An exhaustive crop of bajra was grown before plant crop and removed along with roots at 40 days age.

There were sixteen treatments *viz.*, T_1–Control; T_2–Phosphorus basal; T_3–Phosphorus basal, 45 days; T_4–Phosphorus basal, 45 and 90 days; T_5–Potassium basal; T_6–Potassium basal, 60 days; T_7–Potassium basal, 60, 120 days; T_8–Phosphorus and Potassium basal; T_9–Phosphorus basal + Potassium basal, 60 days; T_{10}–Phosphorus basal + Potassium basal, 60 and 120 days; T_{11}–Phosphorus basal, 45 days + Potassium basal; T_{12}–Phosphorus basal, 45 days + Potassium basal, 60 days; T_{13}–Phosphorus basal, 45 days + Potassium basal, 60. 120 days; T_{14}–Phosphorus basal, 45, 90 days + Potassium basal; T_{15}–Phosphorus basal, 45, 90 days + Potassium basal, 60 days and T_{16}–Phosphorus basal, 45, 90 days + Potassium basal, 60, 120 days. Phosphorus (100 kg P_2O_5 ha^{-1}) and Potassium (120 kg K_2O ha^{-1}) were applied in equal split doses. The treatments were tried in Randomized Block Design (RBD) with three replications.

The variety chosen was 85 A 261 an early maturing, shallow rooted, sucrose rich high yielding variety. Forty thousand three budded sets per hectare were planted in deep furrows spaced at 80 cm apart. Plantings were done during last week of February and harvested during last week of December during both the years. Uniform dose of 112 kg nitrogen ha^{-1} was applied in two equal doses at 45 and 90 days after planting. Phosphorus (100 kg P_2O_5 ha^{-1}) and potassium (120 kg K_2O ha^{-1}) were applied as per the treatments. Irrigations were accorded at weekly interval during formative phase and three weeks interval during maturity phase. All other cultural operations were followed as per the recommendations.

Results and Discussion

Yield attributes *viz.*, number of millable canes, length of millable cane, girth of the cane, single cane weight and cane yield were recorded at harvest and furnished in Tables 53.1 and 53.2.

Table 53.1: Yield Attributes as Influenced by Time of Application of Phosphorus and Potassium

Treatment	*Number of Millable Canes ('000 ha^{-1})*			*Length of Millable Cane (cm)*			*Girth of the Cane (cm)*			*Single Cane Weight (kg)*		
	2001–02	*2002–03*	*Pooled*	*2001–02*	*2002–03*	*Pooled*	*2001–02*	*2002–03*	*Pooled*	*2001–02*	*2002–03*	*Pooled*
T_1: Control	68.40	64.84	66.20	259.0	233.9	246.4	2.06	2.01	2.04	1.149	1.075	1.112
T_2: 100 P_2O_5 basal	72.92	67.40	70.16	266.7	249.3	258.0	2.17	2.15	2.16	1.201	1.128	1.164
T_3: 100 P_2O_5 basal, 45 DAP	73.70	68.92	71.31	268.9	250.4	259.6	2.19	2.17	2.18	1.209	1.147	1.178
T_4: 100 P_2O_5 basal, 90 DAP	74.25	69.98	72.41	273.2	254.9	264.0	2.20	2.19	2.19	1.220	1.154	1.187
T_5: 120 K_2O basal	71.03	67.33	69.18	264.6	248.1	256.3	2.20	2.19	2.19	1.190	1.130	1.160
T_6: 120 K_2O basal, 60 DAP	72.41	68.81	70.61	267.6	251.7	259.6	2.22	2.21	2.21	1.215	1.150	1.182
T_7: 120 K_2O basal, 60, 120 DAP	74.22	69.90	72.06	272.3	253.8	263.0	2.24	2.23	2.23	1.230	1.164	1.197
T_8: 100 P_2O_5 basal + 120 K_2O basal	75.19	71.82	73.51	275.9	259.9	267.9	2.25	2.22	2.23	1.238	1.172	1.205
T_9: 100 P_2O_5 basal + 120 K_2O basal, 60 DAP	76.76	72.92	74.84	278.3	262.3	270.3	2.28	2.24	2.26	1.253	1.187	1.220
T_{10}: 100 P_2O_5 basal + 120 K_2O basal, 60, 120 DAP	77.18	73.22	75.20	282.4	266.0	274.2	2.30	2.26	2.28	1.263	1.198	1.230
T_{11}: 100 P_2O_5 basal, 45 DAP + 120 K_2O basal	76.24	72.40	74.32	280.2	268.0	274.1	2.30	2.24	2.27	1.266	1.202	1.234
T_{12}: 100 P_2O_5 basal, 45 DAP + 120 K_2O basal, 60 DAP	78.58	74.97	76.78	288.9	276.8	282.5	2.33	2.29	2.31	1.286	1.226	1.256
T_{13}: 100 P_2O_5 basal, 45 DAP + 120 K_2O basal, 60, 120 DAP	80.73	76.21	78.47	289.3	279.0	284.1	2.35	2.33	2.34	1.296	1.239	1.267
T_{14}: 100 P_2O_5 basal, 45, 90 DAP + 120 K_2O basal	77.42	73.29	75.35	286.1	274.1	280.1	2.31	2.26	2.28	1.279	1.218	1.248
T_{15}: 100 P_2O_5 basal, 45, 90 DAP + 120 K_2O basal, 60 DAP	80.47	75.43	77.95	293.8	277.6	285.7	2.36	2.30	2.33	1.310	1.249	1.279
T_{16}: 100 P_2O_5 basal, 45, 90 DAP + 120 K_2O basal, 60, 120 DAP	82.88	77.40	80.14	296.1	283.1	289.6	2.39	2.35	2.37	1.321	1.263	1.292
S.Em±	1.50	0.87	1.31	2.5	2.9	2.6	0.02	0.02	0.03	0.017	0.014	0.015
C.D (P = 0.05)	4.32	2.50	3.77	7.3	8.3	7.5	0.06	0.07	0.08	0.051	0.041	0.042

Table 53.2: Cane and Sugar Yield (t/ha) as Influenced by Time of Application of Phosphorus and Potassium

Treatment	Cane Yield (t ha^{-1})			Sugar Yield (t ha^{-1})		
	2001–02	2002–03	Pooled	2001–02	2002–03	Pooled
T_1: Control	77.58	65.70	71.64	9.43	8.39	8.91
T_2: 100 P_2O_5 basal	85.50	70.87	78.18	10.75	9.43	10.08
T_3: 100 P_2O_5 basal, 45 DAP	87.65	72.40	80.02	11.23	9.80	10.51
T_4: 100 P_2O_5 basal, 90 DAP	89.19	74.51	81.85	11.58	10.16	10.87
T_5: 120 K_2O basal	84.38	70.10	77.24	10.97	9.37	10.17
T_6: 120 K_2O basal, 60 DAP	86.18	71.00	78.59	11.30	9.58	10.44
T_7: 120 K_2O basal, 60, 120 DAP	88.39	73.47	80.93	11.67	10.11	10.89
T_8: 100 P_2O_5 basal + 120 K_2O basal	89.42	75.10	82.26	11.50	10.11	10.80
T_9: 100 P_2O_5 basal + 120 K_2O basal, 60 DAP	91.38	77.73	84.55	11.89	10.59	11.24
T_{10}: 100 P_2O_5 basal + 120 K_2O basal, 60, 120 DAP	91.96	78.77	85.36	12.22	10.95	11.58
T_{11}: 100 P_2O_5 basal, 45 DAP + 120 K_2O basal	90.79	77.97	84.38	11.85	10.48	11.16
T_{12}: 100 P_2O_5 basal, 45 DAP + 120 K_2O basal, 60 DAP	95.76	82.60	89.18	12.73	11.47	12.10
T_{13}: 100 P_2O_5 basal, 45 DAP + 120 K_2O basal, 60, 120 DAP	96.08	82.80	89.44	12.99	11.67	12.33
T_{14}: 100 P_2O_5 basal, 45, 90 DAP + 120 K_2O basal	92.19	79.17	85.68	12.23	10.81	11.52
T_{15}: 100 P_2O_5 basal, 45, 90 DAP + 120 K_2O basal, 60 DAP	95.79	82.81	89.30	12.92	11.45	12.18
T_{16}: 100 P_2O_5 basal, 45, 90 DAP + 120 K_2O basal, 60, 120 DAP	97.59	85.05	91.32	13.29	12.08	12.68
S.Em±	1.65	1.43	1.61	0.29	0.39	0.28
C.D (P = 0.05)	4.77	4.10	4.61	0.83	1.01	0.81

Number of Millable Canes

Significant differences in number of millable canes was recorded with different times of application of phosphorus and potassium (Table 53.1). Higher number of millable canes were obtained with application of phosphorus and potassium in three split doses (T_{16}) and found on par with two split doses of phosphorus and potassium (T_{13}), two split doses of phosphorus and three split doses of potassium (T_{13}) and three split doses of P_2O_5 and two split doses of K_2O (T_{15}) and found significantly superior to application of entire dose of P_2O_5 and K_2O (T_8) as basal during both the years and mean over two years. This might be due to more number of tillers and shoots recorded during formative phase and grand growth phases respectively. Ramalingaswamy (1997) and Reddy *et al.* (2000) observed similar increase in number of millable canes with split application of phosphorus and potassium. Individual application of P and K either as basal or in split doses did not improve the millable cane number significantly during both the years and mean over two years.

Length of Millable Cane (cm)

Length of millable cane varied significantly with different times of application of phosphorus and potassium (Table 53.1). Length of millable cane (cm) obtained with split application of phosphorus (upto 90 days age) and potassium upto 120 days age (T_{16}) was higher and found on par with other split application treatments (T_{12}, T_{13} and T_{15}) in both the years as well as pooled over two years. Beneficial effect of late application of phosphorus and potassium in increasing length of millable cane was also reported by Rao (1985) and Kadam *et al.* (1993).

Girth of the Cane (cm)

Application of phosphorus and potassium in three split doses to sugarcane registered higher girth of the cane than application of entire dose of phosphorus and potassium as basal. Individual application of phosphorus and potassium either as basal or in split doses upto 120 days did not found to improve girth of the cane as compared to combined application of P and K.

Single Cane Weight (kg)

Significant variation in single cane weight was recorded with different times of application of phosphorus and potassium (Table 53.2). Single cane weight recorded with control (T_1) plot was significantly lower as compared to rest of the treatments. Significantly higher single cane weight was obtained with three split doses of phosphorus (basal, 45 and 90 DAP) and potassium (basal, 60 and 120 DAP) which was comparable to two split doses of P_2O_5 and K_2O (T_{12}). Application of phosphorus and potassium alone either as basal or split doses did not improve the single cane weight during both the years.

Cane Yield (t/ha)

Significant variation in cane yield was recorded with different times of application of phosphorus and potassium to sugarcane plant crop (Table 53.2). Cane yield was significantly higher with application of phosphorus (0, 45 and 90 DAP) and potassium (0, 60 and 120 DAP) in three split doses (97.59 t ha^{-1} in 2001–02; 85.05 t ha^{-1} in 2002–03 and 91.32 t ha^{-1} in pooled data) which was comparable to two split doses (0, 45 DAP) of phosphorus and two split doses (0.60 DAP) of potassium (T_{12}) or two split doses of phosphorus + three split doses of potassium upto 120 days age (T_{13}) or three splits of P_2O_5 + two splits of potassium (T_{14}) during both the years and found superior to application of entire dose of P_2O_5 and K_2O as basal (T_8). The favourable effect of split doses of phosphorus and potassium on cane yield was mainly due to elevated yield attributes of sugarcane plant crop. Similar increase in cane yield with late application of phosphorus and potassium was observed by Srinivasan and Morachan (1980). Rathi and Singh (1982), Shinde (1986), Yadav and Yaduvamshi (1990), Yadav *et al.* (1991). Kumaraswamy and Rajasekaran (1993) and Ramalingaswamy *et al.* (1999).

Cane yield obtained with control (77.58 t ha^{-1} in 2001–02; 65.70 t ha^{-1} in 2002–03 and 71.64 t ha^{-1} in pooled data was significantly lower compared to rest of the treatment during both the years and pooled over two years.

Sugar Yield (t/ha)

Higher sugar yield (13.29 t ha^{-1} in 2001–02, 12.08 t ha^{-1} in 2002–03 and 12.68 t ha^{-1} in pooled data) was obtained with three split doses of phosphorus and potassium (T_{16}) which was on par with T_{12}, T_{13} and T_{15} in both the years and found significantly superior to application of entire dose of phosphorus and potassium as basal (T_8 : 11.50, 10.11 and 10.80 t ha^{-1} in 2001–02, 2002–03 and pooled data respectively). This might be primarily due to increased cane yield with split application of phosphorus and potassium upto 120 days of the cane crop. These results are in corroboration with Shinde (1986) and Ramalingaswamy *et al.* (1993). Significantly lowest sugar yield was recorded with control plot (9.43, 8.39 and 8.91 t ha^{-1} during 2001–02. 2002–03 and pooled data respectively).

When the yield attributes were correlated with cane and sugar yield (Table 53.3). During 2001–02 season, positive and highly significant correlation coefficients were obtained between number of millable canes (r = 0.984** and r = 0.970** with cane and sugar yield respectively); length of millable cane (r = 0.957** and r = 0.954**), girth of the cane (r = 0.970** and r = 0.9809**) single cane weight (0.797** and 0.779**). During 2002–03 season also positive correlation between yield attributes, cane and sugar yield were recorded which indicates the beneficial effect of late application of phosphorus and potassium in three split doses in improving the yield attributes there by cane productivity.

Table 53.3: Correlation Coefficients ('r' Values) between Yield Attributes with Cane and Sugar Yield of Plant Crop

Character	*Correlation Coefficient ('r' values)*	
	2001–02	*2002–03*
Cane yield with		
Number of millable canes	0.984**	0.992**
Length of millable cane	0.957**	0.986**
Girth of the cane	0.970**	0.930**
Single cane weight	0.797**	0.989**
Sugar yield with		
Number of millable canes	0.970**	0.993**
Length of millable cane	0.954**	0.986**
Girth of the cane	0.980**	0.930**
Single cane weight	0.779**	0.988**

**: Significant at 1 per cent level.

From this study, it can be concluded that, split application of 100 kg P_2O_5 ha^{-1} (basal, 45 and 90 days) and 120 kg K_2O ha^{-1} (basal, 60 and 120 days) was found superior in improving yield attributes and there by cane and sugar yields as compared to entire dose P_2O_5 and K_2O as basal Similarly, two split doses of phosphorous (basal, 45 days) and potassium (basal, 60 days) was also found next best treatment in increasing cane and sugar yield of sugarcane plant crop grown in sandy loam soils.

References

Gillman, G.P., 1973. Studies on some deep sandy soils in cape york Peninsula North Queensland iii. Losses of applied P and S. *Austrialia Journal of Experimental Agriculture and Animal Husbandry*, 13: 418–422.

Kadam, B.S., Akolkar, R.D., Vaidya, B.R. and Jadhav, S.B., 1993. Studies on the effect of time of phosphate application on yield and quality of sugarcane varieties (Suru). In: *Proc. of 43rd Annual Convention of Deccan Sugar Technologists Association*, Pune, Part I: A 21–127.

Kumaraswamy, K. and Rajasekaran, S., 1993. Response of sugarcane crop to phosphorus as influenced by time of application. *Indian Sugar*, 43(4): 233–235.

Ramalingaswamy, K., Jamuna, P. and Veerabhadra Rao, K., 1993. A note on critical limits for soil available phosphorus and potassium in sugarcane growing sandy loam soils of North Coastal Andhra Pradesh in relation to response of sugarcane to phosphorus and potassium fertilization. In: *Proc. of 43rd Annual Convention of Deccan Sugar Technologists Association*, Pune, A: 99–106.

Ramalingaswamy, K., Padma Raju, A. and Chennakesavulu, E., 1997. Effect of phosphorus (through Urea Ammonium Phosphate and 28–28–0) on the yield and quality of sugarcane varieties. *Fertilizer News*, 42(8): 33–35.

Rathi, K.S. and Singh, R.A., 1982. Effect of different dates and time of application of phosphorus on stalk yield of cane ratoon in sandy loam soil of central UP. *Indian Sugar*, 1(11): 823–824.

Reddy, P.R.R., Ravindrababu, P. and Ramachandra Rao, B., 2000. Effect of phosphorus on cane yield and juice quality of sugarcane. *Bharatiya Sugar*, 25(8): 27–29.

Shinde, S.B., 1986. Spraying of phosphate and potash on sugarcane for increasing sugar recovery and yield per acre. Book of Abstracts. *Golden Jubilee of Deccan Sugar Technologists Association*, Pune, A: 49–53.

Srinivasan, T.R. and Morachan, Y.B., 1980. Phosphorus and potassium fertilization on juice quality and cane yield. *Indian Potash Journal*, 5(4): 2–9.

Yadav, D.V. and Yaduvamshi, N.P.S., 1990. Exploring the responses of foliar application vs top dressing of phosphorus in sugarcane. *Annual Report*, Indian Institute of Sugarcane Research, Lucknow, pp. 33.

Yadav, D.V., Yaduvamshi, N.P.S. and Singh, K., 1991. An alternative to basal application of phosphorus in sugarcane. *Fertilizers News*, 36(7): 63–65.

Chapter 54

Biochemical Analysis of *Andrographis paniculata*: The Plant Extract was Used for Reducing Sugar Level in the Human Blood in Empty Stomach

Basukinath Jha and Sanjeev Kumar Choudhary

Introduction and Process

Andrographis paniculata is locally known as 'Kalmegh' or Chiraita. It is a member of the family Acanthaceae Local Vaidyas and Hakims of Inarabaran, Banka, and Katoria used this plant for diabetic patients. The stem and leaves of this plant are deeped in the night in drinking water and extract of plant with drinking water is taken by the diabetic patient in the morning in empty stomach. One diabetic patient in Inarabaran told us that the extract of this plant has positive effect. He told us that 25 days before my blood Sugar level was 371 when I examined blood sugar level at Deoghar yesterday, my blood sugar level slashed to 167. I used this plant extract regularly in the morning in empty stomach. Besides it is used in diabetic patient the extract of this plant is used as blood purifier, febrifuge, vermifuse and in stomach disorder.

Chemical Analysis

The extract of this plant was prepared according to metrology suggested by Ranjan and Loraya (1960) 10μ alcoholic extract of healthy leaves was loaded on Whatman Chromatography paper No. 1 along with standard amino acids. The Chromatogram to run with solvent system containing *n*-butanol : acid : distilled water (4 : 1 : 1 V/V/V top layer). They were dried sprayed with 0.3 per cent ninhydin solution (300my ninhydrin dissolved in 100 ml of n-butanol to which then 3ml of glacial acetic acid added. The chromatography was allowed to air dry. Subsequently they were transferred to oven at 105°C for 5 minutes.

Free Amino Acids

Amino acid to analysis spots of various amino acids were identified by comparing the colour and as R/f value done on the basis of intensities which were by visual observations recorded in 4 categories *i.e.* 3^+, 2^+, + acid + tr.

Bound Amino Acids

The processes were followed for bound amino acids. Result of amino acids analysis has been given in Table 54.1.

Table 54.1: Showing Data of Amino Acids Present in Plant Extract of *A. paniculata* Collected from Inarabaran of Deoghar District

Sport Nos.	*Amino Acids*	*Rf Values*	*Concentration of Amino Acids on Visual Observation*							
			Free Amino Acids				*Bound Amino Acids*			
			T_1	T_2	T_3	*Mean*	T_1	T_2	T_3	*Mean*
1.	Lysine	0.04	+	3+	3+	3+	–	–	–	–
2.	L-Histidine	0.06	–	–	–	–	–	–	–	–
3:	Aspartic acid	0.09	+	+	+	+	+	+	+	+
4.	L-argenine	0.12	+	–	+	+	–	–	–	–
5.	Cysteine	0.16	3+	3+	3+	3+	+	–	+	+
6.	DL-aspargine	0.18	–	–	–	–	–	–	–	–
7.	Proline	0.21	2+	2+	2+	2+	–	–	–	–
8.	L-glutamine	0.23	–	–	–	–	–	–	–	–
9,	Serine	0.24	–	–	–	–	–	–	–	–
10.	Hydroxyproline	0.26	–	–	–	–	–	–	–	–
11.	DL-threonine	0.28	+	+	+	+	+	+	+	+
12.	Glycine	0.31	3+	3+	3+	3+	+	+	2+	+
13.	L-glutamic acid	0.33	–	–	–	–	–	–	–	–
14.	Ornithin	0.34	–	–	–	–	–	–	–	–
15.	Dihydroxy phenylamine	0.37	–	–	–	–	–	–	–	–
16.	DL-alanine	0.45	2+	2+	2+	2+	–	–	–	–
17.	DL-metheonine	0.49	+	+	+	+	–	–	–	–
18.	Amino-*n*-butric acid	0.56	–	–	–	–	–	–	–	–
19.	L-valine	0.61	3+	2+	3+	3+	2+	2+	2+	2+
20.	Tyrosine	0.63	2+	2+	2+	2+	–	–	–	–
21.	Phenylaline	0.67	–	–	–	–	–	–	–	–
22.	DL-tryphtophan	0.71	2+	+	2+	2+	+	2+	2+	2+
23.	Iso-leucine	0.76	–	–	–	–	–	–	–	–
24.	Leucine	0.80	–	–	–	–	–	–	–	–

Protein Analysis

The seeds and another proteins (Water Soluble) were determined qunatitatively by disc electrophoretic method (Ornstein and Davis, 1661) the result of protein analysis of *Androphis parucula* has been in Table 54.2 and finger print (Figure 54.1).

Table 54.2: Showing Data of Protein Analysis of the Plant Extract of *Andrographis paniculata* collected from Inarabaram of Deoghar District

Protein Band Nos.	*Rp Value*	*Visual Observation*
1.	1.3	–
2.	2.5	3+
3.	6.0	3+
4.	0.8	–
5.	3.8	–
6.	5.3	2+
7.	0.4	–
8.	3.0	–
9.	2.0	–
10.	2.3	–

Alkaloid Analysis

Estimation and identification of alkaloids were done according to methods recommended by Clarke (1970). The result of the alkaloid analysis has been given in Table 54.3.

Table 54.3: Biochemical Analysis of Alkaloids of Plant Extract Paniculata Collected from Inarabaran of Deoghar District

Alkaloids	*Chromotographed Rf Value × 100*	*Visual Observation*	*Nature in UV light*	*Reagent Used for the Dectectic*
Cystisine	03	–	Blue	Dragen dorff
Nicotine	07	–	Absorbed	Iodoplatinate
Tomative	08	–	Invisible	Do
Morphine	14	–	Absorbed	Dragen dorff
Solanine	15	–	Invisible	Do
Berbeerine	25	–	Flurescent	Do
Atrophine	37	–	Absorbed	Do
Quinine	56	–	Invisible	Do
Conine				
Unknown extract	17	+	Absorbed	Dragan dorff
	23	3+	Flurescent	Do
	48	2+	Invisible	Iodoplatinate

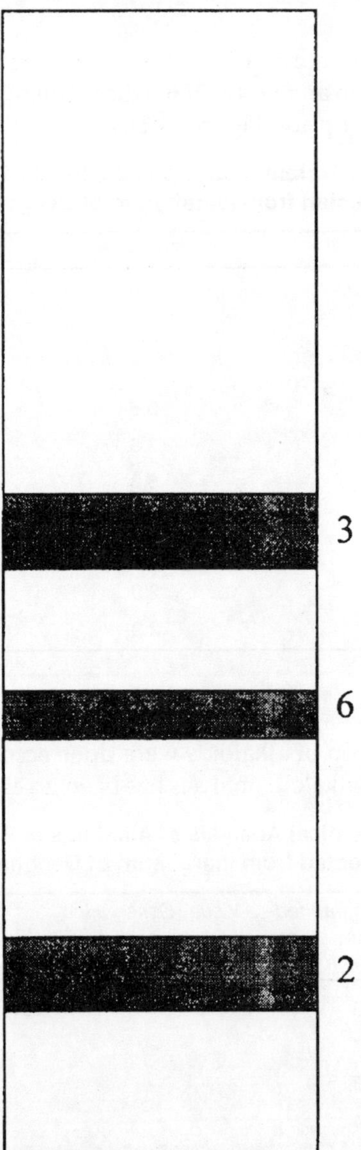

Figure 54.1: Fingerprint of Electrophoretic Protein Analysis of the Plant of Extract of *Andrographs paniculata*

Ultraviolet Measurement

Confirm the presence of alkaloid by measuring UV spectrum of the sample dissolved in NH_2SO_4. Typical maxima values ranges from 250 to 303 nm. Alkaloids with aromatic rings in their structure may be observed by longer wave length. For the purpose Beekman spectrophometer was used noted and colour of emitted light was noted on visual observation of both marker and unknown alkaloids.

Inorganic Analysis

Analysis of detectable chemicals of *Andrographs paniculata* has been given with the help of methods suggested by Bhatia and Parashar (1971). Results of inorganic analysis has been given in Table 54.4.

The plant extract of *Androgrophis paniculata*. In this extract three type of alkaloid could be detected. They were 17, 23, 48 of Rf values.

Table 54.4: Showing Result of Inorganic Analysis *Andropraphis paniculata* from Inarabaran of Deoghar District

Types of the Elements	*Visual Observation*
Nitrogen	2+
Calcium	+
Ferrous (Iron)	3+
Potassium	3+
Phosphorus	3+

Discussion

In nineteenth and twentieth centuries chemical nature of natural products were developed by Weygard (1940); Shemyakin (1965); Miller and Orgel (1972); Joms and Hersh (1974) and Fsicher (1984). Currently, most of the investigation are applying there expertise to examining biochemical nature and biosynthesis of phytochemicals to solving the Stereo-Chemical and mechanistic problems posed. However, seeing the vast member of forms of form of plants relatively little has been made in this regard due to the fact that some of the principles and philosophies are often compounded by lack of the familiarity with biochemical terminology and methods. In spite of that, an even increasing of chemists are being attracted to the challenges of Synthetic Structural, and Stereo chemical and other problems of organic chemistry.

The data presented of medicinal plant *i.e. Androgrophis paniculata* Linn. which is used by diabetes patients in Deoghar district. One cup of the plant juice was used in empty stomach. One patient was silting on the verandah of a local vaidya at Inarabaran. When we enquired about that patient he told us one cup of chiraita was used in empty stomach by diabetic patient, he told us that one month ago my blood sugar level was 371 but when I used chiraita or kalmegh one month ago my sugar level was dropped to 167 on the advice of this Vaidya. I came here to show my report of blood sugar level to Vaidya.

Amino Acids

In *Adrogrophis paniculata* only 12 type of amino acids could detected. Those were Lysine, Aspartic acid, Largenine, cystine praline, DL-theonine, glycine, Alanine out of there Aspartic acid, cysteine, DL threonine, glycine, valine both the forms. While Lysine, Proline, DL-alanine, DL-metheonine, Tyrosine were recorded his free form only. Among the 12 types of amino acid Lysine, Cystein, Glycine and L valine were recorded in high potencial *i.e.* 3+ level while other types of amino acids were recorded at + or 2+ level. Although more than 100 amino acids are known but only 20 amino acids take part in synthesis of proteins.

Protein Analysis

In protein analysis of *Androgrophis paniculata* were recorded the Table 54.2 and Figure 54.1, in protein band No. 2 and 3 were recorded at 3+ level while band No. 6 was recorded in 2+ level. Having Rp value 2.5, 6.0 and 5.3.

The term protein is derived from the Greek noun protein measuring "holding first place" the term was suggested by Berzelius (1838) and first used by G.J. Muller in his book in 1939. Protein has very important in all the living cells. It is essential in diet for another metabolic processes in living organisms. It has varieties of roles, as proteins are of specific nature *viz.* nutritive, enzymatic, hormonal, the making antibodies, medicinal, genetical etc. The way all basic functions of life depends upon specific protein. Indeed we do not know life without proteins. After the works of Neurath (1936-66), Linskens and Tracey (1967), Leggett and Baily (1970), Harbome *et al.* (1971), Arison and Edsall (1973) and Kaziel *et al.* (1993) the details of protein structure and function could be known. According to them protein is a macromolecule of polypeptide chain. Therefore, polypeptide chain assume specific configuration to perform specific functions assigned to them.

Protein constituents of any plant may be very helpful in tracing the phylogeny on the unknown plant. In this regard commendable works have been done by Boulter (1972) MC Leester *et al.* (1983), fair brother (1984) and Naish (1996).

Alkaloids Analysis

The chemistry of alkaloid began with the announcement of isolation of morphine form optimum in 1805 by F. W. Sertuner.

The term alkaloid was proposed by a pharmacist, C.F.W Meissner in 1819. Today more than five thousand alkaloids are known to occur more then 4000 species of Angiosperms. Alkaloids are mostly toxic substances having primary effect on central nervous system are of the organisms.

The noticeable works regarding alkaloids are of Hegnaur (1967), Henery (1969), Clarke (1970) and Luning (1987). The detailed chemistry of alkaloids could be known with the work of above biochemists. In *Andrographis paniculata* three type of alkaloid could be noticed in plant extracts. These were with Rf value 17, 23 and 48 in UV light Rf value 17 was observed while Rf value 23 was with highest potencies on visual observation *i.e.* 3+ while Rf value 48 has lower potencies *i.e.* 2+.

Perhaps these alkaloids lower blood sugar level in human beings.

Inorganic Analysis

The detectable inorganic constituents of *Andrographs paniculata* on methodology suggested by Bhatia and Parasar collected from Inarabaran, of Deoghar District in terms of calcium, phosphorus, potassium, nitrogen and could be detected, this plant was recommended to diabetic patients. Calcium is useful in growth of the bones in human beings and milking cattle and this plant is recommended by local vaidyas and Hakims of tribal belt of Deoghar district to diabetic patients.

In *Andrographs paniculata* phosphorous is detected and visual observation it was detected at 2+ level. Like wise the potassium is detected in this plant. On visual observation it was detected at '+' level only.

In *Andrographs paniculata* nitrogen could be detected. On visual observation it was detected at 3+ level. In this plant iron could be detected and on visual observation it was detected at 3+ level.

Besides diabetes, the plant extract was very much used in high fever, in inflammation, used in Jaundice and in vermifuse. Roof extract was used in hemorrhoids, rubbed after burning on the palm and sold of the feet and used as cardic tonic. The seed extracts were used as paste for leper patients, used in snake bite and skin diseases.

References

Arison, W.A. and Edsall, G.L., 1973. *Structure and Functions of Protein*. Australian Academy of Science, Australia, pp. 7–20.

Berzelium, 1838. *Topics in Stereochemistry*. Academic Press, New York, pp. 733–745.

Bhatia, K.M and Parashar, 1971. *A Plant Physiology: Modern Treaties*, 3rd edn., pp. 188–200.

Boulter, J.M., 1972. *Phylogeny of Some Unknown Plant with the Help of Protein Biochemical*. Cambridge Univ. Press, England, pp. 42–57.

Clarke, E.G.C., 1970. The forensic chemistry of alkaloids. In: *The Alkaloids*, (Ed.) H.F. Manske. Academic Press, New York, 12: 514–590.

Fair Brother, 1984. *Biochemistry*. Academic Press, New York, 2: 836–840.

Harborne, J.B., Morris, J.M. and Miller, S.L., 1971. *Topics of Stereochemistry*. Academic Press, New York, pp. 747–759.

Hegnaur, J.F., 1967. *The Alkaloid of Plants*. Cambridge Univ. Press, England, pp. 49–57.

Henery, J.M., 1969. *The Alkaloid of Plants in Detailed*. Academic Press, New York, pp. 217–223

Keziel, B.A., Bell, A.S.L. and Alworth, W.A., 1993. *The Detailed Structure and Functions of Protein*. Cambridge Univ. Press, England, pp. 213–227.

Leggett, L. and Baily, B.J., 1970. *Structure and Functions of Protein*. Academic Press, New York, pp. 207–212.

Linskens, G.J. and Tracey, W.A., 1967. *Functions of Protein*. Academic Press, New York, pp. 213–219.

Luning, B.A., 1987. *Pharmaceutiical Nature of Alkaloid*. Academic Press, England, pp. 287–300.

McLeester, J.M., Meyer, W.F. and Niconov, G.K., 1983. *The Detailed Structure and Function of Protein*. Academic Press, New York, pp. 257–263.

Meissner, C.F.W., 1819. *Pharma. Sci.*. Cambridge Univ. Press, England, pp. 82–109.

Mulder, G.J., 1939. *The Terminology of Proteins*. Academic Press, New York, pp. 213–237.

Neish, J.F., 1996. *Ann. Rev. of Biochem*. Academic Press, New York, pp. 57–67.

Neurath, W.A., 1936–66. *Protein Structures and Functions*. Academic Press, England, pp. 187–210.

Orstein, L. and Davis, B.J., 1961. *Disc Electrophoresis of Protein by Distillation Products Industries*. Eastman Kodak, Rochester, New York.

Rajan, S. and Loloraya, M.M., 1960. Metabolism of isolated level-I: Change in protein soluble, nitrogenous compound Sugars and organic acids in tobacco leaves. *Clarke F_1 Physiol.*, 95: 714–725.

Chapter 55

Monoculture Growth, Breeding and Fecundity of *Macrobrachium rosenbergii* (De Man) in Haryana

Suday Prasad[1]* and Hardial Singh[2]

[1]ICAR Research Complex for Eastern Region, P.O. B.V. College, Patna – 800 014, Bihar

[2]R.R.C. of Central Institute of Freshwater Aquaculture, PAU, Ludhiana – 141 004, Punjab

ABSTRACT

Giant freshwater prawn *Macrobrachium rosenbergii* seed were cultured in a 0.4 ha pond at a stocking density of 22000 nos during month of May to October 2002. The prawn seed were stocked at a size range of 12–14 mm. The prawn were fed with different grades of a commercial feed (crumbles/pellets) 3–4 times daily @ 20–30 per cent of the estimated biomass. After 5 months of culture the prawns have grown to the size range of 160–205 mm. and weight range of 40–105 g. Berried prawn of the size range of 112 162 mm. and weight 18.32 g to 40.0g were collected during August to October 2002 for a study, breeding and fecundity. In this paper, an effort has been made to study the pond culture growth of this non-resident species. It is also tried to study of breeding and fecundity for establishment of prawn hatchery in north India and to solve the problems to produce scampi seeds in bulk quantities for supply to farmers in the region. Details of the experiment are given in this chapter.

Keywords: Macrobrachium rosenbergii, Scampi growth, Breeding and fecundity.

---

* Corresponding Author; E-mail: sudavprasad@yahoo.co.in

Introduction

Freshwater prawn farming is assuming great importance due to high demand, good price and high returns. Giant freshwater prawn *Macrobrachium rosenbergii* belonging to phylum Arthropoda, order decapoda and family paliminaideae, is available in good quantity in the riverine and lake area of south India. There is a great scope for prawn production in numerous freshwater bodies throughout North India (Prasad and Singh 2006). Tropical climate is congenial for faster growth of scampi. Since profits through scampi farming are much higher as compared to carp culture (Brody et ai, 1980), many fanners in the inland states includhlg Punjab and Haryana have initiated scampi farming. The breeding and fecundity of *M. rosenbergii* has been mentioned briefly by several authors (Ling 1969; Patra 1976; Rao 1986; Jee and Kok 1991; Prasad and Singh 2006). These authors reported that the fecundity of M rosenbergii range from 20 000 to 1, 62, 000. The stipulated objective of the study in the present paper an effort has been done to study in pond Culture, Breeding and fecundity performance of *M. rosenbergii* for establishment of prawn hatchery in north India.

Materials and Methods

The field study site was located at Fatehabad district of Haryana. The Ludhiana Regional Research Centre of CIF A has adopted fish farmer pond in Punjab and Haryana. The adopted farmers' ponds size is around 0.4 ha. Prior to stocking, the pond was treated with lime in order to neutralize the acidity of the pond soil as well as for providing calcium to the prawns, which is essential for moulting. Dried basal portion along with 2 feet stem portion of a locally available bushy plant (*Sacharum munja*) were kept at different locations @ 50 nos./ha to act as artificial shelters/substrates. Prawn seeds were (12–14 mm) procured from the Maharaja Prawn hatchery at Chennai and stocked in the ponds. The prawns were fed with different grades of commercial feed (crumbles/pellets) 3–4 times daily @ 20–30 per cent of the estimated biomass. Monthly sampling was done during the culture period. The growth rate, sex ratio, survival percentage etc. were noted. The culture period ranged from 5 to 6 months. Hydrographical parameters of the pond condition were noted. To study the spawning biology, the berried females were collected from culture pond. They were kept in 80 L water holding capacity in glass aquarium and were provided with good quality 5ppt brackishwater and were also provided with aeration. The hatched zoea was shifted to rearing tanks. the zoeas were provided with live feed as *Artemia nauplii* for further development. For fecundity study the berried prawns were weighed first and they were also weighed again after the release of the eggs. Sample of eggs were taken from berried prawns and weighed, the number of eggs in the sample were counted.

Results and Discussion

During May 2002 the pond was stocked with Prawn seed ranged from 12–14 mm. The prawns were regularly fed with different grades of commercial feed (crumbles/pellets) 3–4 times daily @ 20–30 per cent of the estimated biomass. Sampling were done in regular intervals by using a lift net happa and cast net. The stocking, sampling and harvesting data are summarized in Table 55.1. In the sampling done on July 2002 the size increased to 48.0–92.0 mm and weight increased to 4.50–7.94 g this sampling carried out during October 2002 showed that size and weight increased to 155–193 mm and 30.2–88 g. During harvest all prawns ranged from 160–195 mm and weight ranged from 40.1–95 g.

During the process of scampi culture, a major difference in growth sequence of the males and of the females became very apparent. The estimated survival was 60 per cent at 5 to 6 months of culture. The estimated survival was in general agreement with Singh (2004) Prasad and Singh (2006), who reviewed the experimental culture of Giant freshwater prawn *M. rosenbergii* in North India The estimated production and survival according to him was 1.05–2.2 t/ha and 52–78 per cent in monoculture. Male and female prawns grew at similar rates until the female began to divert much of its energy intake into ovarian development and less into growth. More or less same observations were reported by Brody *et al.*, 1980; Koizumi, 2001; Prasad and Singh, 2006. The hyprographical parameters recorded during the experiment are given in Table 55.2. During the culture period the salinity ranged between 0.1 and 1.0 per cent; the water temperature ranged from 26 to 33.0°C the pH ranged from 7.6 to 8.0 and Transparency ranged from 31.2 to 53.0 cm.

Table 55.1: Details Regarding Giant Freshwater Prawn Culture Growth of *M. rosenbergii* Experiment Carried Out in a 0.4 ha in Pond at Fatehabad During 2002

Source of Seed	*Date of Stocking*	*Stocking Density*	*Total Area*	*Village*
Maharaja Hatchery	02/05/2002	22,000/-	0.4 ha.	Salemeda
Date of Sampling	*Total length mm.*	*Wt. range in gm.*	*Berried*	*Sex ratio F: M*
29/06/2002	30.0–70.0	2.06–4.05	–	7:1
29/06/02	48.0–92.0	4.50–7.94	–	7:1
23/07/02	60.0–130	7.0–20.1	–	6:1
23/08/02	76.0–155	7.4–33.2	Observed	5:1
31/08/02	96.0–176	8.09–41.0	Observed	4:1
16/09/02	120–178	16.01–54.0	Observed	4.6:1
10/10/02	13–180	30.0–65.0	Observed	4.5:1
28/10/02	155–193	30.2–88.0	Observed	4.5:1
November	160–195	40.1–95	Observed	3.6:1

Table 55.2: Physico-chemical Parameters of *M. rosenbergii* Culture Pond at Fatehabad During May 2002 to October 2002

Months	*Salinity*	*Water Temperature (°C)*	*pH*	*Transparency (cm)*
May	0.1	28.6	7.6	51
June	0.1	31.2	7.9	53
July	1.0	32.4	7.9	48
August	1.0	33.0	8.0	31.2
September	0.1	32.8	8.0	32.2
October	0.1	27.4	7.9	39.2
November	0.1	26.0	7.8	43.2

Maturation and Breeding

Scampi reached spawning stage in last week of August 2002. Sexually matured males and females were collected and transported from pond to prawn hatchery at RRC of CIFA Ludhiana. The females measured 140–166 mm and weighted 38–50 g sexually matured females characterized by the fully developed ovaries observed with large yellow colour mass occupied a major portion of the lateral part of the transparent carapace commonly called head. The abdominal pleurae bent slightly inwards, the pleopods become slightly distended and arched out-ward, to form the space between them. Below the abdomen, this formed an enlarged brood chamber to, accommodated large number of eggs being spawned during breeding. The basal segments of the pleopods especially from first three pairs found elongated and provided with soft long setae called ovigearous setae, which bears the eggs. These characters and development observed in females of the prawn species. Once the ovary in females matured the pre-mating molt in females occurred shortly before mating. After pre matting molt the freshly molted female was found weak and. sluggish and searched shelter for hiding to protect itself from predators. It took few hours for the new body covering shell to become sufficiently harden. Perhaps the female prawn secreted a kind of hormones during pre matting molt, which strongly attracted the males for mating. Mating was induced in glass aquarium by introducing a matured male with sexually ripe female. They usually took few minutes to get aquatinted with each others. The maturation fecundity and breeding of *M. rosenbergii* recorded under captive condition from August to October as given in Table 55.3 is similar observed by Ling 1969; Rao 1986; Prasad and Kanaujia 2006; Prasad and Singh 2006, 2007; Prasad 2007.

Sex Ratio and Spawning

For determination of sex ratio the individuals could be broadly grouped as between 30 to 70 mm size, the proportion of females to males was 7:1 the between 48–92 mm the ratio was more or less equal. The proportion of females to males beyond the length of 76–155 mm decreases progressively. The berried female was observed from month of August to harvesting time. After the completion of mating eggs were extruded directly into the brood chamber passing through female genital pore present at the base of third thoracic leg. The eggs were hold in bundles like grapes by some thin, elastic membranous substance that adhered tightly to the fine ovegerous setae of the first to fourth pairs of pleopods. The release of eggs from the ovary to the brood chamber was observed to be in a tubular form, which become rounded after leaving the genital pore. The colour of freshly spawned eggs was yellow, resembling the observations by other authors Ling 1969; Rao 1986; Prasad and Singh 2006; Prasad 2007. The berried female carried the eggs and cared after them till hatching period. During incubation period the variations in important physico-chemical parameters was water temperature ranged between 28.3–30°C, pH 7.6–8.2 and salinity was between 3–5 ppt. During incubation period the pleopods observed to beat back and fourth intermittently to provide aeration for developing embryos. The incubation period recorded in *M. rosenbergii* was 12–16 days. During incubation period the colour of the eggs gradually become lighter when the larvae inside the eggs, and then embryo found fully developed, the colour of egg became gray. More or less similar observations were found by New and Singholka 1985; Prasad 2007. The fecundity was estimated from the berried females collected from the pond sites during different months, different size group and their weight co-related with the egg mass and numbers of eggs by the females are presented in Table 55.3. In the present study the estimated fecundity from 140–166 mm total length of *M. rosenbergii* with 8.9–19.4 g egg mass was in the range of 27,6600–89,121 number. The females with minimum weight 38.9 gm with 8.9 gm egg mass had minimum number of eggs. The number of eggs carried by a female was found to be related to the size of the prawn. Patra 1976; Jee and Kok 1991 mentioned that the relationship of body length with fecundity was significant while it was highly significant between weight of egg mass and fecundity the same has been observed in present experiment.

Table 55.3: Details Regarding Berried Prawn *M. rosenbergii* and their Fecundity

Size Group	*Weight of Prawn (gms)*	*Weight of Egg Mass (gms)*	*Total Number Eggs of Carried by the Female*
140	38	8.9	27,660
148	39	9.2	37,598
150	42	10.1	48,434
153	46	12.1	51,880
157	46	14.0	60,892
160	45	15.2	66,600
164	47	18.1	79,400
164	48	18.8	81,200
166	50	19.4	89,121

The result on growth, breeding and fecundity of this species has indicated the establishment of prawn hatchery and culture in inland region, particularly in north India.

Acknowledgements

The authors gratefully acknowledge the financial support of DBT New Delhi for the Adhoc research project "Establishment of a Nucleus Freshwater Prawn Hatchery in Punjab". Authors wish to express their sincere gratefulness to Dr. S. Ayyappan, DDG (FY) ICAR, New Delhi for his encouragement and support to all staff of RRC of CIFA, Ludhiana.

References

Brody, T., Cohen, D.A. Barnes and Sector, A., 1980. Yield characteristics of the prawn *M. rosenbergii* in temperate zone aquaculture. *Aquaculture*, 12: 375–385.

Jee, A.K. and Kok, L.Y., 1991. Fecundity changes in *Macrobrachium rosenbergii* (de Man) during incubation. *Aquaculture and Fisheries Management*, 22: 1–6.

Koizumi, C., 2001. *Prawn of Japan and the World*. Oxford and IBH Pub., New Delhi, p. 45.

Ling, S.W., 1969. Methods of rearing and culturing *Macrobrachium rosenbergii* (De Man). *FAO Fish Rep.*, 57(3): 607–619.

New, M.B. and Singholka, S., 1985. *Freshwater Prawn Farming: A Manual for the Culture of M. rosenbergii.* FAO Fisheries Technical Paper, 222: 118.

Patra, R.W.R., 1976. The fecundity of *Macrobrachium rosenbergii. Bangladesh Journal of Zoology*, 4: 63–72.

Prasad, S., 2007. *Ganga Jhinga: Ek Parichey*. Daya Publishing House, Delhi, pp. 73–77.

Prasad, S. and Singh, H., 7007. Breeding and larval biology of *Macrobrachium rosenbergii* (de Man) in Punjab. *J. Zoological Society of India*, 6(2): 15–19.

Prasad, S. and Kanaujia, D.R., 2006. Availability and breeding behaviour of the Ganga river Prawn *Macrobrachium gangetium* (Bate) and *M. malcolmsonii* (H. M. Edwards). *Asian Fisheries Science*, 19(4): 377–388.

Prasad, S. and Singh, H., 2006. Brief history and prospects of freshwater prawn farming in North India. *Agricultural Situation in India*, p. 395–398.

Rao, K.J., 1986. Studies on maturation, breeding, fecundity and sex-ratio in *Macrobrachium malcolmsonii* from Kolleru Lake. *J Aquat. Biol.*, 4(2): 62–72.

Singh, H., 2004. Multilocation demonstrations to promote scampi, *Macrobrachium rosenbergii* farming in North India. *Applied Fisheries and Aquaculture*, 4(2): 63–66.

Chapter 56

Ultrasonic Study of Thermodynamic and Mechanical Properties of Liquid Mixture

C.M. Saxena, Archana Saxena and H.C. Srivastava
Department of Chemistry, D.B.S. (P.G.) College, Govind Nagar, Kanpur

ABSTRACT

The Ultrasonic is used in so many fields of Scientific researches in Physics, Chemistry, Biology, medicines and industry. It provides valuable tool to undersigned the nature of molecular interaction among the constituents of the liquid mixture. Young was the first scientist who carried out experimentally thermodynamic and mechanical properties of liquid mixture.

This study is concerned with the evaluation of excess thermodynamic function in order to make critical assessment of various theories of liquid-liquid mixture of non electrolytes *i.e.* these studies are mainly conducted experimentally termed as Sonography. In it frequency of ultrasonic waves in the liquid-liquid mixture is measured. This technique using ultrasonic instruments is in the tremendous use in measuring the rate of flow of blood, through the human body, and getting images of vital organs of the body like Kidney, Liver, Blood vessel, Foetus etc. Frequencies of 10^4 to 10^5 C.P.S. are used for industrial applications.

Introduction

The binary Liquid system Ethylamine + Cyclo hexanol gives the excess values of various parameters represented by 'A' as molar volume (Vm) isentropic compressibility (βs), intermolecular free length (Lf) available volume (Va) and viscosity (η) have been calculated.

The study of equilibrium properties of liquid mixture of non electrolytes has received the attention of various workers for a long time, as it has provided a valuable tool to understand the nature of molecular interactions among the constituents of the liquid mixtures. The first systematic attempt in this directions was made by young by collecting large amount of experimental data on the thermodynamic and mechanical properties liquid mixture since then many investigations (Chaudhary and Katti, 1989; Sharma *et al.*, 1989; Reddy and Jain, 1986; Naroen *et al.*, 1989; Yadav *et al.*, 1989; Ali *et al.*, 1989) have been engaged in the task of collecting more and more data and explaining it in terms of the properties of pure liquid. Quite a systematic theoretical study has also been made on liquid mixture.

Experimental

Ethylamine and cyclohexanol (both are BDH and E Merck reagent) were used without further purification.

The Volume of mixing of binary mixture is given by:

$$Vm = V - X_1 V_1 - X_2 V_2$$

where,

V is molar Volume

V_1 and V_2 are molar volume of pure components and X_1 and X_2 are mole fractions of the components 1 and 2.

In the present study density, ultrasonic Velocity, Viscosity, molar volume, isentropic Compressibility, intermolecular free length and available volume have been reported.

Results and Discussion

The results are shown in Table 56.1–56.12 and Figures 56.1–56.4 the deviations in the properties computed demonstrate that their exist a molecular interaction between unlike molecules of the liquid mixtures. These may be attributed to the change in the adhesive and the cohesive forces.

Table 56.1: Ultrasonic Velocities, Densities, Molar Volumes and their Excess Values for the System Ethylamine+Cyclohexanol (Temp. 25°C)

Mole Fraction of Ethylamine (X_1)	*Ultrasonic Velocity m/sec*	*Density gm/ml*	*Molar Volume (Exp) ml/mole*	*Molar Volume (add) ml/mole*	*Excess Molar Volume ml/mole*
0.0000	1464	0.9472	105.74	105.74	0.00
0.1032	1458	0.9391	100.60	100.49	+0.11
0.2019	1453	0.9306	95.68	95.46	+0.22
0.3007	1449	0.9207	90.79	90.43	+0.36
0.4023	1446	0.9095	85.76	85.26	+0.50
0.5004	1443	0.8978	80.86	80.26	+0.60

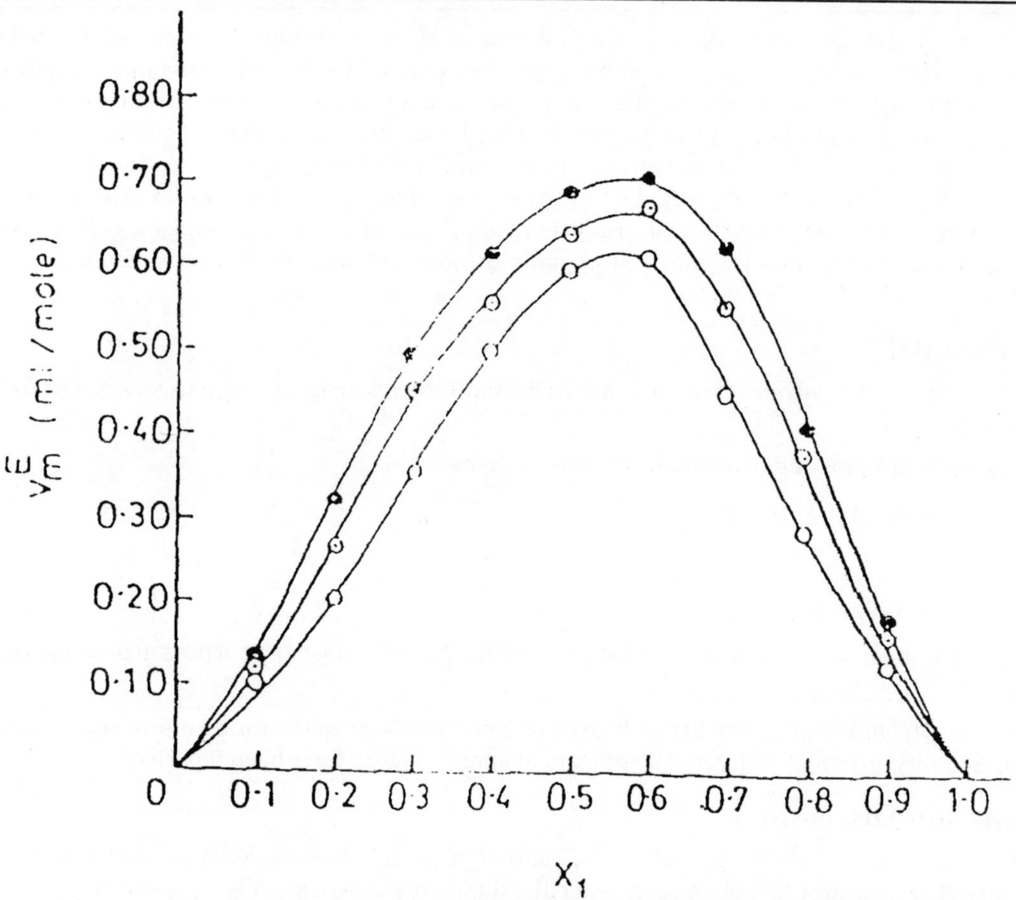

Figure 56.1: Excess Molar Volume V_m^E vs Mole Fraction (X_1) of Ethylamine for Ethylamine+Cyclohexanol at 25°C, 30°C and 35°C

Table 56.2: Isentropic Compressibilities Intermolecular Free Lengths and their Excess Values for the System Ethylamine+Cyclohexanol (Temp. 25°C)

Mole Fraction of Ethylamine (x_i)	*Isentropic Compressibility cm²/dyne × 10¹²*	*Isentropic Compressibility cm²/dyne × 10¹²*	*Excess Isentropic Compressibility cm²/dyne × 10¹²*	*Inter Molecular Free Length (exp) A°*	*Inter Molecular Free Length (add) A°*	*Excess Inter Molecular Free Length A°*
0.0000	49.25	49.25	0.00	0.4386	0.4386	0.0000
0.1032	50.09	50.06	+0.03	0.4423	0.4420	+0.0003
0.2019	50.89	50.83	+0.06	0.4458	0.4452	+0.0006
0.3007	51.73	51.60	+0.13	0.4495	0.4482	+0.0013
0.4023	52.58	52.39	+0.19	0.4532	0.4519	+0.0013
0.5004	53.46	53.16	+0.30	0.4569	0.4552	+0.0017

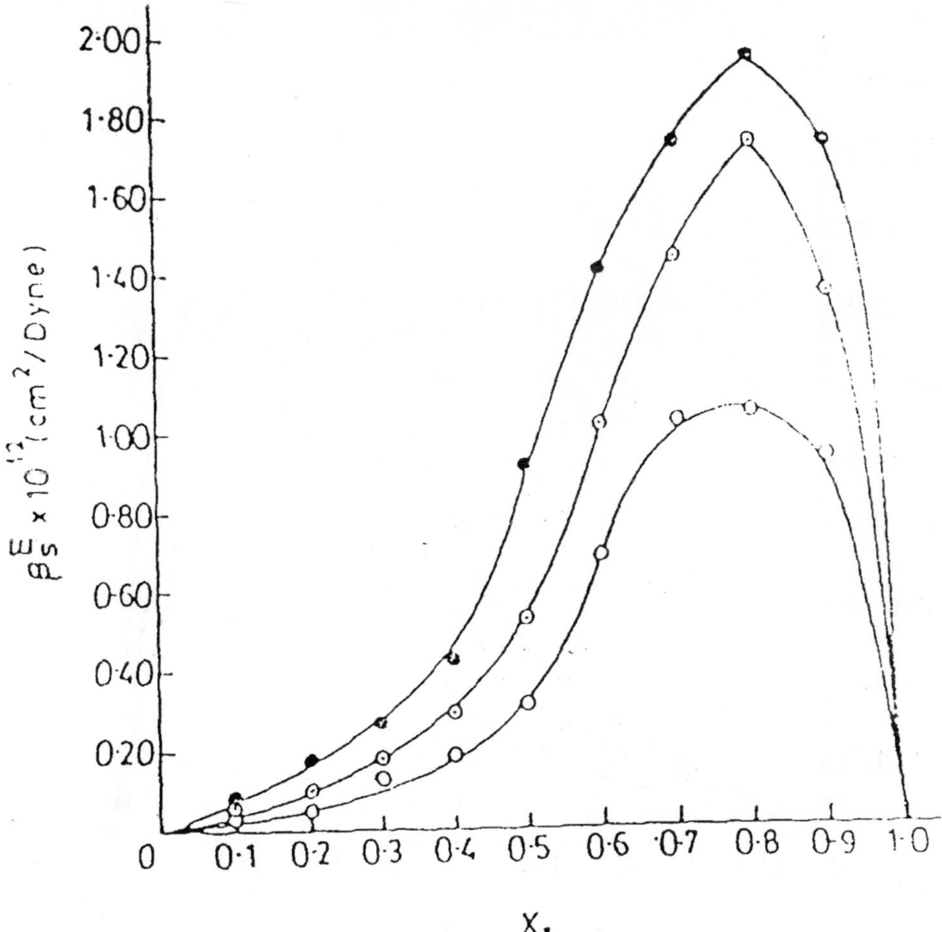

Figure 56.2: Excess Isentropic Compressibility (β_s^E) vs Mole Fraction (X_1) of Ethylamine for Ethylamine+Cyclohexanol at 25°C, 30°C and 35°C

Table 56.3: Available Volumes and their Excess Values for the System Ethylamine+Cyclohexanol (Temp. 25°C)

Mole Fraction of Ethylamine (x_1)	*Available Volume (exp) ml/mole*	*Available Volume (add) ml/mole*	*Excess Available Volume ml/mole*
0.0000	89.87	89.87	0.00
0.1032	89.28	85.53	+3.75
0.2019	87.90	81.40	+6.50
0.3007	85.68	77.17	+8.51
0.4023	82.54	72.90	+9.64
0.5004	79.14	68.88	+10.26

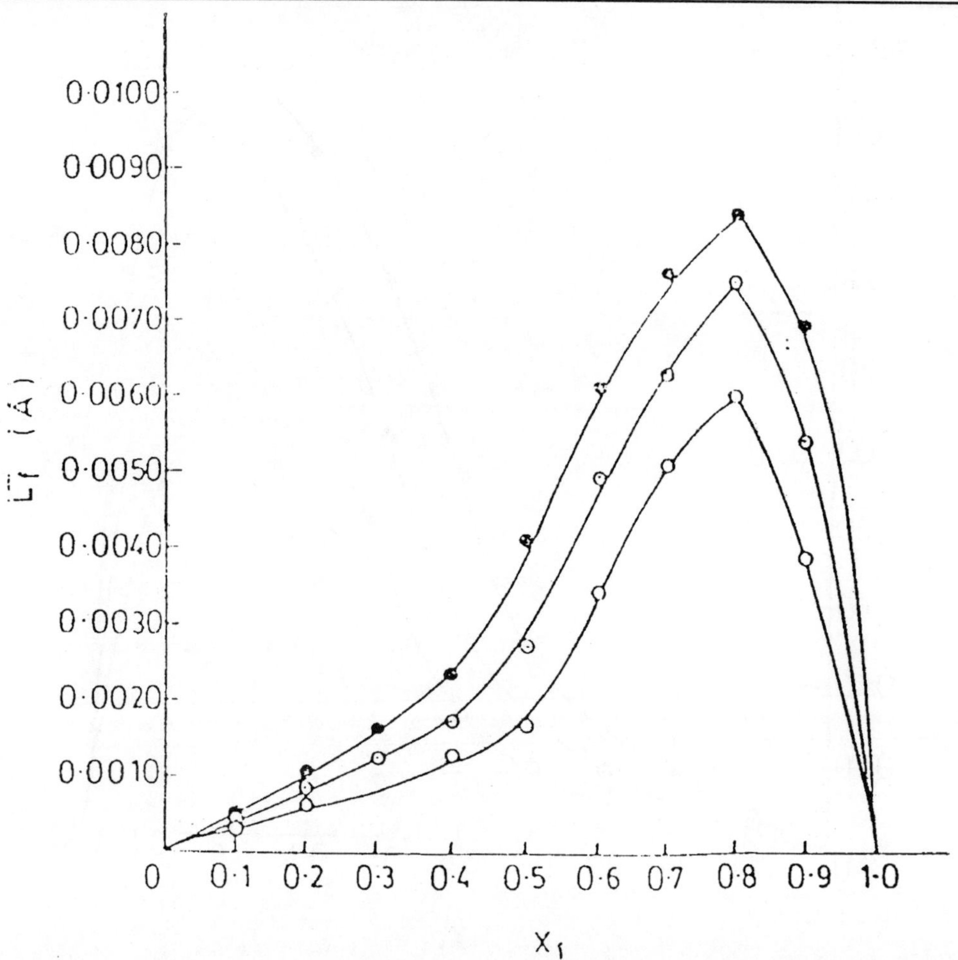

Figure 56.3: Excess Intermolecular Free Length (L_f^E) vs Mole Fraction (X_1) of Ethylamine for Ethylamine+Cyclohexanol at 25°C, 30°C and 35°C

Table 56.4: Viscosities and their Excess Values, Lnη^E and Nissan's Parameter (d) for the System Ethylamine+Cyclohexanol (Temp. 25°C)

Mole Fraction of Ethylamine (x_1)	*Viscosity (exp) cp*	*Viscosity (add) cp*	*Excess Viscosity cp*	*Lnη^E*	*d*
0.0000	29.449	29.449	0.000	0.000	0.0000
0.1032	16.316	26.522	−10.209	−0.249	−2.690
0.2019	11.350	23.720	−12.370	−0.285	−1.786
0.3007	8.073	20.970	−12.897	−0.294	−1.398
0.4023	5,582	18.034	−12.452	−0.331	−1.376
0.5004	3.972	15.250	−11.278	−0.345	−1.380

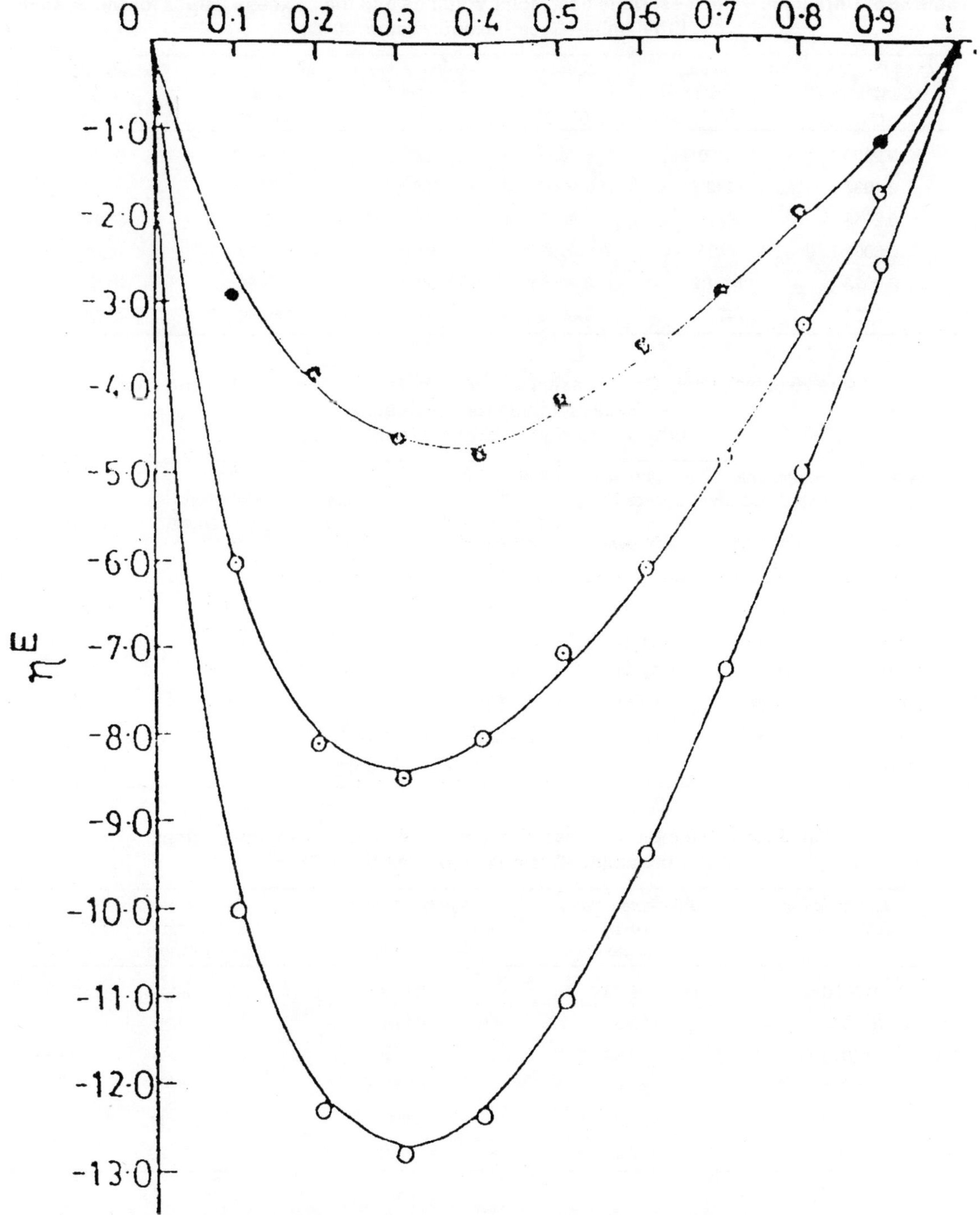

Figure 56.4: Excess Viscosity η^E vs Mole Fraction (X_1) of Ethylamine for Ethylamine+Cyclohexanol at 25°C, 30°C and 35°C

Table 56.5: Ultrasonic Velocities, Densities, Molar Volumes and their Excess Values for the System Ethylamine+Cyclohexanol (Temp. 30°C)

Mole Fraction of Ethylamine (X_1)	*Ultrasonic Velocity m/sec*	*Density gm/ml*	*Molar Volume (exp) ml/mole*	*Molar Volume (add) ml/mole*	*Excess Molar Volume ml/mole*
0.0000	1450	0.9438	106.12	106.12	0.00
0.1032	1443.8	0.9397	100.97	100.85	+0.12
0.2019	1439	0.9267	96.08	95.81	+0.27
0.3007	1435	0.9163	91.23	90.77	+0.46
0.4023	1431.2	0.9055	86.14	85.58	+0.56
0.5004	1427.4	0.8940	81.20	80.56	+0.64

Table 56.6: Isentropic Compressibilities, Intermolecular Free Lengths and their Excess Values for the System Ethylamine+Cyclohexanol (Temp. 30°C)

Mole Fraction of Ethylamine (x_i)	*Isentropic Compressibility* $cm^2/dyne \times 10^{12}$	*Isentropic Compressibility* $cm^2/dyne \times 10^{12}$	*Excess Isentropic Compressibility* $cm^2/dyne \times 10^{12}$	*Inter Molecular Free Length (exp)* A^0	*Inter Molecular Free Length (add)* A^0	*Excess Inter Molecular Free Length* A^0
0.0000	50.39	50.39	0.00	0.4479	0.4479	0.0000
0.1032	51.26	51.22	+0.04	0.4517	0.4513	+0.0004
0.2019	52.11	52.01	+0.10	0.4555	0.4547	+0.0008
0.3007	52.99	52.81	+0.18	0.4593	0.4581	+0.0012
0.4023	53.91	53.62	+0.29	0.4633	0.4616	+0.0017
0.5004	54.93	54.41	+0.52	0.4676	0.4649	+0.0027

Table 56.7: Available Volumes and their Excess Values for the System Ethylamine+Cyclohexanol (Temp. 30°C)

Mole Fraction of Ethylamine (x_1)	*Available Volume (exp) ml/mole*	*Available Volume (add) ml/mole*	*Excess Available Volume ml/mole*
0.0000	99.48	99.48	0.00
0.1032	98.57	94.67	+3.90
0.2019	96.68	90.08	+6.60
0.3007	94.08	95.38	+8.70
0.4023	90.87	80.76	+10.11
0.5004	87.59	76.20	+11.39

Table 56.8: Viscosities and their Excess Values, $Ln\eta^E$ and Nissan's Parameter (d) for the System Ethylamine+Cyclohexanol (Temp. 25°C)

Mole Fraction of Ethylamine (x_1)	*Viscosity (exp) cp*	*Viscosity (add) cp*	*Excess Viscosity cp*	*$Ln\eta^E$*	*d*
0.0000	20.274	20.274	0.0000	0.0000	0.0000
0.1032	12.256	18.279	–6.023	–0.187	–2.020
0.2019	8.358	16.371	–8.013	–0.267	–1.656
0.3007	5.939	14.461	–8.522	–0.267	–1.271
0.4023	4.371	12.497	–8.126	–0.302	–1.255
0.5004	3.412	10.601	–7.188	–0.248	–0.992

Table 56.9: Ultrasonic Velocities, Densities, Molar Volumes and their Excess Values for the System Ethylamine+Cyclohexanol (Temp. 35°C)

Mole Fraction of Ethylamine (X_1)	*Ultrasonic Velocity m/sec*	*Density gm/ml*	*Molar Volume (exp) ml/mole*	*Molar Volume (add) ml/mole*	*Excess Molar Volume ml/mole*
0.0000	1436	0.9404	106.50	106.50	0.00
0.1032	1430	0.9322	101.35	101.22	+0.13
0.2019	1425	0.9227	96.49	96.16	+0.33
0.3007	1421	0.9126	91.60	91.10	+0.50
0.4023	1417	0.9016	86.51	85.89	+0.62
0.5004	1410	0.8901	81.56	80.87	+0.69

Table 56.10: Isentropic Compressibilities, Intermolecular Free Lengths and their Excess Values for the System Ethylamine+Cyclohexanol (Temp. 35°C)

Mole Fraction of Ethylamine (x_i)	*Isentropic Compressibility cm²/dyne × 10^{12}*	*Isentropic Compressibility cm²/dyne × 10^{12}*	*Excess Isentropic Compressibility cm²/dyne × 10^{12}*	*Inter Molecular Free Length (exp) A°*	*Inter Molecular Free Length (add) A°*	*Excess Inter Molecular Free Length A°*
0.0000	51.56	51.56	0.00	0.4570	0.4570	0.0000
0.1032	52.45	52.40	+0.05	0.4609	0.4604	+0.0005
0.2019	53.37	53.19	+0.18	0.4649	0.4639	+0.0010
0.3007	54.26	53.99	+0.27	0.4688	0.4672	+0.0016
0.4023	55.23	54.81	+0.42	0.4730	0.4707	+0.0023
0.5004	56.50	55.60	+0.90	0.4784	0.4743	+0.0041

Table 56.11: Available Volumes and their Excess Values for the System Ethylamine+Cyclohexanol (Temp. 35°C)

Mole Fraction of Ethylamine (x_1)	*Available Volumes (exp) ml/mole*	*Available Volumes (add) ml/mole*	*Excess Available Volumes ml/mole*
0.0000	109.16	109.16	0.00
0.1032	107.68	103.80	+3.88
0.2019	105.53	98.69	+6.84
0.3007	102.47	93.46	+9.01
0.4023	98.94	88.20	+10.64
0.5004	96.85	83.22	+13.60

Table 56.12: Viscosities and their Excess Values, $Ln\eta^E$ and Nissan's Parameter (d) for the System Ethylamine+Cyclohexanol (Temp. 25°C)

Mole Fraction of Ethylamine (x_1)	*Viscosity (exp) cp*	*Viscosity (add) cp*	*Excess Viscosity cp*	$Ln\eta^E$	*d*
0.0000	12.986	12.986	0.0000	0.0000	0.0000
0.1032	8.816	11.731	–2.915	–0.102	–1.102
0.2019	6.682	10.530	–3.848	–0.107	–0.664
0.3007	4.665	9.328	–4.663	–0.190	–0.904
0.4023	3.392	8.092	–4.700	–0.233	.0.968
0.5004	2.623	6.899	–4.276	–0.219	–0.876

The mixture of ethylamine + Cyclohexnol are polar-polar system at 25° C the Ultrasound velocity, density molar volume Viscosity and available volume decrease with the increase in amount of cyclehexanol Tables 56.1, 56.3 and 56.4 and Figures 56.1 and 56.4) while isentropic compressibility intermolecular free length increase under similar condition (Table 56.2 and Figures 56.2 and 56.3) this trend is maintained increase in temperature 25°, 30° and 35°C. The excess values of isentropic compressibility βs^E and Lf^E are positive at all the composition the mixtures and all the three temperature studied.

References

Ali, M. Azar, Parashar, R., Saxena, A. and Lakhanpal, M.L., 1989. *Ind. J. Chem.*, 28A: 512.

Chaudhary, S.K. and Katti, S.S., 1989. *J. Chem Thermodyn.*, 21(6): 561.

Kajuo, Kojima, 1998. *J. Chem. Eng.*, American Chemical Society, 43(5): 687–694.

Naroen, H., Kishore, N. and Suri, S.K., 1989. *Can.J. Chem.*, 67(4): 648.

Paola, B., Lucianoand, L. and Enrico, 1989. *Fluid Phase Eqilib.*, 44(3): 285.

Reddy, S.K. and Jain, D.V.S., 1986. *Thermchim Acta,* 107: 383.

Sharma, S.C., Joshi, I. and Singh, J., 1989. *J. Chem. Thermodyn.*, 211.

Yadav, R.R., Yadav, S.S. and Maurya, S.R., 1989. *Ind. J. Chem.*, 27A: 716.

Chapter 57

The Physico-chemical Characteristics and Nutrient Status of Soils of Tarn Taran District of Punjab, India

Meharban Singh
Farm Advisory Service, Punjab Agricultural University, Kheti Bhawan, Ranjit Avenue, Amritsar – 143 001, Punjab

ABSTRACT

The physico-chemical characteristics and nutrient status of the soils of Tarn Taran district were studied by collecting 90 surface (0–15 cm) soil samples from each of the 8 development blocks. The soil samples were analyzed for texture, pH, electrical conductivity, organic carbon, available phosphorus and available potassium. Soils of the district are mainly loam in texture, normal to alkaline in pH and normal in salt concentration. In general, the nitrogen supplying capacity as indicated by organic carbon status and available phosphorus status of the soil varies from low to medium and available potassium status from medium to high. Low N, P and K status was found in 62.1, 63.5 and 2.4 per cent soils of Tarn Taran district, respectively.

Keywords: Nutrient availability, Soil fertility, Soil testing.

Introduction

To sustain agricultural production and maintain soil productivity, proper management of the basic resource of agriculture *i.e.* soil, is most essential. Study of the soil fertility parameters is important to determine the nutrient needs of crops and to develop the management strategies for maintaining soil productivity. Soil testing is one of the best available tools, to ascertain the physical characteristics and nutrient status of a field so as to assess the fertilizer requirements for a crop or a cropping system or for knowing the reclamation requirements if the soil is saline/sodic in nature. Punjab being an agricultural advanced state and having highest fertilizer use per unit area, needs planning at micro level to further boost agriculture production. Soil fertility status of a particular area can prove highly useful for manufacturers, administrators, planners, researchers and extension workers for proper distribution of chemical fertilizers. So, an attempt has been made to assess the physico-chemical characteristics and fertility status of Tarn Taran district soils for proper fertilizer use and maintenance of soil health.

Materials and Methods

Tarn Taran district lies in the semi-arid and less hot central plain zone of Punjab. It is surrounded by Amritsar district in the north, Kapurthala district in the east and Ferozpur district in the south. The western boundary of Tarn Taran district is international border with Pakistan. After the forming of this district in 2006 after bifurcation of Amritsar district into Tarn Taran and Amritsar districts, the soil test data was reanalyzed (Singh and Brar, 2005). The surface soil samples (0–15 cm) numbering 720 were collected from all the 8 developmental blocks of the district. Three villages, which were far apart from each other and represent the block were selected and 30 soil samples from each village were collected randomly. The soil samples were air dried, ground in a wooden pestle and mortar and sieved through a 2 mm plastic sieve. The samples were analyzed in the soil testing laboratory, PAU, Ludhiana for pH, electrical conductivity (EC) by conductivity bridge in 1 : 2 soil-water suspension, organic carbon (Walkley and Black,1934), available phosphorus (Olsen *et al.*, 1954) and available potassium (Piper, 1966). The tex1ure of the soil was judged by feel method. The samples thus analyzed were grouped into different categories according to the rating limits fixed for Punjab soils (Table 57.1) by Department of Soils, PAU, Ludhiana.

Table 57.1: Rating Limits for Different Soil Characteristics

Parameter	*Categories*				
Texture	Coarse (s + Ls)		Medium (Sl)		Fine (L + Cl)
Soil reaction (pH)	Normal (6.5–8.7)		Marginally Alkaline (8.7–9.3)		Alkaline (>9.3)
Electrical Conductivity (dS/m)	Normal (< 0 8)		Critical (>0.8)		
Organic Carbon (per cent)	Very low (<0.20)	Low (0.20–040)	Medium (0.40–0.75)	High (0.75–0.90)	Very high (>0.90)
Available Phosphorus (P kg ha^{-1})	Very low (<5.0)	Low (5.0–12.5)	Medium (12.5–225)	High (22.5–50.0)	Very high (>50.0)
Available Potassium (K_2O kg h^{-1})	Low (<137.5)		Medium (137.5–337.5)		High (>337.5)

Results and Discussion

Soil Texture

The soils of Tarn Taran district are divided into three categories based on their texture *i.e.* coarse (sand and loamy sand), medium (sandy loam) and fine (loam and finer than loam). Soils of the district, on an average are fine *i.e.*, of loam (74.0 per cent) and clay loam (2.4 per cent) texture. However, sandy loam (19.3 per cent) and loamy sand (4.3 per cent) texture soils are also found in the district (Table 57.3). Loamy sand and sandy loam soils are observed, in the decreasing occurrence, in Chola Sahib (58 per cent), Bhikhiwind (50 per cent), Patti (30 per cent), Tarn Taran (17 per cent), Valtoha (14 per cent), Khadoor Sahib (11 per cent), Gandiwind and Naushera Pannuan (4 per cent each) blocks of the district. Loamy sand soils are reported from Chohla Sahib (18.9 per cent), Tarn Taran (11.1 per cent) and Patti (4.4 per cent) blocks only.

Table 57.2: Per cent Distribution of Soil Samples in Different Categories of Soil Texture, pH and Electrical Conductivity

Block	*Soil Texture*				*pH*			*EC (dS m⁻¹)*	
	Loamy Sand	*Sandy loam*	*Loam loam*	*Clay 8.7*	*6.5–9.3*	*8.8–0.8*	*>9.3*	*<0.4*	*0.4–*
Khadoor Sahib	–	11.1	88.9	–	76.7	23.3	–	100.0	–
Chohla Sahib	18.9	38.9	42.2	–	33.3	57.8	8.9	91.1	8.9
Bhikhiwind	–	50.0	50.0	–	8.9	54.4	36.7	73.3	26.7
Tam Taran	11.1	5.6	83.3	–	32.2	55.6	12.2	93.3	6.7
Gandiwind	–	4.4	95.6	–	17.8	36.7	45.6	65.6	34.4
Patti	4.4	25.6	51.1	18.9	14.4	61.1	24.4	78.9	21.1
Naushera Pannuan	–	4.4	95.6	–	88.9	11.1	–	100.0	–
Valtoha	–	14.4	85.6	–	40.0	43.3	16.7	96.7	3.3
Average	4.3	19.3	74.0	24	39.0	42.9	18.1	87.4	12.6

Soil Reaction

On an average; 39, 42.9 and 18.1 per cent soils of the district were having normal (pH 6.5–8.7), marginally alkaline (pH 8.7–9.3) and alkaline soil reaction (above 9.3) respectively (Table 57.3). Soils of different blocks of the district having pH value above 8.7, varied from 11.1 (Naushera Pannuan) to 91 (Bhikhiwind) percent, which shows wide variation in soil reaction in the district. Alkaline soils (*i.e.* pH > 9.3), which need reclamation by the application of gypsum or any other suitable amendment, are found in Gandiwind (46 per cent), Bhikhiwind (37 per cent), Patti (24 per cent), Valtoha (17 per cent), Tarn Taran (12 per cent) and Chohla Sahib (9 per cent) blocks. The soils of Khadoor Sahib and Naushera Pannuan blocks are found to be relatively free from alkalinity.

Brar *et al.* (1983) studied tour blocks *i.e.* Bhikhiwind, Gandiwind, Tarn Taran and Valtoha of this district and extent of soils having pH > 8.7 was reported as 100, 96, 88 and 8 per cent respectively and soils having pH > 9.3, as 85, 87, 73 and 50 per cent respectively, in these blocks. On an average, for Bhikhiwind, Gandiwind and Tarn Taran blocks, alkaline soils (*i.e.* pH >9.3) have decreased from 81.7 per cent) to 31.7 per cent in about 20 years. The lower percentage of alkaline soils in the present study shows that the effective technology developed by the Department of Soils, P AU, Ludhiana for reclamation of salt affected soils played a key role in lowering the alkalinity in the soils of this area, during the last two decades. However, Valtoha block has shown an increase in soil alkalinity, *i.e.* percentage of soils having pH > 8.7 and > 9.3 have increased from 8 and 5 per cent to 60.0 and 16.7 per cent, respectively, which was probably due to the underground sodic waters in this block.

Table 57.3: Per cent Distribution of Soil Samples in Different Categories of Organic Carbon, Available Phosphorus and Available Potassium

Block	Organic carbon (per cent)					Available phosphorus (kg P ha^{-1})					Available potassium (kg K_2O ha^{-1})		
	V. Low	Low	Medium	High	V. High	V. Low	Low	Medium	High	V. High	V. Low	Medium	High
	< 0.20	0.20–0.40	0.40–0.75	0.75–0.90	>0.90	< 5.0	5.0–12.5	12.5–22.5	22.5–50.0	>50.0	<137.5	137.5–337.5	>337.5
Khadoor Sahib	1.1	8.9	88.9	1.1	–	–	32.2	32.2	26.7	8.9	5.6	52.2	42.2
Chohla Sahib	22.2	56.7	21.1	–	–	–	75.6	21.1	3.3	–	–	65.6	34.4
Bhikhiwind	31.1	55.6	13.3	–	–	–	96.7	3.3	–	–	2.2	76.7	21.1
Tarn Taran	3.3	46.7	47.8	–	2.2	–	86.7	12.2	1.1	–	2.2	38.9	58.9
Gandi wind	–	24.5	72.2	3.3	–	–	62.2	32.2	5.6	–	5.6	42.2	52.2
Patti	11.1	48.9	36.7	3.3	–	–	71.1	27.8	1.1	–	–	16.7	83.3
Naushera Pannuan	8.9	85.6	5.5	–	–	–	58.9	32.2	8.9	–	3.3	71.1	25.6
Valtoha	16.7	75.5	7.8	–	–	–	24.5	73.3	2.2	–	–	62.2	37.8
Average	11.8	50.3	36.7	1.0	0.3	–	63.5	29.3	6.1	1.1	2.4	53.2	44.4

Electrical Conductivity

Soil salinity, as indicated by EC of the soil, was not a problem in the district as no sample showed EC above 0.8 dS m^{-1} (Table 57.3). Brar *et al.* (1983), however, has reported nil to 78 per cent soils having critical (0.81–1.60 dS m^{-1}) to injurious (>1.60 dS m^{-1}) EC levels in different blocks of the district Amritsar including Tarn Taran. The decrease in salinity is due to the leaching of salts during reclamation technology adopted and change in cropping pattern from non rice to rice based cropping sequence during the last two decades. However, slightly higher EC levels *i.e.* between 0.4 to 0.8 dS m^{-1} ranged from nil to 34.4 per cent in different blocks. Gandiwind, Bhikhiwind, Patti and Chohla Sahib blocks showed 34, 27, 21 and 9 per coot soils respectively, having EC between 0.4–0.8 dS m^{-1}.

Available Nitrogen Status

Although the texture and pH of the soil give some indication of the nutrient reserves and their availability in the soil, yet the available nutrient status gives a better estimate of the amount of nutrients which a soil can supply to the crops. Since organic carbon is related to the available nitrogen (Singh and Brar, 1973), the organic carbon content was taken as the index of nitrogen supplying capacity of the soil. The organic carbon percentage of Tarn Taran soils () varied as; very low (11.8 per cent),low (50.3 per cent), medium (36.7 per cent), high (1.0 per cent) and very high (0.3 per cent).

The blocks of Tarn Taran district, for nitrogen availability map, can be grouped into three categories *i.e.* low, medium and high (Table 57.2). According to these categories, soils of Naushera Pannuan, Valtoha, Bhikhiwind and Chola Sahib are low and of Patti and Tarn Taran are medium in nitrogen supplying capacity. Gandiwind, and Khadoor Sahib, blocks, grouped in high category, showed comparatively less percentage of soils having very low or low organic carbon status. There was wide variation in available nitrogen status of different blocks *e.g.*. soils having < 0.4 per cent O.C. were 94.4 per cent in Naushera Pamman and 10.0 per cent in Khadoor Sahib. The comparatively higher deficiency of nitrogen in four blocks of the district seems to be associated with coarse texture or alkalinity of the soils. In the four blocks (*i.e.* Valtoha, Bhikhiwind, Tarn Taran and Gandiwind) studied earlier by Brar *et al.* (1983), the soils having < 0.4 per cent O.C. have decreased from 93, 87, 92 and 78 per cent to 92, 86, 50 and 24 per cent respectively, in the present study. Increase in organic carbon status in, some blocks may be due to the effect of reclamation of alkaline soils, increase in cropping intensity and introduction of paddy on more area which leaves more root mass/crop residue to the soil, resulting in the higher organic matter in the soil. However, this effect was not observed in Valtoha and Bhikhiwind blocks.

Phosphorus

Available phosphorus content of Tarn Taran district soils, according to the classification given in Table 57.1, generally ranged from low (63.5 per cent) to medium (29.3 per cent), but high and very high values of available phosphorus content are also found in some soils. No sample tested very low (*i.e.* < 5.0 kg ha^{-1}) in P content. Soils having very high content of phosphorus (> 50.0 kg ha^{-1}) were observed only in Khadoor Sahib (9 per cent) block. Similarly, the soils having high P (22.5–50.0 kg ha^{-1}) content were also mainly observed in Khadoor Sahib (27 per cent). So, the soils of Khadoor Sahib block are having comparatively high P status, as compared to the other blocks of the district.

For preparing a soil phosphorus status map of the district, categorization can be done on the basis of extent of phosphorus deficient soils. Phosphorus deficient soils (*i.e.* < 12.5 kg P ha^{-1}) are 70–100 per cent in Bhikhiwind, Tarn Taran, Chola Sahib and Patti; 35–70 per cent in Gandiwind and Naushhera Pannuan and < 35 per cent in Khadoor Sahib and Valtoha blocks. As compared to the available phosphorus status in the earlier studies; phosphoms deficiency has increased in three out of the four blocks of Tarn Taran district studied in 1978 (Brar *et al.*, 1983). Percent P deficient soils in Bhikhiwind, Tarn Taran, Gandiwind and Valtoha blocks were 8, 14, 16 and 40 in 1978 (Brar *et al.*, 1983) and 97, 87, 62 and 24 in the present study. The average percent deficient samples in these 4 blocks has increased from 19.5 to 67.5 per cent) in the present study.

Based on temporal analysis of PAU Soil Testing Laboratory data; Benbi *et al.* (2006) has also reported that number of soil samples in low category has increased from 45 per cent in 1981–86 to 70.7 per cent in 1906–2001, for Punjab soils. The decreased consumption of phosphorus fertilizers in rice-wheat system may be the reason for decline in soil P status. The response of crops to applied phosphorus fertilizers varies according to the available P status of the soil and on the season in which crop is grown. So there is need to continuously monitor the P status of these soils for profitable economic response and ensure the optimum use of P based on the soil tests and crop rotations. The fields with very low/low available P status need higher amount of fertilizer P as compared to those testing medium, high/very high in available P; to obtain potential yield of a particular crop.

Potassium

Soils of Tarn Taran are generally medium to high in available potassium content. The available potassium rated as low (137.5 kg K_2O ha^{-1}), medium (137.5–337.5 kg K_2O ha^{-1}) and high (>337.5 kg K_2O ha^{-1}) was observed for 2.4, 53.2 and 44.4 per cent soils of the district. Per cent soils deficient in available potassium in different blocks, in the decreasing order were; Khadoor Sahib and Gandiwind (5.6 per cent); Naushera Pannuan (3.3 per cent); Bhikhiwind and Tarn Taran (2.2 per cent) and Chola Sahib, Patti and Valtoha (Nil). Percent samples in high category varied from 21.1 per cent) in Bhikhiwind to 83.3 per cent in Patti. This shows that though the soils are not deficient in available potassium status in general, yet their content varies much in different blocks of the district.

Earlier, Brar *et al.* (1983) has observed that no soil of Gandiwind and Valtoha blocks was in deficient category. Benbi and Brar (1995.) reported 3.7, 47.8 and 48.5 per cent soils of Amritsar (16 blocks *i.e.* including blocks of Tarn Taran district) in deficient, medium and high categories respectively. Meharban-Singh and Brar (2006) has observed that 9.6 per cent soils of Amritsar district (*i.e.* eight blocks only) were deficient in Potassium and the percentages ranged from 3.3 to 24.4 in different blocks. Slightly higher deficiency of Potassium in Amritsar as compared to Tarn Taran may be due to the fact that Amritsar is nearer the undulating plain and undulating sub mountain region of the state as compared to Tarn Taran. Potassium deficiency observed in some soils of Tarn Taran district may be due to coarse texture of the soils.

The adequate level of available potassium is associated with the prevalence of potassium rich mineral (Illite) in the alluvial plains of Punjab and addition of substantial amounts of potassium through irrigation waters. Benbi *et al.* (2006) reported that potassium deficient soils predominantly occurred in districts of Gurdaspur, Jalandhar and Hoshiarpur (11.6–25.2 per cent) and the per cent deficiency has not aggravated over the years, in Punjab. The response of crops to application of potassium fertilizers is expected in intensively cultivated and heavily leached soils only (Brar and Chhiba, 1994). So, the application of potassium fertilizers to crops, if any, must be on the basis of soil test. Singh and Brar (1977), observed that in loamy sand soil, the available potassium decreased from 160 to 85 kg ha^{-1} after 12 crops in maize-wheat rotation. So, the potassium status of the soil needs continuous monitoring, also.

References

Benbi, D.K. and Brar, J.S., 1995. Available potassium status of soils and delineation of deficient areas. In: *Proc. Workshop on Use of Potassium in Punjab Agriculture,* held at PAU, Ludhiana, India, p. 34–42.

Benbi, D.K., Nayyar, V.K. and Brar, J.S., 2006. The green revolution in Punjab: Impact on soil health. *Indian J. Fertilizers,* 2 4): 57–66.

Brar, J.S. and Chhibba, I.M., 1994. N, P and K status of Punjab soils. *Indian J. Ecol.,* 21(1): 34–38.

Brar, J.S., Singh, Balwinder and Chand, Diwan, 1983. Soil test summaries of soils of Majha tract of Punjab. *J. Res. Punjab Agric. Univ.,* 20(3): 285–292.

Brar, S.P.S., 1978. Fertility status of the Punjab soils. *J. Res. Punjab Agric. Univ.,* 16(3): 272–281.

Singh, Meharban and Brar, J.S., 2005. The physico-chemical characteristics and fertility status of soils of Amritsar district of Punjab, India. *J. Res. Punjab Agric. Univ.,* 42(1): 19–27.

Olsen, S.R., Cole, C.V., Watanbe, F.S. and Dean, L.A., 1954. Estimate of available P by extraction with sodium bicarbonate. *USDA Circ.,* 939: 19.

Piper, C.S., 1966. *Soil and Plant Analysis.* Hans Pub., Bombay.

Singh, B. and Brar, S.P.S., 1977. Dynamics of native and applied potassium in the maize-wheat rotation. *Potash Review (Cereal Crops),* 35th Suits: 1–6.

Singh, R. and Brar, S.P.S., 1973. Correlation of different soil tests with maize response for nitrogen and phosphorus at different fertility levels of the soil. *J. Indian Soc. Soil Science,* 21: 213–217.

Walkley, A. and Black, C.A., 1934. An examination of the digestion method for determining soil organic matter (Chromic acid titration method). *Soil Science,* 37: 28–35.

Chapter 58

Rainfall Prediction for Musi Reservoir Project Raingauge Station in A.P.

P. Srinivas[1]*, G.N. Pradeep Kumar[2] and K. Sunil Raju[3]

[1]Associate Professor, Department of Civil Engineering, Koneru Lakshmaiah College of Engineering, Vaddeswaram, Guntur District - 522 502

[2]Professor, Department of Civil Engineering, S.V.U. College of Engineering, Sri Venkateswara University, Tirupati- 517 502

[3]Assistant Executive Engineer, Vishakapatnam Port Trust, Vishakapatnam - 530 003

ABSTRACT

The primary objective of this study is to develop a time series model to predict the rainfall for Musi Reservoir Rain gauge station for the future years. Water is an important resource for the development of any community and especially for a developing country like India it should be used judiciously. Rainfall is the major means by which fresh water is made available. Effective and efficient operation of water resource systems frequently demands the prediction of hydrological sequences like rainfall. Hence forecasting models are necessary to predict the future rainfall based on past and current information. Rainfall prediction will tell us whether the coming years will be a good, bad or average rainfall years. Accordingly contingency measures can be planned well in advance, tide over difficulties in augmenting water for the use of the teeming population. In this study forecasting model namely Box-Jenkins ARMA model is used to predict the rainfall. Rainfall data of rain gauge station located at Musi Reservoir Project for past 37 years, (1968–2004) was obtained. Auto correlogram and partial auto correlogram drawn

* Corresponding Author; (M) 9866305533; Phone: (08645) 246948 (O), (0863) 2254940 (R); Fax: (08645)247249; E-mails: srinivas7_pasupuleti@yahoo.co.in; vasu77_p@yahoo.com.

for the series indicate that model can be ARMA (2, 2). Auto Regressive Moving Average model was fitted to the data and checked its validity by comparing fitted values with actual rainfall values. Forecast was made for next 6 years (2005–2010). Forecast values are well fitted with actual values. Thus prediction of rainfall is necessary for better planning of water resources and for optimum water management and utilization.

Keywords: *Rainfall prediction, ARMA model, Fitted values.*

Introduction

Water is a precious and important resource for the development of any community. In view of the vital importance of water, the principal constituent for sustenance of all living things in the universe and considering its increasing scarcity for maintaining ecological balance, economic and developmental activities of all kinds, the planning and management of this resource and its optimal, economical and equitable use has become a matter of utmost urgency. Hence, utilization of rainwater which otherwise flows into ocean and not useful, gains significance. Thus forecasting of rainfall is to be made to plan contingency measures well in advance. Makridakis, *et al.* (1978) states that effective and efficient operations of water resource systems frequently demand the prediction of hydrological sequences like rainfall. Hence, forecasting models are necessary to predict the future rainfall based on past and current information. Before being able to forecast hydrological sequences, models have to be found which describe past data adequately. Ideally these models should preserve the properties of the past rainfall data. There are three basic approaches in forecasting hydrological data, namely conceptual, multi regression and time series models. In conceptual models and multiple regression models the data is expressed as a function of important hydrological variables and parameters, the accurate estimation of these variables is a difficult task. On the other hand, in time series models, the rainfall data is evaluated in the form of an equation and future values are predicted using the equation. Water is an important resource for the development of any community and especially for a developing country like India it should be used judiciously. Rainfall data is of utmost importance to hydrologists as it forms the basis of many hydrological studies. Rainfall prediction will tell us whether the coming years will be a good, bad or average rainfall years. Accordingly contingency measures can be planned well in advance, tide over difficulties in augmenting water for the use of the teeming population.

Some of the earlier works in forecasting of Hydrological events are listed below:

A real-time river flow forecasting model, SMAR-AR is used for forecasting the Blue Nile river flows at Eldeim station near the Sudenese-Ethiopia border by Shamseldin *et al.* (1999). Box-Jenkins modeling of time series data can be improved and simplified by adhering to contemporary modeling procedures, proposes by Hipel *et al.* (1977). Srinivas *et al.* (2004) in their article discusses that effective and efficient operation of Water Resource Systems frequently demands the prediction of hydrological sequences like rainfall. Rainfall data of Saidapet raingauge station for past 100 years (1901–2000) is obtained in this study, forecasting model namely Box Jenkins Autoregressive Moving Average (ARMA) Model is used to forecast the rainfall. ARMA Model of order (2, 2) is fitted by using 84 years data (1901–1984) and checked its validity by using last 16 years data (1985–2000), by comparing fitted values with actual rainfall data. Then forecast is made for the next 25 years, (2001–2025). Sharma (2002) in his paper, investigated six models of the autoregressive moving average (ARMA) family for representing the monsoonal river inflow forecast models. The selected models were validated for significance of the residual mean, and for significance of the periodicities in the residuals. Mohan *et al.*

(1991) in their paper highlights that forecasting models are necessary to predict inflows into a reservoir that are helpful in making decisions regarding releases from it. Bertoni *et al.* (1991) in their article, discussed conceptual rainfall runoff model and precipitation forecasting model. Bras (1976) in their paper discusses the on-line hydrologic forecasting application to short-term forecasting of discharges into a river basin and to rainfall over small urban areas. Queen's University Forecasting Method (QUFM) developed by Watt *et al.* (1982) is a general forecasting method since it incorporates both the deterministic and stochastic components, the relative weights of each varying with application. All the earlier models detailed here have used Box-Jenkins models, for forecasting. Each one has their own adjustments and assumptions to suit their study. The above details are taken as guidance for our study. Jenkins (1976) describe a family of linear stochastic models that are commonly referred as either Jenkins or ARMA models. The objective of the present study is to develop a time series model to predict the rainfall for Musi Reservoir project raingauge station for the future years. The study area is Musi Reservoir Project Raingauge station, which is located near Solipet village in Suryapet mandal of Nalgonda district, Andhra Pradesh, India. The Musi Reservoir Project is located between geographical coordinates of East longitude 79°33' and North latitude 17°14'.

Time Series Models

These models use past data as the basis for estimating future results. Past patterns are analyzed and attempts are made to compensate for these patterns in the forecasting process. There are various types of time series models. Some of them are listed below: Single Exponential Smoothing Model, Auto Regressive Moving Average Model (ARMA), Auto Regressive Integrated Moving Average Model (ARIMA), Decomposition Model, Winter's Method, Adaptive Filtering and Census II Method.

Methodology

Box and Jenkins Method

This is one of the most sophisticated of all time series techniques as it is capable of handling almost any pattern of data. It usually provides the most accurate short-range forecasts of any model and even enables statistical tests to be applied to each forecast along with confidence intervals. The objective of this method is to end up with a specific formula for the forecast values that replicates the pattern in the series as closely as possible and also produce accurate results. The phases involved in this process are:

1. Model Identification Phase
2. Model Estimation and Validation Phase
3. Model Forecasting Phase
4. Model Updating Phase

The methodology is given in the flowchart Figure 58.1

Box and Jenkins Model Building Process

1. Plot the original series. Series is said to be stationary if it tends to wander more or less uniformly about mean. Otherwise non-stationary is implied.
2. Non-stationary series can be converted to stationary series by computational process known as regular differencing.

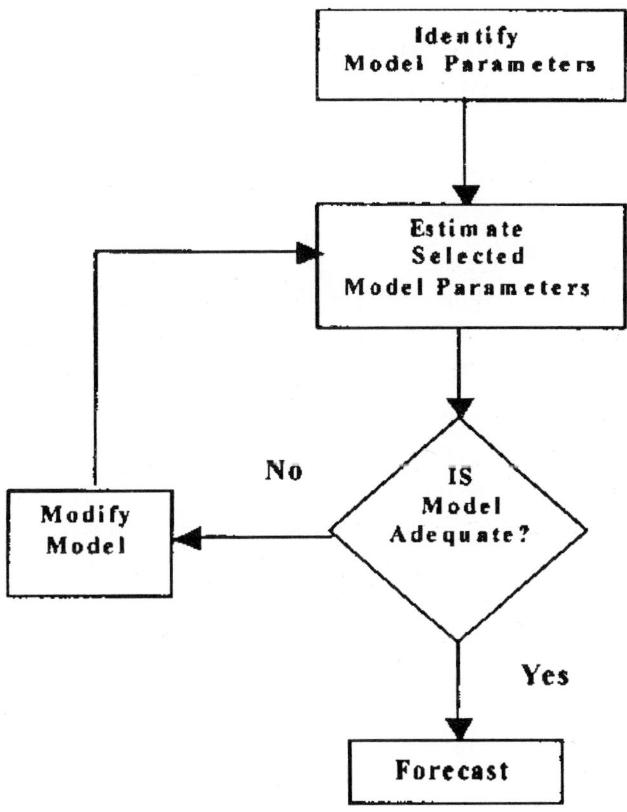

Figure 58.1: Time Series Model Building Process

3. Find the Auto correlations and the Partial auto correlations (Partials) of the original series.
4. Examine the Auto correlations, partials for the patterns, which reveal the type of model (AR/MA/ARMA) and its order.
5. Identify the tentative model and parameters needed to construct the model.
6. The best or optimum values for the selected model parameters are determined so that the model generates fitted values, which are as close as possible to the original series values.
7. Certain diagnostics are provided to check the validity of the model and suggest alternative models.
8. The selected and estimated model is used to generate forecasts of the time series.

Auto Regressive Moving Average Model (ARMA Model)

The ARMA model of order (p, q) is given by the equation as follows:

$$X_t = \Phi_1 X_{t-1} + \Phi_2 X_{t-2} + \cdots + \Phi_p X_{t-p} + e_t - \theta_1 e_{t-1} - \theta_2 e_{t-2} - \cdots - \theta_q e_{t-q} \quad (1)$$

where,

$\Phi_1, \Phi_2, ------ \Phi_p$ are Auto Regressive parameters

$\theta_1, \theta_2, ------- \theta_q$ are Moving Average parameters

e_t is random error

The choice of p and q are determined by Auto correlation coefficients calculated for the data.

Analysis of Data

The analysis of Musi Reservoir Project rainfall data was done for the past 37 years, (1968–2004) are plotted in Figure 58.2. Auto correlations and Partial Auto correlations of the original series were found out for corresponding lags. Autocorrelogram is shown in Figure 58.3. Autocorrelogram and Partial Autocorrelogram clearly indicate that the order of Model can be ARMA (2,2).

Forecast with Box-Jenkins ARMA Model

A Box-Jenkins ARMA model of order (2, 2) is fitted to the time series data. Rainfall data for past 37 years (1968–2004) is obtained. 22 years of data is taken to develop the ARMA model. Last 15 years are used to check the validity of model by comparing the fitted values with the actual rainfall values.

The developed ARMA model of order (2, 2) is given by the equation.

$$X_t = 0.52\, X_{t-1} + 0.27\, X_{t-2} + e_t + 0.25 e_{t-1} + 0.13\, e_{t-2} \tag{2}$$

Rainfall values and fitted values using ARMA (2,2) for years 1968–2004 are shown in Figure 58.4.

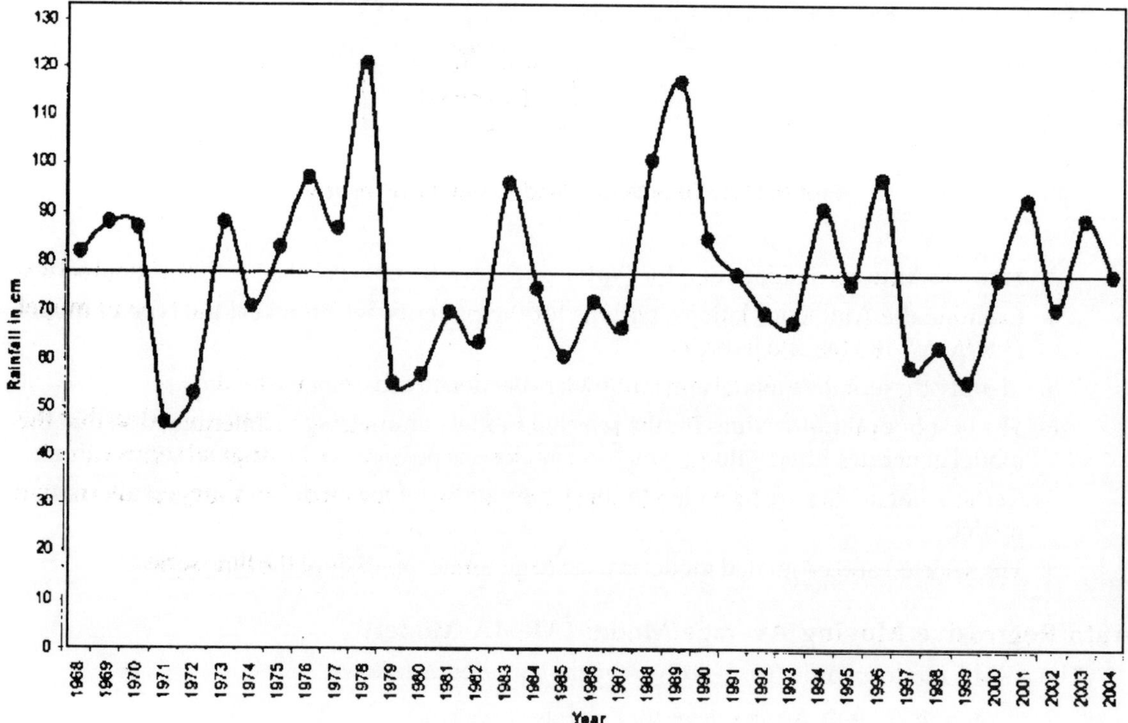

Figure 58.2: Musi Reservoir Project Rainfall Data (1968–2004)

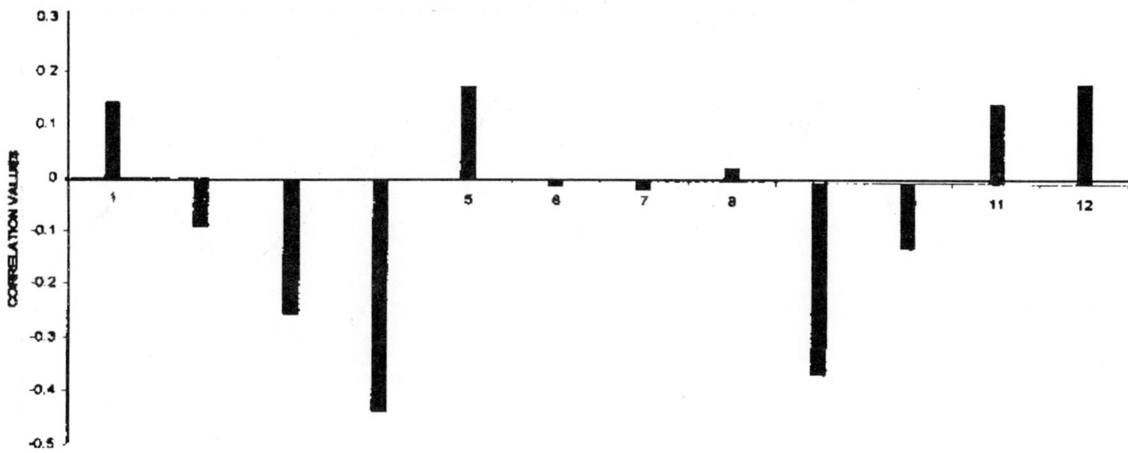

Figure 58.3: Autocorrelogram

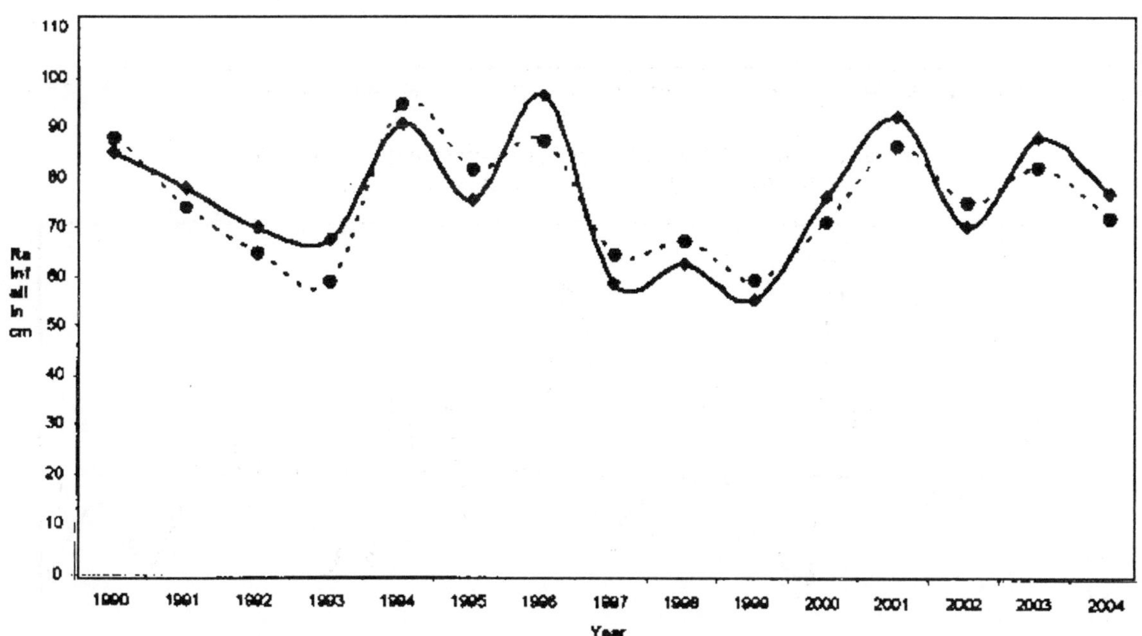

Figure 58.4: Comparison of Fitted Values with Rainfall Values

Results and Discussion

Rainfall data of Musi Reservoir Project raingauge station for the past 37 years, (1968–2004) is obtained for the study. ARMA model of order (2,2) is fitted by using 22 years data (1968–1989) and checked its validity by using 15 years data (1990–2004), by comparing fitted values with actual rainfall data and tabulated in Table 58.1. Then prediction of rainfall is done for the next 6 years, (2005–2010). Forecast values for the rainfall using ARMA (2, 2) for the next 6 years, (2005–2010) is tabulated

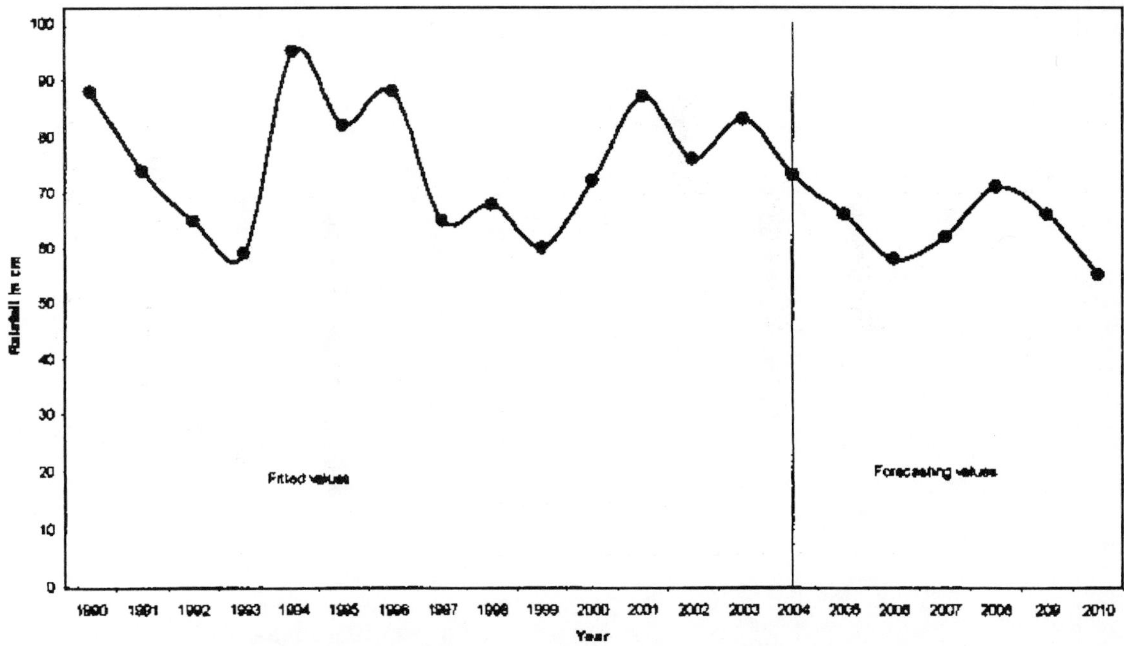

Figure 58.5: Rainfall Data of Musi Reservoir Raingauge Station (Both fitted and forecast values) (1990–2010)

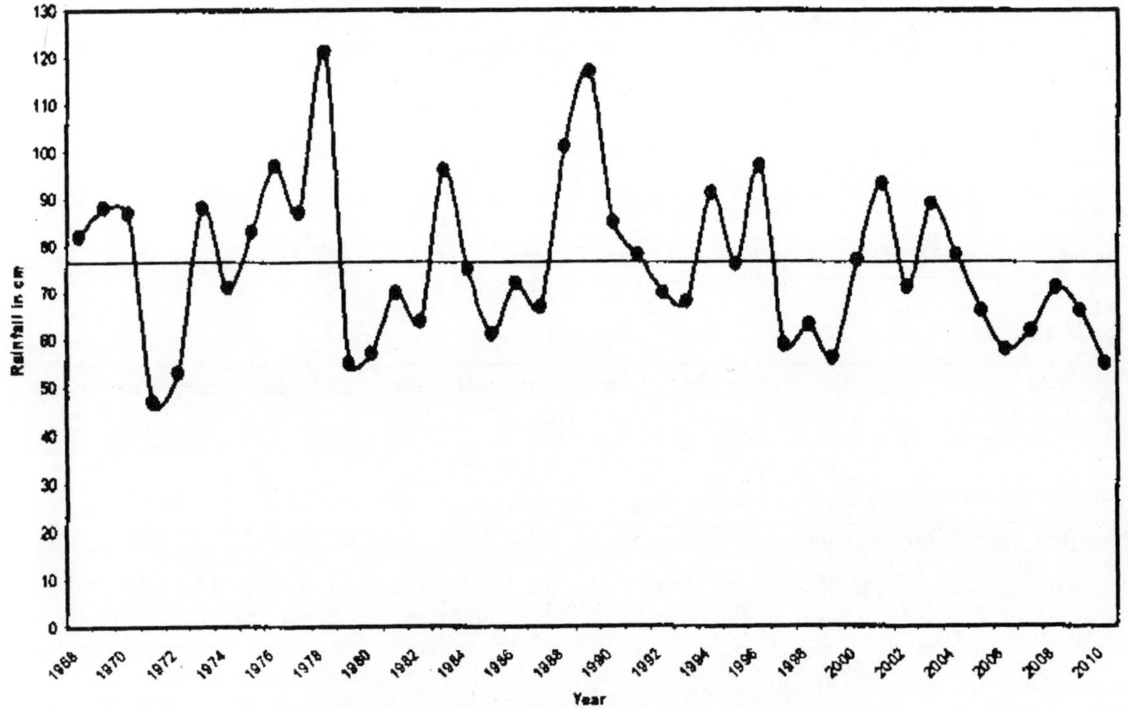

Figure 58.6 Rainfall Data of Musi Reservoir (Actual rainfall data, fitted values, forecast values) (1968–2010)

in Table 58.2. The plot of fitted values and forecast values from 1990 to 2010 were shown in Figure 58.5. Forecast values plotted were matching fairly with the fitted values. Figure 58.6, shows the plot of rainfall including fitted and forecast values for 43 years. 3 year moving average and 5 year moving average for 43 years were computed. From the plots and moving average values, a cyclical trend is not clearly visible but it may be inferred that successive droughts recurred once in a decade. There is not much deviation for the Standard Errors calculated for 37 years and 43 years. It reveals that predicted values were well fitted with actual values.

Table 58.1: Comparison of Fitted Values with Actual Rainfall Values

Year	*Rainfall Value (cm)*	*Fitted Values Using ARMA (2, 2) (cm)*
1990	85	88
1991	78	74
1992	70	65
1993	68	59
1994	91	95
1995	76	82
1996	97	88
1997	59	65
1998	63	68
1999	56	60
2000	77	72
2001	93	87
2002	71	76
2003	89	83
2004	78	73

Table 58.2: Forecast of Rainfall Data for Musi Reservoir Project

Year	*Forecast Values Using ARMA (2,2) (cm)*
2005	66
2006	58
2007	62
2008	71
2009	66
2010	55

Conclusions

The ARMA (2, 2) model was selected as the best fit from the analysis. Box and Jenkins method is best suited for short-range forecasts. By combining this method with linear regression, forecast was made for next Six years. Since, droughts may be expected once in a decade there is a need to have better management. Hence, necessary steps are to be taken for optimum water management and utilization.

References

Bertoni, Juan Carlos, Tucci, Carlos Eduardo and Clarke, R. Thomas, 1991. Rainfall-based real-time flood forecasting. *Journal of Hydrology*, 131: 313–339.

Box, G.E.P. and Jenkins, G.M., 1976. *Time Series Analysis Forecasting and Control*. Holden Day.

Bras, Rafel L., 1976. Short-term forecasting of rainfall and runoff. In: *Real-time Forecasting and Control of Water Resources Systems*. Pergamon Press, pp. 125–137.

Hipel, Keith W., McLeod, A.I. and Lennox, W.C., 1977. Advances in Box-Jenkins modeling: Model construction. *Water Resources Research*, 13: 567–575.

Makridakis, S., Wheelwright, S.C. and Mcgee, V.E., 1978. *Forecasting Methods and Applications*. John Wiley and Sons.

Mohan, S., Raman, H. and Premganesh, G., 1991. A comparative study on models for forecasting Inflows. *Journal of Indian Water Resources Society*, 11(3): 19–22.

Shamseldin., Asaad, Y., Abdoand, Gamal M. and Elzein, Abderhman S., 1999. Real time flood forecasting on the Blue Nile River. *Water International*, 24: 39–45.

Sharma, Arun, 2002. Validation of the monsoonal river inflow forecasting model: A case study. *Journal of Applied Hydrology*, 15: 1–12.

Srinivas, P., Narasiah, M.L.and Thayumanavan, S., 2004. Rainfall forecasting of Saidapet Raingauge station by using Box-Jenkins ARMA model. In: *Proceedings of the International Conference on Advanced Modelling Techniques for Sustainable Management of Water Resources*, pp. 56–60.

Watt, W.E. and Nozdryn, M.J., 1982. *Real-time Flood Forecasting for Flood Damage Reduction*. Water Resources Publication, pp. 551–571.

Chapter 59

Interstitial Fauna in Relation to Water Quality

R.K. Srivastava[1], K.K. Dube[2] and Bazgha Athar[1]
[1]Environmental Research Laboratory,
P.G. Department of Botany and Environmental Sciences,
[2]Department of Zoology
Government Model Science College, Jabalpur – 482 001, M.P.

ABSTRACT

The term "Interstitial fauna" means animal that live between sediment particles in subsurface water and that includes many types of crustaceans, segmented worms, flatworms', rotifers water mites, juvenile stages of aquatic insects and vertebrates. Interstitial spaces that are unique because of the river water that enters the subsurface interstitial zone. When the surface water has dried up in this zone, a lot of organic material is consumed and the nutrients are converted to inorganic ions as NO_3^{-1} or PO_4^{-3}. Much of the water purification occurs because many bacteria, insect larvae and other small creatures live there and aid in purification and may act as biological filter improving water quality and supporting a diverse range of interstitial fauna. The importance of the surface saturated interstitial zone is the ecological functioning and maintenance of water quality of river ecosystem. Water pollution reduces water quality, fauna richness and total fauna is affected. Clean river surface, subsurface and bottom are essential for the health of many fish and aquatic insect communities. The effect of hydrological variations on the vertical distribution of interstitial organisms is the migration of water flow in different places. They also indicate the trophic status of a water body sensitive Ito pollution and many species are recognized as indicators of pollution.

Keywords: Interstitial fauna, Interstitial water, Interstitial spaces river water quality.

Introduction

The "interstitial space" recently has become a subject of general interest in furthering our understanding of lotic ecosystem structure and function of Interstitial space in the subsurface Interstitial zone (0–15 cm deep). The fauna community was described in terms of density taxonomic, richness, Shannon diversity and morphological traits (size, flexibility and body shape of the fauna). The importance of interstitial space, both as a factor partly controlling community structure and as an evolutionary factor acting on macroinvertibral ability to penetrate the substratum interstitial habitat were expected to exhibit the highest species abundance, highest taxonomic richness and lowest relative abundance of species that are adopted to penetrate the substratum. Some invertebrates Migrate deeply within the interstitial space to lescape erosion by scouring (Palmer *et al.*, 1992; Dole Oliver *et al.*, 1997; Gayraud *et al.*, 2000). Recently interested biologists for intestinal fauna tried to identify important features of the substratum that may affect the structure of aquatic fauna communities. Despite its importance, the mechanism relating to interstitial fauna availability (*i.e.* the porosity or the percentage of fine sediment) to the fauna community is poorly understood, probably owing to the difficulty of measuring' this substratum feature thus, the interstitial zone is an important component of river ecosystems. Figures 59.1(a,b and c) hyporheic flow paths can be upwelling and downwelling zone and Nutrient dissolved in surface fauna.

Surface Water

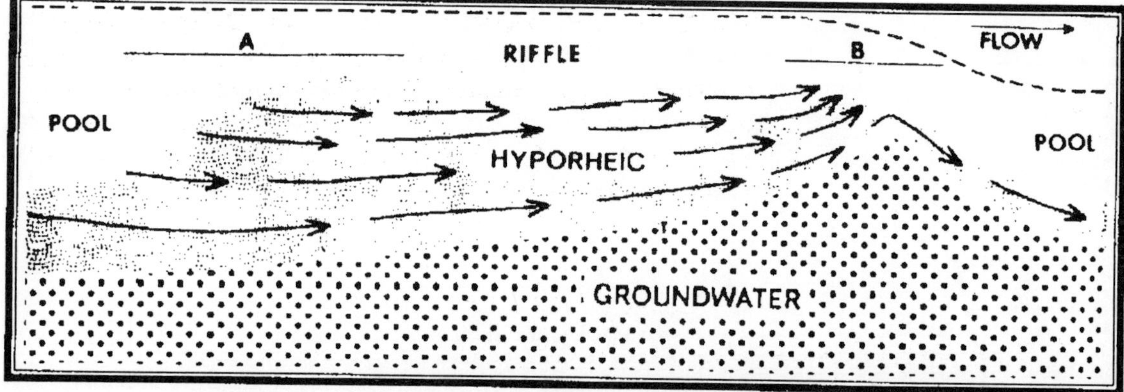

Figure 21.1(*a*): Water Movement through Interstitial Zone of a River

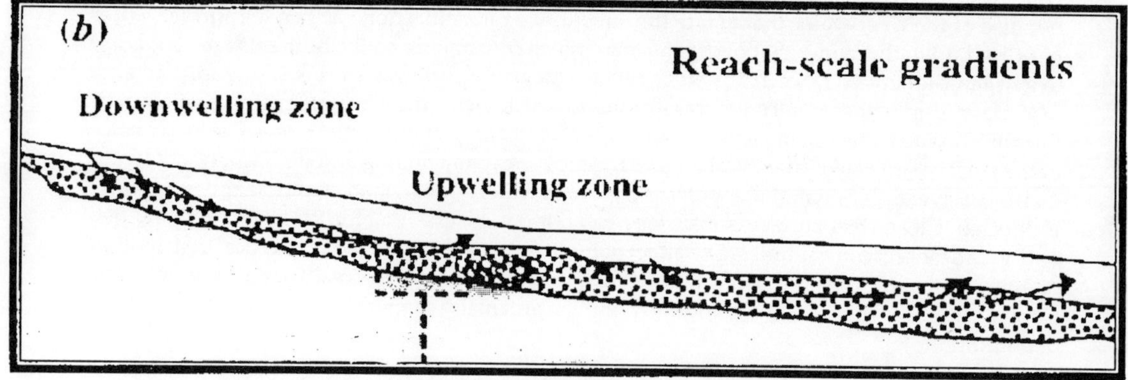

Figure 59.1(*b*): Upwelling and Downwelling Zones Alternate, Generating Gradients

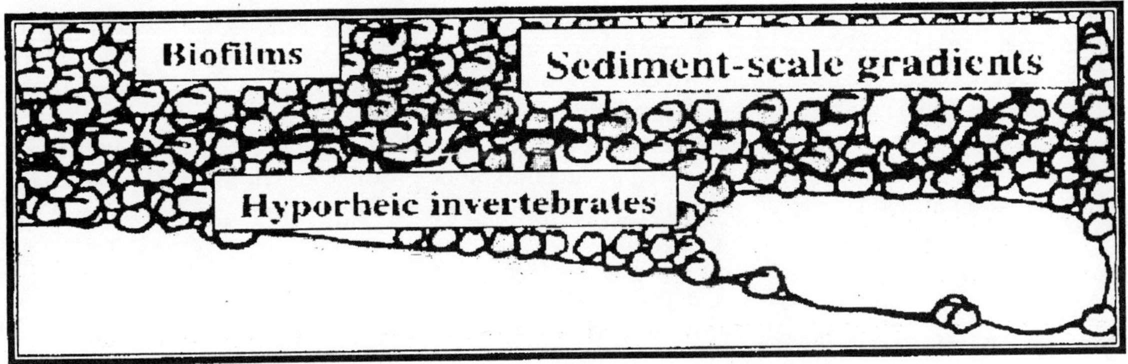

Figure 59.1(*c*): Gradients Along Subsurface Flow Paths in Nutrients, Dissolved Gases, and Subsurface Fauna

The assessment of water quality in a region is an important aspect for any developmental activity of the region. The necessity to determine the quality of a certain water body was given after human population influenced the water courses in a serious manner. The interstitial space of bed-sediments is a major component of the river system as it influences both physical and biological processes and patterns such as exchanges between surface water and ground water of fauna. Many human activities cause interstitial sedimentation surface and subsurface and ground water linkage impacting upon ecological processes in the interstitial zone.

It is very important for the water quality analysis because these are situated in the lower level of the trophic net, showing all changes of the ambient middle. Generally, a high variation of the interstitial fauna density and diversity happens as a result of the changes in the water quality. The interstitial fauna has potential as a source of information for river health. Such studies are important because of increasing human pressures (such as degradation of drainage basins, river regulation or pollution) on natural tropical rivers.

The nature of the river sediments, their interstitial fauna and water quality and the effects of quality changes, even slight but persistent over a period of cumulative effects on the river Claret, *et al.* (1999).A functional classification of interstitial invertebrate of biodiversity using species traits and habitat affinities. The current status of research on the functional significance of the interstitial fauna, addressing the following questions:

1. What hydrological, chemical, and biological processes occur in the interstitial spaces? and how are they interrelated at a range of scales?
2. How do these processes and their interactions influence ecological processes occurring in the surface river?
3. What features determine the functional significance of the interstitial fauna to river ecosystems?
4. What are the promising future research directions in this field, and how do they relate to river management?

Interstitial Fauna Habitats

Many rivers exchange water with their subsurface aquifers at a range of scales (Packman *et al.*,

2000). Ecologists have come to recognize that many river ecosystems possess, surface, subsurface and lateral compartments (Ward, 1989) liked by hydrological exchanges that convey nutrient, organic matters, dissolved oxygen and some invertebrate and vertebrate among these zone. The interstitial zone, the interstitial habitat penetrated by riverine animals, is characterized as being spatially limited to no more than a few meters, in most cases centimetres, away from the river. Indicate that nutrients discharged from the interstitial zone may be crucial to biotic productivity in the river. The surface rivers are transmitted subsurface habitats via alterations of hydrological linkages (Hancock, 2002). Extraction of surface and subsurface water reduces available habitat, alter flashing flow and rates of hydrological exchange between surface and subsurface water and can change, even reverse chemical gradient and conditions in the interstitial zone (Figure 59.2).

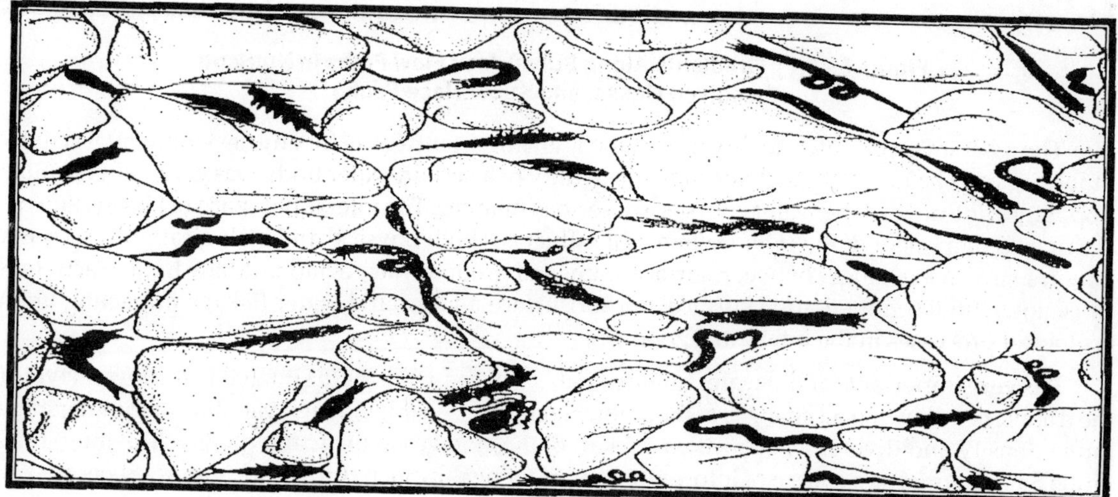

Figure 59.2: Interstitial Habitat Showing Life Between Sand Grains

Size and Sediment Distribution of Interstitial Fauna (Figure 59.3)

Macrofauna

Animals large enough to be seen with the naked eye.

Meiofauna

Meiofauna living between grains of sand have made some fancy adaptations to their harsh environment. Some have hooks on their feet, used to grap the sand. Others have hooked mouthpart, used for locomotion.

Microfauna

The smallest animals in a country, not visible to the naked eye. Microscopic animals such as protozoa and nematodes.

Interstitial Fauna

The interstitial system is truly three-dimensional, often having great vertical extent in the sand. The porous system average about 40 per cent of the total sediment volume. Its inhabitants include

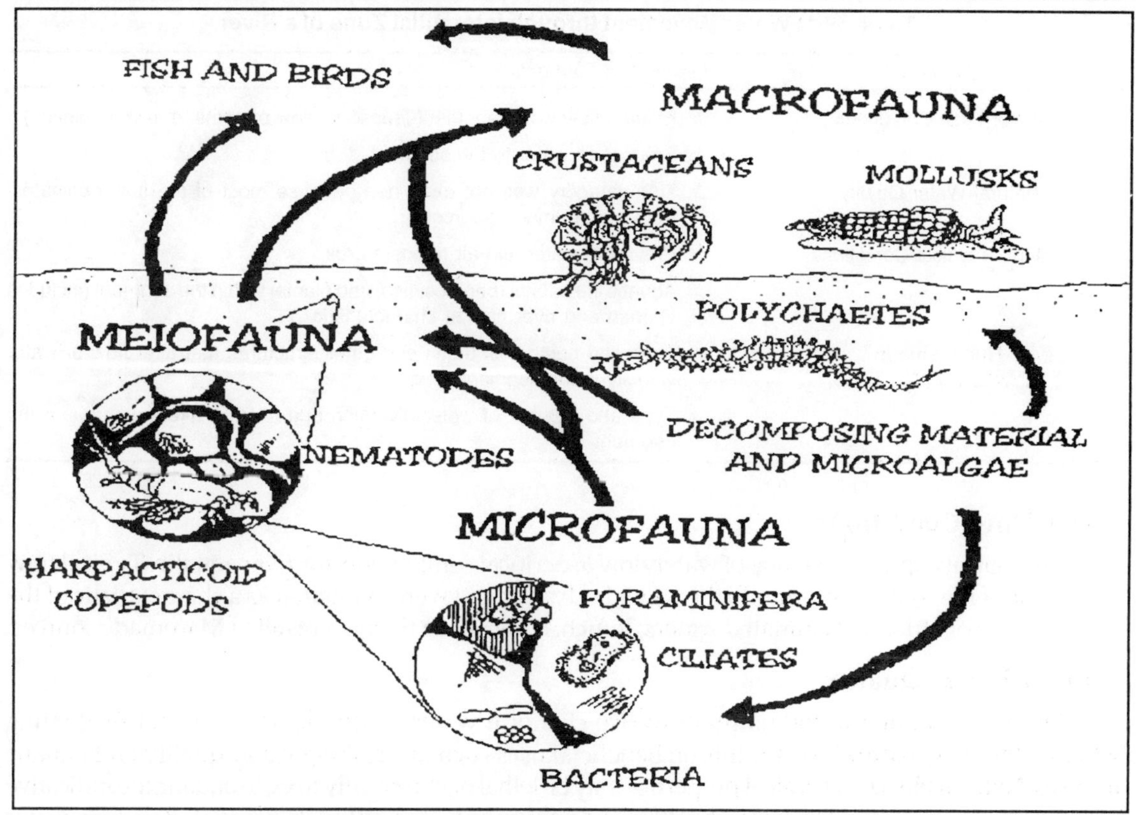

Figure 59.3: Size and Sediment Distribution of Interstitial Fauna

small metazoans forming the meiofauna, protozons, bacteria and diatoms the meiofauna is defined as those metazoans animals passing undamaged though 0.5 to 1.0 mm sieves and trapped on 30 mm screens. On most beaches the interstitial fauna is rich and diverse, even exceeding the macrofauna in biomass in some cases. The dominant taxa of sandy beach, meiofauna are nematodes and harpacticoid copepod with other important groups including turbellarians, oligochaetes, gastrotrichs, ostracods and tardigdades.

Effect of Sediment on Interstitial Fauna

Spates and concomitant increases in both suspended and deposited sediment are a natural occurrence in river. It follows therefore that river organisms must be adapted to withstand occasional increases in suspended and interstitial spaces sediment levels. Sediment deposition has a variety of effects on subsurface: these range from minor interference with feeding (and hence lower productivity) right through to death by smothering. All of these can occur as a result of natural, events. Input from human activities, may be well handled by the river community. Continuous high-level inputs are another matter.

Table 59.1: Water Movement through Interstitial Zone of a River

Category	*Information Elements*
Water Flow Conditions	☆ Effluent flow within segment (seasonal flow patterns, maxima, minima) ☆ Sources of water to the segment
Natural Water Quality	☆ This category was not examined because most of the flow originates' from human-made sources.
Natural Physical Conditions	☆ Prevailing water and air temperatures ☆ Aquatic life habitat (benthos, fish, and plants) within the segment (includes riparian and river habitat characteristics.)
Human-made Physical Alterations	☆ Type and number of dams and other structures that alter flow or water velocity in the segment. ☆ Type and number of water diversion and other intake structures in the segment

Water Flow Conditions

The quantity and consistency of water flow are critical components for river aquatic. Typical flow patterns and the sources of water to the water body segment were evaluated as a characteristic of the water body. For effluent-dominated waters, lmuch, if not all the flow is a result of Man-made sources.

Natural Water Quality

An understanding of the natural physico-chemical water quality is necessary for evaluating whether there are natural constraints on beneficial uses such as drinking water quality and aquatic life uses. Natural physico-chemical properties may be lethal or chronically toxic to aquatic life indicative of a certain use, or a threat to public health as a water supply. For effluent-dominated waters, water quality is largely driven by man-made sources (*i.e.*, treated effluent).

Natural Physical Conditions

The physical aspects of a water body can dictate the types of aquatic life that can be present. The natural physical structure of the river, such as sediment particle size and geology, along with flow conditions, have a large effect on the types of aquatic life. For certain species, the prevailing water temperature is a chief driver affecting distribution and abundance have complex life cycles that require different habitat types to reproduce and persist (*e.g.*., aquatic insects).

Human-made Physical Alterations

Various human-made alterations to the natural river the riparian zone or the flow of water through the segment can have profound effects on the types of aquatic life that can occur and the degree to which a public water supply is feasible or lattainable. Information related to the types of man-made alterations, including barriers, dams, water diversions, intake structures, and physical habitat is therefore, a critical element of this information category.

Ecological and Physical Influences

The interstitial zone is an ecotone between river water and groundwater environments, combining not only biogeochemical but also physical characteristics of both sources. The hyporheic zone (interstitial zone) provides an ideal habitat for a wide array of microbes and invertebrates. Hyporheic ecology is an emerging aspect of hyporheic science, and there are relatively few species of hyporheic fauna for which life histories and adaptation strategies are well documented or understood. Many hyporheic biogeochemical studies (Claret *et al.*, 1997; Edward Richard *et al.*, 1998) have recognized that the resident microbial communities are important in transforming incoming nutrients. Few studies have been conducted to describe and define hyporheic microbial communities. (Bolton *et al.*, 1998; Duff *et al.*, 2000)

Biogeochemical Influences

Although the interstitial zone extent is controlled by surface-water penetration into the subsurface, hyporheic water is generally composed of a mix of surface water and groundwater. From a biogeochemical perspective, groundwater is generally low in dissolved oxygen and enriched in inorganic solutes compared to river water. Thus, biogeochemical gradients exist within the hyporheic zone between two extremes defined by the surface water and groundwater end members. This makes the hyporheic zone a very active location of biogeochemical transformation of nutrients and other dissolved solutes. For example, in many small mountain streams, inorganic nitrogen–as nitrate (NO_3) and ammonium (NH_4)–enters the hyporheic zone in river water, but water the hyporheic zone has far less inorganic nitrogen compounds. *Microbes* and *biofilms* may take up incoming nitrogen compounds or modify them through nitrification and denitrification processes.

Conclusion

Interstitial meiofauna and macrofauna with sediment particle size, disturbance frequency and hydrological exchange at various scales Boulton (2000a and 200b). The subsurface macrofauna, if not, is there a limited subset of additional variables that could improve the association. If there is a good associated, does this extend to matching some of the key ecological roles played by the significance of the interstitial zone as part of many river ecosystem, is generally recognized and its impairment by many human activities is widely acknowledged. However, an infant science, explicit attempts to restore interstitial zone are rare. Historically, this reflected ignorance of the vertical dimension in river corridors but it now may be a reaction to the lack of tools for assessing the condition of the hyporheic zone. Truly holistic approaches to river rehabilitations clearly must seek to recover hydrological connectivity in all three dimensions: longitudinal, lateral and vertical (Kondolf *et al.*, 2006).

References

Bolton, Andrew J., Findlay, Stuart, Marmonier, Pierre, Stanley, Emily H. and Valett, H. Maurice, 1998. The functional significance of the hyporheic zone in streams and rivers. *Annual Review of Ecology and Systematics*, 29: 59–81.

Boulton, A.J., 2000a. The subsurface macrofauna. In: *Streams and Ground Waters*, (Eds.) J.B. Jones and P.J. Mulholland. Academic Press, San Diego, p. 337–362.

Boulton, A., 2000b. River ecosystem health down under: Assessing ecological condition in riverine groundwater zones in Australia. *Ecosys. Health*, 6: 108–118.

Claret, C., Marmonier, P., Boissier, J.M., Fontvieille, D. and Blanc, P., 1997. Nutrient transfer between parafluvial interstitial water and river water: Influence of gravel bar heterogeneity. *Freshwater Biology*, 37: 657 –670.

Claret, C., Marmonier, P., Dole-Olivier, M-J., Creuzdes Chtelliers, M., Boulton, A.J. and Castella, E., 1999. A functional classification of interstitial invertebrates: Supplementing measures of biodiversity using species traits and habitat affinities. *Arch. Hydrobiology*, 145: 385–403.

Dole-Olivier, M-J., Marmonier, P. and Beffy, J.L., 1997. Response of invertebrates to lotic disturbance: Is the hyporheic zone a patchy refugium? *Freshw. Biol.*, 37: 257–276.

Duff, J.H. and Triska, F.J., 2000. Nitrogen biogeochemistry and surface subsurface exchange in streams. In: *Streams and Ground Waters*, (Eds.) J.B. Jones and P.J. Mulholland. Academic Press, San Diego, C.A., pp. 197–220.

Edward, Richard, T., 1998. The hyporheic zone. In: *River Ecology and Management: Lessons from the Pacific Coastal Eco-region*, (Eds.) Robert J. Naiman and Robert E. Bilby. Springer Verlag, New York.

Gayraud, S., Philippe, M. and Maridet, I., 2002.The response of benthic macro-invertebrates to artificial disturbances drift or vertical movements in the gravel bed of two sub-alpine streams? *Arch. Hydrobiology*, 47: 43–446.

Hancock, P.J., 2002. Human impacts on the stream–groundwater exchange zone. *Environ. Manag.*, 29: 763–781.

Kondolf, G.M., Bolton, A.J.,O, Daniel, S., Poole, G.C., Rahel, F.J., Stanley, E.H., Wohl, E., Baing, A., Carlstrome, J., Cristoni, C., Huber, H., Koljonen, S., Louhi, P. and Nakamura, K., 2006. Process-based ecological river restoration: Visualizing three dimensional connectivity and dynamic vectors to recover lost linking. *Ecology and Society*, 11: 1–17.

Palmer, M., Bely, A.E., Berg, K.E., 1992. Response of invertebrates to lotic disturbance: A test of the hyporheic refuge hypothesis. *Oecologia*, 89: 182–194.

Ward, J.V., 1989. The four-dimensional nature of lotic ecosystems. *Journal of the North American Benthol. Society*, 8: 2–8.

Chapter 60

Eco-friendly Management of Bollworms on Cotton through IGR-Pesticide Mixtures*

N. Srinivasa Rao and P. Arjuna Rao**

Department of Entomology, Acharya N.G. Ranga Agricultural University, Agricultural College, Baptla – 522 101, Andhra Pradesh, India

ABSTRACT

The combination treatments containing novaluror + insecticide mixture at half their recommended doses, registered superior performance in containing the bollworms damage and gave high seed cotton yield over the corresponding individual treatments. The combination treatment were found equally good as individual treatments against sucking pests. Further, they were found relatively safe to natural enemies compared to the individual treatments. The descending order of efficacy of the treatments in recording the higher yield was novaluron + beta-cyfluthrin > novaluron + indoxacarb > indoxacarb > beta-cyfluthrin > novaluron + spinosad > spinosad > novaluron > novaluron + NSKE > NSKE.

Keywords: Novaluron, Bollworms, Cotton, Natural enemies.

---

* A part of Ph.D. thesis submitted by N. Srinivasa Rao to Acharya N.G. Ranga Agricultural University, Rajendranagar, Hyderabad

** Corresponding author and Present Mailing Address: Scientist (Entomology), AINP on Rodent Control, APRRI and RARS, Maruteru – 534 122, West Godavari Dist., Andhra Pradesh, E-mail: raoento@yahoo.co.in; Fax No: 08819-247583, (M) 9848646780.

Introduction

The pest spectrum of the cotton is quite complex, bollworms are the major biological constraints in cotton cultivation. The obsession of cotton growers for insecticides usage has brought multifold increase of pesticide resistance on bollworms to many insecticides hitherto recommended for cotton pest control (Armes *et al.*, 1992; Kranthi, 1998). Chitin synthesis inhibitors are more specific towards target insects and are relatively safe to natural enemies compared to other groups of synthetic insecticides (Dhawan, 2000; Patil *et al.*, 2001). But, their poor lethal effects against insect pests make them less preferred by the farmers, besides they potentiate the other groups of insecticides in combination by interfering with the normal deposition of cuticle during and after next moulting. Use of pesticide-IGR mixtures is an effective eco-friendly insecticide resistance management approach against bollworms that have developed multi-fold resistance against several conventional insecticides (Ravi and Verma, 1999; Srinivasan and Uthamasamy, 2001). In this study, an attempt has been made to evaluate the efficacy of a novel chitin synthesis inhibitor, novaluron in combination with different groups of pesticides, commonly used against the bollworms and their safety to natural enemies

Materials and Methods

The experiment was laid in a randomized block design (RBD); with NCS-145 cotton hybrid, replicated thrice with ten treatments including untreated control. The experiment was carried out for two consecutive seasons (*Kharif* 2003–04 and *Kharif* 2004–05) at Agricultural College Farm, Bapatla. The crop was sown on 1st August during both the seasons with a plot size of 42 m^2 (7m × 6m), spacing of 1.05 × 0.6 m between row to row and plant to plant. The crop was raised by practicing all standard agronomic practices.

Application of Treatments

All the treatments including control have received seed treatment with imidacloprid 70 WS @ 5g kg^{-1}. Further, stem application with imidacloprid 200 SL (1 part) + water (20 parts) at 40 days after sowing was given for the control of early season slicking pests. The treatments containing qovaiuron + insecticide at half their recommended doses were imposed at 15 days interval from 60–70 days after sowing, when one of the bollworms reached the Economic Threshold level (ETL) (one larva/plant). Spray fluid was applied @ 700–1000 l ha^{-1} with high volume knapsack hand sprayer. A total of six sprays were imposed during the entire crop growth during both the seasons.

Assessment of Pest and Natural Enemies' Population

The bollworm incidence was assessed based on damaged squares, bolls and locules on five randomly selected plants in each plot at weekly intervals with the commencement of the first spray. The damage due to spotted bollworm, *Earias vitella* (Fab.) and american bollworm, *H. armigera* was assessed in terms of per cent damaged squares and bolls whereas per cent damaged bolls and locules were counted for pink bollworm, *Pectinophora gossypiella* (Saund.). The cumulative per cent damage to the respective reproductive structures was arrived. The incidence of defoliators *viz.*, tobacco caterpillar *Spodoptera litura* (Fab.) (Number of larval plan.), leaf roller, *Sylepta derogata* (Fab.) (Number of leaf rolls per plant) and budworm, *Phycita infusella* Meyr. (Number of damaged tender leaves per plant) were also assessed. The populations of leaf hoppers, aphids and whiteflies were assessed one day before and at weekly intervals after application on three leaves (one leaf each from top, middle and bottom) in each of the five randomly selected plants per plot. The populations of coccinellids and spiders were assessed one day before and at weekly intervals after application on the five randomly selected plants per plot. The seed cotton yield in each plot was weighed separately in each picking and expressed as q ha^{-1}. In all, four pickings were taken during both the seasons. The percentage data was transformed into arcsine data, where as the number data was into square root data and subjected to Analysis of Variance (ANOVA).

Results and Discussions

Bollworms

The combination treatments *viz.*, novaluron–0.005 per cent + beta-cytluthrin 0.00125 per cent, novaluron–0.005 per cent + indoxacarb–0.00725 per cent and novaluron–0.005 per cent + spinosad–0.0075 per cent are more or less equally effective but significantly different with each other and superior over their respective individual treatments in containing the damage to squares and bolls by *E. vitella* (Table 60.1). These findings are in accordance with the findings of Srinivasan and Uthamasamy (2001b). Novaluron–0.005 per cent + indoxacarb–0.00725 per cent and povaluron–0.005 per cent + spinosad–0.0075 per cent were the best chemical treatments in controlling the square land boll damage caused by *H. armigera*.. Beta-cytluthrin and its combination with novaluron were less effective against *H. armigera* compared to the earlier treatments, several folds of resistance in *H. armigera* against synthetic pyrethroids was earlier documented by several workers (Armes *et al.*, 1992; Regupathy *et al.*, 1997; Kranthi, 1998). The present studies indicated that, novaluron–0.005 per cent +beta-cytluthrin–0.00125 per cent and betacytluthirn–0.0025 per cent proved highly effective against the pink bollworm with 14.45 and 16.03 per cent boll and 11.82 and 12.66 per cent locule damage, respectively. Synthetic pyrethroids were rated as the most effective chemicals especially against the pink bollworm due to their long persistent action as reported by several workers (Swamy *et al.*, 2000 and Dhawan and Simwat, 2001). Novaluron–0.005 per cent + indoxacarb–0.00725 per cent and indoxacarb–0.0145 per cent were highly promising in the order of efficacy in controlling *P. gossypiella* followed by novaluron–0.005 per cent + spinosad–0.0075 per cent and spinosad.

Defoliators

The most effective treatments against *S. tilura* were novaluorn–0.005 per cent + beta cyfluthirn–0.00125 per cent (1.0 larvae/plant), novaluron–0.005 per cent + indoxacarb–0.00725 per cent (1.23), indoxacarb–0.0145 per cent (1.25) and beta-cyfluthrin–0.0025 per cent (1.30) novaluron 0.01 per cent (1.48), novaluron–0.005 per cent + spinosad–0.0075 per cent (1.49) and spinosad–0.015 per cent (1.51) being on par with each other with less than 1.60 larvae/plant. The incidence of the leaf roller and budworm were noticed only during 2004–05 season. Among the chemical treatments on non-*Bt* cotton, novaluron–0.005 per cent + beta-cyfluthrin–0.0025 per cent, novaluron–0.005 per cent + indoxacarb–0.0145 per cent and novaluron–0.005 per cent + spinosad–0.015 per cent were equally effective and the best treatments against the leaf roller, *S. derogata*. Novaluron–0.005 per cent + beta-cyfluthrin–0.00125 per cent (2.53), novaluron–0.005 per cent + indoxacarb–0.00725 per cent (2.73), beta-cyfluthrin–0.0025 per cent (2.86) and indoxacarb–0.0145 per cent (3.20) were the best against the bud worm being on par by recording less than 4.0 damaged leaves/plant.

Sucking Pests

Beta-cyfluthrin–0.0025 per cent and indoxacarb–0.0145 per cent were promising in controlling leafhoppers and aphids when applied individually and in combination with novaluron at half the recommended concentration. The other treatments were moderately effective against the sucking pests. The efficacy of beta-cyfluthrin, a second generation synthetic pyrethroid against leafhoppers was in accordance with the findings of Mirikar *et al.* (1995). Similarly the effectiveness of indoxacarb (Dhawan and Simwat, 2000) against the leafhoppers was well established. Application of the synthetic pyrethroid, beta-cyfluthrin–0.0025 per cent individually and in combination with novaluron caused resurgence of whiteflies, which agreed with the findings of Rajak and Diwakar (1987). Novaluron–0.01 per cent was found good enough in controlling the whitefly population. Similar observations were made by Ishaaya *et al.* (1996) who reported that novaluron was highly effective in suppressing the adult emergence of *B. tabaci.*

Table 60.1: Efficacy of IGR-Pesticide Mixtures Against Bollworms on Cotton*

Treatments	*E. vitella*		*H. armigera*		*P. gossyppiella*		*Seed Cotton Yield (Q/ha)*
	Square Damage (%)	*Boll Damage (%)*	*Square Damage (%)*	*Boll Damage (%)*	*Boll Damage (%)*	*Locule Damage (%)*	
T_1–Novaluron	18.50	14.16	10.59	8.56	22.25	18.69	14.32
(0.01%)	(25.47)^{e}	(22.08)^{cd}	(18.99)^{cd}	(17.00)^{bc}	(28.13)^{ef}	(25.60)^{d}	(3.84)^{c}
T_2–Spinosad	16.41	10.85	7.92	6.09	20.38	17.32	16.73
(0.015%)	(23.89)^{d}	(19.22)^{ab}	(16.33)^{ab}	(14.28)^{a}	(26.83)^{def}	(24.58)^{cd}	(4.13)^{b}
T_3–Indoxacarb	14.89	10.36	7.61	6.06	17.69	15.67	17.35
(0.0145%)	(22.70)^{bc}	(18.75)^{ab}	(15.99)^{a}	(14.23)^{a}	(24.87)^{bcd}	(23.31)^{bc}	(4.22)^{ab}
T_4–Beta-cyfluthrin	14.00	10.01	10.48	8.49	16.03	12.66	17.24
(0.0025%)	(21.97)^{ab}	(18.40)^{a}	(18.87)^{cd}	(16.94)^{bc}	(23.60)^{ab}	(20.84)^{a}	(4.21)^{ab}
T_5–NSKE	24.40	16.57	12.71	9.30	26.75	22.05	7.10
(5.0%)	(29.60)^{g}	(24.00)^{d}	(20.88)^{e}	(17.74)c	(31.13)^{g}	(27.99)^{e}	(2.75)^{e}
T_6–Novahiron + Spinosad	15.45	10.68	7.36	5.88	19.20	17.15	16.90
(0.005%) (0.0075%)	(23.14)^{cd}	(19.06)^{ab}	(15.68)^{a}	(14.03)^{a}	(25.98)^{cde}	(24.44)^{cd}	(4.17)^{ab}
T_7–Novaluron + Indoxacarb	14.55	10.08	7.24	5.75	16.43	13.78	17.46
(0.005%) (0.00725%)	(22.42)^{b}	(18.49)^{a}	(15.55)^{a}	(13.86)^{a}	(23.91)^{abc}	(21.78)^{ab}	(4.23)^{ab}
T_8–Novaluron + Beta-cyfluthrin	13.11	9.36	10.03	8.32	14.45	11.82	17.53
(0.005%) (0.00125%)	(21.23)^{a}	(17.79)^{a}	(18.45)^{c}	(16.76)^{bc}	(22.33)^{a}	(20.09)^{a}	(4.23)^{ab}
T_9–Novaluron + NSKE	20.91	14.93	12.22	8.57	23.35	21.16	8.08
(0.005%) (2.5%)	(27.21)^{f}	(22.68)^{cd}	(20.45)^{de}	(17.02)^{bc}	(28.88)^{f}	(27.36)^{e}	(2.92)^{e}
T_{10}–Control	33.32	19.84	16.81	14.08	40.67	36.91	4.04
	(35.25)^{h}	(26.44)^{e}	(24.19)^{f}	(22.01)^{d}	(39.61)^{h}	(37.40)^{f}	(2.12)^{f}
F-test	Sig.	Sig.	Sig.	Sig.	Sig.	Sig.	Sig.
CD (0.05%)	0.972	2.40	1.66	1.36	2.23	1.71	0.19

*: Mean data for two seasons. Figures in parentheses are arcsine transformed values.

Means in a column followed by same superscripts (s) are not significantly different by (p = 0.05) DMRT.

Natural Enemies

Pertaining to the safety of the treatments against natural enemies, NSKE-5 per cent and novaluron–0.005 per cent + NSKE–2.5 per cent and novaluron–0.01 per cent were found highly safe against coccinellids and spiders. The safety of neem compounds (Schmutterer, 1997), as well as chitin inhibitors (Wilkinson *et al.*, 1978) against the coccinellids and spiders are well documented. Spinosad was m6derately toxic to coccinellids and spiders, which is in accordance with the findings of Hendrix *et al.* (1997). Beta-cyfluthrin and incoxacarb were highly toxic to coccinellids and spiders. In general the chemical insecticides *viz.*, spinosad, indoxacarb and beta-cyfluthrin were safe against coccinellids and spiders when applied in combination with novaluron at half the recommended doses compared to individual applications.

Table 60.2: Efficacy of IGR-Pesticide Mixtures Against Defoliators on Cotton*

Treatments	*S. litura*	*S. derogata*	*P. infusella*
	Larvae/Plant	*Leaf Rolls/Plant*	*Dam Leaves/Plant*
T_1–Novaluron (0.01%)	1.48 (1.40)[a]	6.00 (2.51)[cd]	6.33 (2.61)[d]
T_2–Spinosad (0.015%)	1.51 (1.40)[a]	4.73 (2.27)[bc]	4.26 (2.16)[bc]
T_3–Indoxacarb (0.0145%)	1.25 (1.32)[a]	4.66 (2.26)[bc]	3.20 (1.91)[b]
T_4–Beta-cyfluthrin (0.0025%)	1.30 (1.33)[a]	4.53 (2.25)[bc]	2.86 (1.81)[b]
T_5–NSKE (5.0%)	4.14 (2.15)[c]	8.20 (2.94)[e]	7.06 (2.74)[de]
T_6–Novahiron + Spinosad (0.005%) (0.0075%)	1.49 (1.40)[a]	3.40 (1.96)[b]	4.20 (2.14)[bc]
T_7–Novaluron + Indoxacarb (0.005%) (0.00725%)	1.23 (1.30)[a]	3.33 (1.95)[b]	2.73 (1.79)[b]
T_8–Novaluron + Beta-cyfluthrin (0.005%) (0.00125%)	1.00 (1.22)[a]	3.06 (1.88)[b]	2.53 (1.73)[b]
T_9–Novaluron + NSKE (0.005%) (2.5%)	3.14 (1.90)[b]	7.53 (2.83)[de]	6.60 (2.63)[d]
T_{10}–Control	8.82 (3.05)[e]	9.33 (3.33)[f]	9.00 (3.17)[f]
F–test	Sig.	Sig.	Sig.
CD (0.05%)	0.22	1.41	0.43

*: Mean data for two seasons. Figures in parentheses are arcsine transformed values.

Means in a column followed by same superscripts (s) are not significantly different by (p = 0.05) DMRT.

Yield

Among the treatments, novaluron+beta-cyfluthrin (17.53 q ha^{-1}), novaluron + indoxacarb (17.46 q ha^{-1}), indoxacarb (17.35 q ha^{-1}), beta-cyfluthrin (17.24 q ha^{-1}) novaluron + spinosad (16.90 q ha^{-1}) and spinosad (16.73 q ha^{-1}) were statistically at par with each other in recording the highest seed cotton yield. The superior performance of the combination treatments against bollworms resulted in increased yields over and above the other treatments. The lowest yields obtained in NSKE–5 per cėnt (7.1.0 q ha^{-1}) and novaluron–0.005 per cent + NSKE–2.5 per cent (8.80 q ha^{-1}) treatments over other treatments may be attributed to poor efficacy of these treatments against the bollworms.

Table 60.3: Efficacy of IGR-Pesticide Mixtures Against Sucking Pests on Cotton*

Treatments	*No./3 Leaves*		*No./3 Leaves*		*No./3 Leaves*	
	Before Spray	*After Spray*	*Before Spray*	*After Spray*	*Before Spray*	*After Spray*
T_1–Novaluron (0.01%)	6.33	5.40 (2.63)[c]	21.98	15.60 (4.01)[cd]	3.19	1.31 (1.34)[a]
T_2–Spinosad (0.015%)	6.05	5.00 (2.57)[bc]	22.76	14.41 (3.86)[bc]	3.36	2.56 (1.77)[cd]
T_3–Indoxacarb (0.0145%)	5.87	3.54 (2.31)[a]	21.54	11.39 (3.44)[ab]	3.29	2.21 (1.64)[bc]
T_4–Beta-cyfluthrin (0.0025%)	5.40	3.33 (2.24)[a]	22.23	9.10 (3.09)[a]	3.39	5.16 (2.37)[f]
T_5–NSKE (5.0%)	5.66	4.87 (2.69)[b]	21.05	17.58 (4.24)[d]	3.36	2.81 (1.81)[d]
T_6–Novahiron + Spinosad (0.005%) (0.0075%)	5.56	5.40 (2.68)[c]	20.81	19.68 (4.44)[dc]	3.50	2.40 (1.70)[bcd]
T_7–Novaluron + Indoxacarb (0.005%) (0.00725%)	5.33	4.73 (2.42)[b]	20.67	12.29 (3.57)[b]	3.33	2.06 (1.60)[b]
T_8–Novaluron + Beta-cyfluthrin (0.005%) (0.00125%)	5.00	4.33 (2.34)[b]	22.10	11.78 (3.50)[ab]	3.13	4.61 (2.26)[f]
T_9–Novaluron + NSKE (0.005%) (2.5%)	5.96	5.00 (2.63)[bc]	21.90	16.52 (4.12)[cd]	3.06	2.67 (1.78)[cd]
T_{10}–Control	5.59	9.26 (3.01)[de]	19.40	29.82 (5.50)[f]	3.00	3.94 (2.10)[e]
F–test	–	Sig.	–	Sig.	–	Sig.
CD (0.05%)		0.08		0.43		0.14

* Mean data for two seasons. Figures in parentheses are square root transformed values (√X).

Means in a column followed by same superscript (s) are not significantly different by (p=0.05) DMRT.

Table 60.4: Effect of IGR-Pesticide Mixtures Against Natural Enemies*

Treatments	*Coccinellids No./Plant*		*Spiders No./Plant*	
	Before Spray	*After Spray*	*Before Spray*	*After Spray*
T_1–Novaluron (0.01%)	1.51	0.92 (2.07)[c]	3.92	3.86 (2.08)[bc]
T_2–Spinosad (0.015%)	1.48	0.84 (1.94)[c]	4.16	3.68 (2.04)[bcd]
T_3–Indoxacarb (0.0145%)	1.66	0.52 (1.60)[e]	3.83	2.98 (1.86)[ef]
T_4–Beta-cyfluthrin (0.0025%)	1.43	0.34 (1.29)[c]	3.76	2.71 (1.79)[f]
T_5–NSKE (5.0%)	1.46	1.42 (2.44)[b]	3.99	3.95 (2.10)[b]

Contd...

Table 60.4–Contd...

Treatments	*Coccinellids*		*Spiders*	
	No./Plant		*No./Plant*	
	Before Spray	*After Spray*	*Before Spray*	*After Spray*
T_6–Novahiron + Spinosad (0.005%) (0.0075%)	1.54	0.98 (2.21)[bc]	3.88	3.49 (1.99)[d]
T_7–Novaluron + Indoxacarb (0.005%) (0.00725%)	1.46	0.86 (1.95)[c]	3.68	3.51 (2.00)[cd]
T_8–Novaluron + Beta-cyfluthrin (0.005%) (0.00125%)	1.33	0.76 (1.86)[dc]	3.81	3.06 (1.88)[e]
T_9–Novaluron + NSKE (0.005%) (2.5%)	1.53	1.20 (2.23)[bc]	456	3.86 (2.08)[bc]
T_{10}–Control	1.66	2.46 (2.83)[a]	4.08	5.33 (2.41)[a]
F–test	–	Sig. ·	–	Sig.
CD (0.05%)		0.15		0.08

* Mean data for two seasons. Figures in parentheses are square root transformed values (√X).

Means in a column followed by same superscript (s) are not significantly different by (p=0.05).

The present findings on the potentiation of chemical insecticides by chitin synthesis inhibitor, novaluron in controlling bollworms on non-*Bt.* cotton was of significance for four reasons. Firstly, the potentiation effect observed in mixtures may be useful in insecticide resistance management strategy. According to Ravi and Verma (1999) and Srinivasan and Uthamasamy, (2001a), the use of insect growth regulators in combination with insecticides forms a key approach in insecticide resistance management (IRM) programme. The second area of significance is that, these combinations involving low dose of toxic compounds are less harmful to natural enemies, answers the queries of several environmentalists. Thirdly, application of new and highly effective chemicals as mixtures would directly contribute to prolonging the life of new chemistries. Lastly, this technique will also be useful in improving the field efficacy of relatively less toxic compounds such as novaluron.

References

Armes, N.J., Jadhav, D.R., Bond, G.S. and King, A.B.S., 1992. Insecticide resistance in *H. armigera* in South India. *Pesticide Sci.*, 34: 355–364.

Dhawan, A.K., 2000. Impact of some new insecticides on natural enemy complex of cotton ecosystem. *Pestology*, 24(5): 8–14.

Dhawan, A.K. and Simwat, G.S., 2000. Evaluation of indoxacarb for the control of bollworm complex and its impact on population of sucking pests and beneficial arthropods in cotton ecosystem. *Pestology*, 24(7): 15–20.

Dhawan, A.K. and Simwat, G.S., 2001. Evaluation of beta-cyfluthrin for the control of bollworm complex on cotton. *Pesticide Res. Journal*, 13(2): 178–187.

Hendrix, W.H., Huckaba, R., Nead, B., Peterson, L., Porteous, D. and Thompson, G., 1997. Tracer insect control–1996 EUP results. In: *Proceedings of Beltwide Cotton Conferences*, New Orleans, LA, USA, January 6–10, 2: 1086–1087.

Ishaaya, I., Yablonski, S., Meandelson, Z., Mansour, Y. and Horowitz, A.R., 1996. Novaluron (MCW–275): A novel benzophenyl urea, suppressing developing stages of lepidopteran, whitefly and leafminer pests. In: *Bringhton Crop Protection Conference: Pests and Diseases*, 3: 1013–1020.

Kranthi, K.R., 1998. Insecticide resistance mechanisms in Indian *Helicoverpa armigera*. In: *Proceedings of World Cotton Conference–II*, Athens, Greece, pp. 689–696.

Mirikar, A.V., Borle, M.N., Borikar, P.S. and Jahagirkar, J.E., 1995. Efficacy of mixture of some synthetic pyrethroids with urea on sucking pests of cotton. *Pestology*, 19(3): 14–16.

Patil, B.V., Bheemanna, M., Prasad, A.R.B. and Goudar, S.B., 2001. Integration of insect growth regulators in the management of cotton bollworm, *Helicoverpa armigera* (Hubner). *Pestology*, 25(9): 10–12.

Rajak, R.C. and Diwakar, M.C., 1987. Resurgence of cotton whitefly in India and its integrated management. *Pl. Prot. Bull. of India*, 39(3): 13–14.

Ravi, G. and Verma, S., 1999. Pesticide–IGR mixtures: An effective eco-friendly approach against *Helicoverpa armigera* (Hub.). *Pestology (Spl.)*, p. 235–294.

Regupathy, A., Kranthi, K.R., Ashokan, G., Ardtes, N.J., Rao, N.V., Sounderarajan, R.P. and Kussel, D.A., 1997. Mechanism of insecticide resistance in field population of *Helicoverpa armigera* in India. In: *Proceedings of International Conference on Integrated Approach Combating Resistance*, ICAR, Rothamsted, Harpenden, Herts, U.K., 117 p.

Schmutterer, H., 1997. Side effects of neem (*Azadirachta indica* A. Juss.) products on insect pathogens and natural enemies of spider mites and insects. *J. of Appi. Entomol.*, 121(2): 121–128.

Srinivasan, R. and Uthamasamy, S., 2001a. Bio-efficacy of difluoenzuron, insecticides and diflubenzuron + insecticide mixtures on the incidence of American bollworm, *Helicoverpa armigera* (Hubner) in cotton. *Pestology*, 25(11): 13–15.

Srinivasan, R. and Uthamasamy, S., 2001b. Interactions of diflubenzuron, insecticides amidiflubenzuron + insecticide mixtures on spotted bollworm, *Earias vitella* Fab. and tobacco caterpillar *Spodoptera litura* Fab. in cotton. *Pestology*, 25 (9): 7–9.

Wilkinson, J.D., Biever, K.D., Ignoffo, C.M., Pons, W.J., Morrison, R.K. and Seys, 1978. Evaluation of diflubenzuron formulations on selected insect parasitoids and predators. *J. of Georgia Entomological Society*, 13: 227–236.

Chapter 61

Larvicidal and Ovicidal Effect of a Neem Product on Aedine and Culex Mosquitoes

Kiran Joseph, R.S. Mohanraj*, B. Dhanakkodi and K. Sathya
Department of Zoology, Kongunadu Arts and Science College (Autonomous), Coimbatore – 641 029, Tamil Nadu, India

ABSTRACT

An *in vitro* study was carried out to understand the effect of a neem based insecticide on the larvae and egg hatchability of Aedine and Culex mosquitoes. The LC_{50}/24 hour values against the different instar stages indicated larvicidal effect of the formulation. Hatchability of mosquito eggs was considerably reduced when treated with the botanical insecticide.

Keywords: Aedes aegypti, Culex quinquefasciatus, Neem product, Toxicity, Hatchability.

Introduction

For 260 million years, the two-winged insect has been on the face of this earth. It has outlived the dinosaurs and faced all the rough weather from volcanoes to human onslaught and emerged as a true nemesis to the human race. No other insect may have caused us so much trouble as this tiny perpetrator whom we encounter everyday. This common day foe is none other than the Mosquito.

* Corresponding Author.

Aedes aegypti is the only known potential vector of dengue and urban yellow fever (Mekuria, 1991; Mazzarri and Georghiou, 1995). This species of mosquito was shown to be a competent laboratory vector of Chikungunya (CHIK) virus (Jupp and McIntosh, 1990). *Aedes aegypti* has also been noted to transmit filariasis and encephalitis (Dorland's illustrated Medical Dictionary, 1982). The *Culex quinquefasciatus*, a known vector of Bancroftian filariasis, is an important pest in all the tropical towns and cities in Asia (Bowers *et al.*, 1995).

Among the various control measures for mosquitoes, use of synthetic insecticides is still the most convenient and indispensable method all over the world. The extensive application of synthetic pesticides has however caused irrepairable hazards to the environment and humans. Moreover, insecticide resistance has also become the greatest problem in the control of many medically important mosquito species. *Ae. aegypti* in particular has developed resistance to a variety of organochloride, organophosphorus, carbamate and pyrethroid insecticides (Kumar *et al.*, 1991; Mekuria *et al.*, 1991; Mazzarri and Georghiou, 1995; Rawlins and Wan, 1995). Neem products are also relatively advantageous over synthetic insecticides in being eco-friendly and non-toxic to non-target organisms including humans. Purohit *et al.* (1989) demonstrated that plant products are easily biodegradable. It was also emphasized that they do not leave poisonous residues and insects do not develop resistance. Oroumchi and Lorra (1993) stated that neem products may be a source of inexpensive and environmentally sound pesticides.

Neem derivatives obtained from different parts of the plant were documented to produce diverse toxic and biological effects, on a variety of insect species (Kaur *et al.*, 2001; Ramarethinam *et al.*, 2002; Mohanraj and Dhanakkodi, 2004, 2005 a, 2005b, 2006, 2007 a, 2007b; Neloilya, 2007). However, investigations on the influence of neem derivatives on development aspects of mosquitoes are only a few that too fragmentary.

Reports on the toxicity of neem as mosquitocide (adulticide and larvicide) are available. Attri and Prasad (1980) observed the toxic effects of neem oil extractive on larvae of *Cx. fatigans.* Chavan and Nikam (1988) demonstrated that petroleum ether extract and alkanes from neem leaves were toxic to *Cx. pipiens fatigans.* Neem leaf extracts of polar and non-polar solvents (Deshmuhk and Renapurkar, 1992; Obomanu *et al.*, 2006) and neem oil (Das *et al.*, 1995) have been tested against larvae of *Culex* mosquito. The susceptibility of *Cx. quinquefasciatus* to acetone extract of neem seed coat (Sagar and Sehgal, 1997), NSK and neem gum extracts (Babu and Murugan, 1998) and Nimbex, Nimbecidine, Neem gold, Salotrap (Thamil Solai *et al.*, 1999) and that of *Ae. aegypti* to acetone extract fraction of the fruit of *Melia volkensii* and *Azadirachta indica* (Awala *et al.*, 1998; Umar *et al.*, 2007), resinous exudate from the tender leaves of neem (Babu *et al.*, 2000) and Econeem, a neem based insecticide (Mohanraj *et al.*, 2000) has been found. Larvicidal effect of neem products on the malarial vector, *Anopheles* have also been reported earlier (Aliero, 2003; Vatandoost and Vaziri, 2004; Okumu *et al.*, 2007). However, conducting more investigations on toxicity of different forms of neem to mosquitoes is therefore the need of the hour.

Hatchability of eggs is one of the biological features that ascertain the renewal of a species. The ovicidal effect produced by any means on egg hatching is likely to reduce the population of harmful insects. Neem derivatives were found to lessen the hatching of eggs of many species of insects (Ekesi, 2000; Sundar *et al.*, 2000, Kaur *et al.*, 2001). Although, ovicidal activities of various chemicals on mosquito eggs including *Aedes aegypti* were reported earlier (Judson, 1967; Miura *et al.*, 1976; Fournet *et al.*, 1993; Sahgal and Pillai, 1993), reports on the ovicidal effects of neem substances on mosquito eggs are meagre. It has been stated that no information is available regarding ovicidal activity of AZ and related products in mosquitoes Su and Mulla (1998a). This raises an ample chance to carry out investigations on this aspect.

Materials and Methods

Animals

Aedine and *Culex* eggs were collected, hatched, reared and maintained for many generations in the laboratory. The eggs and larvae obtained from this stock were used for different experiments.

Test Compound

A neem based insecticide, EC, containing 0.15 per cent w/w azadirachtin (1500 ppm).

Different concentrations of the neem formulation was prepared in parts per million (ppm) basis, based on the concentration of azadirachtin through serial dilution with unchlorinated filtered water.

Bioassay Test

Bioassay tests were carried out for testing the efficacy of the neem formulation *Aedes aegypti* and *Culex quinquefasciatus* at different stages of development *viz.*, I, II, III and IV instars and pupae. Instructions of WHO (1960) as detailed by Pampana (1963) for conducting bioassay experiment with mosquito larvae were carefully followed.

Different concentrations of the test compounds were prepared using unchlorinated water. Clean plastic tubs of 500 ml capacity were used as test containers. Batches of 20 larvae were exposed to 200 ml of a particular concentration of test solution. The larvae were collected with an eye dropper placed onto filter paper strips and immediately transferred to test cup containing test solution according to Cilek *et al.* (1991).

Mortality rates of larvae were recorded after 24 hours. Five or more concentrations of a test compound giving between 0 and 100 per cent mortality for larvae at different instar stages were tested. Parallel controls were maintained. Two replicates were done at each concentration. In recording the percent mortality for each concentration, the moribund and dead larvae in both replicates were combined.

For computing LC_{50}, the data were subjected to Finney's method of probit analysis as detailed by Regupathy and Dhamu (1990).

Determination of Effect of Neem Formulation on Hatching of *Aedes aegypti, Culex quinquefasciatus* Eggs

Twenty eggs of aedine were exposed to a particular concentration of the neem product tested. The hatchability was recorded until 96 hours from the initial time of the experiment. The time was fixed so, because it was demonstrated that the completion of embryogeny occurs within 4 days (Judson and Gojrati, 1967). In case of *Culex,* the egg rafts with known number of eggs were carefully exposed to different concentrations.

Hatching rate was calculated on the basis of non-hatchability of eggs according to Sahgal and Pillai (1993). To ensure non-hatchability, the eggs from any test container were collected after 96 hours. Unhatched and decapped eggs were separated and counted using dissection microscope. Five replications were conducted at each concentration of neem product. The data were statistically examined using student 't' test.

Results and Discussion

LC_{50}/24 hour values of the neem formulation to I instar larvae was 63.095 ppm (*Aedes*) 25.118 ppm (*Culex*) and this was found to gradually increase with the age of larvae. The LC_{50} values of the neem formulation to the II, III and IV instar stages were 199.526, 316.277, 501.187 ppm of (*Aedes*); 39.810, 79.432, 158.489 ppm (*Culex*) respectively. Pupae showed the highest resistance to the neem formulation as is evident from the relatively higher LC_{50} values 794.328 ppm (*Aedes*) 501.187 ppm (*Culex*) (Tables 61.1 and 61.2).

Table 61.1: LC_{50}/24 Hour Values of a Neem Formulation to the Pre-adult Stages (I, II, III, IV Instar and Pupae) of *Aedes aegypti*

Stages of Development (Instars)	*Number of Larva/ Trial*	*LC_{50}/ 24 Hour Log Dose*	*Confidence Interval*		*Variance*	*Chi-Square*	*Degree of Freedom*
			LL	*UL*			
I	20	1.8 (63.095 ppm)	1.664 (46.131 ppm)	1.936 (86.297 ppm)	0.00486	4.964	3(16.266)
II	20	2.3 (199.526ppm)	2.2362 (172.266 ppm)	2.3638 (231.100 ppm)	0.00106	1.915	3(16.266)
III	20	2.5 (316.277ppm)	2.4283 (268.101 ppm)	2.5717 (372.992 ppm)	0.00134	0.2741	3(16.266)
IV	20	2.7 (501.187 ppm)	2.6692 (466.874 ppm)	2.7308 (538.021 ppm)	0.000247	4.531	3(16.266)
Pupa	20	2.9 (794.328 ppm)	2.8562 (718.124 ppm)	2.9438 (878.617 ppm)	0.000501	7.8021	3(16.266)

Table 61.2: LC_{50}/24 Hour Values of a Neem Formulation to the Pre-adult Stages (I, II, III, IV Instar and Pupae) of *Culex quinquefasciatus*

Stages of Development (Instars)	*Number of Larva/ Trial*	*LC_{50}/ 24 Hour Log Dose*	*Confidence Interval*		*Variance*	*Chi-Square*	*Degree of Freedom*
			LL	*UL*			
I	20	1.4 (25.118 ppm)	1.3996 (25.095 ppm)	1.4003 (25.136 ppm)	0.0003749	4.032	3 (16.266)
II	20	1.6 (39.810 ppm)	1.5048 (31.974 ppm)	1.6952 (49.567 ppm)	0.002370	6.171	3 (16.266)
III	20	1.9 (79.432 ppm)	1.812 (64.868 ppm)	1.988 (97.274 ppm)	0.002022	3.871	3 (16.266)
IV	20	2.2 (158.489 ppm)	2.115 (129.270 ppm)	2.2885 (194.312 ppm)	0.002051	2.264	3 (16.266)
Pupa	20	2.7 (501.187 ppm)	2.6716 (469.461 ppm)	2.7284 (535.056 ppm)	0.0002122	0.091	3 (16.266)

Average number of eggs hatched in control medium was 18.4 out of 20 (92 per cent). This was reduced to the lowest level of 0 among eggs of *Aedes* and *Culex* placed in medium of 400 ppm, neem formulation (Tables 61.3 and 61.4).

Table 61.3: Alteration in the Hatchability of *Aedes aegypti* Eggs Exposed to Different Concentrations of a Neem Formulation and Control

Parameters	*Control*	*Concentrations (ppm)*				
		100	*150*	*200*	*300*	*400*
Number of eggs introduced	20	20	20	20	20	20
Mean number of eggs hatched*	18.4	14.0	9.6	5.2	3.0	0
S.D.	±0.894	±0.836	±0.836	±0.836	±0.707	0
Pe c ent hatchability	92	70#	48#	26#	15#	0
Per cent reduction over control		23.91	47.82	71.73	83.69	100

*: Mean (±SD) of 5 replicates.

#: Significantly different from control; P < 0.001 per cent.

Table 61.4: Alteration in the Hatchability of *Culex quinquefasciatus* Eggs Exposed to Different Concentration of a Neem Formulation and Control

Parameters	*Control*	*Concentrations (ppm)*				
		100	*150*	*200*	*300*	*400*
Number of eggs introduced/trial	20	20	20	20	20	20
Mean number of eggs hatched*	18.4	15.8	11.0	7.7	3.4	0
SD	±0.894	±0.386	±0.942	±0.833	±0.547	
Percent hatchability	92	79#	55#	38#	17#	0
Per cent reduction over control		14.13	40.21	58.15	81.52	100

*: Mean (±SD) of 5 replicates.

#: Significantly different from control; P < 0.001 per cent.

Attri and Prasad (1980) demonstrated that neem oil extractive acted as strong toxicant to larval mosquito. Zebitz (1984) recorded 78.2 ppm as LC_{50} value of aqueous neem seed kernel extracts to fourth instar larvae of *Ae. aegypti.* Neem oil was found to kill mosquito larvae in respect to the concentration (Rawat, 1994). Mortality of 50 per cent and more occurred in larval *Aedes aegypti* (Sinniah *et al.*, 1994) and *Culex quinquefasciatus* (Amorose, 1995) exposed to relatively low concentrations (0.02 per cent as LC_{100}/24 hour as well as 0.99 and 1.2 ppm as LC_{50}/24 hour to 3rd and 4th stage larvae) of neem extract and neem oil. Acetone extract of neem seed coat at 40 ppm caused 100 per cent mortality of the first instar larvae of *Aedes aegypti* (Sagar and Sehgal, 1997). In a previous study on the toxicity of a neem formulation azadirachtin against larval stages of *Aedes aegypti* (Mohanraj *et al.*, 2000) the LC_{50}/ 24 hour values recorded against I, II, III, IV instar and pupal stages were 7.94, 12.365, 14.565, 14.642 and 17.38 ppm, respectively.

The difference in the values between the present and previous studies may be due to the difference in the composition of pesticides. The toxic effect may mainly be due to the active component azadirachtin present in neem. This compound was demonstrated to be the major active constituent of neem and was well known for its toxic, antifeedant, growth regulating effect on insects (Schmutterer, 1990). Xie *et al.* (1995) confirmed that azadirachtin was largely responsible for toxic actions of neem on insects. The species and formulation–dependent differences in the larvicidal activity of neem products was noticed (Mulla *et al.*, 1997 cit. Su and Mulla, 1999a). Death of mosquito larvae was probably due to the poisoning effect of neem ingredients (Friend, 1998) and deficiency of dissolved oxygen in the neem treated water (Aliero, 2003).

In the current study, a considerable increase in the LC_{50} values with the age of larvae was observed, *i.e.*, early instar larvae were more sensitive to the neem product than late instars. It was recognized that fourth stage larvae and pupae of mosquitoes were more tolerant to toxicants, than early instars (Jebanesan *et al.*, 1999). Low sensitivity with increase in age among different pre adult stages of *Aedesaegypti* exposed to various neem extracts as well as formulations (Mohanraj *et al.*, 2000). The decreased susceptibility may be attributed to increased size and weight of older larvae (Helson and Surgenoner, 1983).

Mohanraj *et al.* (2000) observed a considerable dose-based reduction in the hatchability of eggs treated with a neem formulation. One ppm of AZ found to produce hundred percent mortality in eggs (exposed for 4 hour) of *Culex tarsalis* and *Culex quinquefasciatus* (Su and Mulla, 1998a). Ovicidal activity of growth regulators/inhibitors (Miura *et al.*, 1976) and insecticides (Saghal and Pillai, 1993) in mosquitoes of different species was reported. Severe reduction in hatching of eggs from Lepidopteran and Hymenopteran insects on treatment with aqueous extracts of black pepper, neem seed and garlic bulb at concentrations 5, 10 and 15 per cent and also with their volatiles was recorded (Ekesi, 2000).

Boradent and Free (1984) based on a study on the ovicidal activity of insect growth regulators, Diflubenzuron and Bay SIR 8514 opined that the chemicals diffuse into eggs and affect the vital physiological and biochemical processes associated with embryonic development, thereby inhibiting eclosion of eggs. Penetration of ovicides through chorion was demonstrated earlier (Smith and Salkerd, 1966). The mode of ovicidal action of an IGR (Dimilin) was explained as the inhibition of chitin synthesis, thus interfering with the formation of endocuticular deposition (Miura *et al.*, 1976). A similar mode of action may be suggested for neem products as they were observed to affect chitin synthesis (Sharma and Chander sheikher, 1999a).

It is therefore suggested that the neem derivatives can also be considered for mosquito control operation, besides their use against agricultural pests.

References

Aliero, B.L., 2003. Larvicidal effects of aqueous extracts of *Azadirachta indica* (neem) on the larvae of *Anopheles* mosquito. *African Journal of Biotechnology*, 2: 325–327.

Amorose, T., 1995. Larvicidal efficacy of neem (*Azadirachta indica*) oil and defatted cake on *Culex quinquefasciatus* Say. *Geobios*, 22: 169–173.

Attri, B.S. and Prasad, R., 1980. Neem oil extractive: An effective mosquito larvicide. *Indian J. Ent.*, 42: 371–374.

Awala, P., Mwangi, R.M. and Irungu, L.W., 1998. Larvicidal activity of a granular formation of a *Melia volkensii* (Gurke) acetone extract against *Aedes aegypti* L. *Insect Sci. Applic.*, 18: 225–228.

Babu, R. and Murugan, K., 1998. Interactive effect of neem seed kernel and neem gum extracts on the control of *Culex quinquefasciatus* Say. *Neem Newsletter*, 15: 9–10.

Babu, R., Murugan, K. and Mohan, P.S., 2000. Larvicidal effect of resinous exudate from the tender leaves of *Azadirachta indica. Neem Newsletter*, 17: 1–3.

Boradbent, A.B. and Free, D.J., 1984. Effects of dyflobenzuron and Bay SIR 8514 on the oriental fruit moth (Lepidotera : Olethreutidae) and the oblique banded leaf roller (Lepidotera : Tortricidae). *J. Econ. Entomol.*, 77: 194–197.

Bowers, W.S., Sener, H., Evans, F., Bingal, A. and Erdogon, J., 1995. Activity of Turkish medical plant against mosquitoes *Aedes aegypti* and *Anopheles gambiae. Insect Sci. Applic.*, 16: 339–342.

Chavan, S.R. and Nikam, S.T., 1988. Investigation of alkanes from neem leaves and their mosquito larvicidal activity. *Pesticides*, 22: 32–33.

Cilek, J.W., Webb, J.D. and Knapp, F.W., 1991. Residual concentration and efficacy of three temphos formulations for control of larval *Aedes aegypti. J. Am. Mosq. Control Assoc.*, 7: 310–312.

Das, T.K., Somchoudhury, A.K., Saha, K. and Sarkar, P.K., 1995. Persistant toxicity of modern insecticides, germicides and plant extracts against mosquito larvae. *Pestology*, 19: 5–12.

Deshmukh, P.B. and Renapurkar, D.M., 1992. Insect repellent activity of *Azadirachta indica* Juss against mosquito *Culex pipiens fatigans. Pestology*, 16: 33–36.

Dorland's Illustrated Medical Dictionary, 1982, 26th Edition. W.B. Saunder Company, Igaku-Shoin/ Saunders, London, p. 35.

Ekesi, S., 2000. Effect of volatiles and crude extracts of different plant materials on egg viability of *Maruca vitrata* and *Clavigralla tomentosicollis. Phytoparasitica*, 28: 305–310.

Fournet, F., Sannier, C. and Monteny, N., 1993. Effects of the insect growth regulators OMS 2017 and diflubenzuron on the reproductive potential of *Aedes aegypti. J. Am. Mosq. Control Assoc.*, 9: 426–430.

Friend, J.A., 1998. Twenty years of horticulture research with neem against diamond backmoth, *Plutella xylostella* L. *Pestology*, 22: 33–40.

Helson, B.V. and Surgeoner, G.A., 1983. Effect of temperature and stage of development on susceptibility of *Aedes euedes* and *Aedes stimulans* (Diptera : Culicidae) larvae to temephos. *Can. Ent.*, 115: 623–628.

Jebanesan, A., Venkatachalam, M.R. and Jagadeesan, G., 1999. Toxicity evaluation of certain neem based bio–insecticides against the larval mosquitoes of *Culex quinquefasciatus* Say. *Pestology*, 23: 72–75.

Judson, C.L., 1967. Alternation of feeding behaviour and fertility in *Aedes aegypti* by the chemosterilant apholate. *Entomol. Exp. Appl.*, 10: 387–394.

Judson, C.L. and Gojrati, H.A.N., 1967. The effects of various oxygen tensions on embryogeny and larval responses of *Aedes aegypti. Entomol. Exp. Appl.*, 10: 181–188.

Jupp, P.G. and McIntosh, B.M., 1990. *Aedes furcifer* and other mosquitoes as vectors of Chikungunya virus at mica, north-eastern transvaal, South Africa. *J. Am. Mosq. Control Assoc.*, 12: 415–419.

Kaur, J.J., Rao, D.K., Sehgal, S.S. and Seth, R.K., 2001. Effect of hexane extract of neem seed kernel on development and reproductive behaviour of *Spodoptera litura* Fabr. *Ann. Pl. Protec. Sci.*, 9: 171–178.

Kumar, S., Thomas, S. and Pillai, M.K.K., 1991. Involvement of mono-oxygenases as a major mechanism of deltamethrin resistance in larvae of three species of mosquitoes. *Indian J. Exp. Biol.*, 29: 379–384.

Mazzarri, M.B. and Georghiou, P., 1995. Characterization of resistance to organophosphate, carbomate and pyrethroid insecticides in field populations of *Aedes aegypti* from Venezuela. *J. Am. Mosq. Control Assoc.*, 11: 315–322.

Mekuria, Y., Gwinn, T.A., Williams, D.C. and Tidwell M.A., 1991. Insecticide susceptibility of *Aedes aegypti* from Santo Domingo, Dominican Republic. *J. Am. Mosq. Control Assoc.*, 7: 69–72.

Miura, T., Schaefer, H., Takahashi, R.M. and Mulligan, F.S., 1976. Effects of the insects growth inhibitor, dimilin on hatching of mosquito egg. *J. Econ. Entomol.*, 69: 655–658.

Mohanraj, R.S. and Dhanakkodi, B., 2004. Effect of neem extracts on ovarian length and protein of *Aedes aegypti. J. Exp. Zool., India*, 7: 125 – 128.

Mohanraj, R.S. and Dhanakkodi, B., 2005a. Ovicidal effects of neem products on mosquito eggs. *J. Ecotoxicol. Environ. Monit.*, 15: 73–77.

Mohanraj, R.S. and Dhanakkodi, B., 2005b. Growth inhibiting effect of crude neem seed kernel extract of *Aedes aegypti. Indian J. Ent.*, 67(3): 241–243.

Mohanraj, R.S. and Dhanakkodi, B., 2006. Influence of neem extracts of ovipositional activity of *Aedes aegypti. Pestology*, 30: 38–41.

Mohanraj, R.S. and Dhanakkodi, B., 2007a. Delayed effect of neem extracts on fitness parameters of *Aedes aegypti. Journal of Current Sciences*, 10: 161–165.

Mohanraj, R.S. and Dhanakkodi, B., 2007b. Interference of botanical extracts on oviposition of *Aedes aegypti. Adv. Bio. Sci.*, 6: 11–17.

Mohanraj, R.S., Saraswathy, S., Thangavel, K., Logaswamy, S. and Dhanakkodi, B., 2000. Effect of neem formulation on pre-adult stages and egg hatchability of *Aedes aegypti. Neem Newsletter*, 17: 4--6.

Neoliya, N.K., Singh, D. and Sangwan, R.S., 2007. Azadirachtin-based insecticides induce alteratioin in *Helicoverpa armigera* Hub. Head polypeptides. *Current Science*, 92: 94–99.

Obomanu, F.G., Ogbalu, O.K., Gabriel, U.U., Fekarurhobo, G.K. and Adediran, B.I., 2006. Larvicidal properties of *Lepiadagathis alopecuroides* and *Azadirachta indica* on *Anopheles gambiae* and *Culex quinquefasciatus. African Journal of Biotechnology*, 5: 761–765.

Okumu, F.O., Knols, B.G.J. and Fillinger, U., 2007. Larvicidal effects of a neem (*Azadirachta indica*) oil formulation on the malaria vector *Anopheles gambiae. Malaria Journal*, 6: 63–73.

Oroumchi, S. and Lorra, C., 1993. Investigation on the effects of aqueous extracts of neem and Chinaberry on development and mortality of the alfalfa weevil *Hypera postica* Gyllenh (Col., Curculionidae). *J. Appl. Ent.*, 116: 345–351.

Pampana, E., 1963. *A Textbook of Malaria Eradication.* Oxford University Press, London, pp. 459–462.

Purohit, P., Jyotsna, D. and Srimannarayana, G., 1989. Anti-feedant activity of indigenous plant extracts against larvae of castor semi-looper. *Pesticides*, 25: 23–26.

Ramarethinam, S., Loganathan, S., Marimuthu, S. and Murugesan, N.V., 2002. Studies on the evaluation of neem oil based EC formulation (0.03 per cent Azadirachtin) on the semilooper caterpillar, *Achaeajanata* L. (Lepidoptera : Noctuidae) infesting castor plant, *Ricinus communis* L. *Pestology*, 26: 9–14.

Rawat, S., 1994. Neem products and their pesticidal characteristics. *Everyman's Sciences.* December–January, 1993–94.

Rawlins, S.C. and Wan, J.O.H., 1995. Resistance in some caribbean populations of *Aedes aegypti* to several insecticides. *J. Am. Mosq. Control Assoc.*, 11: 59–65.

Regupathy, A. and Dhamu, K.P., 1990. *Statistic Workbook for Insecticide Toxicology.* Surya Desktop Publishers, India, pp. 4–7, 16–22.

Sagar, S.K. and Sehgal, S.S., 1997. Toxicity of neem seed coat extract against mosquitoes. *Indian J. Ent.*, 59: 215–223.

Sahgal, A. and Pillai, M.K.K., 1993. Ovicidal activity of permethrin and deltamethrin on mosquitoes. *Entomon.*, 17: 149–154.

Schmutterer, H., 1990. Properties and potential of natural pesticides from the neem tree, *Azadirachta indica. Ann. Rev. Ent.*, 35: 271–297.

Sharma, P.L. and Chander Sheikher, 1999a. Toxic and morphogenic effects of neem leaf extracts on *Helicoverpa armigera* Hubner. *Pest Management and Economic Zoology*, 5: 95–100.

Sinniah, B., Sinniah, D. and Ibrahim, J., 1994. Effect of neem oil on mosquito larvae. *Mosq. Borne. Dis. Bull.*, 11: 90–94.

Smith, E.H. and Salkerd, E.H., 1966. The use and action of ovicides. *A. Rev. Ent.*, 11: 331–368.

Su, T.Y. and Mulla, M.S., 1998a. Ovicidal activity of neem products (Azadirachtin) against *Culex tarsalis* and *Culex quinquefasciatus* (Diptera : Culicidae). *J. Am. Mosq. Control Assoc.*, 14: 204–209.

Su, T.Y. and Mulla, M.S., 1999a. Oviposition bioassay responses of *Culex tarsalis* and *Culex quinquefaciatus* to neem products containing azadirachtin. *Entomol. Exp. Appl.*, 91: 337 – 345.

Sundar, M.S., Krishnayya, P.V. and Rao, P.A., 2000. Evaluation of lethal concentrations of azadirachtin based proprietary formulation for larvicidal and ovicidal effect against *Spodoptera litura* Fab. *J. Ent. Res.*, 24: 147–150.

Thamil Solai, P.K., Selvaraj Pandian, R. and Charles Manoharan, A., 1999. Toxicity evaluation of the pesticides Salotrap, Neemgold, Nimbecidine and Nimbex against the larvae and pupae of *Culex quinquefasciatus* Say. In: *Biopesticides in Insect Pest Management*, (Eds.) S. Ignacimuthu and Alok Sen. Phoneix Publishing House Pvt. Ltd., New Delhi, pp. 46–49.

Umar, A., Kela, S.L. and Ogidi, J.A., 2007. Effects of extraction solvents on the toxicity of *Azadirachta indica* A. Juss (Neem) seed kernel extracts to *Aedes aegypti* (Diptera : Culicidae) larvae. *Int. Jor. P. App. Scs.*, 1: 32–38.

Vatandoost, H. and Haziri, V.M., 2004. Larvicidal activity of a neem tree extract (Neemarin) against mosquito larvae in the Islamic Republic of Iran. *Eastern Mediterranean Health Journal*, 10: 573–581.

Xie, Y.S., Fields, P.G. and Isman, M.B., 1995. Repellency and toxicity of azadirachtin and neem concentrates to three stored-product beetles. *J. Econ. Entomol.*, 80: 1024–1031.

Zebitz, C.P.W., 1984. Effect of some crude and azadirachtin enriched neem (*Azadirachta indica*) seed kernel extracts on larvae of *Aedes aegypti. Entomol. Exp. Appl.*, 35: 11–16.

Chapter 62

Identification of Droughts in Anantapur District, Andhra Pradesh Utilizing DBMS Approach

P. Srinivas[1]*, G.N. Pradeep Kumar[2] and T. Tajmul Basha[3]

[1]Associate Professor, Department of Civil Engineering, Koneru Lakshmaiah College of Engineering, Vaddeswaram, Guntur District – 522 502

[2]Professor, Department of Civil Engineering, S.V.U. College of Engineering, Sri Venkateswara University, Tirupati– 517 502

[3]Assistant Engineer, Panchayat Raj Engineering Department, Somajiguda, Hyderabad – 500 082

ABSTRACT

The agricultural productivity depends upon chiefly the occurrence and distribution of rainfall in a particular region. But due to non-uniform distribution of rainfall and prolonged dry spells during monsoon season and crop period, the dry land agriculture in arid and semi arid regions is becoming difficult. The identification of dry spells and wet spells to find out drought occurring conditions are complex because of the fact that, it requires analyzing the vast hydrological data in a systematic order. Anantapur is one of the drought-affected districts in Andhra Pradesh. Due to prolonged dry spells and ill distributed rainfall the district underwent a metamorphosis from drought to desert prone area. Keeping this in view, the present study is carried out to identify the droughts in Anantapur district by analyzing the daily rainfall and evaporation data for a period of 21 years from 1979 to 2000 by the application of DBMS approach by developing

* Corresponding Author: (M) 9866305533; Phone: (08645) 246948 (O), (0863) 2254940 (R); Fax: (08645)247249; E-mails: srinivas7_pasupuleti@yahoo.co.in; vasu77_p@yahoo.com

programs using Microsoft Visual FoxPro Software. From the data analysis, dry days, dry spells, wet days, wet spells, monthly, seasonal and annual rainfall, driest and wettest months were obtained. The interpretation of the results yields that it is possible to identify the likely hood of occurrence of dry spells and wet spells. The onset of monsoons was late and consequently the late sowing of crops which results in crop failure. The identification of dry spells is therefore helpful in agricultural planning, reservoir operations, releasing of water to canals for irrigation and for planning Cloud seeding operations.

Keywords: *DBMS, Drought, Wet spell, Wet period.*

Introduction

Meteorology as well as Hydrology are concerned with the atmospheric and land phases of hydrologic cycle with the emphasis on the interrelationship involved. The water resources of a region mainly depends upon the precipitation, which in turn depends upon the meteorology parameters such as temperature, evaporation etc., where as industrialization, deforestation and modem living leading to the phenomenon called global warming. The warming of the earth's surface could cause large scale climatic changes, especially affects rainfall pattern, which greatly alters both agriculture and natural vegetation. World wide decrease in forest area is also affecting the climate, which has direct physical affects on water balance of the earth by altering the temperature, monsoon pattern and thus affecting agricultural production. Water is one of the Nature's wonderful and most precious gift to the earth, the most crucial for sustaining life and is required in almost all the activities of man. All the civilizations have flourished along the water bodies in the world. Water is the fundamental basic need for sustaining human economic activities. Providing water in the desired quantity and quality, and at the right time and place, has been a constant endeavor of all civilizations, and no other natural resource has had such an overwhelming influence on human history and life. As the human population increases, as people express their desire for a better standard of living, and as economic activities continue to expand in scale and diversity, the demand on fresh water resources will continue to grow. However, the water resources in a region are limited considering the future demands. Moreover, the rainfall is confined to the monsoon season and is unevenly distributed both in space and time even during the monsoon season. Hence the conservation and use of water therefore, forms one of the main elements in the development planning.

Study Area

Anantapur District is the study area, which is one of the drought-affected areas of Andhra Pradesh, located in the middle of the peninsular India, which renders it the driest part of the state. Being far away from the East coast, it does not enjoy the full benefits of South-West monsoon. Being cut off by the high Western Ghats, the North-East monsoon is also prevented from penetrating and quenching the thirst of these perched soils. It is therefore seen that, the entire area of Anantapur district is deprived of both the monsoons and is subjected to frequent droughts. The predominant causes for drought are undependable and ill distributed rainfall with prolonged dry spells between successive rains during the crop period, intensive wet spells of short duration causing soil erosion, and no significant perennial water resources in the district. These causes have resulted the district to undergo a metamorphosis from a drought to desert prone area. The district falls between the Northern latitudes of 13°04′ and 15°15′ and Eastern longitudes of 76°50′ and 78°30′. The normal annual rainfall in the district is 520mm where as the normal rainfall of Andhra Pradesh as a whole is 925mm. The number of rainy days on an average in a year is only 36. The normal rainfall during the South-West monsoon (June to September) is 296 mm which forms about 57 per cent of the total rainfall for the year. The rainfall during North-East monsoon (October–December) is 140 mm which forms 27 per cent of the rainfall for the year. Some times, intense storms occur even in the drier regions because of severe cyclonic activity in the Bay of Bengal. The gross cultivated area is only 55 per cent of the total geographical area (19163 Km^2), while the average area cultivated more than once is only 6 per cent of the total area. The percentage of the gross irrigated area both from surface and ground water to the culturable area works out to be only 11.7 per cent, which is very low. Groundnut and Sunflower are the major crops grown in dry land agriculture in Anantapur district. Owing to prolonged dry spells, the yield of the crop is greatly reduced, resulting in the weak economic position of the cultivators in this area.

Necessity of the Study

The identification of dry spells and wettest months is helpful in crop planning, pre and post agricultural activities in Semi-arid areas such as Anantapur District. After the estimation of the likelihood of occurrence of prolonged dry spells, it is possible to select the type of crops that are to be raised in a particular area, which can with stand the moisture stress. Sufficient moisture is necessarily required for the crop, during the critical periods. The critical periods may vary from one crop to other. Unless the required water is not supplied, the crop starts wilting. If the prolonged dry spells occur during the critical stages of crop, then the crop will not be able to survive and results in decrease in the yield, even after sufficient rainfall occur after reaching the ultimate wilting point of the crop. This is because of the greater moisture stress to the plant. In such situations, the knowledge of likelihood of occurrence of dry spells will greatly help to protect the crop from wilting, by supplying the minimum moisture required for the survival of the crop. This can be achieved by following various methods even in dry land agriculture, such as construction of farm ponds and utilizing the stored water using modem methods of application of water to the crops. The identification of wet spells is very much helpful in reservoir operation and planning and for releasing water to the canals for irrigating crops and for planning and carrying out cloud seeding operations.

Objectives of the Study

The various objectives of the study are:

1. Identification of Droughts in Anantapur district by analyzing the daily rainfall and evaporation data for a period of 21 years.
2. Analysing data by the application of DBMS approach using Microsoft Visual Fox Pro software and obtaining Dry spells, Wet spells, Driest and Wettest months and classify Dry spells.

Literature Review

Rainfall pattern can be represented in two ways namely trend and cycles. Cycles refer to the natural variation while trend refers to man induced pattern such as increasing or decreasing. Topography plays an important role on Indian rainfall. The Western Ghats and North Eastern mountain ranges are monumental examples. Colaba raingauge station and Santacruz raingauge station within Mumbai show an increasing trend in rainfall with Santacruz rainfall higher than Colaba rainfall by about 300 mm, which may be due to cutting of hillocks in wind ward direction of S-W monsoon on the Eastern and North-Eastern side of Santhacruz observatory on the runway (Jeevananda, 2000). The geographical position of Anantapur District in Andhra Pradesh is in the middle of the peninsular

India, located in the rain shadow region and the leeward side of Western Ghats, renders it the driest part of the state with a normal annual rainfall as low as 520 mm, the second lowest in India after Jaisalmer of Rajasthan (Rao, 2002). Crop production under rain fed conditions in the arid and semiarid regions is often affected by droughts during the monsoon season because of prolonged dry spells associated with break monsoon conditions. Some times, heavy rainfall occurs even in the drier regions because of severe cyclonic activity in the Bay of Bengal. In Anantapur, the maximum amount of rainfall received during a wetspell was more than 100 mm even during the years with seasonal rainfall of about 250 mm, although it was less than 100 mm in most cases during the years with seasonal rainfall of up to 450 mm. During the years with seasonal rainfall exceeding 450 mm a single wet spell can contribute more than 150 mm to seasonal rainfall. Rainfall of more than 250 mm during a wet spell was observed in the early part of the season (July–August) during the years with seasonal rainfall exceeding 600 mm (Singh *et al.*). Though the crop loss due to unfavourable rainfall cannot be avoided fully, it can be minimized with the accurate and timely weather forecasting. Medium range (3–10 days) forecasts are greatly helpful to the agriculture for taking decisions and implementing the same. (Krishna Murthy *et al.*, 1949). Out of 14 famine occurrences, Anantapur district experienced as many as 11 times. The trend of famine in the district evidently indicate that the south-western part of the district is recurrently subjected to famine compared to Northern and Eastern parts of the District. Ananthapur area is one among the south-western parts of India subjected to famine 10 times (Baburao *et al.*, 1999). Dry spell and Wet spell water availability index is calculated using threshold values of weekly rain fall and probability of dry spells over three weeks and Agro-climatic classification was carried out for assessment of the crop potential of Karnataka state in India (Khambate, 1992).

Database Management Systems (DBMS)

A Database Management System (DBMS) is a software designed to assist in maintaining and utilizing large collections of data. Database Management System consists of a collection of interrelated data and efficiently to use in retrieving and storing information for the manipulation. A DBMS provide users with an abstract view of the large data. A database is a collection of data, typically describing the activities of one or more related events. For example the hydrological database might contain information about the following.

1. Entities such as rainfall, evaporation, humidity, temperature, wind and velocity.
2. Relationships between entities such as rainfall, evaporation and humidity.

A DBMS allows a user to define the data to be stored in terms of data model, which is a collection of high level data description constructs that hide many low level storage details. A database system provides two different types of languages one to specify the database schema and the other to express database queries and updates. The relational data model is based on a collection of tables. The user of the data base system may query these tables, insert new tables, delete tables and update (modify) tables. The overall design of the database is specified by a set of definitions that are expressed by a Data-Definition Language (DDL). A Data-Manipulation Language (DML) is a language that enables users to access or manipulate data. There are several languages for expressing these operations numerous relational-data base products are now commercially available. Data base products for personal computers include MS access, Dbase and Microsoft Visual Fox Pro.

Methodology

General

In the present study, daily rainfall and evaporation data for Anantapur District for a period of 21 years from 1979 to 2000 was obtained and analysed using Microsoft Visual Foxpro software employing personal computer.

Calculation of Dry Spells and Wet Spells

Dry Days and Dry Spell

In order to calculate the dry days the daily rainfall and evaporation amount in mm were taken in to consideration. The condition for a dry day in that if

Rainfall (mm) ≤ evaporation (mm)

If there are more than three dry days occurred successively, then the entire period is considered as dry spell. The dry days (d.d : count) and the dry spells were calculated between 1st March (as starting date) of a particular year and 28th of February (as ending date) of the next year as the Rabi season end with February normally. In leap year 29th of February is considered as ending date.

Wet Days and Wet Spells

The condition for a wet day is that if

Rain fall (mm) > Evaporation (mm)

If there are one or more wet days occurred successively, then the entire period during which the above condition is satisfied was considered as wet spell. The wet days (w.d count) and the wet spells were calculated between the starting date (1st January) and the end (31st December) of a particular year.

Application of Microsoft Visual FoxPro Software

Analysis of daily rainfall and evaporation data for a period of 21 years from 1979 to 2000 can be done by application of data base management system approach by developing programs using Microsoft Visual FoxPro Software. Programs are written in Visual FoxPro to display main menu, data entry screen, dry days and wet days, identification of dry spells, wet spells and displaying Answer table. Various FoxPro display Screens are shown in Figures 62.1–62.4.

Analysis of Data

Dry Days and Dry Spells Analysis

Dry days and Dry spells were computed by analyzing the data with Microsoft Visual FoxPro Software. The dry days and dry spells greater than 3 days duration were computed. It was indicated that, on an average about 14 days spells were occurred per year. The minimum number of dry spells were 10 during 1994–95 and the minimum number of drys spells were 18 during 1983–84. It was also observed that, the dry spell having maximum duration *i.e.* 122 days was occurred during *Rabi* Season in 1995–96. Mostly the dry spells having greater than 60 dry days were occurred during *Rabi* Season in almost all the 21 years period of study. During monsoon season *i.e.* from July to November, the maximum dry spell having 65 days duration was occurred in 1989–90. During hot weather season, the maximum dry spells with more than 90 days duration were occurred in 1984, 1986, 1990, 1991, 1992, 1993.

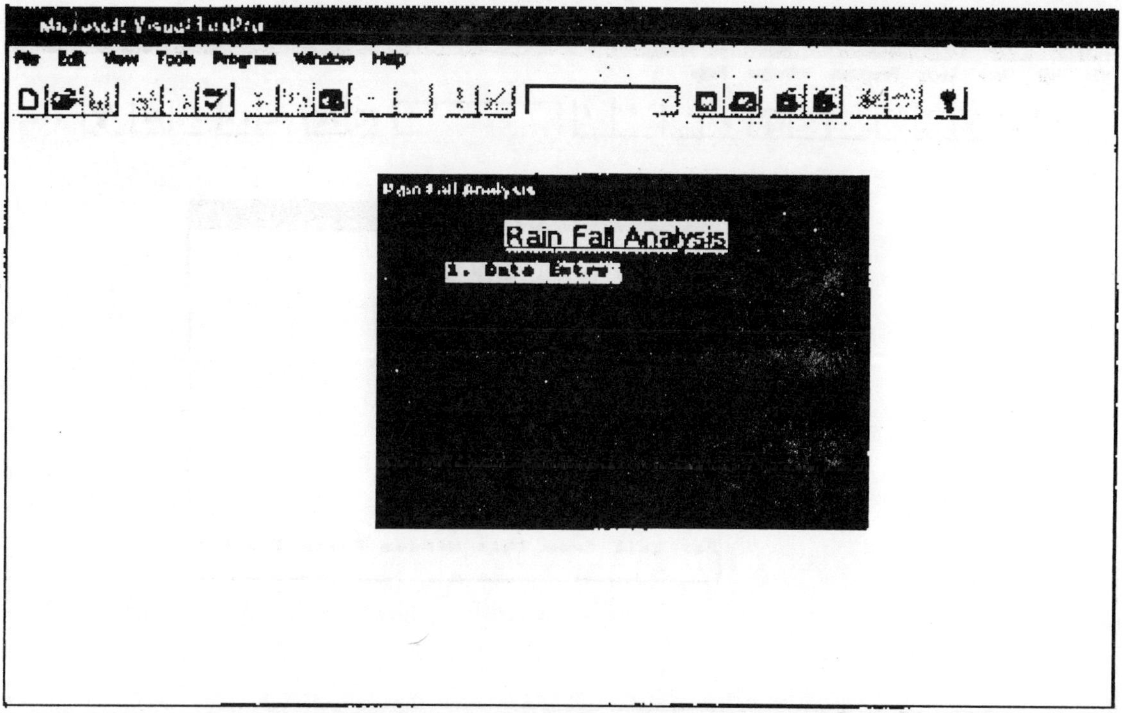

Figure 62.1: Microsoft Visual FoxPro Main Menu Screen

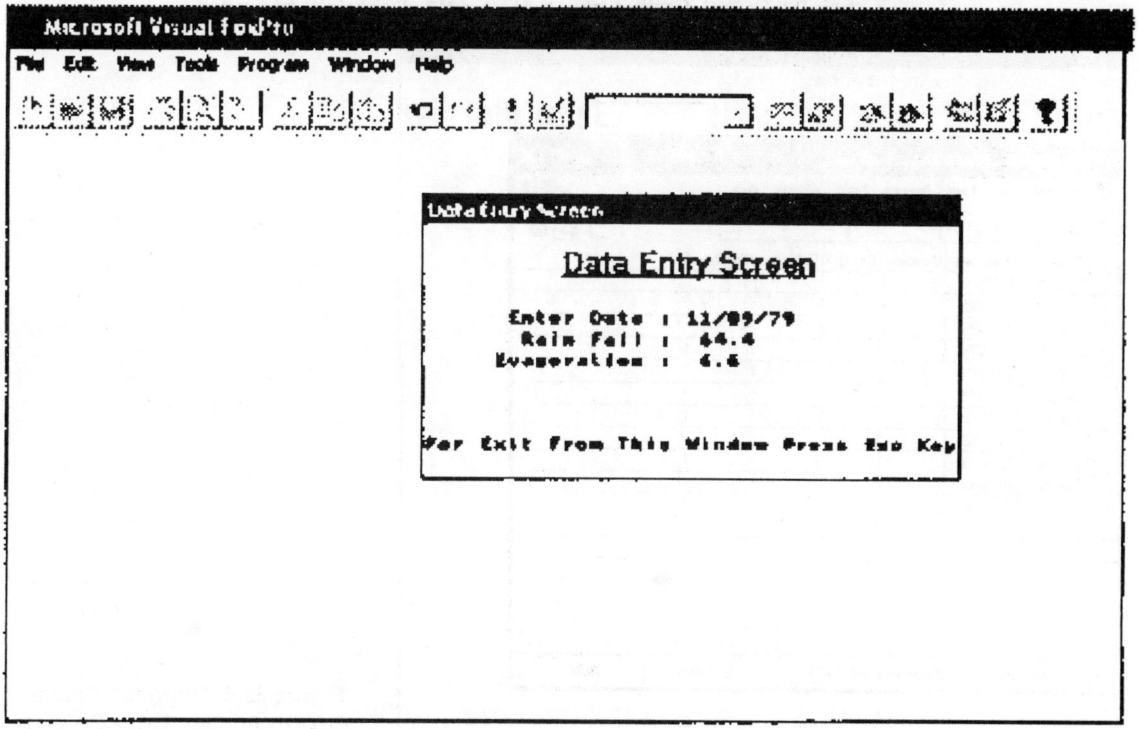

Figure 62.2: Microsoft Visual FoxPro Data Entry Screen

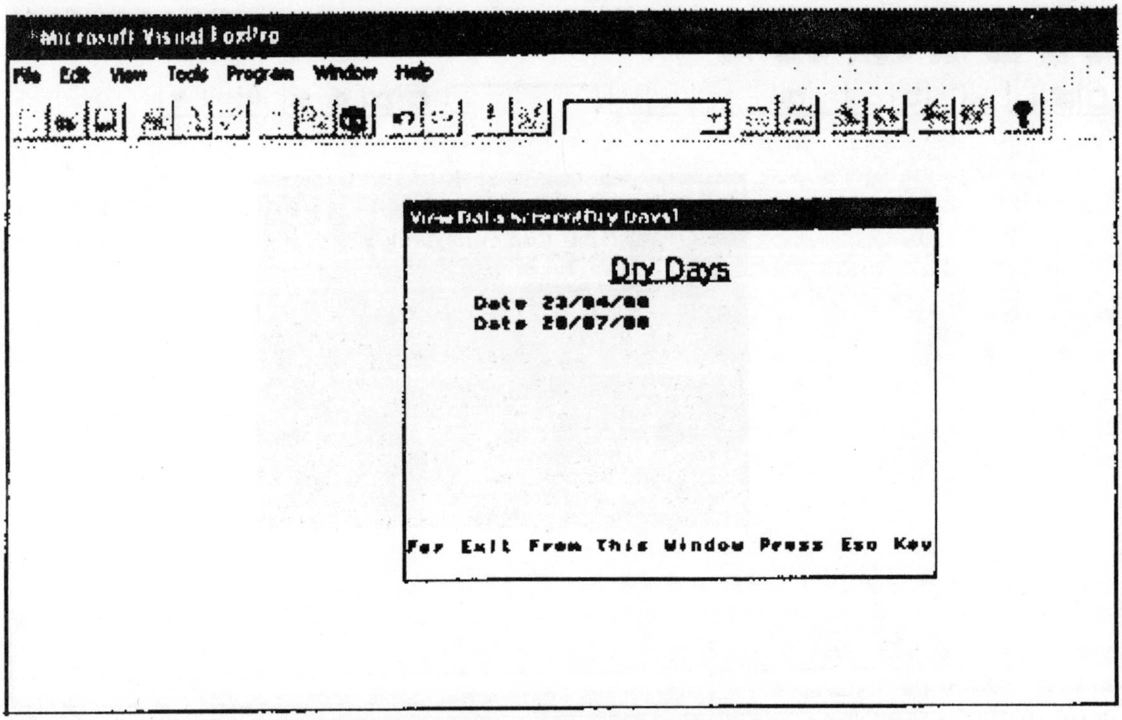

Microsoft Visual FoxPro

File Edit View Tools Program Window Help

View Data Screen(Dry Days)

Dry Days

Date 23/04/00

Date 20/07/00

For Exit From This Window Press Esc Key

Figure 62.3: Microsoft Visual FoxPro View Data Screen for Dry Days

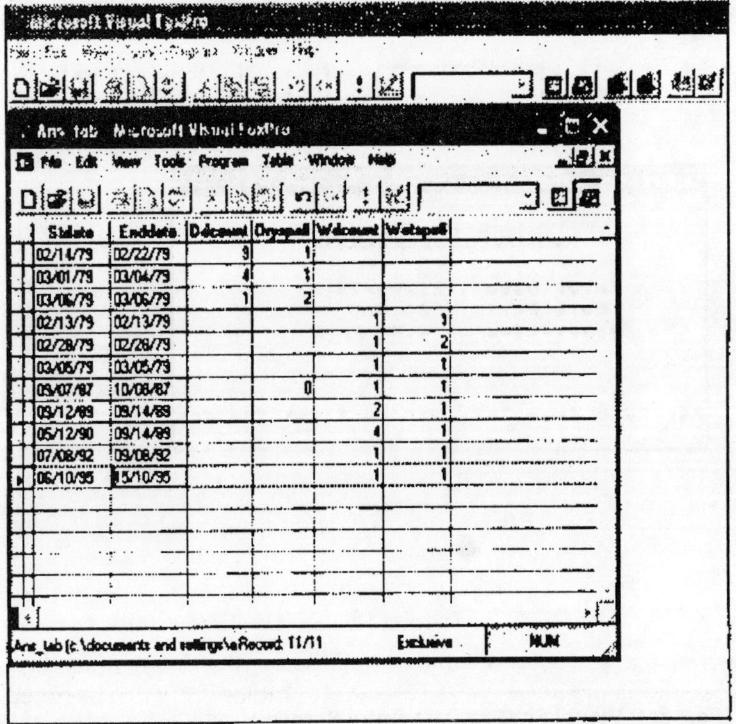

Stdate	Enddate	Dcount	Dryspell	Wdcount	Wetspell
02/14/79	02/22/79	9	1		
03/01/79	03/04/79	4	1		
03/06/79	03/06/79	1	2		
02/13/79	02/13/79			1	1
02/28/79	02/28/79			1	2
03/05/79	03/05/79			1	1
09/07/87	10/08/87		0	1	1
09/12/89	09/14/89			2	1
05/12/90	09/14/89			4	1
07/08/92	09/08/92			1	1
06/10/95	15/10/95			1	1

Figure 62.4: Microsoft Visual FoxPro Answer Table Screen

Wet Days and Wet Spells Analysis

Wet days and Wet spells were computed by analyzing the data with Microsoft Visual FoxPro Software. It was observed from the above results of wet days and wet spells that mostly one day duration wet spells were occurred in all the 21 years period of study. The wet spell having maximum duration *i.e.* 7 days was occurred during the August Month in 1988. The number of wet spells was varying from minimum of 10 in 1980–81 to maximum of 23 in 1996–97. The number of wet days per year were varying from 15 to 35 and on an average, the number of wet days were 23 per year. The average number of wet spells per year was worked out to be 18. The number of wet days were minimum of 15 in 1980–81 and maximum of 35 in 1988–89.

Monthly Dry Days and Wet Days Analysis

The monthly dry days and monthly wet days during South West and North East monsoon seasons starting from July to November for the period of 21 years from 1979–2000 were obtained from the analysis. It is indicated that, September and October were the wettest months. This indicates that the maximum number of wet days were occurred in September and the minimum number of wet days were occurred in June.

Monthly and Seasonal Rainfall Computation and Analysis

The monthly and seasonal rainfall were calculated from the daily rainfall data for a period of 21 years from 1979 to 2000.The careful observation of the data reveals that, most of the rainfall events in Anantapur district occurred during monsoon season from July to November. In the 21 years rainfall data, the highest was 937.0 during 1998 and the lowest was 223.0 in 1980. The average annual rainfall at Anantapur district was worked out to be 547.5 mm. The average monthly rainfall over a period of 21 years indicated that the highest rainfall of 127.6 mm occurred in September.

Computation and Analysis of Year Wise Dry Days, Wet Days and Total Rainfall

The yearly Dry days, Wet days and the Total rainfall were calculated. The results indicates that, when the number of wet days are less than 22, the amount of rainfall is also less than the average annual rainfall (547.5 mm) except in one or two cases over the 21 years period of study. When the number of rainy days are greater than or equal to 30, the annual rainfall amount was greater than or equal to the average annual rainfall and the rainfall was distributed in all the monsoon months. During such periods the agriculture yield was not much affected.

Dry Spells Classification

The number of dry spells of different duration's during monsoon period over the 21 years period of study and the total number of dry spells greater than 10 days duration was obtained.

The dry spell duration was classified into five categories such as, Cat. 1: 1 0 to 20 days, Cat. 2: 21 to 30 days, Cat. 3: 31 to 40 days, Cat. 4: 41 to 50 days and Cat. 5: greater than 50 days.

The maximum number of dry spells (> 10 days) were maximum *i.e.*, 8 in 1982–83, 85–86 and 92–93 and the corresponding rainfall was 604.3, 404.4 and 433.8 respectively. The number of dry spells (> 10 days) were minimum *i.e.*, 3 in 1979–80 and in 1988–89 and the corresponding rainfall were 610 and 927.6 mm respectively.

Computation and Analysis of Driest and Wettest Months during 1979–2000

The interpretation of the results from the tables, the months with maximum number of rainy days and maximum number of dry days were calculated and tabulated in the Table 62.1. It can be noted from the Table 62.1 that in 15 years out of 21 years, September was seen to be the wettest month and the next was October. The driest months were June and July in 15 years out of 21 years. From the statistics of the rainfall data and from the interpretation of the results, it was observed that the onset of monsoons were late in almost all the years over the period of study, due to which the sowing of crops was not done with in the correct time, resulting in reduction of yield. It was also observed that the annual rainfall was even less than the normal annual rainfall in most of the years and once in four years, the annual rainfall exceeded the normal annual rainfall, resulting in drought conditions prevailing in Anantapur district. Because of the number of rainy days which are also non uniformly distributed during the S–W and N–E monsoon period, and the prolonged dry spells during monsoons, the yield of crops were affected and this has resulted in frequent droughts in Anantapur district.

Table 62.1: Driest and Wettest Months During 1979–2000

Year	*Total Rainfall (mm)*	*No. of Rainy Days (Wet days)*	*No. of Dry Spells> 10 Days Duration*	*Max. Wet days Months*	*Maximum Dry Days Months*
1979–80	610	23	3	September, November	June, July, August
80–81	223	15	6	September, November	June, July, August
81–82	526.7	32	5	September, November	June, July, August
82–83	604.3	28	8	September, November	August
83–84	665.5	30	6	August, September	July, November
84–85	273.2	18	5	July, October	June, August, November
85–86	404.4	28	8	July, September, October	July, August, November
86–87	423.5	19	5	Sept, October	July, November
87–88	505.5	19	6	August, October	July, Sept
88–89	927.6	35	3	July, September	June, Nov
89–90	826.0	25	7	July, September	August, October, November
90–91	539.6	26	5	August, October	July
91–92	433.8	19	6	June, October, November	July, August
92–93	498.7	25	8	September, October, November	June, August
93–94	699.6	32	7	August, September, October	June, July
94–95	395.1	19	4	October, November	July, August, September
95–96	762.5	24	6	August, October	June, July
96–97	853.9	32	4	June, September, October	August, November
97 –98	502.4	17	5	September	July, October, November
98–99	937.0	25	5	September, October	June
1999–2000	472.6	21	7	August, September	June, July

Conclusion

It was observed that the onset of monsoons were late in almost all the years during the period of study and consequently the late sowing of crops which results in decrease in yield. It can be concluded that, the annual rainfall was even less than normal annual rainfall in most of the years and once in four years, the annual rainfall exceeded the normal annual rainfall over the 21 years period of study, which results in drought conditions prevailing in Anantapur district. It was also concluded from the analysis that when ever in a particular year, the annual rainfall is less than the normal annual rainfall, wet days are less than 23, dry spells are more than 5 during monsoon season and the dry spells with more than 30 days duration occurs during June, July and August, then it can be identified as drought year.

The average numbers of rainy days were less than 30 days and which are also non uniformly distributed during the S–W and N–E monsoon resulting in prolonged dry spells to occur in Anantapur district. The knowledge of likelihood of occurrence of dry spells will greatly help in protection of crop from wilting by supplying minimum quantity of stored water from farm ponds, check dams etc., using modem methods of application of water. The identification of dry spells and wettest months will be helpful in crop planning and in pre and post agricultural operations. The identification of dry spells is therefore helpful in agricultural planning, reservoir operations, releasing of water to canals for irrigation and for planning Cloud seeding operations.

References

Baburao, P. and Seshubabu, K., 1999. Environment impact on water resources and sustainability in and around Anantapur. In: *Proc. National Workshop on Integrated Rural Development through Participatory Approach,* The District Administration, Anantapur, pp. 132–139.

Huda, A.K.S., Rahman, M.S., Egan, J.P. and Holloway, R.E., 1993. Assessing and managing cropping risks in low rainfall areas of southern Australic using agro-climatic data. Techical Paper, Department of Agriculture, South Australia No. 34, Xii, pp. 191.

Jeevananda Reddy, P., 2000. Relevance of pollution monitoring in environmental context. In: *Proc. of National Seminar on Analytical Techniques in Chemistry,* S.V. University, Tirupathi, pp. 152–163.

Khambate, N.N., 1993. Use of Markov chain model in agro-climatic classification for assessment of crop potential. *J. Maharastra Agricultural University,* 18(2): 174–177.

Khambeta, N.N., 1992. Agro-climatic classification for assessment of the crop potential of Karnataka. *Mousam,* 43(1): 91–98.

Krishnamurthy, S.K., Padmalatha, Y. and Yellamanda Reddy, T., 1999. Role of weather forecasting in agriculture. In: *Proc. National Workshop on Integrated Rural Development through Participatory Approach,* The District administration, Anantapur, pp. 19–25.

Ranade, D.H., Vishwakarma, S.K. and Paradkar, V.K., 2000. Probability analysis of wet and dry spells for agricultural planning at Chhindwara. *Indian Journal of Agricultural Sciences,* 70(10): 719–721.

Rao, K.V.M., 2002. Rainfall variation on ground water table. In: *Proc. of National Conference on Groundwater Management and Rural Development,* Anjuman Engineering College, Bhatkal, pp. 9–10.

Chapter 63

Investigation on Water Quality Parameters on Some Selected Freshwater Bodies of Imphal District

L. Joymati Devi, M. Suresh Singh and W. Mema Devi
Department of Zoology, G.P. Women's College, Imphal

ABSTRACT

The water samples from fresh water bodies of 10 different ponds of Imphal district were studied. Some of the water samples was found suitable for drinking as well as other purposes but some area was found beyond permissible standard limit when compared with value for drinking water prescribed by WHO, BIS and ICMR.

Keywords: Pond water, Monitoring of water quality, BOD, COD, TDS etc.

Introduction

Water is a unique liquid which is regarded to be the most important substance for the sustenance of life for all the living organisms of the biosphere. The Imphal valley is an oval shaped depression located in the central part of Manipur. Water resources in Manipur are provided mainly by the river, lakes, ponds, ground water, rain water etc., but pollution of water is one of the most serious problems confronting the modern human society, specially Imphal valley, where congested people but lack of supplying tap water. Water quality is affected by various factors like its uses in agriculture, industry, recreation and flow through dams by Jayanthi (1993), Ayed (2002). It is an established fact that water quality is closely linked to the surrounding environment and the prevalent land used (APHA, 1975).

Table 63.1: Water Quality Standards for Drinking Water

Parameter	pH	Alkalinity	No_2	TDS	DO	COD	Ca	Fe	Cl^{-1}	F^-	K^+	NO_3	Cr	SO_4	Col
WHO	7–8.5	120	3.0	500	6.0	10.0	75	0.3	250	1–1.5	5–10	50	0.05	200	<4
BIS	6.5–8.5	200		500	3.0		75	0.3	250		150		45		
ICMR	7–8.5			500	4–6										

Table 63.2: Physico-chemical and Microbiological Analysis of Freshwater Samples of Imphal District

Study Sites	pH	TDS	DO	COD	Alkalinity	Ca Hardness	Fe	Cl^-	SO_4	Nitrate	Nitrite	Fluoride	K^+	Cr	Zn	No. of Coliners
Kanglapat	6.5	115.0	4.2	ND	125	70	0.05	200	8.6	4.0	1.15	1	5.2	ND	ND	10
Bijoygo vinda Thangapat	6.3	156.3	5.8	7.4	104	73	0.07	170	2.5	3.0	4	1.1	6.7	ND	ND	4
Ningthem-pukhri	6.2	209.3	4.1	2.5	135	75	0.08	225	12.5	2.4	3.8	1.6	8.9	ND	0.1	3
Lalambung Pukhri	7.2	400	5.6	ND	107	64	0.1	180	ND	0.78	1.45	1.4	10.6	ND	ND	ND
Khuyathong Pukhri	7	320	6	1.0	121	43	0.35	130	25	1.5	1.25	1.5	11	ND	ND	ND
Chingmeirong Pukhria-chouba	6.8	160	5.5	8.6	106	25	0,01	140	27	4.8	2.1	1.3	5.2	ND	0.05	ND
Mantripukhri	6.9	480	4.2	3.5	118	83	0,07	178	1.8	5.6	3.2	0.9	4.3	ND	ND	14
Kwakeithel Pukhria-chouba	7.1	440	4.3	ND	115	125	0,08	192	ND	1.4	2.5	1.2	6.7	ND	ND	8
M.R. Ground Pukhri	7.6	516	6.6	ND	103	46	0.3	205	2.8	2.1	1.6	1.6	7.8	ND	ND	2
Kongba Pukhreia-chouba	6.1	267	I 3.4	3.4	102	62	0.2	220	110	3.4	1.3	1.1	8.4	ND	ND	ND

*: The results is the mean of five replication for five months.

In Manipur specially in Imphal, there is no significant work on water quality of fresh water bodies except by Rajeshori (2005) on Nambul river. Therefore, the present work was undertaken to investigate about the physico-chemical parameters of selected fresh water bodies of Imphal valley, which can give an useful information of water quality to the local users of the surrounding area.

Materials and Methods

Water samples of 10 different fresh water bodies of Imphal districts were collected from March 2007 upto July 2007 (50 samples collected over a period of five months).The samples collected were analyzed in the laboratory, pH determination was carried out by electrometric method using a digital pH Meter. Calcium hardness was determined by EDTA titrimetric method. Nitrate, nitrite and fluorides were analyzed by UV Spectrophotometer, while chloride, alkalinity and acidity were analyzed by Atomic Absorption Spectrophotometer (APHA, 1975).

Results and Discussion

The results obtained for the analysis of water samples of 10 different fresh water bodies of Imphal District of over 5 months are shown in Table 63.2. The Table 63.1 shows drinking water standard prescribed by the Indian Council of Medical Research (ICMR, 1975). So far pH range of the study area is concerned it was found to be varied between 6.1 to 7.6 as against the desirable range but for drinking water is 6.5 to 8.5(BIS).This investigation shows that some of water samples had pH less than the permissible limit, which may increase the corrosive and activity of water on concrete. In this investigation TDS of water sample ranged between 115 to 516 mg/L, whereas desirable limit of TDS is 500 mg/L. In some area such as M.R. ground pond, there is slight high concentration of TDS, but not as much only 16 mg/L. The concentration of calcium in water sample range from 25 to 125 mg/L, that means some of the samples had higher values than the permissible limit. But, this metal is not problematic but it can increase total hardness of water. This results agree with the works of Jayalakshmi *et al.* (2005) at Karnataka. When we compared with BIS the concentration of Fe is 0.03, in this experiment most of them except Khoyathong pond have 0.35 which have slightly higher in permissible limit *i.e.* 0.3. The primary concern about iron in drinking water is its objectionable taste, the nitrate concentration varied between 0.01 to 4.91 mg/L. The concentration of nitrate is high in Mantripukhri pond in which it reaches upto 5.6, which is slightly greater than permissible limit.

Occurrence of nitrite in water is an intermediate product in the biological breakdown of ammonia or the reduction of nitrate. Because of biological decomposition of nitrogenous organic matter such as sea wage and imimal waste. During this investigation some samples were found to be highly contaminated with coliforms, while some other samples were suitable for human consumption. According to WHO, water is considered safe for drinking if it contains four or lesser coliners per 100 ml of water. Nitrate nitrogen concentration was minimum in some ponds such as in Chingmeirong and Mantripukhri ponds, these concentration enters in a aquatic body or soil as well as artificially fertilized soil, through rainfall and sewage (Lal, 2005).

From the above investigation it can be concluded that more on basic information and data on fresh water aquatic habitat are required to be generated in order to evolve effective and appropriate strategies for management of our natural and man made reservoirs.

Acknowledgement

The author greatly acknowledged to the Principal, G.P. Women's College for providing laboratory facilities during the course of investigation.

References

APHA, AWWA, WDCF, 1975. *Standard Methods for the Examination of Water and Wastewater*, 14th edn. American Public Health Association, Washington, D.C.

Ayed, P.K., 2002. Ecological studies on two fresh water bodies near Mysore, Karnataka. *Ph.D. Thesis*, University of Mysore, Mysore.

Jayalakshmi, O., Belagali, S.L., Ramaswamy, S.N., Das, R.S. and Janardhana, G.R., 2005. Investigation on water quality characteristics in selected areas of Karnataka state, India. *Indian J. Environ. and Ecoplan.*, 10(2): 405–508.

Jayanthi, S.K., 1993. Studies on industrial water pollution of Cauvery river near Belugola, Mysore. *Ph.D. Thesis*, University of Mysore, Mysore.

Lal, R.B., 2005. Effect of heavy metal toxicity and exposure on human health. *Indian J. Environ. and Ecoplan.*, 10(2): 533–536.

Rajeshwari, R.K., 2005. Investigation on physico-chemical condition of major effluent points of Nambul River, Manipur, India. *Indian J. Environ. and Ecoplan.*, 10(2): 537–540.

Chapter 64

Treatment of Polluted Water Containing Heavy Metal Ions in Basic Medium

Deepika Bansal, R.K. Mittal and C.P. Singh Chandel*
Department of Chemistry, University of Rajasthan, Jaipur – 302 004, Rajasthan

ABSTRACT

This article throws light on a cost-effective technology for the removal of heavy metal ions like Al^{+3}, Cr^{+3}, Fe^{+3}, Mn^{+2}, Zn^{+2}, Ni^{+2} and Co^{+2} from the polluted water discharged by industries. The removal of these metal ions from polluted water was investigated in the basic medium. The effect of pH was found to be most important for their precipitation. It was found that Al^{+3}, Cr^{+3} and Fe^{+3} ions precipitate in the form of their hydroxides. Mn^{+2}, Zn^{+2}, Ni^{+2} and Co^{+2} ions precipitate in the form of their sulphides with the help of Na_2S.

Introduction

In Rajasthan, one of the main causes of industrial pollution is the discharge of effluents containing heavy metal ions in places like "Amani Shah Ka Nala" of Jaipur. The polluted water of the Nala is used by farmers for growing vegetables. The metal ions pass on to the vegetables and act as poison on consumption by people. Heavy metal ions have atomic density greater than 6g/cm^3 (Ghaffar, 2006). These metal ions include Al^{+3}, Cr^{+3}, Fe^{+3}, Mn^{+2}, Zn^{+2}, Ni^{+2} and Co^{+2}. They are highly toxic for biological organisms. Hence, they present a threat to human health and the environment. Therefore, these heavy metal ions are to be removed from water sources before using for agriculture. According to World Health Organization (WHO), the metal ions of most immediate concern are Zn, Fe, Mn, Al, Ni and Co (WHO, 1984).

---

* Corresponding Author; E-mail: chandelcps@rediffmail.com.

A number of conventional technologies have been employed for the treatment of wastewater contaminated with heavy metal. ions. Aklil *et al.* (2004) used calcined phosphate as adsorbent for removal of heavy metal ions from water. Sand was used as an adsorbent by Awan *et al.* (2003) for the removal of heavy metal ions. Removal of heavy metal ions by using alumina or chitosan as adsorbent was studied by Luisa *et al.* (2003). Activated carbon was used as adsorbent by Reed *et al.* (2002). Some of the conventional techniques for the removal of metal ions from industrial wastewater include chemical precipitation. adsorption, solvent extraction, membrane separation, ion exchange, electrolytic techniques, coagulation, sedimentation, filtration, membrane process, biological process and chemical reaction (Blanco *et al.*, 1999; Blanchard *et al.*, 1984; Gloaguen and Morvan, 1997; Jeon and Park, 2001; Kim *et al.*, 1998; Lee and Jung, 1998; Mofa, 1995; Lujan *et al.*, 1994; Gardea-Torresdey and Henning, 1996). Removal of heavy metal ions from industrial wastewater using H_2O_2 was studied by Badmus and Anyata (2006). Saifuddin and Kumaran (2005) used chitosan coated oil palm shell charcoal for the removal of heavy metal ions from water. There appears to be no report in literature for the removal of heavy metal ions by ammonium hydroxide and by sodium sulphide in basic medium. In view of this it was considered worthwhile to undertake the present study.

Materials and Methods

Determination of Aluminium Ions

A solution of $AlCl_3$ was prepared by dissolving 1.0 gm of it in 100 ml distilled water. Two grams of ammonium chloride was added in Ammonium hydroxide solution (5N) and Al^{+3} solution was added to it. The reaction mixture was allowed to stand for one hour. A white gelatinous precipitate of $Al(OH)_3$ appeared. This precipitate was filtered, washed, dried and weighed. The pH of the reaction mixture was 7.6.

Determination of Chromium Ions

A solution of $CrCl_3$ was prepared by dissolving 1.0 gm of it in 100 ml distilled water. Two grams of ammonium chloride was added in Ammonium hydroxide solution (5N) and Cr^{3+} solution was added to it. The reaction mixture was allowed to stand for one hour. A green precipitate of $Cr(OH)_3$ appeared. This precipitate was filtered, washed, dried and weighed. The pH of the reaction mixture was 8.8.

Determination of Iron Ions

A solution of $FcCl_3$ was prepared by dissolving 1.0 gm of it in 100 ml distilled water. Two grams of ammonium chloride was added in Ammonium hydroxide solution (5N) and Fe^{3+} solution was added to it. The reaction mixture was allowed to stand for one hour. A brown precipitate of $Fe(OH)_3$ appeared. This precipitate was filtered, washed, dried and weighed. The pH of the reaction mixture was 7.4.

Determination of Manganese Ions

A solution of $MnCl_2.4H_2O$ was prepared by dissolving 0.1 gm of it in 100 ml distilled water. 40 ml ammonium hydroxide solution (5N) was added in the above solution. After that Sodium sulphide solution (100 ppm) was added in excess to the above solution. The reaction mixture was allowed to stand for one-hour. A light pink precipitate of MnS appeared. This precipitate was filtered, washed, dried and weighed. The pH of the reaction mixture was 7.9.

Determination of Zinc Ions

A solution of $ZnSO_4$ was prepared by dissolving 0.1 gm of it in 100 ml distilled water. 40 ml ammonium hydroxide solution (5N) was added in the above solution. Sodium sulphide solution (100 ppm) was then added in excess to the above solution. The reaction mixture was allowed to stand for one hour. A white precipitate of ZnS appeared. This precipitate was filtered, washed, dried and weighed. The pH of the reaction mixture was 8.5.

Determination of Nickel Ions

A solution of $NiCl_2$ was prepared by dissolving 0.1 gm of it in 100 ml distilled water. 40 ml ammonium hydroxide solution (5N) was added in the above solution. After that Sodium sulphide solution (100 ppm) was added in excess to the above solution. The reaction mixture was allowed to stand for one hour. A black precipitate of NiS appeared. This precipitate was filtered, washed, dried and weighed. The pH of the reaction mixture was 9.4.

Determination of Cobalt Ions

A solution of $CoCl_2$ was prepared by dissolving 0.1 gm of it in 100 ml distilled water. 40 ml ammonium hydroxide solution (5N) was added in the above solution. After that Sodium sulphide solution (100 ppm) was added in excess to the above solution. The reaction mixture was allowed to stand for one hour. A black precipitate of CoS appeared. This precipitate was filtered, washed, dried and weighed. The pH of the reaction mixture was 9.7.

Results and Discussion

The results of these estimations are given in the Tables 64.1 and 64.2.

Table 64.1

Name of the Metal Ion	*Volume of the Metal Ion Solution Taken*	*Amount of NH_4Cl Added*	*Volume of NH_4OH- 5N Added*	*pH*	*Weight of Precipitate found in gm*	*Theoretical Weight of Precipitate in gm*
Al^{+3}	100 ml	2 gm	10 ml	7.6	0.63 gm	0.58 gm
Cr^{+3}	100 ml	2 gm	10 ml	8.8	0.42 gm	0.38 gm
Fe^{+3}	100 ml	2 gm	10 ml	7.4	0.70 gm	0.65 gm

Table 64.2

Name of the Metal Ion	*Volume of the Metal Ion Solution Taken*	*Volume of Na_2S (100 ppm)*	*Volume of NH_4OH- 5N Added*	*pH*	*Weight of Precipitate found in gm*	*Theoretical Weight of Precipitate in gm*
Mn^{+2}	100 ml	50 ml	40 ml	7.9	0.046	0.043
Zn^{+2}	100 ml	50 ml	40 ml	8.5	0.056	0.060
Ni^{+2}	100 ml	50 ml	40 ml	9.4	0.041	0.038
Co^{+2}	100 ml	50 ml	40 ml	9.7	0.042	0.038

It was observed during the present investigations that the metal ions precipitate almost quantitatively in the form of hydroxides or in the form of sulphides in basic medium. Hence, this method can be employed for the-removal of heavy metal ions discharged by industries into water bodies like "Amani Shah Ka' Nala". The toxicity level in vegetables grown by the polluted water has been found to be very high for human consumption. The great hazard of human health can be protected if the water is freed from heavy metal ions before using for irrigation.

References

Aklil, A., Mouflih, M. and Sebti, S., 2004. Removal of heavy metals from water using calcined, phosphate as adsorbent. *Journal of Hazardous Materials*, 112(3): 183–190.

Awan, M. and Qazi, Ishtiaq, 2003. Removal of heavy metals through adsorption using sand. *Journal of Environmental Sciences*, 15(3): 413–416.

Badmus, M.A.O. and Anyata, B.D., 2006. Removal of heavy metal from industrial wastewater using H_2O_2. *African Journal of Biotechnology*, 6(3): 238–242.

Blanchard, G., Maunaye, M. and Martin, G., 1984. Removal of heavy metals from water by means of natural Zeolites. *Wat. Res.*, 18: 1501.

Blanco, A., Sanz, B. and Llama, M.J., 1999. Biosorption of heavy metals to immobilized *Phormidium laminosum* biomass. *J. Biotechnol.*, 6(9): 227–240.

Gardea-Torresdey, J. and Henning, J., 1996. Ability of silica-immobilized *Medicago sativa* to remove copper ions from solution. *J. Hazard Mater.*, 57: 29–39.

Ghaffar, Abdul, 2006. Removal and stabilization of chromium metal ions from industrial effluents. *Electronic J. Environ. Agric. Food Chem.*, 5(2): 1286–1295.

Gloaguen, V. and Morvan, H., 1997. Removal of heavy metal ions from aqueous solution by modified barks. *J. Environ. Sci. Health*, A32: 901.

Jeon, C. and Park, J., 2001. Removal of heavy metals in plating wastewater using Carboxylated alginic acid. *Korean J. Chem. Eng.*, 18: 955.

Kim, S., Jeung, S. and Moon, H., 1998. Removal and recovery of heavy metal ions in fixed and semi-fluidized beds. *Korean J. Chem. Eng.*, 15: 37.

Lee, S. and Jung, C., 1998. Removal of heavy metals from aqueous solution by apple residues. *Process Biochem.*, 33: 205.

Luisa, Cervera, Cannen, M. and Arnal, M., 2003. Removal of heavy metals by using adsorption on alumina or chitosan. *Analytical and Bioanalytical Chemistry*, 375(6): 820–825.

Lujan, J., Damall, D. and Stark, P., 1994. Metal ion binding by algae and higher plant tissues: A phenomenological study of solution pH dependence. *Solvent Extr. Ion Exch.*, 12: 803–816.

Mofa, A., 1995. Plants proving their worth in toxic metal cleanup. *Science*, 269: 302–305.

Reed, Brian E., 2002. Removal of heavy metals by activated carbon. *Environmental Separation of Heavy Metals*, p. 205–264.

Saifuddin, M. and Kumaran, Palanisamy, 2005. Removal of heavy metal from industrial wastewater using chitosan coated oil palm shell charcoal. *Electronic Journal of Biotechnology*, 8(1).

Chapter 65

Studies on Environmental Degradation of Two Wetlands in and Around Guwahati City, Assam, India

P.K. Das[1] and Aruna Bhattacharyya[2]

[1]**Department of Geological Sciences, Gauhati University, Guwahati – 781 014, Assam**

[2]**Department of Geology, Dimoria College, Khetri, Assam**

ABSTRACT

Solabeel and Deepar beel are two notable wetlands of Guwahati city. These two beels have been identified as a large water storage beels. The heels once provided habitat for a large number of flora and fauna and received rain water from nearby hills and plains. Previously the both beels accommodates huge water and could control flood of Gwahati area, but nowadays due to huge encroachment, dumping of solid wastes, the morphology of the beels have changed and reduces in size. The quality of the water and soil of the beels has exceeds tolerable limit destroying the aquatic habitat.

To evaluate the nature of pollution and its effect, a systematic study has been a carried out from different locations of the water and soil samples and the different parameters have been determined.

Huge population explosion, rapid urbanization, unauthorized occupation of the beel areas, lack of social awareness for environmental protection and sort of proper planning, the area of wetlands is gradually decreasing and has changed the ecosystem of the beels. Some remedial measures have been suggested to protect the beels for future generations.

Keywords: Environmental degradation, Wetlands, Guwahati city.

Introduction

The term 'wetland' is generally applied to seasonally or permanently waterlogged areas of land. Wetlands include lakes, rivers, estuaries and marshes. Wetland habitats are vibrant with a diversity of living beings.

Wetlands are transitional areas between aquatic and terrestrial ecosystems where the water table is usually at or near the surface. Wetlands are among the most productive life support systems in the world and are of immense socio-economic and ecological importance to mankind.

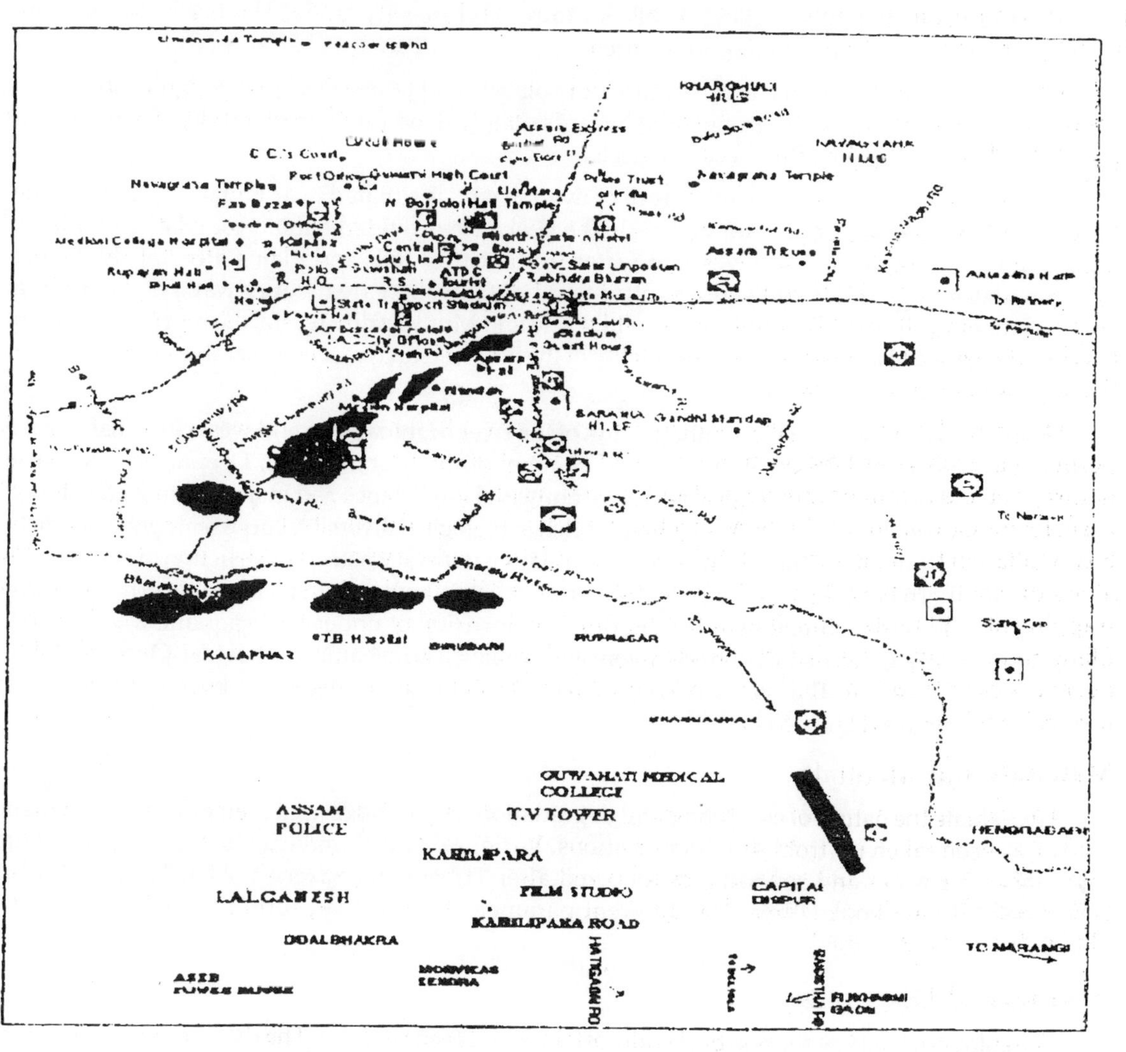

Figure 65.1: Location Map of the Solabeel Area of Guwahati City (Road map is downloaded from Internet)

Wetlands play critical role in the improvement of water quality through the airest pollutants (organic and inorganic) from direct release to the natural water bodies and by removal of sediments, production of oxygen, recycling of nutrients and treatment of wastewater by removing nitrogen phosphorus, bacteria and viruses. The wetlands act as habitat for fish and shellfish and for birds, mammals, reptiles and amphibians. Unfortunately, biodiversity and the ecological functions of wetlands are not widely recognized.

Guwahati, the premier city of North-east India, has covered an area of more than 300 sq.km. of the Brahmaputra valley and occurs in the northern fringe of Assam and Meghalaya plateau. The morpho-tectonic features under the influence of fluvio-denudational process of erosion has given rise to the various land forms like hills, valleys, beels, swamps, and marshy lands. The beels, swamps and marshy lands are the main wetlands of the area.

There are few wetlands in Guwahati city. But Solabeel and Deeparbeel play a significant role for restoring the ecosystem and environmental hazards mainly flood problem of the city. Therefore, this type of study is confined within these two beels.

Solabeel, one of the notable wetlands of Guwahati city, is situated near Paltanbazar, which has been identified as a large water storage beel. The beel once provided habitat for a large number of aquatic flora and fauna and receives rain water from nearest hills and plains. Previously the beel accommodates huge water and could control flood of Guwahati area, but nowadays due to huge encroachment by the people, dumping of solid waste and Municipality waste, the morphology of the beel has changed and reduces in size. The quality of the beel water has exceeds tolerable limit destroying the aquatic habitat (Figure 65.2).

Deeparbeel is situated on the southern bank of the river Brahmaputra and west Guwahati area. A permanent freshwater lake on an abandoned channel of the Brahmaputra, Deepar beel is a large natural wetland of immense biological and environmental importance and major storm water storage basin for the Guwahati city. Moreover wetland is rich in biological diversity. Large scale encroachment, heavy siltation from the surrounding hills, accumulation of toxic wastes, construction of railway line along the southern boundary of the beel, industrial development within its periphery, hunting and trapping of wild birds, unregulated fishing practice, invasion of undesirable aquatic weeds, earth-filling and soil cutting etc. disturb the ecosystem and waging a grim battle for survival. Once spreading over an area of 41 sq.km., the heel now barely covers 5 sq.km. and its depth has been reduced to 2.25 meters from 5 metres (Figure 65.3).

Materials and Methods

To evaluate the nature of environmental degradation and pollution and their effects, a systematic study has been taken up from different locations. Various physical and chemical parameters were undertaken for water and soil samples followed after Tribedi and Saxena (1987), Trivedi and Raj (1996) and ISI handbook (1986). The different parameters like colour, turbidity, pH, BOD, COD, alkalinity etc. were studied.

Results and Discussion

The colour of water of the Solabeel is almost dark and faulty selling. The dark colour of the water is due to solid waste deposited there and not easily biodegradable and decrease the photosynthetic activity and other self natural purification of the beel water (Dara, 1995). PH of the water ranges from 6.2–6.5 and it is slightly acidic in nature. Biological activity remains usually restricted to a narrow pH range of 6.5 to 8.5 (Goel, 1996). The hardness ranges from 3.6–5.4. The Dissolved Oxygen (DO), Biological

Figure 65.2: Major Portion of Solabeel Showing Wetland

Figure 65.3: Photograph of Deeparbeel (Carrying water and soil samples)

Oxygen Demand (BOD), Chemical Oxygen Demand (COD) shows maximum pollution has taken place (Table 65.1). The soil colour of the beel is almost dark brown. The result of the chemical analysis of soil is shown in Table 65.2.

Table 65.1: Water Analysis of Solabeel

Sample	Colour	pH	Total Alkalinity	Total Hardness mg/l	DO mg/l	BOD mg/1	COD mg/l	Chloride mg/l
1	Black	6.5	5	0.6	Nil	40.0	83.6	1.4
2	Greenish with black	6.5	6.5	3.6	Nil	38.0	53.2	1.0
3	Green dark	6.4	6.4	4.8	Nil	38.0	68.4	1.1
4	Greenish	6.5	6.5	4.8	Nil	40.0	60.6	1.2
5	Normal	6.2	6.2	5.4	Nil	36.0	33.3	1.1

Table 65.2: Chemical Analysis of Soil Near Solabeel

Sample	Color	pH	Water Holding Capacity
1. Northern part of Beel	Dark brown	6.6	0.23 per cent
2. Near Fish Market	Dark brown	7.5	0.23 per cent
3. Eastern side of Beel	Dark brown	6.9	0.24 per cent

The results of chemical analysis of water and soil of Deepar beel were shown in Tables 65.3 and 65.4. The results indicate that the water and soil are less polluted than the water and soil of the Sola beel. Moreover, environmental degradation is also less than the Sola beel. Still, it losses its unique ecosystem balance gradually.

Table 65.3: Water Analysis of Deeparbeel

Sample	Colour	pH	Conductance μmho km	Turbidity	Total Alkalinity mg/l	Total Hardness mg/l	DO mg/l	BOD mg/l	COD mg/l	Chloride mg/l
1	Colurless	6.3	217	Nil	1.7	3.0	6.9	20.8	68.0	
2	Colurless	6.5	226	Nil	1.5	2.6	5.7	29.0	49.0	
3	Colurless	6.0	247	Nil	1.9	4.1	6.2	32.5	53.0	
4	Colurless	6.1	212	Nil	2.5	4.8	6.5	26.9	63.0	
5	Colurless	6.2	229	Nil	2.2	2.4	6.6	36.0	69.0	1.1

Table 65.4: Chemical Analysis of Soil Near Deeparbeel

Sample	Color	pH	Water Holding Capacity
1. Near Borholabeel	Dark brown	6.6	0.23 per cent
2. Azara side	Dark brown	7.5	0.23 per cent

Remedial Measures

1. Scientific management of waste-utilize sewage waste as substitute of wood, cattle feed, fertilizer and fuel production *e.g.* CH_4 gas released due to bacterial growth can be used as a fuel, use of sewage water for fish farming etc. These type of practices should be made to conserve Solabeel and Deeparbeel.
2. Clearance of the water by hyacinth and reclamation of beel water should be done. The Government of Assam has already taken this step in 1995 according to the decision of the "State Wetland Committee".
3. Sewage and water treatment plant should be established by Municipal body.
4. Social awareness will be helpful for saving the beels.
5. Pollution Control Board must enforce laws to save the beels.
6. Fencing around the beel should be completed immediately.
7. New settlement and further encroachment should be stopped.
8. Plantation can be done around the beels.
9. Recreation place can be made around the beels.
10. The beels may converted to a Government Fishery which will help state's economic development.
11. NGO's should come forward to restore the ecosystem of the beels.

References

Dara, S.S., 1995. *A Textbook for Environmental Chemistry and Pollution Control*, New Delhi.

Dey, S.K., 1996. *Environmental Chemistry*, New Delhi.

Goel, P.V., 1996. *Water Pollution Effects and Control.*

Goswami, I.D. and Barman, D.K., 1998. Geo-environmental impact on the geohydro-morphological setting of Guwahati city and its environment. In: *International Conference on Disaster Management*, Guwahati, Assam, India, April 23–26.

I.S.I., 1996. *Methods of Sampling and Test (Physical and Chemical) for Water used in Industry*. Manak Bhawan, Bahadur Shah Jafar Marg, New Delhi.

Saxena, M.M., 1987. *Environmental Analysis (Water, Soil and Air)*. Agri-Botanical Publication.

Tribedi, P.K., 1987. *Practical Methods in Ecology and Environmental Sciences*. Environmental Publication.

Index

G

H

I